# HANDBOOK ON
# LINEAR ALGEBRA AND NUMERICAL METHODS
# (SOLVED EXAMPLES)

By

Dr. A. KAMESWARA RAO

# SYLLABUS

**Unit I: Solving systems of linear equations, Eigen values and Eigen vectors:**
Rank of a matrix by echelon form and normal form Solving system of homogeneous and nonhomogeneous equations linear equations Gauss Elimination for solving system of equations Eigen values and Eigen vectors and their properties.

**Unit-II: Cayley-Hamilton theorem and Quadratic forms:**
Cayley - Hamilton theorem (without proof) Finding inverse and power of a matrix by Cayley- Hamilton theorem Reduction to Diagonal form Quadratic forms and nature of the quadratic forms Reduction of quadratic form to canonical forms by orthogonal transformation Complex matrices

**UNIT III: Iterative methods:**
Introduction Bisection method Method of false position Iteration method Newton-Raphson method (One variable and simultaneous Equations) Jacobi and Gauss-Seidel methods for solving system of equations.

**UNIT IV: Interpolation:**
Introduction Errors in polynomial interpolation Finite differences Forward differences Backward differences Central differences Relations between operators formulae for interpolation Interpolation with unequal intervals Newtons's divided difference formula.

**UNIT V: Numerical integration and solution of ordinary differential equations:**
Trapezoidal rule-1/3rd and 3/8th rule-Solution of ordinary differential equations by Taylor's series-Picard's method of successive approximations-Euler's method- Runge-Kutta method (second and fourth order)

Dr. K. L. Sai Prasad, M.Sc., Ph.D. (B.H.U.)
Head-Department of Mathematics
& In-Charge Examinations
G.V.P. College of Engineering for Women
Member: IMS, ISCA, TSI
Phone: +91 9989084548
email: klsprasad@yahoo.com

## FOREWORD

Mathematics is not about numbers, equations, computations, or algorithms:
it is about understanding

—**William Paul Thurston**

A good textbook that is lucid in its presentation and easy to understand is the need of the day. The present book 'Handbook on Linear Algebra and Numerical Methods (Solved Examples)' authored by Dr. A. Kameswara Rao experienced faculty member of Department of Mathematics is ideal, both for the students and teachers. A conscious attempt has been made by the author to present a large number of examples in a simple manner. There is no doubt that the solved examples together with well-structured exercises will help the student to enrich his/her comprehension of the subject.

**(Dr. K. L. Sai Prasad)**
**04-05-2024**

# PREFACE

There is no dearth of books on Engineering Mathematics but the students find it difficult to solve most of the problems in the exercises in the absence of an adequate number of solved problems. An outstanding and distinguishing feature of the book is large number of typical solved examples followed by well-graded problems. Many examples and problems have been selected from reputed text books and the recent papers of various Engineering examinations conducted by different Technical Universities.

Sufficient care has been taken by the author to present all example problems in five units in a lucid manner. The present form of the book is suitable to all B. Tech. students of all Engineering colleges.

All efforts have been made to keep the book free from errors. Answers to all exercises have been re-checked. Suggestions for improvement will be highly appreciated and gratefully acknowledged.

-AUTHOR

# ABOUT THE AUTHOR

**Dr. A. Kameswara Rao** is an Assistant Professor, Department of Mathematics, GVP College of Engineering for Women, Visakhapatnam. He has received his doctorate degree from the prestigious Andhra University, Visakhapatnam in 2010. During his vast sixteen years of teaching experience, he taught various Mathematics courses for undergraduate students.

His enthusiasm for research still keeps him as an active researcher. He has published about twenty two research papers in various international journals and guiding one research scholars at present.

His dedication towards his profession, passion for teaching and research, his work culture and love and affection towards students, made him a role model as a teacher.

# HANDBOOK ON
# LINEAR ALGEBRA AND NUMERICAL METHODS
(Solved Examples)

## ABOUT THE BOOK

The author has endeavoured to present the problems in a comprehensive and lucid manner

- An outstanding and distinguishing feature of the book is the large number of typical solved examples followed by well graded exercises for practice.

- Many examples and problems have been selected from recent papers of various Engineering examinations conducted by different Technical Universities.

- A careful selection of objective-type questions with answers is given at the end of the each chapter.

# ACKNOWLEDGEMENTS

First, I would like to thank Dr. Rajkumar Goswami, Principal and Dr. G. Sudheer, Vice-Principal of the college for giving me all necessary support. At the outset I would like to extend by deep gratitude to Dr. K. L. Sai Prasad, the Head of the department of Basic Sciences & Humanities for his consistent encouragement in bringing out this volume.

I also extend my thanks to all my department colleagues Dr. A. Suseelatha, Dr. T. Poornakantha, Ms. B. Bharathi, Ms. Ranjani and Mr. V.S.S.V.D. Prakash and also my fellow mates of the Institution for their support extended in all respects from time to time

# Table of Contents

Table of Contents     iii

**1 Solving System of Linear Equations, Eigen values and Eigen vectors**    1
- 1.0.1 Sub Matrix .................................... 1
- 1.0.2 Minor of a Matrix ............................. 1
- 1.1 Rank of a Matrix .................................. 2
  - 1.1.1 Echelon form of a Matrix ..................... 3
  - 1.1.2 Normal Form of a Matrix ..................... 11
  - 1.1.3 Rank of the Matrix by $PAQ$ form ............. 20
  - 1.1.4 Gauss-Jordan Method to find Inverse of a Matrix ....... 25
- 1.2 System of Linear Equations (Non-Homogeneous) ............. 29
- 1.3 Conditions for Solution of Linear Systems ................. 29
- 1.4 Homogeneous System of Linear Equations .................. 42
- 1.5 Gauss Elimination Method ............................. 54
- 1.6 Gauss-Jordan Method ................................. 61
- 1.7 Applications ......................................... 65
  - 1.7.1 Kirchoff's Laws .............................. 65
- 1.8 Eigenvalues and Eigenvectors .......................... 72
  - 1.8.1 Properties of Eigenvalues and Eigenvectors ........ 99
  - 1.8.2 Algebraic Multiplicity and Geometric Multiplicity ..... 110
  - 1.8.3 Applications of Eigenvalues and Eigenvectors ....... 114

**2 Cayley-Hamilton Theorem and Quadratic forms**    117
- 2.1 Cayley-Hamilton Theorem ............................. 117

| | | | |
|---|---|---|---|
| 2.2 | Diagonalization of a Matrix | | 146 |
| 2.3 | Quadratic Forms | | 171 |
| | 2.3.1 | General Form of a Quadratic Form | 173 |
| | 2.3.2 | Matrix Form of a Quadratic Form | 174 |
| | 2.3.3 | Rank of the Quadratic form | 174 |
| | 2.3.4 | Linear Transformation of a Quadratic Form | 177 |
| | 2.3.5 | Orthogonal Transformation | 177 |
| | 2.3.6 | Canonical Form or Normal Form of a Quadratic Form | 177 |
| | 2.3.7 | Rank, Index and Signature of the Quadratic Form | 178 |
| | 2.3.8 | Nature of the Quadratic Form | 178 |
| 2.4 | Method of Reduction of Quadratic Form to Canonical Form | | 179 |
| | 2.4.1 | Diagonalization | 180 |
| | 2.4.2 | Lagrange's Reduction Method | 192 |
| | 2.4.3 | Orthogonal Transformation | 197 |
| 2.5 | Complex Matrices | | 228 |
| | 2.5.1 | Hermitian Matrix | 229 |
| | 2.5.2 | Skew-Hermitian Matrix | 230 |
| | 2.5.3 | Unitary Matrix | 240 |

# 3 Iterative Methods — 248

| | | | |
|---|---|---|---|
| 3.1 | Introduction | | 248 |
| | 3.1.1 | Root of an Equation | 248 |
| | 3.1.2 | Intermediate Value Theorem | 248 |
| | 3.1.3 | Absolute Error and Relative Error | 249 |
| 3.2 | Bisection Method | | 249 |
| 3.3 | Regula-False Method | | 262 |
| 3.4 | Iteration Method | | 273 |
| 3.5 | Newton-Raphson Method | | 284 |
| | 3.5.1 | Convergence of Newton-Raphson Method | 285 |
| | 3.5.2 | Some Deductions from the Newton-Raphson formula | 293 |
| 3.6 | Solution of Non-Linear Simultaneous Equations | | 296 |
| 3.7 | Gauss-Jacobi Iteration Method | | 305 |

|   |     |                                                            |     |
|---|-----|------------------------------------------------------------|-----|
|   | 3.8 | Gauss-Seidel Iteration Method ............................ | 314 |

# 4 Interpolation     331

- 4.1 Finite Differences .................................... 331
  - 4.1.1 Forward Differences ............................ 331
  - 4.1.2 Backward Differences ........................... 337
  - 4.1.3 Central Differences ............................. 340
  - 4.1.4 Differences of a Polynomial ..................... 342
- 4.2 Other Difference Operators ............................ 343
- 4.3 Relations Between the Operators ...................... 345
  - 4.3.1 Examples ....................................... 346
  - 4.3.2 Missing Term Techniques ........................ 368
- 4.4 Newton's Forward Difference Interpolation Formula .... 373
- 4.5 Newton's Backward Difference Interpolation Formula ... 390
- 4.6 Gauss's Forward Interpolation formula ................. 397
- 4.7 Gauss's Backward Interpolation formula ................ 404
- 4.8 Interpolation with Unequal Intervals .................. 410
  - 4.8.1 Lagrange's Interpolation formula ................ 410
  - 4.8.2 Inverse Interpolation ........................... 428
  - 4.8.3 Divided Differences ............................. 429
  - 4.8.4 Newton's Divided Difference Formula ............. 435

# 5 Numerical Integration and Solution of Ordinary Differential Equations     451

- 5.1 Numerical Integration ................................. 451
- 5.2 Numerical Solution of Ordinary Differential Equations .. 481
  - 5.2.1 Taylor's Series Method .......................... 481
  - 5.2.2 Picard's Method of Successive Approximation ..... 504
  - 5.2.3 Euler's Method .................................. 517
  - 5.2.4 Examples ........................................ 518
  - 5.2.5 Modified Euler's Method ......................... 529
  - 5.2.6 Examples ........................................ 531
  - 5.2.7 Runge-Kutta Methods ............................. 543

5.2.8 R-K Second Order Method . . . . . . . . . . . . . . . . . . 543
5.2.9 Runge-Kutta Fourth-Order Method . . . . . . . . . . . . . 547

# Chapter 1

# Solving System of Linear Equations, Eigen values and Eigen vectors

## 1.0.1 Sub Matrix

**Definition:** A Matrix obtained by deleting some rows or columns or both of a given matrix is called its sub-matrix.

**Example:** Let $A = \begin{pmatrix} 1 & 5 & 6 & 7 \\ 8 & 9 & 10 & 5 \\ 3 & 4 & 5 & -1 \end{pmatrix}$. Then $\begin{pmatrix} 1 & 5 & 6 \\ 8 & 9 & 10 \end{pmatrix}$ is a sub-matrix of $A$ obtained by deleting third row and $4^{th}$ column from $A$.

## 1.0.2 Minor of a Matrix

Let $A$ be an $m \times n$ matrix. The determinant of a square sub-matrix of $A$ is called a minor of the matrix. If the order of the square sub-matrix is $t$ then its determinant is called minor of order $t$.

**Example:** Let $A = \begin{pmatrix} 2 & 1 & 1 \\ 3 & 1 & 2 \\ 1 & 2 & 3 \\ 5 & 6 & 7 \end{pmatrix}$ be a matrix. We have $B = \begin{pmatrix} 2 & 1 \\ 3 & 1 \end{pmatrix}$ is a sub-matrix of order 0 2. $\det(B) = 2 - 3 = -1$ is a minor of order 2.

## 1.1 Rank of a Matrix

Let $A$ be an $m \times n$ matrix. If $A$ is a null matrix, we define its rank to be 0(zero). If $A$ is a non-zero matrix, we say that $r$ is the rank of $A$ if

(i) every $(r+1)$ th order minor of $A$ is 0 (zero) and

(ii) there exists at least one $r$ th order minor of $A$ which is not zero.

Rank of $A$ is denoted by $\rho(A)$.

**Note:** (1) Every matrix will have a rank.

(2) Rank of a matrix is unique.

(3) $\rho(A) \geq 1$ when $A$ is a non-zero matrix.

(4) If $A$ is a matrix of order $m \times n$, rank of $A = \rho(A) \leq \min(m, n)$.

(5) Rank of the identity matrix $I_n$ is $n$.

**Example:** Find the rank of the matrix $A = \begin{pmatrix} -1 & 0 & 6 \\ 3 & 6 & 1 \\ -5 & 1 & 3 \end{pmatrix}$.

**Solution:** We have $\det A = -1(18-1) - 0(9+5) + 6(3+30)$
$$= -17 - 0 + 6(33) = 181 \neq 0.$$

We have minor of order $3 \neq 0$

$$\therefore \rho(A) = 3.$$

**Example:** Find the rank of the matrix $\begin{pmatrix} 1 & 2 & 3 \\ 3 & 4 & 5 \\ 4 & 5 & 6 \end{pmatrix}$.

**Solution:** Let $A$ be the given matrix. Then

$$\det A = \begin{vmatrix} 1 & 2 & 3 \\ 3 & 4 & 5 \\ 4 & 5 & 6 \end{vmatrix} = 1(24-25) - 2(18-20) + 3(15-16) = -1 + 4 - 3 = 0.$$

$\therefore$ Rank $A \neq 3$. So it must be less than 3.

Consider a minor of order $2 = \begin{vmatrix} 1 & 2 \\ 3 & 4 \end{vmatrix} = 4 - 6 - 2 \neq 0$

Hence there is at least a minor of order 2 which is not zero.

∴ Rank of $A = 2$.

**Example:** Find the rank of the matrix $\begin{pmatrix} 1 & 2 & 3 & 4 \\ 5 & 6 & 7 & 8 \\ 8 & 7 & 0 & 5 \end{pmatrix}$.

**Solution:** Here the matrix is of order $3 \times 4$. Its rank $\leq \min(3, 4) = 3$.

∴ Highest order of the minor will be 3.

Let us consider the minor $\begin{pmatrix} 1 & 2 & 3 \\ 5 & 6 & 7 \\ 8 & 7 & 0 \end{pmatrix}$.

Its determinant $= 24 \neq 0$.

∴ The order of the highest order non-zero minor of $A$ is 3.
Hence the rank of the given matrix is 3.

**Example:** Find the rank of the matrix $\begin{pmatrix} 1 & 1 & 1 \\ 1 & 1 & 1 \\ 1 & 1 & 1 \end{pmatrix}$.

**Ans:** 1.

## 1.1.1 Echelon form of a Matrix

A matrix is said to be in Echelon form if it satisfies the following two condition:

(1) Zero rows if any occurs then they should be below the non-zero rows.

(2) The number of zeros before the non-zero element increases with the row number.

**Note:** Every matrix can be transformed to Echelon form by applying elementary row transformation. In Echelon form the rank of the matrix is equal to the number of non-zero rows.

**Examples:** $\begin{pmatrix} 9 & 1 & 2 & 5 \\ 0 & 2 & 0 & 5 \\ 0 & 0 & 1 & 2 \\ 0 & 0 & 0 & 0 \end{pmatrix}$, $\begin{pmatrix} 0 & 1 & 2 & 3 & 4 \\ 0 & 0 & 1 & 0 & 2 \\ 0 & 0 & 0 & 7 & 2 \\ 0 & 0 & 0 & 0 & 0 \end{pmatrix}$, $\begin{pmatrix} 1 & 0 & 0 & 0 \\ 0 & 1 & 0 & 0 \\ 0 & 0 & 1 & 0 \\ 0 & 0 & 0 & 1 \end{pmatrix}$ are the matrices in Echelon form.

**Note:** $\rho(A) = \rho(A^T)$.

**Example:** Reduce the matrix to echelon form and hence find its rank $\begin{pmatrix} 1 & 2 & 3 & 0 \\ 2 & 4 & 3 & 2 \\ 3 & 2 & 1 & 3 \\ 6 & 8 & 7 & 5 \end{pmatrix}$.

**Solution:** Let $A = \begin{pmatrix} 1 & 2 & 3 & 0 \\ 2 & 4 & 3 & 2 \\ 3 & 2 & 1 & 3 \\ 6 & 8 & 7 & 5 \end{pmatrix}$ (Apr-2017), (Jan-2020), (Apr-2022)

$R_2 \to R_2 - 2R_1, R_3 \to R_3 - 3R_1, R_4 \to R_4 - 6R_1$, we have

$$\sim \begin{pmatrix} 1 & 2 & 3 & 0 \\ 0 & 0 & -3 & 2 \\ 0 & -4 & -5 & 3 \\ 0 & -4 & -11 & 5 \end{pmatrix}$$

$R_4 \to R_4 - R_3$, we have

$$\sim \begin{pmatrix} 1 & 2 & 3 & 0 \\ 0 & 0 & -3 & 2 \\ 0 & -4 & -5 & 3 \\ 0 & 0 & -3 & 2 \end{pmatrix}$$

$R_4 \to R_4 - R_2$, we have

$$\sim \begin{pmatrix} 1 & 2 & 3 & 0 \\ 0 & 0 & -3 & 2 \\ 0 & -4 & -5 & 3 \\ 0 & 0 & 0 & 0 \end{pmatrix}$$

$R_2 \leftrightarrow R_3$, we have

$$\sim \begin{pmatrix} 1 & 2 & 3 & 0 \\ 0 & -4 & -5 & 3 \\ 0 & 0 & -3 & 2 \\ 0 & 0 & 0 & 0 \end{pmatrix}$$

This matrix is in Echelon form

No. of non-zero rows $=3$

Rank of $A = \rho(A) = 3$.

**Example:** Find the rank of the matrix

$$A = \begin{pmatrix} 1 & 4 & 3 & -2 & 1 \\ -2 & -3 & -1 & 4 & 3 \\ -1 & 6 & 7 & 2 & 9 \\ -3 & 3 & 6 & 6 & 12 \end{pmatrix}$$

(Jan-2020), (Aug-2022)

**Solution:** Let $A = \begin{pmatrix} 1 & 4 & 3 & -2 & 1 \\ -2 & -3 & -1 & 4 & 3 \\ -1 & 6 & 7 & 2 & 9 \\ -3 & 3 & 6 & 6 & 12 \end{pmatrix}$

$R_2 \to R_2 + 2R_1$, $R_3 \to R_3 + R_1$, $R_4 \to R_4 - 3R_2$, we have

$$\sim \begin{pmatrix} 1 & 4 & 3 & -2 & 1 \\ 0 & 5 & 5 & 0 & 5 \\ 0 & 10 & 10 & 0 & 10 \\ 0 & 15 & 15 & 0 & 15 \end{pmatrix}$$

$R_3 \to R_3 - 2R_2$, $R_4 \to R_4 - 3R_2$, we have

$$\sim \begin{pmatrix} 1 & 4 & 3 & -2 & 1 \\ 0 & 5 & 5 & 0 & 5 \\ 0 & 0 & 0 & 0 & 0 \\ 0 & 0 & 0 & 0 & 0 \end{pmatrix}$$

$R_2 \to \frac{R_2}{5}$, we have

$$\sim \begin{pmatrix} 1 & 4 & 3 & -2 & 1 \\ 0 & 1 & 1 & 0 & 1 \\ 0 & 0 & 0 & 0 & 0 \\ 0 & 0 & 0 & 0 & 0 \end{pmatrix}$$

This matrix is in Echelon form

No. of non-zero rows $=2$

Rank of $A = \rho(A) = 2$.

**Example:** Find the rank of the matrix $\begin{pmatrix} 2 & 1 & 3 & 5 \\ 4 & 2 & 1 & 3 \\ 8 & 4 & 7 & 13 \\ 8 & 4 & -3 & -1 \end{pmatrix}$.

**Solution:** Let $A = \begin{pmatrix} 2 & 1 & 3 & 5 \\ 4 & 2 & 1 & 3 \\ 8 & 4 & 7 & 13 \\ 8 & 4 & -3 & -1 \end{pmatrix}$. **(Nov-2021), (Mar-2022)**

$R_2 \to R_2 - 2R_1$, $R_3 \to R_3 - 4R_1$, $R_4 \to R_4 - 4R_1$, we have

$$\sim \begin{pmatrix} 2 & 1 & 3 & 5 \\ 0 & 0 & -5 & -7 \\ 0 & 0 & -5 & -7 \\ 0 & 0 & -15 & -21 \end{pmatrix}.$$

$R_3 \to R_3 - R_2$, $R_4 \to R_4 - 3R_2$, we have

$$\sim \begin{pmatrix} 2 & 1 & 3 & 5 \\ 0 & 0 & -5 & -7 \\ 0 & 0 & 0 & 0 \\ 0 & 0 & 0 & 0 \end{pmatrix}.$$

This matrix is in Echelon form

No. of non-zero rows $= 2$

Rank of $A = \rho(A) = 2$.

**Example:** Find the rank of the matrix $\begin{pmatrix} 2 & -4 & 3 & -1 & 0 \\ 1 & -2 & -1 & -4 & 2 \\ 0 & 1 & -1 & 3 & 1 \\ 4 & -7 & 4 & -4 & 5 \end{pmatrix}$. **(Aug-2022)**

**Solution:** Let $A = \begin{pmatrix} 2 & -4 & 3 & -1 & 0 \\ 1 & -2 & -1 & -4 & 2 \\ 0 & 1 & -1 & 3 & 1 \\ 4 & -7 & 4 & -4 & 5 \end{pmatrix}$.

$R_1 \leftrightarrow R_2$, we have

$$\sim \begin{pmatrix} 1 & -2 & -1 & -4 & 2 \\ 2 & -4 & 3 & -1 & 0 \\ 0 & 1 & -1 & 3 & 1 \\ 4 & -7 & 4 & -4 & 5 \end{pmatrix}.$$

$R_2 \to R_2 - 2R_1$, $R_4 \to R_4 - 4R_1$, we have

$$\sim \begin{pmatrix} 1 & -2 & -1 & -4 & 2 \\ 0 & 0 & 5 & 7 & -4 \\ 0 & 1 & -1 & 3 & 1 \\ 0 & 1 & 8 & 12 & -3 \end{pmatrix}$$

$R_4 \to R_4 - R_3$, we have

$$\sim \begin{pmatrix} 1 & -2 & -1 & -4 & 2 \\ 0 & 0 & 5 & 7 & -4 \\ 0 & 1 & -1 & 3 & 1 \\ 0 & 0 & 9 & 9 & -4 \end{pmatrix}$$

$R_4 \to 5R_4 - 9R_2$, we have

$$\sim \begin{pmatrix} 1 & -2 & -1 & -4 & 2 \\ 0 & 0 & 5 & 7 & -4 \\ 0 & 1 & -1 & 3 & 1 \\ 0 & 0 & 0 & -18 & 16 \end{pmatrix}$$

$R_2 \leftrightarrow R_3$, we have

$$\sim \begin{pmatrix} 1 & -2 & -1 & -4 & 2 \\ 0 & 1 & -1 & 3 & 1 \\ 0 & 0 & 5 & 7 & -4 \\ 0 & 0 & 0 & -18 & 16 \end{pmatrix}$$

This matrix is in Echelon form

No. of non-zero rows $=4$

Rank of $A = \rho(A) = 4$.

**Example:** Find the rank of the matrix $\begin{pmatrix} 10 & -2 & 3 & 10 \\ 2 & 10 & 2 & 4 \\ -1 & -2 & 10 & 1 \\ 2 & 3 & 4 & 9 \end{pmatrix}$.

**Solution:** Let $A = \begin{pmatrix} 10 & -2 & 3 & 10 \\ 2 & 10 & 2 & 4 \\ -1 & -2 & 10 & 1 \\ 2 & 3 & 4 & 9 \end{pmatrix}$

$R_1 \leftrightarrow R_2$, we have

$$\sim \begin{pmatrix} -1 & -2 & 10 & 1 \\ 2 & 10 & 2 & 4 \\ 10 & -2 & 3 & 10 \\ 2 & 3 & 4 & 9 \end{pmatrix}$$

$R_2 \to R_2 + 2R_1,\ R_3 \to R_3 + 10R_1,\ R_4 \to R_4 + 2R_1$, we have

$$\sim \begin{pmatrix} -1 & -2 & 10 & 1 \\ 0 & 6 & 22 & 6 \\ 0 & -22 & 103 & 20 \\ 0 & -1 & 24 & 11 \end{pmatrix}$$

$R_2 \to R_2 + 6R_4,\ R_3 \to R_3 - 22R_4$, we have

$$\sim \begin{pmatrix} -1 & -2 & 10 & 1 \\ 0 & 0 & 166 & 72 \\ 0 & 0 & -425 & -222 \\ 0 & -1 & 24 & 11 \end{pmatrix}$$

$R_2 \leftrightarrow R_4$, we have

$$\sim \begin{pmatrix} -1 & -2 & 10 & 1 \\ 0 & -1 & 24 & 11 \\ 0 & 0 & -425 & -222 \\ 0 & 0 & 166 & 72 \end{pmatrix}$$

$R_4 \to 425R_4 + 166R_3$, we have

$$\sim \begin{pmatrix} -1 & -2 & 10 & 1 \\ 0 & -1 & 24 & 11 \\ 0 & 0 & -425 & -222 \\ 0 & 0 & 0 & 67452 \end{pmatrix}$$

$R_3 \to \frac{R_3}{-425},\ R_4 \to \frac{R_4}{67452}$, we have

$$\sim \begin{pmatrix} -1 & -2 & 10 & 1 \\ 0 & -1 & 24 & 11 \\ 0 & 0 & 2 & \frac{222}{425} \\ 0 & 0 & 0 & 1 \end{pmatrix}.$$

This matrix is in Echelon form

No. of non-zero rows $=4$

Rank of $A = \rho(A) = 4$.

**Example:** Find the value of $k$ such that the rank of $\begin{pmatrix} 1 & 2 & 3 \\ 2 & k & 7 \\ 3 & 6 & 10 \end{pmatrix}$ is 2.

**Solution:** Let $A = \begin{pmatrix} 1 & 2 & 3 \\ 2 & k & 7 \\ 3 & 6 & 10 \end{pmatrix}$.

Since the rank of matrix $A$ is 2 so the minor of 3rd order must be zero i.e., $|A| = 0$

$\Rightarrow 1(10k - 42) - 2(20 - 21) + 3(12 - 3k) = 0$

$\Rightarrow 10k - 42 + 2 + 36 - 9k = 0$

$\Rightarrow k - 4 = 0$

$\Rightarrow k = 4$.

**Example:** For what value $k$, the matrix $\begin{pmatrix} 4 & 4 & -3 & 1 \\ 1 & 1 & -1 & 0 \\ k & 2 & 2 & 2 \\ 9 & 9 & k & 3 \end{pmatrix}$ has rank 3.

**Solution:** Let $A = \begin{pmatrix} 4 & 4 & -3 & 1 \\ 1 & 1 & -1 & 0 \\ k & 2 & 2 & 2 \\ 9 & 9 & k & 3 \end{pmatrix}$

(Apr-2018)

$R_1 \leftrightarrow R_2$, we have

$$\sim \begin{pmatrix} 1 & 1 & -1 & 0 \\ 4 & 4 & -3 & 1 \\ k & 2 & 2 & 2 \\ 9 & 9 & k & 3 \end{pmatrix}$$

$R_2 \to R_2 - 4R_1$, $R_3 \to R_3 - kR_1$, $R_4 \to R_4 - 9R_1$, we have

$$\sim \begin{pmatrix} 1 & 1 & -1 & 0 \\ 0 & 0 & 1 & 1 \\ 0 & 2-k & 2+k & 2 \\ 0 & 0 & 9+k & 3 \end{pmatrix}$$

The given matrix is of order 4. If its rank is 3 then we must have $\det A = 0$

$$1 \cdot \begin{vmatrix} 0 & 1 & 1 \\ 2-k & 2+k & 2 \\ 0 & 9+k & 3 \end{vmatrix} = 0$$

$\Rightarrow 0[\ ] - 1[(2-k)3 - 0] + 1[(2-k)(9+k) = 0] = 0$

$\Rightarrow -3(2-k) + (2-k)(9+k) = 0$

$\Rightarrow -6 + 3k = 18 + 2k - 9k - k^2$

$\Rightarrow -k^2 - 4k + 12 = 0$

$\Rightarrow k^2 + 4k - 12 = 0$

$\Rightarrow (k-2)(k+6) = 0$

$\Rightarrow k = 2$ or $-6$.

### EXERCISE

1. Reduce the following matrices into Echelon form and find its rank.

(i) $\begin{pmatrix} 3 & 1 & 4 & 6 \\ 2 & 1 & 2 & 4 \\ 4 & 2 & 5 & 8 \\ 1 & 1 & 5 & 2 \end{pmatrix}$

(ii) $\begin{pmatrix} 2 & -1 & 3 & 4 \\ 0 & 3 & 4 & 1 \\ 2 & 3 & 7 & 5 \\ 2 & 5 & 11 & 6 \end{pmatrix}$  (Apr-2018)

(iii) $\begin{pmatrix} -1 & 2 & 1 & 8 \\ 2 & 1 & -1 & 0 \\ 3 & 2 & 1 & 7 \end{pmatrix}$

(iv) $\begin{pmatrix} 0 & 1 & 2 & -2 \\ 4 & 0 & 2 & 6 \\ 2 & 1 & 3 & 1 \end{pmatrix}$

(v) $\begin{pmatrix} 2 & 3 & -1 & -1 \\ 1 & -1 & -2 & -4 \\ 3 & 1 & 3 & -2 \\ 6 & 3 & 0 & -7 \end{pmatrix}$ (Dec-2020)  (vi) $\begin{pmatrix} 1 & 1 & 1 & 1 \\ 1 & 2 & 3 & -4 \\ 2 & 3 & 5 & -5 \\ 3 & -4 & -5 & 8 \end{pmatrix}$ (Apr-2022)

(vii) $\begin{pmatrix} 2 & 3 & 4 & -1 \\ 5 & 2 & 0 & -1 \\ -4 & 5 & 12 & -1 \\ 2 & 4 & 0 & 3 \end{pmatrix}$ (Apr-2022)  (viii) $\begin{pmatrix} 2 & 3 & 4 & 5 \\ 3 & 4 & 5 & 6 \\ 4 & 5 & 6 & 7 \\ 5 & 6 & 7 & 8 \end{pmatrix}$ (Apr-2022)

**Ans:** (i) 3  (ii) 3  (iii) 3  (iv) 3  (v) 3  (vi) 4

2. Find the values of $\lambda$ and $\mu$ if the rank of $\begin{pmatrix} 1 & -2 & 3 & 1 \\ 2 & 1 & -1 & 2 \\ 6 & -2 & \lambda & \mu \end{pmatrix}$ is 2.

**Ans:** $\lambda = 4, \mu = 6$.

3. Find the value of $k$ such that the rank of the matrix $\begin{pmatrix} 1 & 2 & -1 & 3 \\ 4 & 1 & 2 & 1 \\ 3 & -1 & 1 & 2 \\ -1 & 2 & 0 & k \end{pmatrix}$ is 3.

**Ans:** $k = \frac{-11}{3}$.

4. Find for what values of $a$ such that the rank of the matrix $A$ is 2, where $A = \begin{pmatrix} 1 & 1 & -1 & 1 \\ 1 & -1 & a & -1 \\ 3 & 1 & 0 & 1 \end{pmatrix}$.

**Ans:** $a = 2$.

### 1.1.2 Normal Form of a Matrix

The normal form of a matrix $'A'$ of rank $r$ is one of the forms $I_r$, $\begin{bmatrix} I_r & 0 \\ 0 & 0 \end{bmatrix}$, $\begin{bmatrix} I_r & 0 \end{bmatrix}$, $\begin{bmatrix} I_r \\ 0 \end{bmatrix}$, where $I_r$ is the identity matrix of order $r$. This form can be obtained by the application of both row and column operations on any given matrix $A$.

**Example:** Find the rank of the matrix using normal form $\begin{pmatrix} 8 & 1 & 3 & 6 \\ 0 & 3 & 2 & 2 \\ -8 & -1 & -3 & 4 \end{pmatrix}$.

**Solution:** Let $A = \begin{pmatrix} 8 & 1 & 3 & 6 \\ 0 & 3 & 2 & 2 \\ -8 & -1 & -3 & 4 \end{pmatrix}$

$R_3 \to R_3 + R_1$, we have

$$\sim \begin{pmatrix} 8 & 1 & 3 & 6 \\ 0 & 3 & 2 & 2 \\ 0 & 0 & 0 & 10 \end{pmatrix}$$

$C_1 \to \frac{C_1}{8}$, we have

$$\sim \begin{pmatrix} 1 & 1 & 3 & 6 \\ 0 & 3 & 2 & 2 \\ 0 & 0 & 0 & 10 \end{pmatrix}$$

$C_2 \to C_2 - C_1$, $C_3 \to C_3 - 3C_1$, $C_4 \to C_4 - 6C_1$, we have

$$\sim \begin{pmatrix} 1 & 0 & 0 & 0 \\ 0 & 3 & 2 & 2 \\ 0 & 0 & 0 & 10 \end{pmatrix}$$

$C_3 \to 3C_3 - 2C_2$, $C_4 \to 3C_4 - 2C_3$, we have

$$\sim \begin{pmatrix} 1 & 0 & 0 & 0 \\ 0 & 3 & 0 & 0 \\ 0 & 0 & 0 & 30 \end{pmatrix}$$

$R_3 \to \frac{R_3}{30}$, $R_2 \to \frac{R_2}{3}$, we have

$$\sim \begin{pmatrix} 1 & 0 & 0 & 0 \\ 0 & 1 & 0 & 0 \\ 0 & 0 & 0 & 1 \end{pmatrix}$$

$C_3 \leftrightarrow C_4$, we have

$$\sim \begin{pmatrix} 1 & 0 & 0 & 0 \\ 0 & 1 & 0 & 0 \\ 0 & 0 & 1 & 0 \end{pmatrix}$$

$$= \begin{bmatrix} I_3 & 0 \end{bmatrix}$$

∴ The rank of the matrix=Order of the identity matrix in normal form=3

**Example:** Find the rank of the matrix using normal form $\begin{pmatrix} 1 & 2 & 3 & 0 \\ 2 & 4 & 3 & 2 \\ 3 & 2 & 1 & 3 \\ 6 & 8 & 7 & 5 \end{pmatrix}$.

**Solution:** Let $A = \begin{pmatrix} 1 & 2 & 3 & 0 \\ 2 & 4 & 3 & 2 \\ 3 & 2 & 1 & 3 \\ 6 & 8 & 7 & 5 \end{pmatrix}$ (Jan-2020)

$R_2 \to R_2 - 2R_1$, $R_3 \to R_3 - 3R_1$, $R_4 \to R_4 - 6R_1$, we have

$$\sim \begin{pmatrix} 1 & 2 & 3 & 0 \\ 0 & 0 & -3 & 2 \\ 0 & -4 & -8 & 3 \\ 0 & -4 & -11 & 5 \end{pmatrix}$$

$C_2 \to C_2 - 2C_1$, $C_3 \to C_3 - 3C_1$, we have

$$\sim \begin{pmatrix} 1 & 0 & 0 & 0 \\ 0 & 0 & -3 & 2 \\ 0 & -4 & -8 & 3 \\ 0 & -4 & -11 & 5 \end{pmatrix}$$

$R_4 \to R_4 - R_3$, we have

$$\sim \begin{pmatrix} 1 & 0 & 0 & 0 \\ 0 & 0 & -3 & 2 \\ 0 & -4 & -8 & 3 \\ 0 & 0 & -3 & 2 \end{pmatrix}$$

$R_4 \to R_4 - R_2$, we have

$$\sim \begin{pmatrix} 1 & 0 & 0 & 0 \\ 0 & 0 & -3 & 2 \\ 0 & -4 & -8 & 3 \\ 0 & 0 & 0 & 0 \end{pmatrix}$$

$C_2 \to \frac{C_2}{-4}$, we have

$$\sim \begin{pmatrix} 1 & 0 & 0 & 0 \\ 0 & 0 & -3 & 2 \\ 0 & 1 & -8 & 3 \\ 0 & 0 & 0 & 0 \end{pmatrix}$$

$C_3 \to C_3 + 8C_2$, $C_4 \to C_4 - 3C_2$, we have

$$\sim \begin{pmatrix} 1 & 0 & 0 & 0 \\ 0 & 0 & -3 & 2 \\ 0 & 1 & 0 & 0 \\ 0 & 0 & 0 & 0 \end{pmatrix}$$

$C_4 \to 3C_4 + 2C_3$, we have

$$\sim \begin{pmatrix} 1 & 0 & 0 & 0 \\ 0 & 0 & -3 & 0 \\ 0 & 1 & 0 & 0 \\ 0 & 0 & 0 & 0 \end{pmatrix}$$

$C_3 \to \frac{C_3}{-3}$, we have

$$\sim \begin{pmatrix} 1 & 0 & 0 & 0 \\ 0 & 0 & 1 & 0 \\ 0 & 1 & 0 & 0 \\ 0 & 0 & 0 & 0 \end{pmatrix}$$

$R_2 \leftrightarrow R_3$, we have

$$\sim \begin{pmatrix} 1 & 0 & 0 & 0 \\ 0 & 1 & 0 & 0 \\ 0 & 0 & 1 & 0 \\ 0 & 0 & 0 & 0 \end{pmatrix}$$

$$= \begin{pmatrix} I_3 & 0 \\ 0 & 0 \end{pmatrix}$$

∴ The rank of the matrix $A$=Order of the identity matrix in normal form=3

**Example:** Find the rank of the matrix using normal form $\begin{pmatrix} 6 & 1 & 3 & 8 \\ 4 & 2 & 6 & -1 \\ 10 & 3 & 9 & 7 \\ 16 & 4 & 12 & 15 \end{pmatrix}$.

**Solution:** Let $A = \begin{pmatrix} 6 & 1 & 3 & 8 \\ 4 & 2 & 6 & -1 \\ 10 & 3 & 9 & 7 \\ 16 & 4 & 12 & 15 \end{pmatrix}$

$C_2 \leftrightarrow C_1$, we have

$$\sim \begin{pmatrix} 1 & 6 & 3 & 8 \\ 2 & 4 & 6 & -1 \\ 3 & 10 & 9 & 7 \\ 4 & 16 & 12 & 15 \end{pmatrix}$$

$R_2 \to R_2 - 2R_1, R_3 \to R_3 - 3R_1, R_4 \to R_4 - 4R_1$, we have

$$\sim \begin{pmatrix} 1 & 6 & 3 & 8 \\ 0 & -8 & 0 & -17 \\ 0 & -8 & 0 & -17 \\ 0 & -8 & 0 & -17 \end{pmatrix}$$

$C_2 \to C_2 - 6C_1, C_3 \to C_3 - 3C_1, C_4 \to C_4 - 8C_1$, we have

$$\sim \begin{pmatrix} 1 & 0 & 0 & 0 \\ 0 & -8 & 0 & -17 \\ 0 & -8 & 0 & -17 \\ 0 & -8 & 0 & -17 \end{pmatrix}$$

$R_3 \to R_3 - R_2, R_4 \to R_4 - R_2$, we have

$$\sim \begin{pmatrix} 1 & 0 & 0 & 0 \\ 0 & -8 & 0 & -17 \\ 0 & 0 & 0 & 0 \\ 0 & 0 & 0 & 0 \end{pmatrix}$$

$C_2 \to \frac{C_2}{-8}, C_4 \to \frac{C_4}{-17}$, we have

$$\sim \begin{pmatrix} 1 & 0 & 0 & 0 \\ 0 & 1 & 0 & 1 \\ 0 & 0 & 0 & 0 \\ 0 & 0 & 0 & 0 \end{pmatrix}$$

$C_4 \to C_4 - C_2$, we have

$$\sim \begin{pmatrix} 1 & 0 & 0 & 0 \\ 0 & 1 & 0 & 0 \\ 0 & 0 & 0 & 0 \\ 0 & 0 & 0 & 0 \end{pmatrix}$$

$$= \begin{pmatrix} I_2 & 0 \\ 0 & 0 \end{pmatrix}$$

∴ The rank of the matrix $A$=Order of the identity matrix in normal form=2

**Example:** Find the rank of the matrix using normal form $\begin{pmatrix} 2 & 1 & 2 & 4 \\ 4 & 2 & 5 & 8 \\ 3 & 1 & 4 & 6 \\ 1 & 1 & 2 & 2 \end{pmatrix}$.

**Solution:** Let $A = \begin{pmatrix} 2 & 1 & 2 & 4 \\ 4 & 2 & 5 & 8 \\ 3 & 1 & 4 & 6 \\ 1 & 1 & 2 & 2 \end{pmatrix}$

$R_1 \leftrightarrow R_4$, we have

$$\sim \begin{pmatrix} 1 & 1 & 2 & 2 \\ 4 & 2 & 5 & 8 \\ 3 & 1 & 4 & 6 \\ 2 & 1 & 2 & 4 \end{pmatrix}$$

$R_2 \to R_2 - 4R_1$, $R_3 \to R_3 - 3R_1$, $R_4 \to R_4 - 2R_1$, we have

$$\sim \begin{pmatrix} 1 & 1 & 2 & 2 \\ 0 & -2 & -3 & 0 \\ 0 & -2 & -2 & 0 \\ 0 & -1 & -2 & 0 \end{pmatrix}$$

$C_2 \to C_2 - C_1$, $C_3 \to C_3 - 2C_1$, $C_4 \to C_4 - 2C_1$, we have

$$\sim \begin{pmatrix} 1 & 0 & 0 & 0 \\ 0 & -2 & -3 & 0 \\ 0 & -2 & -2 & 0 \\ 0 & -1 & -2 & 0 \end{pmatrix}$$

$R_2 \to R_2 - 2R_4$, $R_3 \to R_3 - 2R_4$, we have

$$\sim \begin{pmatrix} 1 & 0 & 0 & 0 \\ 0 & 0 & 1 & 0 \\ 0 & 0 & -2 & 0 \\ 0 & -1 & -2 & 0 \end{pmatrix}$$

$R_3 \to R_3 - 2R_2$, we have

$$\sim \begin{pmatrix} 1 & 0 & 0 & 0 \\ 0 & 0 & 1 & 0 \\ 0 & 0 & 0 & 0 \\ 0 & -1 & -2 & 0 \end{pmatrix}$$

$R_4 \to R_4 + 2R_2$, we have

$$\sim \begin{pmatrix} 1 & 0 & 0 & 0 \\ 0 & 0 & 1 & 0 \\ 0 & 0 & 0 & 0 \\ 0 & -1 & 0 & 0 \end{pmatrix}$$

$R_3 \to (-1)R_3$, we have

$$\sim \begin{pmatrix} 1 & 0 & 0 & 0 \\ 0 & 0 & 1 & 0 \\ 0 & 0 & 0 & 0 \\ 0 & 1 & 0 & 0 \end{pmatrix}$$

$R_2 \leftrightarrow R_4$, we have

$$\sim \begin{pmatrix} 1 & 0 & 0 & 0 \\ 0 & 1 & 0 & 0 \\ 0 & 0 & 0 & 0 \\ 0 & 0 & 1 & 0 \end{pmatrix}$$

$R_3 \leftrightarrow R_4$, we have

$$\sim \begin{pmatrix} 1 & 0 & 0 & 0 \\ 0 & 1 & 0 & 0 \\ 0 & 0 & 1 & 0 \\ 0 & 0 & 0 & 0 \end{pmatrix}$$

$$= \begin{pmatrix} I_3 & 0 \\ 0 & 0 \end{pmatrix}$$

∴ The rank of the matrix $A$=Order of the identity matrix in normal form=3

**Example:** Find the rank of the matrix using normal form for the matrix $\begin{pmatrix} 2 & -2 & 0 & 6 \\ 4 & 2 & 0 & 2 \\ 1 & -1 & 0 & 3 \\ 1 & -2 & 1 & 2 \end{pmatrix}$

**Solution:** Let $A = \begin{pmatrix} 2 & -2 & 0 & 6 \\ 4 & 2 & 0 & 2 \\ 1 & -1 & 0 & 3 \\ 1 & -2 & 1 & 2 \end{pmatrix}$ (Aug-2022)

$R_3 \leftrightarrow R_1$, we have

$$\sim \begin{pmatrix} 1 & -1 & 0 & 3 \\ 4 & 2 & 0 & 2 \\ 2 & -2 & 0 & 6 \\ 1 & -2 & 1 & 2 \end{pmatrix}$$

$R_2 \to R_2 - 4R_1$, $R_3 \to R_3 - 2R_1$, $R_4 \to R_4 - R_1$, we have

$$\sim \begin{pmatrix} 1 & -1 & 0 & 3 \\ 0 & 6 & 0 & -10 \\ 0 & 0 & 0 & 0 \\ 0 & -1 & 1 & -1 \end{pmatrix}$$

$C_2 \to C_2 + C_1$, $C_4 \to C_4 - 3C_1$, we have

$$\sim \begin{pmatrix} 1 & 0 & 0 & 0 \\ 0 & 6 & 0 & -10 \\ 0 & 0 & 0 & 0 \\ 0 & -1 & 1 & -1 \end{pmatrix}$$

$C_3 \to C_3 + C_2$, $C_4 \to C_4 - C_2$, we have

$$\sim \begin{pmatrix} 1 & 0 & 0 & 0 \\ 0 & 6 & 6 & -16 \\ 0 & 0 & 0 & 0 \\ 0 & -1 & 0 & 0 \end{pmatrix}$$

$C_3 \to \frac{C_3}{6}$, $C_4 \to \frac{C_4}{-16}$, we have

$$\sim \begin{pmatrix} 1 & 0 & 0 & 0 \\ 0 & 6 & 1 & 1 \\ 0 & 0 & 0 & 0 \\ 0 & -1 & 0 & 0 \end{pmatrix}$$

$C_4 \to C_4 - C_3$, $C_4 \to (-1)C_4$, we have

$$\sim \begin{pmatrix} 1 & 0 & 0 & 0 \\ 0 & 6 & 1 & 0 \\ 0 & 0 & 0 & 0 \\ 0 & 1 & 0 & 0 \end{pmatrix}$$

$R_2 \to R_2 - 6R_4$, we have

$$\sim \begin{pmatrix} 1 & 0 & 0 & 0 \\ 0 & 0 & 1 & 0 \\ 0 & 0 & 0 & 0 \\ 0 & 1 & 0 & 0 \end{pmatrix}$$

$R_2 \leftrightarrow R_4$, we have

$$\sim \begin{pmatrix} 1 & 0 & 0 & 0 \\ 0 & 1 & 0 & 0 \\ 0 & 0 & 0 & 0 \\ 0 & 0 & 1 & 0 \end{pmatrix}$$

$R_3 \leftrightarrow R_4$, we have

$$\sim \begin{pmatrix} 1 & 0 & 0 & 0 \\ 0 & 1 & 0 & 0 \\ 0 & 0 & 1 & 0 \\ 0 & 0 & 0 & 0 \end{pmatrix}$$

$$= \begin{pmatrix} I_3 & 0 \\ 0 & 0 \end{pmatrix}$$

∴ The rank of the matrix $A$=Order of the identity matrix in normal form=3.

## EXERCISE

Reduce the following matrices into normal form and find its tank:

$$\begin{pmatrix} 1 & 0 & -3 & 2 \\ 0 & 1 & 4 & 5 \\ 1 & 3 & 2 & 0 \\ 1 & 1 & -2 & 0 \end{pmatrix}, \begin{pmatrix} 2 & -1 & 3 & 4 \\ 0 & 3 & 4 & 1 \\ 2 & 3 & 7 & 5 \\ 2 & 5 & 11 & 6 \end{pmatrix}, \begin{pmatrix} 2 & 1 & 3 & 5 \\ 4 & 2 & 1 & 3 \\ 8 & 4 & 7 & 13 \\ 8 & 4 & -3 & -1 \end{pmatrix}, \begin{pmatrix} -2 & -1 & -3 & -1 \\ 1 & 2 & 3 & -1 \\ 1 & 0 & 1 & 1 \\ 0 & 1 & 1 & -1 \end{pmatrix}$$

$$\begin{pmatrix} 3 & -2 & 0 & -1 \\ 0 & 2 & 2 & 1 \\ 1 & -2 & -3 & 2 \\ 0 & 1 & 2 & 1 \end{pmatrix} \text{(Apr-2018)} \quad \begin{pmatrix} 1 & 2 & 2 & 4 \\ 2 & 3 & 4 & 6 \\ 3 & 5 & 6 & 10 \\ -1 & 1 & -2 & -2 \end{pmatrix} \text{(Dec-2020)}$$

$$\begin{pmatrix} 1 & 2 & 3 & -2 \\ 2 & -2 & 1 & 3 \\ 3 & 0 & 4 & 1 \end{pmatrix} \text{(Dec-2020)} \quad \begin{pmatrix} 0 & 1 & -3 & -1 \\ 1 & 0 & 1 & 1 \\ 3 & 1 & 0 & 2 \\ 1 & 1 & -2 & 0 \end{pmatrix} \text{(Dec-2020), (Mar-2022)}$$

$$\begin{pmatrix} 2 & -4 & 3 & -1 & 0 \\ 1 & -2 & -1 & -4 & 2 \\ 0 & 1 & -1 & 3 & 1 \\ 4 & -7 & 4 & -4 & 5 \end{pmatrix} \text{(Jan-2020), (Aug-2021), (Dec-2021)}$$

$$\begin{pmatrix} 0 & 1 & 2 & -2 \\ 4 & 0 & 2 & 6 \\ 2 & 1 & 3 & 3 \end{pmatrix} \text{(Sep-2021)} \quad \begin{pmatrix} 1 & 4 & 3 & -2 & 1 \\ -2 & -3 & -1 & 4 & 3 \\ -1 & 6 & 7 & 2 & 9 \\ -3 & 3 & 6 & 6 & 12 \end{pmatrix} \text{(Aug-2022)}$$

**Ans:** (i) **4** (ii) **2** (iii) **2** (iv) **2** (v) **4** (vi) **3** (vii) **3** (viii) **2** (ix) **4** (x) **2**

### 1.1.3 Rank of the Matrix by $PAQ$ form

Consider $A_{m \times n} = I_{m \times m} A_{m \times n} I_{n \times n}$. Applying elementary row operations on $A$ and on prefactor $I_{m \times m}$ and applying elementary column operations on $A$ and postfactor $I_{n \times n}$ such that $A$ on the LHS is reduces to normal form, then $I_{m \times m}$ is reduced to $P_{m \times m}$ and $I_{n \times n}$ is reduces to $Q_{n \times n}$.

$$\therefore \ N = PAQ.$$

Here $P$ and $Q$ are non-singular matrices and not unique.

**Example:** Find two non-singular matrices $P$ and $Q$ such that the normal form of $A$ is $PAQ$, where $A = \begin{pmatrix} 2 & -1 & 3 \\ 1 & 1 & 1 \\ 1 & -1 & 1 \end{pmatrix}$. Hence find its rank.

**Solution:** Given $A = \begin{pmatrix} 2 & -1 & 3 \\ 1 & 1 & 1 \\ 1 & -1 & 1 \end{pmatrix}$.

Consider,
$$A_{3\times 3} = I_{3\times 3} A I_{3\times 3}$$

$$\begin{pmatrix} 2 & -1 & 3 \\ 1 & 1 & 1 \\ 1 & -1 & 1 \end{pmatrix} = \begin{pmatrix} 1 & 0 & 0 \\ 0 & 1 & 0 \\ 0 & 0 & 1 \end{pmatrix} A \begin{pmatrix} 1 & 0 & 0 \\ 0 & 1 & 0 \\ 0 & 0 & 1 \end{pmatrix}$$

$R_2 \leftrightarrow R_1$, we have

$$\begin{pmatrix} 1 & 1 & 1 \\ 2 & -1 & 3 \\ 1 & -1 & 1 \end{pmatrix} = \begin{pmatrix} 0 & 1 & 0 \\ 1 & 0 & 0 \\ 0 & 0 & 1 \end{pmatrix} A \begin{pmatrix} 1 & 0 & 0 \\ 0 & 1 & 0 \\ 0 & 0 & 1 \end{pmatrix}$$

$R_2 \to R_2 - 2R_1$, $R_3 \to R_3 - R_1$, we have

$$\begin{pmatrix} 1 & 1 & 1 \\ 0 & -3 & 1 \\ 0 & -2 & 0 \end{pmatrix} = \begin{pmatrix} 0 & 1 & 0 \\ 1 & -2 & 0 \\ 0 & -1 & 1 \end{pmatrix} A \begin{pmatrix} 1 & 0 & 0 \\ 0 & 1 & 0 \\ 0 & 0 & 1 \end{pmatrix}$$

$C_2 \to C_2 - C_1$, $C_3 \to C_3 - C_1$, we have

$$\begin{pmatrix} 1 & 0 & 0 \\ 0 & -3 & 1 \\ 0 & -2 & 0 \end{pmatrix} = \begin{pmatrix} 0 & 1 & 0 \\ 1 & -2 & 0 \\ 0 & -1 & 1 \end{pmatrix} A \begin{pmatrix} 1 & -1 & -1 \\ 0 & 1 & 0 \\ 0 & 0 & 1 \end{pmatrix}$$

$C_2 \to C_2 + 3C_3$, we have

$$\begin{pmatrix} 1 & 0 & 0 \\ 0 & 0 & 1 \\ 0 & -2 & 0 \end{pmatrix} = \begin{pmatrix} 0 & 1 & 0 \\ 1 & -2 & 0 \\ 0 & -1 & 1 \end{pmatrix} A \begin{pmatrix} 1 & -4 & -1 \\ 0 & 1 & 0 \\ 0 & 3 & 1 \end{pmatrix}$$

$C_2 \to \frac{C_2}{-2}$ we have

$$\begin{pmatrix} 1 & 0 & 0 \\ 0 & 0 & 1 \\ 0 & 1 & 0 \end{pmatrix} = \begin{pmatrix} 0 & 1 & 0 \\ 1 & -2 & 0 \\ 0 & -1 & 1 \end{pmatrix} A \begin{pmatrix} 1 & 2 & -1 \\ 0 & \frac{-1}{2} & 0 \\ 0 & \frac{-3}{2} & 1 \end{pmatrix}$$

$R_2 \leftrightarrow R_3$, we have

$$\begin{pmatrix} 1 & 0 & 0 \\ 0 & 1 & 0 \\ 0 & 0 & 1 \end{pmatrix} = \begin{pmatrix} 0 & 1 & 0 \\ 0 & -1 & 1 \\ 1 & -2 & 0 \end{pmatrix} A \begin{pmatrix} 1 & 2 & -1 \\ 0 & \frac{-1}{2} & 0 \\ 0 & \frac{-3}{2} & 1 \end{pmatrix}$$

$$\Rightarrow I_3 = PAQ$$

where $P = \begin{pmatrix} 0 & 1 & 0 \\ 0 & -1 & 1 \\ 1 & -2 & 0 \end{pmatrix}, Q = \begin{pmatrix} 1 & 2 & -1 \\ 0 & \frac{-1}{2} & 0 \\ 0 & \frac{-3}{2} & 1 \end{pmatrix}$ are non-singular matrices.

∴ Rank of $A = \rho(A) =$ order of the identity matrix in the normal form=3.

**Example:** Find two non-singular matrices $P$ and $Q$ such that the normal form of $A$ is $PAQ$, where $A = \begin{pmatrix} 1 & 3 & 6 & -1 \\ 1 & 4 & 5 & 1 \\ 1 & 5 & 4 & 3 \end{pmatrix}$. Hence find its rank.

**Solution:** Given $A = \begin{pmatrix} 1 & 3 & 6 & -1 \\ 1 & 4 & 5 & 1 \\ 1 & 5 & 4 & 3 \end{pmatrix}$.

Consider,

$$A_{3\times 4} = I_{3\times 3} A I_{4\times 4}$$

$$\begin{pmatrix} 1 & 3 & 6 & -1 \\ 1 & 4 & 5 & 1 \\ 1 & 5 & 4 & 3 \end{pmatrix} = \begin{pmatrix} 1 & 0 & 0 \\ 0 & 1 & 0 \\ 0 & 0 & 1 \end{pmatrix} A \begin{pmatrix} 1 & 0 & 0 & 0 \\ 0 & 1 & 0 & 0 \\ 0 & 0 & 1 & 0 \\ 0 & 0 & 0 & 1 \end{pmatrix}$$

$R_2 \to R_2 - R_1, R_3 \to R_3 - R_1$, we have

$$\begin{pmatrix} 1 & 3 & 6 & -1 \\ 0 & 1 & -1 & 2 \\ 0 & 2 & -2 & 4 \end{pmatrix} = \begin{pmatrix} 1 & 0 & 0 \\ -1 & 1 & 0 \\ -1 & 0 & 1 \end{pmatrix} A \begin{pmatrix} 1 & 0 & 0 & 0 \\ 0 & 1 & 0 & 0 \\ 0 & 0 & 1 & 0 \\ 0 & 0 & 0 & 1 \end{pmatrix}$$

$C_2 \to C_2 - 3C_1$, $C_3 \to C_3 - 6C_1$, $C_4 \to C_4 + C_1$, we have

$$\begin{pmatrix} 1 & 0 & 0 & 0 \\ 0 & 1 & -1 & 2 \\ 0 & 2 & -2 & 4 \end{pmatrix} = \begin{pmatrix} 1 & 0 & 0 \\ -1 & 1 & 0 \\ -1 & 0 & 1 \end{pmatrix} A \begin{pmatrix} 1 & -3 & -6 & 1 \\ 0 & 1 & 0 & 0 \\ 0 & 0 & 1 & 0 \\ 0 & 0 & 0 & 1 \end{pmatrix}$$

$R_3 \to R_3 - 2R_2$, we have

$$\begin{pmatrix} 1 & 0 & 0 & 0 \\ 0 & 1 & -1 & 2 \\ 0 & 0 & 0 & 0 \end{pmatrix} = \begin{pmatrix} 1 & 0 & 0 \\ -1 & 1 & 0 \\ -1 & -2 & 1 \end{pmatrix} A \begin{pmatrix} 1 & -3 & -6 & 1 \\ 0 & 1 & 0 & 0 \\ 0 & 0 & 1 & 0 \\ 0 & 0 & 0 & 1 \end{pmatrix}$$

$C_3 \to C_3 + C_2$, $C_4 \to C_4 - 2C_2$, we have

$$\begin{pmatrix} 1 & 0 & 0 & 0 \\ 0 & 1 & 0 & 0 \\ 0 & 0 & 0 & 0 \end{pmatrix} = \begin{pmatrix} 1 & 0 & 0 \\ -1 & 1 & 0 \\ -1 & -2 & 1 \end{pmatrix} A \begin{pmatrix} 1 & -3 & -9 & 7 \\ 0 & 1 & 1 & -2 \\ 0 & 0 & 1 & 0 \\ 0 & 0 & 0 & 1 \end{pmatrix}$$

$$\begin{pmatrix} I_2 & 0 \\ 0 & 0 \end{pmatrix} = PAQ$$

where $P = \begin{pmatrix} 1 & 0 & 0 \\ -1 & 1 & 0 \\ -1 & -2 & 1 \end{pmatrix}$, $Q = \begin{pmatrix} 1 & -3 & -9 & 7 \\ 0 & 1 & 1 & -2 \\ 0 & 0 & 1 & 0 \\ 0 & 0 & 0 & 1 \end{pmatrix}$ are non-singular matrices.

$\therefore$ Rank of $A = \rho(A) =$ order of the identity matrix in the normal form $= 2$.

**Example:** Reduce the matrix $A = \begin{pmatrix} 1 & 2 & 3 & -2 \\ 2 & -2 & 1 & 3 \\ 3 & 0 & 4 & 1 \end{pmatrix}$ into $PAQ$ form, and hence find the rank of the matrix.

(Apr-2018)

**Solution:** Given $A = \begin{pmatrix} 1 & 2 & 3 & -2 \\ 2 & -2 & 1 & 3 \\ 3 & 0 & 4 & 1 \end{pmatrix}$.

Consider,

$$A_{3\times 4} = I_{3\times 3} A I_{4\times 4}$$

$$\begin{pmatrix} 1 & 2 & 3 & -2 \\ 2 & -2 & 1 & 3 \\ 3 & 0 & 4 & 1 \end{pmatrix} = \begin{pmatrix} 1 & 0 & 0 \\ 0 & 1 & 0 \\ 0 & 0 & 1 \end{pmatrix} A \begin{pmatrix} 1 & 0 & 0 & 0 \\ 0 & 1 & 0 & 0 \\ 0 & 0 & 1 & 0 \\ 0 & 0 & 0 & 1 \end{pmatrix}$$

$R_2 \to R_2 - 2R_1, R_3 \to R_3 - 3R_1$, we have

$$\begin{pmatrix} 1 & 2 & 3 & -2 \\ 0 & -6 & -5 & 7 \\ 0 & -6 & -5 & 7 \end{pmatrix} = \begin{pmatrix} 1 & 0 & 0 \\ -2 & 1 & 0 \\ -3 & 0 & 1 \end{pmatrix} A \begin{pmatrix} 1 & 0 & 0 & 0 \\ 0 & 1 & 0 & 0 \\ 0 & 0 & 1 & 0 \\ 0 & 0 & 0 & 1 \end{pmatrix}$$

$C_2 \to C_2 - 2C_1, C_3 \to C_3 - 3C_1, C_4 \to C_4 + 2C_1$, we have

$$\begin{pmatrix} 1 & 0 & 0 & 0 \\ 0 & -6 & -5 & 7 \\ 0 & -6 & -5 & 7 \end{pmatrix} = \begin{pmatrix} 1 & 0 & 0 \\ -2 & 1 & 0 \\ -3 & 0 & 1 \end{pmatrix} A \begin{pmatrix} 1 & -2 & -3 & 2 \\ 0 & 1 & 0 & 0 \\ 0 & 0 & 1 & 0 \\ 0 & 0 & 0 & 1 \end{pmatrix}$$

$R_3 \to R_3 - R_2$, we have

$$\begin{pmatrix} 1 & 0 & 0 & 0 \\ 0 & -6 & -5 & 7 \\ 0 & 0 & 0 & 0 \end{pmatrix} = \begin{pmatrix} 1 & 0 & 0 \\ -2 & 1 & 0 \\ -1 & -1 & 1 \end{pmatrix} A \begin{pmatrix} 1 & -2 & -3 & 2 \\ 0 & 1 & 0 & 0 \\ 0 & 0 & 1 & 0 \\ 0 & 0 & 0 & 1 \end{pmatrix}$$

$C_3 \to 6C_3 - 5C_2, C_4 \to 6C_4 + 7C_2$, we have

$$\begin{pmatrix} 1 & 0 & 0 & 0 \\ 0 & -6 & 0 & 0 \\ 0 & 0 & 0 & 0 \end{pmatrix} = \begin{pmatrix} 1 & 0 & 0 \\ -2 & 1 & 0 \\ -1 & -1 & 1 \end{pmatrix} A \begin{pmatrix} 1 & -2 & -8 & -2 \\ 0 & 1 & -5 & 7 \\ 0 & 0 & 6 & 0 \\ 0 & 0 & 0 & 6 \end{pmatrix}$$

$R_2 \to \frac{R_2}{-6}$, we have

$$\begin{pmatrix} 1 & 0 & 0 & 0 \\ 0 & 1 & 0 & 0 \\ 0 & 0 & 0 & 0 \end{pmatrix} = \begin{pmatrix} 1 & 0 & 0 \\ \frac{-1}{3} & \frac{-1}{6} & 0 \\ -1 & -1 & 1 \end{pmatrix} A \begin{pmatrix} 1 & -2 & -8 & -2 \\ 0 & 1 & -5 & 7 \\ 0 & 0 & 6 & 0 \\ 0 & 0 & 0 & 6 \end{pmatrix}$$

$$\Rightarrow \begin{pmatrix} I_2 & 0 \\ 0 & 0 \end{pmatrix} = PAQ$$

where $P = \begin{pmatrix} 1 & 0 & 0 \\ \frac{-1}{3} & \frac{-1}{6} & 0 \\ -1 & -1 & 1 \end{pmatrix}$, $Q = \begin{pmatrix} 1 & -2 & -8 & -2 \\ 0 & 1 & -5 & 7 \\ 0 & 0 & 6 & 0 \\ 0 & 0 & 0 & 6 \end{pmatrix}$ are non-singular matrices.

$\therefore$ Rank of $A = \rho(A)$ = order of the identity matrix in the normal form = 2.

**EXERCISE**

1. Find two non-singular matrices $P$ and $Q$ such that $PAQ$ is in the normal form where $A = \begin{pmatrix} 1 & 1 & 2 \\ 1 & 2 & 3 \\ 0 & -1 & -1 \end{pmatrix}$.

   **Ans: 2.**

2. If $A = \begin{pmatrix} 1 & -1 & -1 & 2 \\ 4 & 2 & 2 & -1 \\ 2 & 2 & 0 & -2 \end{pmatrix}$, find two non-singular matrices $P$ and $Q$ such that $PAQ$ is in the normal form.

   **Ans: 3.**

### 1.1.4 Gauss-Jordan Method to find Inverse of a Matrix

Suppose $A$ is a non-singular matrix of order $n$. We write $A = I_n A$.

Now, we apply elementary row operations only to the matrix $A$ and the prefactor $I_n$ of the LHS. We will do this till, we get an equation of the form

$$I_n = BA.$$

Then obviously, $B$ is the inverse of $A$.

**Example:** Find the inverse of $A$ by Gauss-Jordan method where $A = \begin{pmatrix} 1 & 2 & 3 \\ 2 & 4 & 5 \\ 3 & 5 & 6 \end{pmatrix}$.

**Solution:** Consider $A = IA$ and apply elementary row operations on LHS $A$ and prefactor $I$ on RHS until the LHS matrix is an identity matrix.

$$\begin{pmatrix} 1 & 2 & 3 \\ 2 & 4 & 5 \\ 3 & 5 & 6 \end{pmatrix} = \begin{pmatrix} 1 & 0 & 0 \\ 0 & 1 & 0 \\ 0 & 0 & 1 \end{pmatrix} A$$

$R_2 \to R_2 - 2R_1$, $R_3 \to R_3 - 3R_1$, we have

$$\begin{pmatrix} 1 & 2 & 3 \\ 0 & 0 & -1 \\ 0 & -1 & -3 \end{pmatrix} = \begin{pmatrix} 1 & 0 & 0 \\ -2 & 1 & 0 \\ -3 & 0 & 1 \end{pmatrix} A$$

$R_3 \to R_3 - 3R_2$, we have

$$\begin{pmatrix} 1 & 2 & 3 \\ 0 & 0 & -1 \\ 0 & -1 & 0 \end{pmatrix} = \begin{pmatrix} 1 & 0 & 0 \\ -2 & 1 & 0 \\ 3 & -3 & 1 \end{pmatrix} A$$

$R_1 \to R_1 + 3R_2$, we have

$$\begin{pmatrix} 1 & 2 & 0 \\ 0 & 0 & -1 \\ 0 & -1 & 0 \end{pmatrix} = \begin{pmatrix} -5 & 3 & 0 \\ -2 & 1 & 0 \\ 3 & -3 & 1 \end{pmatrix} A$$

$R_1 \to R_1 + 2R_3$, we have

$$\begin{pmatrix} 1 & 0 & 0 \\ 0 & 0 & -1 \\ 0 & -1 & 0 \end{pmatrix} = \begin{pmatrix} 1 & -3 & 2 \\ -2 & 1 & 0 \\ 3 & -3 & 1 \end{pmatrix} A$$

$R_2 \to R_2(-1)$, $R_3 \to R_3(-1)$, we have

$$\begin{pmatrix} 1 & 0 & 0 \\ 0 & 0 & 1 \\ 0 & 1 & 0 \end{pmatrix} = \begin{pmatrix} 1 & -3 & 2 \\ 2 & -1 & 0 \\ -3 & 3 & 1 \end{pmatrix} A$$

$R_2 \leftrightarrow R_3$, we have

$$\begin{pmatrix} 1 & 0 & 0 \\ 0 & 1 & 0 \\ 0 & 0 & 1 \end{pmatrix} = \begin{pmatrix} 1 & -3 & 2 \\ -3 & 3 & 1 \\ 2 & -1 & 0 \end{pmatrix} A$$

Thus,

$$A^{-1} = \begin{pmatrix} 1 & -3 & 2 \\ -3 & 3 & 1 \\ 2 & -1 & 0 \end{pmatrix}.$$

**Example:** Find the inverse of $A$ by Gauss-Jordan method where $A = \begin{pmatrix} -1 & -3 & 3 & -1 \\ 1 & 1 & -1 & 0 \\ 2 & -5 & 2 & -3 \\ -1 & 1 & 0 & 1 \end{pmatrix}.$

**Solution:** Consider $A = IA$ and apply elementary row operations on LHS $A$ and prefactor $I$ on RHS until the LHS matrix is an identity matrix.

$$\begin{pmatrix} -1 & -3 & 3 & -1 \\ 1 & 1 & -1 & 0 \\ 2 & -5 & 2 & -3 \\ -1 & 1 & 0 & 1 \end{pmatrix} = \begin{pmatrix} 1 & 0 & 0 & 0 \\ 0 & 1 & 0 & 0 \\ 0 & 0 & 1 & 0 \\ 0 & 0 & 0 & 1 \end{pmatrix} A$$

$R_1 \leftrightarrow R_2$, we have

$$\begin{pmatrix} 1 & 1 & -1 & 0 \\ -1 & -3 & 3 & -1 \\ 2 & -5 & 2 & -3 \\ -1 & 1 & 0 & 1 \end{pmatrix} = \begin{pmatrix} 0 & 1 & 0 & 0 \\ 1 & 0 & 0 & 0 \\ 0 & 0 & 1 & 0 \\ 0 & 0 & 0 & 1 \end{pmatrix} A$$

$R_2 \to R_2 + R_1$, $R_3 \to R_3 - 2R_1$, $R_4 \to R_4 + R_1$, we have

$$\begin{pmatrix} 1 & 1 & -1 & 0 \\ 0 & -2 & 2 & -1 \\ 0 & -7 & 4 & -3 \\ 0 & 2 & -1 & 1 \end{pmatrix} = \begin{pmatrix} 0 & 1 & 0 & 0 \\ 1 & 1 & 0 & 0 \\ 0 & -2 & 1 & 0 \\ 0 & 1 & 0 & 1 \end{pmatrix} A$$

$R_1 \to 2R_1 + R_2$, $R_4 \to R_4 + R_2$, we have

$$\begin{pmatrix} 2 & 0 & 0 & -1 \\ 0 & -2 & 2 & -1 \\ 0 & -7 & 4 & -3 \\ 0 & 0 & 1 & 0 \end{pmatrix} = \begin{pmatrix} 1 & 3 & 0 & 0 \\ 1 & 1 & 0 & 0 \\ 0 & -2 & 1 & 0 \\ 1 & 2 & 0 & 1 \end{pmatrix} A$$

$R_3 \to 2R_3 - 7R_2$, we have

$$\begin{pmatrix} 2 & 0 & 0 & -1 \\ 0 & -2 & 2 & -1 \\ 0 & 0 & -6 & 1 \\ 0 & 0 & 1 & 0 \end{pmatrix} = \begin{pmatrix} 1 & 3 & 0 & 0 \\ 1 & 1 & 0 & 0 \\ -7 & -11 & 2 & 0 \\ 1 & 2 & 0 & 1 \end{pmatrix} A$$

$R_1 \to R_1 + R_4,\ R_2 \to R_2 + 2R_4, R_3 \to R_3 - R_4$, we have

$$\begin{pmatrix} 2 & 0 & 0 & 0 \\ 0 & -6 & 0 & 0 \\ 0 & 0 & -6 & 0 \\ 0 & 0 & 0 & 1 \end{pmatrix} = \begin{pmatrix} 0 & 4 & 2 & 6 \\ -6 & -6 & 6 & 12 \\ -6 & -12 & 0 & -6 \\ -1 & 1 & 2 & 6 \end{pmatrix} A$$

$R_1 \to \frac{R_1}{2},\ R_2 \to \frac{R_2}{-6},\ R_3 \to \frac{R_3}{-6}$, we have

$$\begin{pmatrix} 1 & 0 & 0 & 0 \\ 0 & 1 & 0 & 0 \\ 0 & 0 & 1 & 0 \\ 0 & 0 & 0 & 1 \end{pmatrix} = \begin{pmatrix} 0 & 2 & 1 & 3 \\ 1 & 1 & -1 & -2 \\ 1 & 2 & 0 & 1 \\ -1 & 1 & 2 & 6 \end{pmatrix} A$$

Thus,
$$A^{-1} = \begin{pmatrix} 0 & 2 & 1 & 3 \\ 1 & 1 & -1 & -2 \\ 1 & 2 & 0 & 1 \\ -1 & 1 & 2 & 6 \end{pmatrix}.$$

**Example:** Find the inverse of $A$ by Gauss-Jordan method where $A = \begin{pmatrix} 1 & 1 & 3 \\ 1 & 3 & -3 \\ -2 & -4 & -4 \end{pmatrix}$.

**Ans:** $A^{-1} = \begin{pmatrix} 3 & 1 & 3/2 \\ -5/4 & -1/4 & -3/4 \\ -1/4 & -1/4 & -1/4 \end{pmatrix}.$

## 1.2 System of Linear Equations (Non-Homogeneous)

Let us consider the following system of $m$ linear equations in $n$ unknowns $x_1, x_2, \cdots, x_n$.

$$\left.\begin{array}{c} a_{11}x_1 + a_{12}x_2 + \cdots + a_{1n}x_n = b_1 \\ a_{21}x_1 + a_{22}x_2 + \cdots + a_{2n}x_n = b_2 \\ \cdots \qquad \cdots \qquad \cdots \qquad \cdots \\ a_{m1}x_1 + a_{m2}x_2 + \cdots + a_{mn}x_n = b_m \end{array}\right\} \quad \cdots \cdots \quad (1)$$

In the matrix notation, these equations can be put in the form

$$AX = B \quad \cdots \cdots \quad (2)$$

where $A = \begin{pmatrix} a_{11} & a_{12} & \cdots & a_{1n} \\ a_{21} & a_{22} & \cdots & a_{2n} \\ \cdots & \cdots & \cdots & \cdots \\ a_{m1} & a_{m2} & \cdots & a_{mn} \end{pmatrix}$, $X = \begin{pmatrix} x_1 \\ x_2 \\ \vdots \\ x_n \end{pmatrix}$, $B = \begin{pmatrix} b_1 \\ b_2 \\ \vdots \\ b_m \end{pmatrix}$.

**Augmented Matrix:** The augmented matrix $[A|B]$ or $[A:B]$ of system (1) is obtained by augmenting $A$ by column $B$.

$$[A|B] = \begin{pmatrix} a_{11} & a_{12} & \cdots & a_{1n} & b_1 \\ a_{21} & a_{22} & \cdots & a_{2n} & b_2 \\ \cdots & \cdots & \cdots & \cdots & \vdots \\ a_{m1} & a_{m2} & \cdots & a_{mn} & b_m \end{pmatrix}.$$

## 1.3 Conditions for Solution of Linear Systems

**Consistent:** If the ranks of $A$ and augmented matrix $[A|B]$ are equal then the system is said to be consistent, otherwise inconsistent.

There are following conditions for exist the solution of any system of linear equations:

**Case (i):** If $\rho(A) = \rho[A|B] = r = n$, where $n$ is the number of unknowns, then the system has a unique solution.

**Case (ii):** If $\rho(A) = \rho[A|B] = r < n$, then the system has infinitely many solutions in terms of remaining $n - r$ unknowns which are arbitrary.

**Example:** Solve the given system of equations if consistent $x + 2y + z = 3$, $2x + 3y + 2z = 5$, $3x - 5y + 5z = 2$, $3x + 9y - z = 4$.

**Solution:** The given system of equations can be put in the matrix form as

$$\begin{pmatrix} 1 & 2 & 1 \\ 2 & 3 & 2 \\ 3 & -5 & 5 \\ 3 & 9 & -1 \end{pmatrix} \begin{pmatrix} x \\ y \\ z \end{pmatrix} = \begin{pmatrix} 3 \\ 5 \\ 2 \\ 4 \end{pmatrix}.$$

$$\Rightarrow AX = B,$$

where $A = \begin{pmatrix} 1 & 2 & 1 \\ 2 & 3 & 2 \\ 3 & -5 & 5 \\ 3 & 9 & -1 \end{pmatrix}$, $X = \begin{pmatrix} x \\ y \\ z \end{pmatrix}$ and $B = \begin{pmatrix} 3 \\ 5 \\ 2 \\ 4 \end{pmatrix}$.

The augmented matrix is given by

$$[A|B] = \begin{pmatrix} 1 & 2 & 1 & | & 3 \\ 2 & 3 & 2 & | & 5 \\ 3 & -5 & 5 & | & 2 \\ 3 & 9 & -1 & | & 4 \end{pmatrix}$$

$R_2 \to R_2 - 2R_1$, $R_3 \to R_3 - 3R_1$, $R_4 \to R_4 - 3R_1$, we have

$$\sim \begin{pmatrix} 1 & 2 & 1 & | & 3 \\ 0 & -1 & 0 & | & -1 \\ 0 & -11 & 2 & | & -7 \\ 0 & 3 & -4 & | & -5 \end{pmatrix}$$

$R_3 \to R_3 - 11R_2$, $R_4 \to R_4 + 3R_2$, we have

$R_4 \to R_4 + 2R_3$, we have

$$\sim \begin{pmatrix} 1 & 2 & 1 & | & 3 \\ 0 & -1 & 0 & | & -1 \\ 0 & 0 & 2 & | & 4 \\ 0 & 0 & -4 & | & -8 \end{pmatrix}$$

$$\sim \begin{pmatrix} 1 & 2 & 1 & | & 3 \\ 0 & -1 & 0 & | & -1 \\ 0 & 0 & 2 & | & 4 \\ 0 & 0 & 0 & | & 0 \end{pmatrix} \quad \ldots \ldots \text{(i)}$$

From this, we find that

$\rho(A) = \rho[A|B]$=No. of non-zero rows=3=No. of unknowns.

Therefore, the given system is consistent and has a unique solution.

From (i), we observe that the given equations are equivalent to the following equations

$$x + 2y + z = 3, \quad -y = -1, \quad 2z = 4.$$

From these, we get $y = 1, z = 2$ and $x = 3 - 2y - z = -1$.

Thus, $x = -1, y = 1, z = 2$ is the unique solution of the given system.

**Example:** Show that the following system of equations **(Dec-2021)**

$$3x + 3y + 2z = 1$$
$$x + 2y = 4$$
$$10y + 3z = -2$$
$$2x - 3y - z = 5$$

is consistent and solve it.

**Solution:** The given system of equations can be put in the matrix form as

$$\begin{pmatrix} 3 & 3 & 2 \\ 1 & 2 & 0 \\ 0 & 10 & 3 \\ 2 & -3 & -1 \end{pmatrix} \begin{pmatrix} x \\ y \\ z \end{pmatrix} = \begin{pmatrix} 1 \\ 4 \\ -2 \\ 5 \end{pmatrix}.$$

$$\Rightarrow AX = B,$$

where $A = \begin{pmatrix} 3 & 3 & 2 \\ 1 & 2 & 0 \\ 0 & 10 & 3 \\ 2 & -3 & -1 \end{pmatrix}$, $X = \begin{pmatrix} x \\ y \\ z \end{pmatrix}$ and $B = \begin{pmatrix} 1 \\ 4 \\ -2 \\ 5 \end{pmatrix}$.

The augmented matrix is given by

$$[A|B] = \left(\begin{array}{ccc|c} 3 & 3 & 2 & 1 \\ 1 & 2 & 0 & 4 \\ 0 & 10 & 3 & -2 \\ 2 & -3 & -1 & 5 \end{array}\right)$$

$R_2 \leftrightarrow R_1$, we have

$$\sim \left(\begin{array}{ccc|c} 1 & 2 & 0 & 4 \\ 3 & 3 & 2 & 1 \\ 0 & 10 & 3 & -2 \\ 2 & -3 & -1 & 5 \end{array}\right)$$

$R_2 \to R_2 - 3R_1$, $R_4 \to R_4 - 2R_1$, we have

$$\sim \left(\begin{array}{ccc|c} 1 & 2 & 0 & 4 \\ 0 & -3 & 2 & -11 \\ 0 & 10 & 3 & -2 \\ 0 & -7 & -1 & -3 \end{array}\right)$$

$R_2 \to R_2 + 2R_4$, $R_3 \to R_3 + 3R_4$, we have

$$\sim \left(\begin{array}{ccc|c} 1 & 2 & 0 & 4 \\ 0 & -17 & 0 & -17 \\ 0 & -11 & 0 & -11 \\ 0 & -7 & -1 & -3 \end{array}\right)$$

$R_2 \to \frac{R_2}{-17}$, $R_3 \to \frac{R_3}{-11}$, $R_4 \to \frac{R_4}{-1}$, we have

$$\sim \left(\begin{array}{ccc|c} 1 & 2 & 0 & 4 \\ 0 & 1 & 0 & 1 \\ 0 & 1 & 0 & 1 \\ 0 & 7 & 1 & 3 \end{array}\right)$$

$R_3 \to R_3 - R_2, R_4 \to R_4 - 7R_2$, we have

$$\sim \left(\begin{array}{ccc|c} 1 & 2 & 0 & 4 \\ 0 & 1 & 0 & 1 \\ 0 & 0 & 0 & 0 \\ 0 & 0 & 1 & -4 \end{array}\right)$$

$R_3 \leftrightarrow R_4$, we have

$$\sim \left(\begin{array}{ccc|c} 1 & 2 & 0 & 4 \\ 0 & 1 & 0 & 1 \\ 0 & 0 & 1 & -4 \\ 0 & 0 & 0 & 0 \end{array}\right) \quad \ldots\ldots \text{(i)}$$

From this, we find that

$$\rho(A) = \rho[A|B] = \text{No. of non-zero rows} = 3 = \text{No. of unknowns}.$$

Therefore, the given system is consistent and has a unique solution.

From (i), we observe that the given equations are equivalent to the following equations

$x + 2y = 4$, $y = 1$ and $z = -4$.

From these, we get $y = 1, z = -4$ and $x = 4 - 2y = 2$.

Thus, $x = 2, y = 1, z = -4$ is the unique solution of the given system.

**Example:** Find whether the following equations are consistent, if so, solve them
$x + y + 2z = 4$, $2x - y + 3z = 9$, $3x - y - z = 2$.

**Solution:** The given system of equations can be put in the matrix form as

$$\begin{pmatrix} 1 & 1 & 2 \\ 2 & -1 & 3 \\ 3 & -1 & -1 \end{pmatrix} \begin{pmatrix} x \\ y \\ z \end{pmatrix} = \begin{pmatrix} 4 \\ 9 \\ 2 \end{pmatrix}.$$

$$\Rightarrow AX = B,$$

where $A = \begin{pmatrix} 1 & 1 & 2 \\ 2 & -1 & 3 \\ 3 & -1 & -1 \end{pmatrix}$, $X = \begin{pmatrix} x \\ y \\ z \end{pmatrix}$ and $B = \begin{pmatrix} 4 \\ 9 \\ 2 \end{pmatrix}$.

The augmented matrix is given by

$$[A|B] = \begin{pmatrix} 1 & 1 & 2 & | & 4 \\ 2 & -1 & 3 & | & 9 \\ 3 & -1 & -1 & | & 2 \end{pmatrix}$$

$R_2 \to R_2 - 2R_1$, $R_3 \to R_3 - 3R_1$, we have

$$\sim \begin{pmatrix} 1 & 1 & 2 & | & 4 \\ 0 & -3 & -1 & | & 1 \\ 0 & -4 & -7 & | & -10 \end{pmatrix}$$

$R_3 \to R_3 - 7R_2$, we have

$$\sim \begin{pmatrix} 1 & 1 & 2 & | & 4 \\ 0 & -3 & -1 & | & 1 \\ 0 & 17 & 0 & | & -17 \end{pmatrix}$$

$R_3 \to \frac{R_3}{17}$, we have

$$\sim \begin{pmatrix} 1 & 1 & 2 & | & 4 \\ 0 & -3 & -1 & | & 1 \\ 0 & 1 & 0 & | & -1 \end{pmatrix}$$

$R_2 \to R_2 + 3R_3$, we have

$$\sim \begin{pmatrix} 1 & 1 & 2 & | & 4 \\ 0 & 0 & -1 & | & -2 \\ 0 & 1 & 0 & | & -1 \end{pmatrix}$$

$R_2 \leftrightarrow R_3$, we have

$$\sim \begin{pmatrix} 1 & 1 & 2 & | & 4 \\ 0 & 1 & 0 & | & -1 \\ 0 & 0 & -1 & | & -2 \end{pmatrix} \quad \ldots\ldots \text{(i)}$$

From this, we find that

$\rho(A) = \rho[A|B]$=No. of non-zero rows=3=No. of unknowns.

Therefore, the given system is consistent and has a unique solution.

From (i), we observe that the given equations are equivalent to the following equations

$$x + y + 2z = 4,\; y = -1 \text{ and } -z = -2.$$

From these, we get $y = -1, z = 2$ and $x = 4 - y - 2z = 1$.

Thus, $x = 1, y = -1, z = 2$ is the unique solution of the given system.

**Example:** Find whether the following equations are consistent, if so, solve them
$2x - y - z = 2$, $x + 2y + z = 2$, $4x - 7y - 5z = 2$.  (Apr-2017)

**Solution:** The given system of equations can be put in the matrix form as

$$\begin{pmatrix} 2 & -1 & -1 \\ 1 & 2 & 1 \\ 4 & -7 & -5 \end{pmatrix} \begin{pmatrix} x \\ y \\ z \end{pmatrix} = \begin{pmatrix} 2 \\ 2 \\ 2 \end{pmatrix}.$$

$$\Rightarrow AX = B,$$

where $A = \begin{pmatrix} 2 & -1 & -1 \\ 1 & 2 & 1 \\ 4 & -7 & -5 \end{pmatrix}$, $X = \begin{pmatrix} x \\ y \\ z \end{pmatrix}$ and $B = \begin{pmatrix} 2 \\ 2 \\ 2 \end{pmatrix}$.

The augmented matrix is given by

$$[A|B] = \begin{pmatrix} 2 & -1 & -1 & | & 2 \\ 1 & 2 & 1 & | & 2 \\ 4 & -7 & -5 & | & 2 \end{pmatrix}$$

$R_1 \leftrightarrow R_2$, we have

$$\sim \begin{pmatrix} 1 & 2 & 1 & | & 2 \\ 2 & -1 & -1 & | & 2 \\ 4 & -7 & -5 & | & 2 \end{pmatrix}$$

$R_2 \to R_2 - 2R_1$, $R_3 \to R_3 - 4R_1$ we have

$$\sim \begin{pmatrix} 1 & 2 & 1 & | & 2 \\ 0 & -5 & -3 & | & -2 \\ 0 & -15 & -9 & | & -6 \end{pmatrix}$$

$R_3 \to R_3 - 3R_2$, we have

$$\sim \begin{pmatrix} 1 & 2 & 1 & | & 2 \\ 0 & -5 & -3 & | & -2 \\ 0 & 0 & 0 & | & 0 \end{pmatrix} \quad \ldots\ldots \text{(i)}$$

From this, we find that
$$\rho(A) = \rho[A|B] = \text{No. of non-zero rows} = 2 < 3 = \text{No. of unknowns}.$$
Therefore, the given system is consistent and has infinitely many solutions.
The number of unknowns that can be chosen arbitrarily is $n - r = 3 - 2 = 1$.
From (i), we observe that the given equations are equivalent to the following equations
$$x + 2y + z = 2, \quad -5y - 3z = -2$$
Choose $z = k$ (arbitrary), we get from these
$$y = \frac{2-3k}{5}$$
$$x = 2 - 2y - z = 2 - 2(\tfrac{2-3k}{5}) - k = \tfrac{k+6}{5}.$$
Thus,
$$x = \frac{k+6}{5}, \quad y = \frac{2-3k}{5}, \quad z = k$$
where $k$ is arbitrary, gives all solutions of the given system.

**Example:** Are the following system of equations consistent? If so, solve them,
$x + y + 2z + w = 5,\ 2x + 3y - z - 2w = 2,\ 4x + 5y + 3z = 7.$

**Solution:** The given system of equations can be put in the matrix form as
$$\begin{pmatrix} 1 & 1 & 2 & 1 \\ 2 & 3 & -1 & -2 \\ 4 & 5 & 3 & 0 \end{pmatrix} \begin{pmatrix} x \\ y \\ z \end{pmatrix} = \begin{pmatrix} 5 \\ 2 \\ 7 \end{pmatrix}.$$
$$\Rightarrow AX = B,$$
where $A = \begin{pmatrix} 1 & 1 & 2 & 1 \\ 2 & 3 & -1 & -2 \\ 4 & 5 & 3 & 0 \end{pmatrix}, X = \begin{pmatrix} x \\ y \\ z \end{pmatrix}$ and $B = \begin{pmatrix} 5 \\ 2 \\ 7 \end{pmatrix}.$

The augmented matrix is given by
$$[A|B] = \left(\begin{array}{cccc|c} 1 & 1 & 2 & 1 & 5 \\ 2 & 3 & -1 & -2 & 2 \\ 4 & 5 & 3 & 0 & 7 \end{array}\right)$$

$R_2 \to R_2 - 2R_1$, $R_3 \to R_3 - 4R_1$, we have

$$\sim \begin{pmatrix} 1 & 1 & 2 & 1 & | & 5 \\ 0 & 1 & -5 & -4 & | & -8 \\ 0 & 1 & -5 & -4 & | & -13 \end{pmatrix}$$

$R_3 \to R_3 - R_2$, we have

$$\sim \begin{pmatrix} 1 & 1 & 2 & 1 & | & 5 \\ 0 & 1 & -5 & -4 & | & -8 \\ 0 & 0 & 0 & 0 & | & -5 \end{pmatrix}$$

This is in Echelon form. Therefore, $\rho(A) \neq \rho[A|B]$. Hence, the given system is inconsistent and has no solution.

**Example:** Find the values of $\lambda$ and $\mu$ for which the system

$$x + y + z = 6$$

$$x + 2y + 3z = 10$$

$$x + 2y + \lambda z = \mu$$

has (i) no solution (ii) unique solution (iii) infinite number of solutions.

**Solution:** The given system of equations can be put in the matrix form as

$$\begin{pmatrix} 1 & 1 & 1 \\ 1 & 2 & 3 \\ 1 & 2 & \lambda \end{pmatrix} \begin{pmatrix} x \\ y \\ z \end{pmatrix} = \begin{pmatrix} 6 \\ 10 \\ \mu \end{pmatrix}.$$

$$\Rightarrow AX = B,$$

where $A = \begin{pmatrix} 1 & 1 & 1 \\ 1 & 2 & 3 \\ 1 & 2 & \lambda \end{pmatrix}$, $X = \begin{pmatrix} x \\ y \\ z \end{pmatrix}$ and $B = \begin{pmatrix} 6 \\ 10 \\ \mu \end{pmatrix}$.

The augmented matrix is given by

$$[A|B] = \begin{pmatrix} 1 & 1 & 1 & | & 6 \\ 1 & 2 & 3 & | & 10 \\ 1 & 2 & \lambda & | & \mu \end{pmatrix}$$

$R_2 \to R_2 - R_1$, $R_3 \to R_3 - R_1$, we have

$$\sim \begin{pmatrix} 1 & 1 & 1 & | & 6 \\ 0 & 1 & 2 & | & 4 \\ 0 & 1 & \lambda - 1 & | & \mu - 6 \end{pmatrix}$$

$R_3 \to R_3 - R_2$, we have

$$\sim \begin{pmatrix} 1 & 1 & 1 & | & 6 \\ 0 & 1 & 2 & | & 4 \\ 0 & 0 & \lambda - 3 & | & \mu - 10 \end{pmatrix}$$

(i) **For no solution:** $\rho(A) \neq \rho(A|B)$

It is only possible only when $\lambda = 3$ and $\mu \neq 10$.

(ii) **For unique solution:** $\rho(A) = \rho(A|B) =$ number of unknowns

It is only possible only when $\lambda \neq 3$ and for any $\mu$.

(iii) **For infinite number of solutions:** $\rho(A) = \rho(A|B) <$ number of unknowns

It is only possible only when $\lambda = 3$ and $\mu = 10$.

**Example:** Find the values of $a$ and $b$ for which the equations $x + y + z = 3$, $x + 2y + 2z = 6$, $x + ay + 3z = b$ have

(i) no solution

(ii) infinite number of solutions

(iii) unique solution.

**Solution:** The given system of equations can be put in the matrix form as

$$\begin{pmatrix} 1 & 1 & 1 \\ 1 & 2 & 2 \\ 1 & a & 3 \end{pmatrix} \begin{pmatrix} x \\ y \\ z \end{pmatrix} = \begin{pmatrix} 3 \\ 6 \\ b \end{pmatrix}.$$

$$\Rightarrow AX = B,$$

where $A = \begin{pmatrix} 1 & 1 & 1 \\ 1 & 2 & 2 \\ 1 & a & 3 \end{pmatrix}$, $X = \begin{pmatrix} x \\ y \\ z \end{pmatrix}$ and $B = \begin{pmatrix} 3 \\ 6 \\ b \end{pmatrix}$.

The augmented matrix is given by

$$[A|B] = \begin{pmatrix} 1 & 1 & 1 & | & 3 \\ 1 & 2 & 2 & | & 6 \\ 1 & a & 3 & | & b \end{pmatrix}$$

$R_2 \to R_2 - R_1$, $R_3 \to R_3 - R_1$, we have

$$\sim \begin{pmatrix} 1 & 1 & 1 & | & 3 \\ 0 & 1 & 1 & | & 3 \\ 0 & a-2 & 1 & | & b-3 \end{pmatrix}$$

$R_3 \to R_3 - R_2$, we have

$$\sim \begin{pmatrix} 1 & 1 & 1 & | & 3 \\ 0 & 1 & 1 & | & 3 \\ 0 & a-3 & 0 & | & b-9 \end{pmatrix}$$

(i) **For no solution:** If $a = 3$ and $b \neq 9$, then $\rho(A)=2$ and $\rho[A|B] = 3$. Since $\rho(A) \neq \rho[A|B]$, the system is inconsistent. Therefore, the system has no solution, when $a = 3$ and $b \neq 9$.

(ii) **For unique solution:** If $a \neq 3$ and $b$ has any value, then $\rho(A)=3$ and $\rho[A|B] = 3$. Since $\rho(A) = \rho[A|B] = $ number of unknowns, the system is consistent. Therefore, the system has unique solution, when $a \neq 3$ and $b$ has any value.

(iii) **For infinite number of solutions:** If $a = 3$ and $b = 9$, then $\rho(A) = 2 = \rho[A|B]$. Since $\rho(A) \neq \rho[A|B]$ <number of unknowns=3, the system is inconsistent. Therefore, the system will have infinite number of solutions, with $n - r = 3 - 2 = 1$ arbitrary variables.

**Example:** Find the value of $\lambda$ for which the system

$$x + y + z = 1$$
$$4x + y + 10z = \lambda^2$$
$$2x + y + 4z = \lambda$$

has a solution. Solve the system in each possible case.

**Solution:** The given system of equations can be put in the matrix form as

$$\begin{pmatrix} 1 & 1 & 1 \\ 4 & 1 & 10 \\ 2 & 1 & 4 \end{pmatrix} \begin{pmatrix} x \\ y \\ z \end{pmatrix} = \begin{pmatrix} 1 \\ \lambda \\ \lambda^2 \end{pmatrix}.$$

$$\Rightarrow AX = B,$$

where $A = \begin{pmatrix} 1 & 1 & 1 \\ 4 & 1 & 10 \\ 2 & 1 & 4 \end{pmatrix}$, $X = \begin{pmatrix} x \\ y \\ z \end{pmatrix}$ and $B = \begin{pmatrix} 1 \\ \lambda \\ \lambda^2 \end{pmatrix}$.

The augmented matrix is given by

$$[A|B] = \begin{pmatrix} 1 & 1 & 1 & | & 1 \\ 4 & 1 & 10 & | & \lambda \\ 2 & 1 & 4 & | & \lambda^2 \end{pmatrix}$$

$R_2 \to R_2 - 4R_1$, $R_3 \to R_3 - 2R_1$, we have

$$\sim \begin{pmatrix} 1 & 1 & 1 & | & 1 \\ 0 & -3 & 6 & | & \lambda^2 - 4 \\ 0 & -1 & 2 & | & \lambda - 2 \end{pmatrix}$$

$R_3 \to 3R_3 - R_2$, we have

$$\sim \begin{pmatrix} 1 & 1 & 1 & | & 1 \\ 0 & -3 & 6 & | & \lambda^2 - 4 \\ 0 & 0 & 0 & | & \lambda^2 - 3\lambda + 2 \end{pmatrix} \quad \cdots \cdots \quad \text{(i)}$$

Since the given system is consistent $\Rightarrow \rho(A) = \rho[A|B]$

$$\Rightarrow \lambda^2 - 3\lambda + 2 = 0$$

$$\Rightarrow \lambda = 1, 2.$$

Thus, the given system is consistent only in the cases of $\lambda = 1$ and $\lambda = 2$.

For $\lambda = 1$, the system corresponding to (i) is

$$x + y + z = 1$$

$$-3y + 6z = -3$$

These equations yield $z = k_1, y = 1 + 2k_1, x = -3k_1$, where $k_1$ is arbitrary.
Thus, when $\lambda = 1$,

$$x = -3k_1, \quad y = 1 + 2k_1, \quad z = k_1$$

give all solutions of the given system.

For $\lambda = 2$, the system corresponding to (i) is

$$x + y + z = 1$$
$$-3y + 6z = 0$$

These equations yield $z = k_2, y = 2k_2, x = 1 - 3k_2$, where $k_2$ is arbitrary.
Thus, when $\lambda = 1$,

$$x = 1 - 3k_2, \quad y = 2k_2, \quad z = k_2$$

give all solutions of the given system.

## EXERCISE

1. If consistent, solve the system of equations $x+y+z+t = 4, x-z+2t = 2, y+z-3t = -1, x+2y-z+t = 3$. **(Apr-2018)**
   Ans: $\rho(A) = \rho(A|B) = 4, x = 1, y = 1, z = 1, t = 1$.

2. Solve the system of equations $10x - y - z = 13; x + 10y + z = 36; -x - y + 10z = 35$.
   Ans: $x = 2, y = 3, z = 4$.

3. Test the consistency and solve $x + y + z = 6; x - y + 2z = 5; 2x - 2y + 3z = -7$.
   Ans: $x = -20, y = 9, z = 17$. **(Jan-2020), (Mar-2022)**

4. Test the consistency and solve $x + y + z = 6; x + 2y + 3z = 14; x + 4y + 9z = 36$.
   Ans: $x = 1, y = 2, z = 3$. **(Jan-2020)**

5. Test the consistency of the system $x + y + z = 6; x - y + 2z = 5; 3x + y + z = 8$ and hence solve. **(Sep-2021)**
   Ans: $x = 1, y = 2, z = 3$.

6. Test the consistency and solve $5x+3y+7z = 4; 3x+26y+2z = 9; 7x+2y+10z = 5$.
   Ans: $z = k, y = \dfrac{3+k}{11}, x = \dfrac{-16k+7}{11}$. **(Jan-2020), (Aug-2022)**

7. Solve the equations $2x - y + 3z - 9 = 0; x + y + z = 6; x - y + z - 2 = 0$.

**Ans:** $x = 1, y = 2, z = 3$.

8. Show that the system $x + 2y - 5z = -9, 3x - y + 2z = 5, 2x + 3y - z = 3, 4x - 5y + z = -3$ is consistent and solve it. **(Dec-2020)**

**Ans:** $x = \frac{1}{2}, y = \frac{3}{2}, z = \frac{5}{2}$.

9. Test for Consistency the set of equations and solve them if they are consistent:
$x + 2y + 2z = 2; 3x - 2y - z = 5; 2x - 5y + 3z = -4; x + 4y + 6z = 0$.

**Ans:** $x = 2, y = 1, z = -1$.

10. Test the following system for consistency and if consistent solve it $u + 2v + 2w = 1, 2u + v + w = 2, 3u + 2v + 2w = 3, v + w = 0$.

**Ans:** $u = 1, v = -k, w = k$.

11. Find the values of $a$ and $b$ such that the system of equations $2x + 3y + 5z = 9$, $7x + 3y - 2z = 8, 2x + 3y + az = b$ have (i) No solution (ii) A unique solution (iii) Infinite number of solutions. **(Aug-2022)**

12. Find the values of $a$ and $b$ such that the system of equations $x + y + z = 6$, $x + 2y + 3z = 10, x + 2y + az = b$ have (i) No solution (ii) A unique solution (iii) Infinite number of solutions. **(Aug-2022)**

## 1.4 Homogeneous System of Linear Equations

Consider the following system of $m$ simultaneous linear equations in $n$ unknowns:

$$\left.\begin{array}{c} a_{11}x_1 + a_{12}x_2 + \cdots + a_{1n}x_n = b_1 \\ a_{21}x_1 + a_{22}x_2 + \cdots + a_{2n}x_n = b_2 \\ \cdots \quad \cdots \quad \cdots \quad \cdots \\ a_{m1}x_1 + a_{m2}x_2 + \cdots + a_{mn}x_n = b_m \end{array}\right\} \quad \cdots \cdots \quad (1)$$

In system (1), all $b_1 = b_2 = \cdots = b_m = 0$, then the system is homogeneous.

**Result:**

1. If $r = n$; i,e., the rank of the coefficient matrix is equal to number of unknowns then the system has a trivial solution ($x_1 = x_2 = \cdots = x_n = 0$).
2. If $r < n$; i.e., the rank of the coefficient matrix is smaller than the number of unknowns then the system has non-trivial solution.
3. For non-trivial solution always $|A| = 0$.

**Example:** Solve the system of equations $x+y+w = 0$, $y+z = 0$, $x+y+z+w = 0$, $x+y+2z = 0$.

**Solution:** The given system of equations can be written as $AX = B$, where

$$A = \begin{pmatrix} 1 & 1 & 0 & 1 \\ 0 & 1 & 1 & 0 \\ 1 & 1 & 1 & 1 \\ 1 & 1 & 2 & 0 \end{pmatrix}, X = \begin{pmatrix} x \\ y \\ z \\ w \end{pmatrix}, B = \begin{pmatrix} 0 \\ 0 \\ 0 \\ 0 \end{pmatrix}.$$

Let us apply elementary row operations on the coefficient matrix, we have

$$A = \begin{pmatrix} 1 & 1 & 0 & 1 \\ 0 & 1 & 1 & 0 \\ 1 & 1 & 1 & 1 \\ 1 & 1 & 2 & 0 \end{pmatrix}$$

$R_3 \to R_3 - R_1$, $R_4 \to R_4 - R_1$, we have

$$\sim \begin{pmatrix} 1 & 1 & 0 & 1 \\ 0 & 1 & 1 & 0 \\ 0 & 0 & 1 & 0 \\ 0 & 0 & 2 & -1 \end{pmatrix}$$

$R_4 \to R_4 - 2R_3$, we have

$$\sim \begin{pmatrix} 1 & 1 & 0 & 1 \\ 0 & 1 & 1 & 0 \\ 0 & 0 & 1 & 0 \\ 0 & 0 & 0 & -1 \end{pmatrix}$$

This matrix is in Echelon form and it has four non-zero rows.

$\therefore \rho(A) = 4 =$ No. of unknowns.

The given system has a unique trivial solution.

$\therefore x = 0, y = 0, z = 0, w = 0$.

**Example:** Determine whether the following equations will have a non-trivial solution, if so solve them $x+y-2z+3w = 0$, $x-2y+z-w = 0$, $4x+y-5z+8w = 0$, $5x-7y+2z-w = 0$. **(Aug-2021), (Mar-2022)**

**Solution:** The given system of equations can be written as $AX = B$, where

$$A = \begin{pmatrix} 1 & 1 & -2 & 3 \\ 1 & -2 & 1 & -1 \\ 4 & 1 & -5 & 8 \\ 5 & -7 & 2 & -1 \end{pmatrix}, X = \begin{pmatrix} x \\ y \\ z \\ w \end{pmatrix}, B = \begin{pmatrix} 0 \\ 0 \\ 0 \\ 0 \end{pmatrix}.$$

Let us apply elementary row operations on the coefficient matrix, we have

$$A = \begin{pmatrix} 1 & 1 & -2 & 3 \\ 1 & -2 & 1 & -1 \\ 4 & 1 & -5 & 8 \\ 5 & -7 & 2 & -1 \end{pmatrix}$$

$R_2 \to R_2 - R_1$, $R_3 \to R_3 - 4R_1$, $R_4 \to R_4 - 5R_1$, we have

$$\sim \begin{pmatrix} 1 & 1 & -2 & 3 \\ 0 & -3 & 3 & -4 \\ 0 & -3 & 3 & -4 \\ 0 & -12 & 12 & -16 \end{pmatrix}$$

$R_3 \to R_3 - R_2$, $R_4 \to R_4 - 4R_2$, we have

$$\sim \begin{pmatrix} 1 & 1 & -2 & 3 \\ 0 & -3 & 3 & -4 \\ 0 & 0 & 0 & 0 \\ 0 & 0 & 0 & 0 \end{pmatrix} \quad \cdots \cdots \text{ (i)}$$

This matrix is in Echelon form and it has two non-zero rows.

$$\therefore \rho(A) = 2 < 4 \text{ (No. of unknowns)}.$$

Therefore, the given system has a non-trivial solution in which $n - r = 4 - 2 = 2$ unknowns can be chosen arbitrarily.

From (i), we find that the given system is equivalent to the following system:

$$x + y - 2z + 3w = 0$$

$$-3y + 3z - 4w = 0$$

Let $z = k_1$ and $w = k_2$, we have

$$y = \frac{3z-4w}{3} = \frac{3k_1-4k_2}{3}$$

$$x = 2z - 3w - y = 2k_2 - 3k_1 - \frac{3k_1-4k_2}{3} = \frac{3k_1-5k_2}{3}, \text{ where } k_1 \text{ and } k_2 \text{ are any arbitrary constants.}$$

**Example:** Determine $b$ such that the system of homogeneous equations $2x + y + 2z = 0$, $x + y + 3z = 0$, $4x + 3y + bz = 0$ has trivial and non-trivial solutions. Find the non-trivial solution. **(Dec-2020)**

**Solution:** The given system of equations can be written as $AX = B$, where

$$A = \begin{pmatrix} 2 & 1 & 2 \\ 1 & 1 & 3 \\ 4 & 3 & b \end{pmatrix}, X = \begin{pmatrix} x \\ y \\ z \end{pmatrix}, B = \begin{pmatrix} 0 \\ 0 \\ 0 \end{pmatrix}.$$

Let us apply elementary row operations on the coefficient matrix, we have

$$A = \begin{pmatrix} 2 & 1 & 2 \\ 1 & 1 & 3 \\ 4 & 3 & b \end{pmatrix}$$

$R_2 \leftrightarrow R_1$, we have

$$\sim \begin{pmatrix} 1 & 1 & 3 \\ 2 & 1 & 2 \\ 4 & 3 & b \end{pmatrix}$$

$R_2 \to R_2 - 2R_1$, $R_3 \to R_3 - 4R_1$, we have

$$\sim \begin{pmatrix} 1 & 1 & 3 \\ 0 & -1 & -4 \\ 0 & -1 & b-12 \end{pmatrix}$$

$R_3 \to R_3 - R_2$, we have

$$\sim \begin{pmatrix} 1 & 1 & 3 \\ 0 & -1 & -4 \\ 0 & 0 & b-8 \end{pmatrix} \quad \cdots \cdots \text{ (i)}$$

This matrix is in Echelon form

(i) If $b \neq 8$, then $\rho(A) = $ No. of unknowns $= 3$. i.e., $|A| \neq 0$.

The given system has a trivial solution if $b \neq 8$.

(ii) If $b = 8$, $\rho(A) = 2 < 3$(No. of unknowns), i.e., $|A| = 0$.

Therefore, the given system has a non-trivial solution if $b = 8$.

Since $n - r = 3 - 2 = 1$, one unknown in the given system can be chosen arbitrarily.

From (i), the given system is equivalent to the following equations:
$$x + y + 3z = 0$$
$$y + 4z = 0$$

Let $z = k \Rightarrow y = -4k$ and $x = -y - 3z = 4k - 3k = k$.

$\therefore$ Non-trivial solution of the given system is $x = k$, $y = -4k$, $z = k$ (where $k$ is any arbitrary constant).

**Example:** Show that the system of equations
$$2x_1 - 2x_2 + x_3 = \lambda x_1, \; 2x_1 - 3x_2 + 2x_3 = \lambda x_2, \; -x_1 + 2x_2 = \lambda x_3$$
can possess a non-trivial solution only if $\lambda = 1$ and $\lambda = -3$ and obtain the general solution in each case.

**Solution:** The above system equations can be written in the matrix form as
$$\begin{pmatrix} 2-\lambda & -2 & 1 \\ 2 & -3-\lambda & 2 \\ -1 & 2 & -\lambda \end{pmatrix} \begin{pmatrix} x_1 \\ x_2 \\ x_3 \end{pmatrix} = \begin{pmatrix} 0 \\ 0 \\ 0 \end{pmatrix}$$

The coefficient matrix $A = \begin{pmatrix} 2-\lambda & -2 & 1 \\ 2 & -3-\lambda & 2 \\ -1 & 2 & -\lambda \end{pmatrix}$.

For non-trivial solution, $|A| = 0$
$$\begin{vmatrix} 2-\lambda & -2 & 1 \\ 2 & -3-\lambda & 2 \\ -1 & 2 & -\lambda \end{vmatrix} = 0$$
$\Rightarrow (2-\lambda)[(-3-\lambda)(-\lambda) - 4] + 2[-2\lambda + 2] + 1[4 - 3 - \lambda] = 0$
$\Rightarrow (2-\lambda)[3\lambda + \lambda^2 - 4] + 2[-2\lambda + 2] + (1 - \lambda) = 0$
$\Rightarrow \lambda^3 + \lambda^2 - 5\lambda + 3 = 0$

$\Rightarrow \lambda = 1, 1, -3$.

The given system has a non-trivial solution for $\lambda = 1$ and $-3$.
When $\lambda = 1$, the coefficient matrix is

$$A = \begin{pmatrix} 1 & -2 & 1 \\ 2 & -4 & 2 \\ -1 & 2 & -1 \end{pmatrix}$$

We find that

$$R_2 \to R_2 - 2R_1, R_3 \to R_3 + R_1 \sim \begin{pmatrix} 1 & -2 & 1 \\ 0 & 0 & 0 \\ 0 & 0 & 0 \end{pmatrix} \quad \ldots\ldots \text{(i)}$$

This is in Echelon form. $\rho(A) = 1 < 3$(No. of unknowns). Since $n - r = 3 - 1 = 2$, two unknowns in the given system can be chosen arbitrarily.
From (i), the given system is equivalent to the following equation:

$$x_1 - 2x_2 + x_3 = 0.$$

Let $x_3 = k_1$, $x_2 = k_2$, then $x_1 = 2x_2 - x_3 = 2k_2 - k_1$.
$\therefore$ The solution when $\lambda = 1$ is $x_1 = 2k_2 - 1, x_2 = k_2, x_3 = k_1$.
For $\lambda = -3$, the coefficient matrix is

$$A = \begin{pmatrix} 5 & -2 & 1 \\ 2 & 0 & 2 \\ -1 & 2 & 3 \end{pmatrix}$$

We find that

$$R_2 \leftrightarrow R_3 \sim \begin{pmatrix} -1 & 2 & 3 \\ 2 & 0 & 2 \\ 5 & -2 & 1 \end{pmatrix}$$

$$R_2 \to R_2 + 2R_1, R_3 \to R_3 + 5R_1 \sim \begin{pmatrix} -1 & 2 & 3 \\ 0 & 4 & 8 \\ 0 & 8 & 16 \end{pmatrix}$$

$$R_3 \to R_3 - 2R_2 \sim \begin{pmatrix} -1 & 2 & 3 \\ 0 & 4 & 8 \\ 0 & 0 & 0 \end{pmatrix} \quad \cdots \cdots \text{(ii)}$$

This is in Echelon form. $\rho(A) = 2 < 3$(No. of unknowns). Since $n - r = 3 - 2 = 1$, one unknown in the given system can be chosen arbitrarily.

From (ii), the given system is equivalent to the following equations:
$$-x_1 + 2x_2 + 3x_3 = 0$$
$$4x_2 + 8x_3 = 0$$

Let $x_3 = k$, then $x_2 = -2k$ and $x_1 = 2x_2 + 3x_3 = -k$.

$\therefore$ Solution when $\lambda = -3$ is $x_1 = -k, x_2 = -2k, x_2 = k$.

**Example:** Show that the only real value of $\lambda$ for which the system **(Sep-2021)**

$$x + 2y + 3z = \lambda x$$
$$3x + y + 2z = \lambda y$$
$$2x + 3y + z = \lambda z$$

has a non-trivial solutions is 6. Solve the system for this value of $\lambda$.

**Solution:** The given equations can be rewritten as

$$(1 - \lambda)x + 2y + 3z = 0$$
$$3x + (1 - \lambda)y + 2z = 0$$
$$2x + 3y + (1 - \lambda)z = 0$$

The coefficient matrix $A = \begin{pmatrix} 1-\lambda & 2 & 3 \\ 3 & 1-\lambda & 2 \\ 2 & 3 & 1-\lambda \end{pmatrix}$.

For non-trivial solution, $|A| = 0$.

$$\begin{vmatrix} 1-\lambda & 2 & 3 \\ 3 & 1-\lambda & 2 \\ 2 & 3 & 1-\lambda \end{vmatrix} = 0$$

$\Rightarrow (1-\lambda)[(1-\lambda)^2 - 6] - 2[3 - 3\lambda - 4] + 3[9 - 2 + 2\lambda] = 0$

$\Rightarrow (1-\lambda)[1 + \lambda^2 - 2\lambda - 6] + 2[3\lambda + 1] + 3[7 + 2\lambda] = 0$

$\Rightarrow \lambda^3 - 3\lambda^2 - 15\lambda - 18 = 0$

$\Rightarrow (\lambda - 6)(\lambda^2 + 3\lambda + 3) = 0$

$\Rightarrow \lambda = 6 \text{ or } \lambda = \dfrac{-3 \pm \sqrt{3}i}{2}$.

This shows that the only real value of $\lambda$ for which $|A| = 0$ is $\lambda = 6$.

Thus, $\lambda = 6$ is the only real value of $\lambda$ for which the given system has a non-trivial solution.

When $\lambda = 6$, the coefficient matrix is

$$A = \begin{pmatrix} -5 & 2 & 3 \\ 3 & -5 & 2 \\ 2 & 3 & -5 \end{pmatrix}$$

We find that $R_1 \to R_1 + R_2 + R_3$, $\sim \begin{pmatrix} 0 & 0 & 0 \\ 3 & -5 & 2 \\ 2 & 3 & -5 \end{pmatrix}$

$R_1 \leftrightarrow R_3$, we have

$$\sim \begin{pmatrix} 2 & 3 & -5 \\ 3 & -5 & 2 \\ 0 & 0 & 0 \end{pmatrix}$$

$R_2 \to 2R_2 - 3R_1$, we have

$$\sim \begin{pmatrix} 2 & 3 & -5 \\ 0 & -19 & 19 \\ 0 & 0 & 0 \end{pmatrix} \quad \cdots \cdots \text{(i)}$$

This is in Echelon form. $\rho(A) = 2 < 3$(No. of unknowns). Since $n - r = 3 - 2 = 1$, one unknown in the given system can be chosen arbitrarily.

From (i), the given system is equivalent to the following system of equations:
$$2x + 3y - 5z = 0$$
$$-19y + 19z = 0.$$

Let $z = k$, then $-19y + 19k \Rightarrow y = k$.
$$2x + 3k - 5k = 0 \Rightarrow x = k.$$

$\therefore \quad x = k, y = k, z = k$ ($k$ is any arbitrary constant).

**Example:** Solve the following system of equations for all values of $k$
$$2x + 3ky + (3k+4)z = 0$$
$$x + (k+4)y + (4k+2)z = 0$$
$$x + 2(k+1)y + (3k+4)z = 0.$$

**Solution:** The matrix form of the given system of equations is $AX = O$

where $A = \begin{pmatrix} 2 & 3k & 3k+4 \\ 1 & k+4 & 4k+2 \\ 1 & 2(k+1) & 3k+4 \end{pmatrix}$ is the coefficient matrix and $X = \begin{pmatrix} x \\ y \\ z \end{pmatrix}$;

$O = \begin{pmatrix} 0 \\ 0 \\ 0 \end{pmatrix}$.

If the given system of equations possesses a non-zero solution. then the coefficient matrix is singular, i.e., $|A| = 0$

$$\begin{vmatrix} 2 & 3k & 3k+4 \\ 1 & k+4 & 4k+2 \\ 1 & 2(k+1) & 3k+4 \end{vmatrix} = 0$$

$R_1 \leftrightarrow R_2$, we have $\begin{vmatrix} 1 & k+4 & 4k+2 \\ 2 & 3k & 3k+4 \\ 1 & 2k+2 & 3k+4 \end{vmatrix} = 0$

$R_2 \to R_2 - 2R_1, R_3 \to R_3 - R_1, \begin{vmatrix} 1 & k+4 & 4k+2 \\ 0 & k-8 & -5k \\ 0 & k-2 & -k+2 \end{vmatrix} = 0$

$\Rightarrow (k-2) \begin{vmatrix} 1 & k+4 & 4k+2 \\ 0 & k-8 & -5k \\ 0 & k-2 & -k+2 \end{vmatrix} = 0$

Expanding with first column, we have

$$(k-2)\big[(k-8)(-1) + 5k\big] = 0$$
$$\Rightarrow (k-2)\big[-k+8+5k\big] = 0$$
$$\Rightarrow (k-2)(4k+8) = 0$$
$$\Rightarrow (k-2)(k+2) = 0$$
$$\Rightarrow k = \pm 2.$$

Now three cases arise.

**Case 1:** Suppose $k \ne \pm 2$. Then $\det A \ne 0 \Rightarrow$ Rank of $A = 3$.

We have number of variables is 3.

The give system of equations possesses a zero solution, i.e., trivial solution.

$\therefore x = y = z = 0$ is the only solution for the given system of equations.

**Case 2:** When $k = 2$, the equivalent matrix of $A$ is

$A \sim \begin{pmatrix} 1 & 6 & 10 \\ 0 & -6 & -10 \\ 0 & 0 & 0 \end{pmatrix}$; which is in Echelon form.

Here the rank of $A = 2 =$ the number of non-zero rows of equivalent $A$ i.e., $r = 2 < n$, i.e., the number of unknowns, so that the given system of equations an infinite number of solutions.

The number of independent solutions $n - r = 3 - 2 = 1$.

The equivalent matrix equation is
$$\begin{pmatrix} 1 & 6 & 10 \\ 0 & -6 & -10 \\ 0 & 0 & 0 \end{pmatrix} \begin{pmatrix} x \\ y \\ z \end{pmatrix} = \begin{pmatrix} 0 \\ 0 \\ 0 \end{pmatrix}.$$

The equations can be written as:
$$x + 6y + 10z = 0$$
$$-6y - 10z = 0.$$

Choosing $z = c$, then $-6y - 10c \Rightarrow y = -\frac{5}{3}c$.

$x + 10c - 10c = 0 \Rightarrow x = 0$.

$\therefore x = 0, y = -\frac{5}{3}c, z = c$, where $c$ is any arbitrary constant, gives the general solution of the given system of equations.

**Case 3:** When $k = -2$, the equivalent matrix of $A$ is
$$A \sim \begin{pmatrix} 1 & 2 & -6 \\ 0 & -10 & -10 \\ 0 & -4 & 4 \end{pmatrix}$$

Applying $R_3 \to 5R_3 - 2R_2$, we have
$$A \sim \begin{pmatrix} 1 & 2 & -6 \\ 0 & -10 & -10 \\ 0 & 0 & 0 \end{pmatrix}; \text{ which is in Echelon form.}$$

Here the rank of $A = 2 =$ the number of non-zero rows of equivalent $A$ i.e., $r = 2 < n$, i.e., the number of unknowns.

So that the given system has an infinite number of solutions. Of these $n - r = 3 - 2 = 1$ is linearly independent and the remaining are depending upon it.

The equivalent matrix equation is
$$\begin{pmatrix} 1 & 2 & -6 \\ 0 & -10 & 10 \\ 0 & 0 & 0 \end{pmatrix} \begin{pmatrix} x \\ y \\ z \end{pmatrix} = \begin{pmatrix} 0 \\ 0 \\ 0 \end{pmatrix}.$$

The equations can be written as:
$$x + 2y - 6z = 0$$

$$-10y + 10z = 0.$$

Choosing $z = c$, then $-10y + 10c \Rightarrow y = c$.

$x + 2c - 6c = 0 \Rightarrow x = 4c$.

$\therefore x = 4c, y = c, z = c$, where $c$ is any arbitrary constant, gives the general solution of the given system of equations.

## EXERCISE

1. Solve the system $2x - y + 3z = 0, 3x + 2y + z = 0$ and $x - 4y + 5z = 0$.
Ans: $x = -k, y = k, z = k$, $k$ is a arbitrary constant.

2. Find all solutions of the system $x + 2y - z = 0, 2x + y + z = 0$ and $x - 4y + 5z = 0$.
Ans: $x = -k, y = k, z = k$, $k$ is a arbitrary constant.

3. Solve the system of equations $4x + 3y - z = 0, 3x + 4y + z = 0, x - y - 2z = 0$.
(Nov-2021)

4. Determine the values of $\lambda$ for which the following set of equations may possess non-trivial solutions and solve them in each case:

$$3x_1 + x_2 - \lambda x_3 = 0$$
$$4x_1 - 2x_2 - 3x_3 = 0$$
$$2\lambda x_1 + 4x_2 + \lambda x_3 = 0$$

Ans: When $\lambda = 1$, $\begin{pmatrix} x_1 \\ x_2 \\ x_3 \end{pmatrix} = \begin{pmatrix} -k \\ k \\ -2k \end{pmatrix}$. When $\lambda = -9$, $\begin{pmatrix} x_1 \\ x_2 \\ x_3 \end{pmatrix} = \begin{pmatrix} \frac{-3s}{2} \\ \frac{-9s}{2} \\ s \end{pmatrix}$.

5. Solve the system of equations $x + 2y + (2+k)z = 0, 2x + (2+k)y + 4z = 0, 7x + 3y + (18+k)z = 0$ for all values of $k$.
(Jan-2020)

Ans: $k = 1$ and $k = \dfrac{4}{3}$.

Case 1: When $k \neq 1$ and $k \neq \dfrac{4}{3}$
$x = 0 = y = z$, is the only solution.

Case 2: When $k = 1$.

$x = c, y = -2c, z = c$.

**Case 3:** When $k = \dfrac{4}{3}$.

$x = \dfrac{14}{3}c, y = -4c, z = c$.

6. Determine whether the following equations will have a non-trivial solution, if so solve them $4x + 2y + z + 3w = 0, 6x + 3y + 4z + 7w = 0, 2x + y + w = 0$.

**Ans:** $x = t, y = -2t - s, z = -s, w = s$.      **(Nov-2017), (Jan-2020)**

or $x = a, y = b, z = 2a + b, w = -2a - b$.

7. Solve the system of equations $4x+2y+z+w = 0, 6x+3y+4z+7w = 0, 2x+y+w = 0$.

**(Aug-2022)**

8. Solve the following homogeneous system of liner equations:

$$x + 3y - 2z = 0$$

$$2x - y + 4z = 0$$

$$x - 11y + 14z = 0$$

**Ans:** $x = -(10/7)k, \ y = (8/7)k, \ z = k$

9. Test for consistency and solve $x + y = 0, y + z = 0, z + x = 0$.

10. Solve the system of equations $x + 3y - 5z = 0, 3x - y + 5z = 0, 3x + 2y - z = 0$.

**(Aug-2022)**

11. Solve the system of equations $3x + 8y + 2z = 0, 2x + y + 3z = 0, 5x - y + z = 0$.

**(Aug-2022)**

## 1.5   Gauss Elimination Method

Gauss elimination method is an exact method which solves a given system of equations in $n$ unknowns by transforming the coefficient matrix into an upper triangular matrix by applying elementary row operations and then solve for the unknowns by back substitution.

**Example:** Solve the system of equations $3x + y + 2z = 3, \ 2x - 3y - z = -3$,

$x + 2y + z = 4$ by Gauss elimination method.

**Solution:** In Gauss elimination method, the coefficient matrix $A$ is reduced to an upper triangular matrix by applying elementary row operation.

The given system of equations can be written as $AX = B$, where

$$A = \begin{pmatrix} 3 & 1 & 2 \\ 2 & -3 & -1 \\ 1 & 2 & 1 \end{pmatrix}, X = \begin{pmatrix} x \\ y \\ z \end{pmatrix}, B = \begin{pmatrix} 3 \\ -3 \\ 4 \end{pmatrix}.$$

The augmented matrix $[A|B] = \begin{pmatrix} 3 & 1 & 2 & | & 3 \\ 2 & -3 & -1 & | & -3 \\ 1 & 2 & 1 & | & 4 \end{pmatrix}$

$R_3 \leftrightarrow R_1$, we have

$$\sim \begin{pmatrix} 1 & 2 & 1 & | & 4 \\ 2 & -3 & -1 & | & -3 \\ 3 & 1 & 2 & | & 3 \end{pmatrix}$$

$R_2 \to R_2 - 2R_1$, $R_3 \to R_3 - 3R_1$, we have

$$\sim \begin{pmatrix} 1 & 2 & 1 & | & 4 \\ 0 & -7 & -3 & | & -11 \\ 0 & -5 & -1 & | & -9 \end{pmatrix}$$

$R_2 \to R_2 - 3R_3$, we have

$$\sim \begin{pmatrix} 1 & 2 & 1 & | & 4 \\ 0 & 8 & 0 & | & 16 \\ 0 & -5 & -1 & | & -9 \end{pmatrix}$$

$R_2 \to \frac{R_2}{2}$, $R_3 \to (-1)R_3$, we have

$$\sim \begin{pmatrix} 1 & 2 & 1 & | & 4 \\ 0 & 1 & 0 & | & 2 \\ 0 & 5 & 1 & | & -9 \end{pmatrix}$$

$R_3 \to R_3 - 5R_2$, we have

$$\sim \begin{pmatrix} 1 & 2 & 1 & | & 4 \\ 0 & 1 & 0 & | & 2 \\ 0 & 0 & 1 & | & -1 \end{pmatrix}$$

Here, coefficient matrix is an upper triangular matrix form.

By Gauss elimination method, we have
$$x + 2y + z = 4$$
$$y = 2$$
$$z = -1$$
By back substitution, we have
$$x = 4 - 2y - z = 4 - 4 + 1 = 1.$$
Therefore, $x = 1, y = 2, z = -1$ is the solution of the given system.

**Example:** Solve the system of equations $x + y + z = 8$, $2x + 3y + 2z = 19$, $4x + 2y + 3z = 23$ by Gauss elimination method. **(Apr-2018)**

**Solution:** In Gauss elimination method, the coefficient matrix $A$ is reduced to an upper triangular matrix by applying elementary row operation.

The given system of equations can be written as $AX = B$, where
$$A = \begin{pmatrix} 1 & 1 & 1 \\ 2 & 3 & 2 \\ 4 & 2 & 3 \end{pmatrix}, X = \begin{pmatrix} x \\ y \\ z \end{pmatrix}, B = \begin{pmatrix} 8 \\ 19 \\ 23 \end{pmatrix}.$$

The augmented matrix $[A|B] = \begin{pmatrix} 1 & 1 & 1 & | & 8 \\ 2 & 3 & 2 & | & 19 \\ 4 & 2 & 3 & | & 23 \end{pmatrix}$

$R_2 \to R_2 - 2R_1, R_3 \to R_3 - 4R_1$, we have
$$\sim \begin{pmatrix} 1 & 1 & 1 & | & 8 \\ 0 & 1 & 0 & | & 3 \\ 0 & -2 & -1 & | & -9 \end{pmatrix}$$

$R_3 \to R_3 + 2R_2$, we have
$$\sim \begin{pmatrix} 1 & 1 & 1 & | & 8 \\ 0 & 1 & 0 & | & 3 \\ 0 & 0 & -1 & | & -3 \end{pmatrix}$$

Here, coefficient matrix is an upper triangular matrix form.

By Gauss elimination method, we have
$$x + y + z = 8$$

$$y = 3$$
$$-z = -3$$

By back substitution, we have
$$z = 3, \ y = 3 \text{ and } x = 8 - y - z = 8 - 3 - 3 = 2.$$
Therefore, $x = 2, y = 3, z = 3$ is the solution of the given system.

**Example:** Express the following system in matrix form and solve by Gauss elimination method $2x_1 + x_2 + 2x_3 + x_4 = 6, \ 6x_1 - 6x_2 + 6x_3 + 12x_4 = 36, \ 4x_1 + 3x_2 + 3x_3 - 3x_4 = -1, \ 2x_1 + 2x_2 - x_3 + x_4 = 10.$ **(Dec-2021)**

( or) Solve the system of equations by Gauss-elimination method $2x + y + 2z + w = 6, \ 6x - 6y + 6z + 12w = 36, \ 4x + 3y + 3z - 3w = -1, \ 2x + 2y - z + w = 10.$

**(Apr-2022)**

**Solution:** In Gauss elimination method, the coefficient matrix $A$ is reduced to an upper triangular matrix by applying elementary row operation.

The given system of equations can be written as $AX = B$, where

$$A = \begin{pmatrix} 2 & 1 & 2 & 1 \\ 6 & -6 & 6 & 12 \\ 4 & 3 & 3 & -3 \\ 2 & 2 & -1 & 1 \end{pmatrix}, \ X = \begin{pmatrix} x_1 \\ x_2 \\ x_3 \\ x_4 \end{pmatrix}, \ B = \begin{pmatrix} 6 \\ 36 \\ -1 \\ 10 \end{pmatrix}.$$

The augmented matrix $[A|B] = \left( \begin{array}{cccc|c} 2 & 1 & 2 & 1 & 6 \\ 6 & -6 & 6 & 12 & 36 \\ 4 & 3 & 3 & -3 & -1 \\ 2 & 2 & -1 & 1 & 10 \end{array} \right)$

$R_2 \to \frac{R_2}{6}$, we have

$$\sim \left( \begin{array}{cccc|c} 2 & 1 & 2 & 1 & 6 \\ 1 & -1 & 1 & 2 & 6 \\ 4 & 3 & 3 & -3 & -1 \\ 2 & 2 & -1 & 1 & 10 \end{array} \right)$$

$R_1 \leftrightarrow R_2$, we have

$$\sim \begin{pmatrix} 1 & -1 & 1 & 2 & | & 6 \\ 2 & 1 & 2 & 1 & | & 6 \\ 4 & 3 & 3 & -3 & | & -1 \\ 2 & 2 & -1 & 1 & | & 10 \end{pmatrix}$$

$R_2 \to R_2 - 2R_1$, $R_3 \to R_3 - 4R_1$, $R_4 \to R_4 - 2R_1$, we have

$$\sim \begin{pmatrix} 1 & -1 & 1 & 2 & | & 6 \\ 0 & 3 & 0 & -3 & | & -6 \\ 0 & 7 & -1 & -11 & | & -25 \\ 0 & 4 & -3 & -3 & | & -2 \end{pmatrix}$$

$R_2 \to \frac{R_2}{3}$, we have

$$\sim \begin{pmatrix} 1 & -1 & 1 & 2 & | & 6 \\ 0 & 1 & 0 & -1 & | & -2 \\ 0 & 7 & -1 & -11 & | & -25 \\ 0 & 4 & -3 & -3 & | & -2 \end{pmatrix}$$

$R_3 \to R_3 - 7R_2$, $R_4 \to R_4 - 4R_2$, we have

$$\sim \begin{pmatrix} 1 & -1 & 1 & 2 & | & 6 \\ 0 & 1 & 0 & -1 & | & -2 \\ 0 & 0 & -1 & -4 & | & -11 \\ 0 & 0 & -3 & 1 & | & 6 \end{pmatrix}$$

$R_4 \to R_4 - 3R_3$, we have

$$\sim \begin{pmatrix} 1 & -1 & 1 & 2 & | & 6 \\ 0 & 1 & 0 & -1 & | & -2 \\ 0 & 0 & -1 & -4 & | & -11 \\ 0 & 0 & 0 & 13 & | & 39 \end{pmatrix}$$

Here, coefficient matrix is an upper triangular matrix form.

By Gauss elimination method, we have

$$x_1 - x_2 + x_3 + 2x_4 = 6$$
$$x_2 - x_4 = -2$$
$$-x_3 - 4x_4 = -11$$
$$13x_4 = 39$$

By back substitution, we have

$$x_4 = \frac{39}{13} = 3$$
$$x_3 = 11 - 4x_4 = 11 - 12 = -1$$
$$x_2 = -2 + x_4 = -2 + 3 = 1$$
$$x_1 = 6 + x_2 - x_3 - 2x_4 = 6 + 1 + 1 - 6 = 2$$

Therefore, $x_1 = 2, x_2 = 1, x_3 = -1$ and $x_4 = 3$ is the solution of the given system.

**Example:** Express the following system in matrix form and solve by Gauss elimination method $2x_1 + x_2 + 2x_3 + x_4 = 6$, $6x_1 - 6x_2 + 6x_3 + 12x_4 = 36$, $4x_1 + 3x_2 + 3x_3 - 3x_4 = 1$, $2x_1 + 2x_2 - x_3 + x_4 = 10$. **(Dec-2021)**

**Solution:** In Gauss elimination method, the coefficient matrix $A$ is reduced to an upper triangular matrix by applying elementary row operation.

The given system of equations can be written as $AX = B$, where

$$A = \begin{pmatrix} 2 & 1 & 2 & 1 \\ 6 & -6 & 6 & 12 \\ 4 & 3 & 3 & -3 \\ 2 & 2 & -1 & 1 \end{pmatrix}, X = \begin{pmatrix} x_1 \\ x_2 \\ x_3 \\ x_4 \end{pmatrix}, B = \begin{pmatrix} 6 \\ 36 \\ 1 \\ 10 \end{pmatrix}.$$

The augmented matrix $[A|B] = \begin{pmatrix} 2 & 1 & 2 & 1 & | & 6 \\ 6 & -6 & 6 & 12 & | & 36 \\ 4 & 3 & 3 & -3 & | & -1 \\ 2 & 2 & -1 & 1 & | & 10 \end{pmatrix}$

$R_2 \to R_2 - 3R_1$, $R_3 \to R_4 - 2R_1$, $R_4 \to R_4 - R_1$, we have

$$\sim \begin{pmatrix} 2 & 1 & 2 & 1 & | & 6 \\ 0 & -9 & 0 & 9 & | & 18 \\ 0 & 1 & -1 & -5 & | & -11 \\ 0 & 1 & -3 & 0 & | & 4 \end{pmatrix}$$

$R_2 \to \frac{R_2}{-9}$, we have

$$\sim \begin{pmatrix} 2 & 1 & 2 & 1 & | & 6 \\ 0 & 1 & 0 & -1 & | & -2 \\ 0 & 1 & -1 & -5 & | & -11 \\ 0 & 1 & -3 & 0 & | & 4 \end{pmatrix}$$

$R_3 \to R_3 - R_2$, $R_4 \to R_4 - R_2$, we have

$$\sim \begin{pmatrix} 2 & 1 & 2 & 1 & | & 6 \\ 0 & 1 & 0 & -1 & | & -2 \\ 0 & 0 & -1 & -4 & | & -9 \\ 0 & 0 & -3 & 1 & | & 6 \end{pmatrix}$$

$R_4 \to R_4 - 3R_3$, we have

$$\sim \begin{pmatrix} 2 & 1 & 2 & 1 & | & 6 \\ 0 & 1 & 0 & -1 & | & -2 \\ 0 & 0 & -1 & -4 & | & -9 \\ 0 & 0 & 0 & 13 & | & 33 \end{pmatrix}$$

Here, coefficient matrix is an upper triangular matrix form.

By Gauss elimination method, we have

$$2x_1 + x_2 + 2x_3 + x_4 = 6$$
$$x_2 - x_4 = -2$$
$$-x_3 - 4x_4 = -9$$
$$13x_4 = 33$$

By back substitution, we have

$$x_4 = \frac{33}{13}$$
$$x_3 = 9 - 4x_4 = 9 - \frac{132}{13} = \frac{-15}{13}$$
$$x_2 = -2 + x_4 = -2 + \frac{33}{13} = \frac{7}{13}$$
$$2x_1 = 6 - x_2 - 2x_3 - x_4 = 6 - \frac{7}{13} + \frac{30}{13} - \frac{33}{13} = \frac{68}{13}$$
$$\Rightarrow x_1 = \frac{34}{13}.$$

Therefore, $x_1 = \frac{34}{13}$, $x_2 = \frac{7}{13}$, $x_3 = \frac{-15}{13}$ and $x_4 = \frac{33}{13}$ is the solution of the given system.

1. Solve the equations $3x + y + 2z = 3, 2x - 3y - z = -3, x + 2y + z = 4$ by Gauss elimination method.

    **Ans:** $x = 1, y = 2, z = -1$.

2. Solve $x - y + 2z = 4, 3x + y + 4z = 6, x + y + z = 1$ using Gauss elimination method. **(Dec-2020)**

3. Solve the system of equations $x + 2y + 3z = 1, 2x + 3y + 8z = 2, x + y + z = 3$ using Gauss elimination method. **(Dec-2020)**

    **Ans:** $x = \dfrac{9}{2}, y = -1, z = -\dfrac{1}{2}$.

4. Solve the equations $x + y + z - w = 2, 7x + y + 3z + w = 12, 8x - y + z - 3w = 5, 10x + 5y + 3z + 2w = 20$ by Gauss elimination method.

    **(Jan-2020), (Nov-2021), (Mar-2022), (Apr-2022)**

    **Ans:** $x = 1, y = 1, z = 1, w = 1$

4. Solve the equations $2x - 6y + 8z = 24, 5x + 4y - 3z = 2, 3x + y + 2z = 16$ by Gauss elimination method. **(Aug-2021), (Aug-2022)**

    **Ans:** $x = 1, y = 3, z = 5$.

5. Solve the system equations by Gauss elimination method $5x + y + z + w = 4, x + 7y + z + w = 12, x + y + 6z + w = -5, x + y + z + 4w = -6$. **(Apr-2022)**

6. Solve the system equations by Gauss elimination method $2x - y + 2z + 6w = 4, 6x + y + 6z + 12w = 2, 4x + y + 3z - 3w = -1, 2x + 2y - z + w = 1$. **(Apr-2022)**

4. Solve the system of equations by Gauss elimination method: $x - 3y + 7z = 2, 2x + 4y - 3z = -1, -3x + 7y + 2z = 3$. **(Aug-2022)**

## 1.6 Gauss-Jordan Method

This method is modification of Gauss elimination. In this method, the coefficient matrix $A$ is reduced to diagonal form. Then solution can be obtained directly without any back substitution.

**Example:** Solve the following system of equations using Gauss-Jordan method:

$x + y + z = 3$, $2x - y + 3z = 16$, $3x + y - z = -3$.

**Solution:** In Gauss-Jordan method, the coefficient matrix $A$ is reduced to diagonal form by applying elementary row operation.

The given system of equations can be written as $AX = B$, where

$$A = \begin{pmatrix} 1 & 1 & 1 \\ 2 & -1 & 3 \\ 3 & 1 & -1 \end{pmatrix}, X = \begin{pmatrix} x \\ y \\ z \end{pmatrix}, B = \begin{pmatrix} 3 \\ 16 \\ -3 \end{pmatrix}.$$

The augmented matrix $[A|B] = \begin{pmatrix} 1 & 1 & 1 & | & 3 \\ 2 & -1 & 3 & | & 16 \\ 3 & 1 & -1 & | & -3 \end{pmatrix}$

$R_2 \to R_2 - 2R_1$, $R_3 \to R_3 - 3R_1$, we have

$$\sim \begin{pmatrix} 1 & 1 & 1 & | & 3 \\ 0 & -3 & 1 & | & 10 \\ 0 & -2 & -4 & | & -12 \end{pmatrix}$$

$R_3 \to 3R_3 - 2R_2$, we have

$$\sim \begin{pmatrix} 1 & 1 & 1 & | & 3 \\ 0 & -3 & 1 & | & 10 \\ 0 & 0 & -14 & | & -56 \end{pmatrix}$$

$R_4 \to \frac{R_4}{-14}$, we have

$$\sim \begin{pmatrix} 1 & 1 & 1 & | & 3 \\ 0 & -3 & 1 & | & 10 \\ 0 & 0 & 1 & | & 4 \end{pmatrix}$$

$R_2 \to R_2 - R_3$, $R_1 \to R_1 - R_3$, we have

$$\sim \begin{pmatrix} 1 & 1 & 0 & | & -1 \\ 0 & -3 & 0 & | & 6 \\ 0 & 0 & 1 & | & 4 \end{pmatrix}$$

$R_2 \to \frac{R_2}{-3}$, we have

$$\sim \begin{pmatrix} 1 & 1 & 0 & | & -1 \\ 0 & 1 & 0 & | & -2 \\ 0 & 0 & 1 & | & 4 \end{pmatrix}$$

$R_1 \to R_1 - R_2$, we have

$$\sim \begin{pmatrix} 1 & 0 & 0 & | & 1 \\ 0 & 1 & 0 & | & -2 \\ 0 & 0 & 1 & | & 4 \end{pmatrix}$$

Here the coefficient matrix is in diagonal form.

By Gauss-Jordan method, we have $x = 1, y = -2, z = 4$.

**Example:** Solve the equations $10x_1 + x_2 + x_3 = 12$, $x_1 + 10x_2 - x_3 = 10$ and $x_1 - 2x_2 + 10x_3 = 9$ by Gauss-Jordan method.

**Solution:** In Gauss-Jordan method, the coefficient matrix $A$ is reduced to diagonal form by applying elementary row operation.

The given system of equations can be written as $AX = B$, where

$$A = \begin{pmatrix} 10 & 1 & 1 \\ 1 & 10 & -1 \\ 1 & -2 & 10 \end{pmatrix}, X = \begin{pmatrix} x_1 \\ x_2 \\ x_3 \end{pmatrix}, B = \begin{pmatrix} 12 \\ 10 \\ 9 \end{pmatrix}.$$

The augmented matrix $[A|B] = \begin{pmatrix} 10 & 1 & 1 & | & 12 \\ 1 & 10 & -1 & | & 10 \\ 1 & -2 & 10 & | & 9 \end{pmatrix}$

$R_1 \leftrightarrow R_2$, we have

$$\sim \begin{pmatrix} 1 & 10 & -1 & | & 10 \\ 10 & 1 & 1 & | & 12 \\ 1 & -2 & 10 & | & 9 \end{pmatrix}$$

$R_2 \to R_2 - 10R_1, R_3 \to R_3 - R_2$, we have

$$\sim \begin{pmatrix} 1 & 10 & -1 & | & 10 \\ 0 & -99 & 11 & | & -88 \\ 0 & -12 & 11 & | & -1 \end{pmatrix}$$

$R_2 \to \frac{R_2}{11}$, we have

$$\sim \begin{pmatrix} 1 & 10 & -1 & | & 10 \\ 0 & -9 & 1 & | & -8 \\ 0 & -12 & 11 & | & -1 \end{pmatrix}$$

$R_3 \to 3R_3 - 4R_2$, we have

$$\sim \begin{pmatrix} 1 & 10 & -1 & | & 10 \\ 0 & -9 & 1 & | & -8 \\ 0 & 0 & 29 & | & 29 \end{pmatrix}$$

$R_3 \to \frac{R_3}{29}$, we have

$$\sim \begin{pmatrix} 1 & 10 & -1 & | & 10 \\ 0 & -9 & 1 & | & -8 \\ 0 & 0 & 1 & | & 1 \end{pmatrix}$$

$R_1 \to R_1 + R_3, R_2 \to R_2 - R_3$, we have

$$\sim \begin{pmatrix} 1 & 10 & 0 & | & 11 \\ 0 & -9 & 0 & | & -9 \\ 0 & 0 & 1 & | & 1 \end{pmatrix}$$

$R_2 \to \frac{R_2}{-9}$, we have

$$\sim \begin{pmatrix} 1 & 10 & 0 & | & 11 \\ 0 & 1 & 0 & | & 1 \\ 0 & 0 & 1 & | & 1 \end{pmatrix}$$

$R_1 \to R_1 - 10R_2$, we have

$$\sim \begin{pmatrix} 1 & 0 & 0 & | & 1 \\ 0 & 1 & 0 & | & 1 \\ 0 & 0 & 1 & | & 1 \end{pmatrix}$$

Here the coefficient matrix is in diagonal form.
By Gauss-Jordan method, we have $x_1 = 1, x_2 = 1, x_3 = 1$.

### EXERCISE

1. Solve the system of equations $x + y + z = 8, 2x + 3y + 2z = 19, 4x + 2y + 3z = 23$ by Gauss-Jordan method.

   **Ans:** $x = 2, y = 3, z = 3$.

2. Solve the equations $10x + y + z = 12, 2x + 10y + z = 13, x + y + 5z = 7$ by Gauss-Jordan method. **(Apr-2018)**

   **Ans:** $x = 1, y = 1, z = 1$.

3. Solve the equations $x + y + z - w = 2, 7x + y + 3z + w = 12, 8x - y + z - 3w =$

$5, 10x + 5y + 3z + 2w = 20$ by Gauss-Jordan method.

4. Using Gauss-Jordan method, solve the system $2x + y + z = 10, 3x + 2y + 3z = 18, x + 4y + 9z = 16$.

**Ans:** $x = 7, y = -9, z = 5$.

## 1.7 Applications

### 1.7.1 Kirchoff's Laws

**Kirchoff's Current Law (KCL):** At any node of a circuit, the sum of the inflowing currents equals to sum of the out flowing currents.

**Kirchoff's Voltage Law (KVL):** In any closed loop, the sum of all voltage drops equals the impressed electromagnetic force.

The following physical quantities are measured in the electrical circuit.

(i). **Current:** Current is denoted by $i$, measured in Amps ($A$).

(ii). **Resistance:** Resistance is denoted by $R$, measured in Ohms ($\Omega$).

(ii). **Electrical Potential difference:** It is denoted by $V$, measured in Volts ($V$).

**Example:** There is a circuit (electrical network) in figure given below.

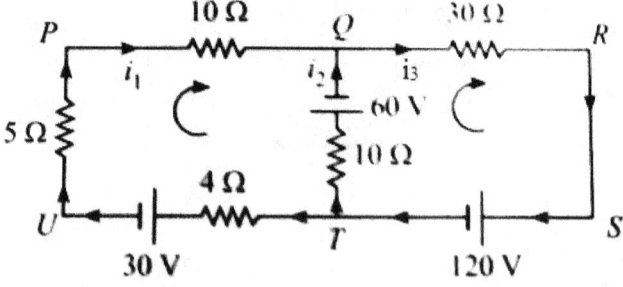

Find the currents $i_1, i_2$ and $i_3$ respectively.

**Solution:** Using KCL, current at node $P$

$$i_1 + i_3 = i_2 \Rightarrow i_1 - i_2 + i_3 = 0 \quad \cdots \cdots \quad (1)$$

Applying KVL to the loop PQTUP, we have

$$10i_2 + 20i_1 = 80 \Rightarrow 20i_1 + 10i_2 = 80 \quad \cdots\cdots \quad (2)$$

Applying KVL to the loop QRSTQ, we have

$$15i_3 + 10i_2 + 10i_3 = 90 \Rightarrow 10i_2 + 25i_3 = 90 \quad \cdots\cdots \quad (3)$$

To find the currents $i_1, i_2$ and $i_3$, solve equations (1), (2) and (3) using matrix method.

The matrix is
$$\begin{pmatrix} 1 & -1 & 1 \\ 20 & 10 & 0 \\ 0 & 10 & 25 \end{pmatrix} \begin{pmatrix} i_1 \\ i_2 \\ i_3 \end{pmatrix} = \begin{pmatrix} 0 \\ 80 \\ 90 \end{pmatrix}.$$

The augmented matrix is $\begin{pmatrix} 1 & -1 & 1 & | & 0 \\ 20 & 10 & 0 & | & 80 \\ 0 & 10 & 25 & | & 90 \end{pmatrix}$

$R_2 \to R_2 - 20R_1, \sim \begin{pmatrix} 1 & -1 & 1 & | & 0 \\ 0 & 30 & -20 & | & 80 \\ 0 & 10 & 25 & | & 90 \end{pmatrix}$

$R_2 \to R_2 - 3R_3, \sim \begin{pmatrix} 1 & -1 & 1 & | & 0 \\ 0 & 0 & -95 & | & -190 \\ 0 & 10 & 25 & | & 90 \end{pmatrix}$

$R_2 \leftrightarrow R_3, \sim \begin{pmatrix} 1 & -1 & 1 & | & 0 \\ 0 & 10 & 25 & | & 90 \\ 0 & 0 & -95 & | & -190 \end{pmatrix}$

Here the coefficient matrix is in upper triangular form. By Gauss elimination method,

$$i_1 - i_2 + i_3 = 0$$
$$10i_2 + 25i_3 = 90$$
$$-95i_3 = -190$$

By back substitution, we have

$$-95i_3 = -190 \Rightarrow i_3 = 2$$
$$10i_2 + 25i_3 = 90 \Rightarrow i_2 = 4$$
$$i_1 - i_2 + i_3 = 0 \Rightarrow i_1 = 2$$

$\therefore i_1 = 2$ Amp, $i_2 = 4$ Amp, $i_3 = 2$ Amp.

**Example:** Find the currents in the following circuit:

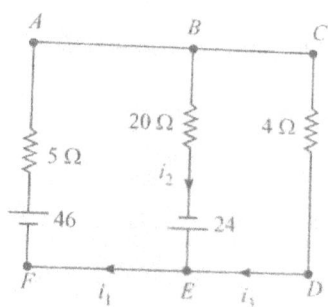

**Solution:** Using KCL, current at node $E$

$$i_2 + i_3 = i_1 \Rightarrow -i_1 + i_2 + i_3 = 0 \quad \cdots \cdots \quad (1)$$

Applying KVL to the loop ABEFA, we have

$$5i_1 + 20i_2 = 46 + 24 \Rightarrow 5i_1 + 20i_2 = 70 \quad \cdots \cdots \quad (2)$$

Applying KVL to the loop BEDCB, we have

$$20i_2 - 4i_3 = 24 \quad \cdots \cdots \quad (3)$$

To find the currents $i_1, i_2$ and $i_3$, solve equations (1), (2) and (3) using matrix method. The matrix is

$$\begin{pmatrix} -1 & 1 & 1 \\ 5 & 20 & 0 \\ 0 & 20 & -4 \end{pmatrix} \begin{pmatrix} i_1 \\ i_2 \\ i_3 \end{pmatrix} = \begin{pmatrix} 0 \\ 70 \\ 24 \end{pmatrix}.$$

The augmented matrix is $\left( \begin{array}{ccc|c} -1 & 1 & 1 & 0 \\ 5 & 20 & 0 & 70 \\ 0 & 20 & -4 & 24 \end{array} \right)$

$R_2 \to R_2 + 5R_1, R_3 \to \frac{R_3}{4}, \sim \left( \begin{array}{ccc|c} -1 & 1 & 1 & 0 \\ 0 & 25 & 5 & 70 \\ 0 & 5 & -1 & 6 \end{array} \right)$

$$R_2 \to R_2 - 5R_3, \sim \begin{pmatrix} -1 & 1 & 1 & | & 0 \\ 0 & 0 & 10 & | & 40 \\ 0 & 5 & -1 & | & 6 \end{pmatrix}$$

$$R_2 \leftrightarrow R_3, \sim \begin{pmatrix} -1 & 1 & 1 & | & 0 \\ 0 & 5 & -1 & | & 6 \\ 0 & 0 & 10 & | & 40 \end{pmatrix}$$

Here the coefficient matrix is in upper triangular form. By Gauss elimination method,

$$-i_1 + i_2 + i_3 = 0$$
$$5i_2 - i_3 = 6$$
$$10i_3 = 40$$

By back substitution, we have

$$10i_3 = 40 \Rightarrow i_3 = 4$$
$$5i_2 - i_3 = 6 \Rightarrow i_2 = 2$$
$$-i_1 + i_2 + i_3 = 0 \Rightarrow i_1 = 6$$

$\therefore i_1 = 6$ Amp, $i_2 = 2$ Amp, $i_3 = 4$ Amp.

**Example:** Find the currents in the following circuit:

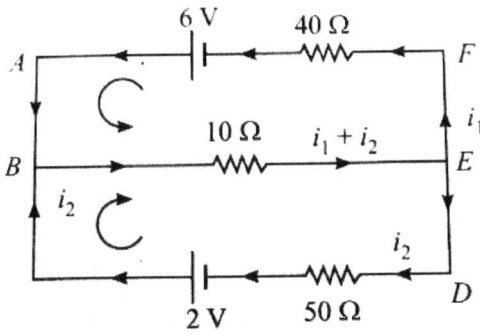

**Solution:** Using KVL to the loop ABEFA, we have

$$10(i_1 + i_2) + 40i_1 = 6 \Rightarrow 50i_1 + 10i_2 = 6 \quad \cdots \cdots \quad (1)$$

Using KVL to the loop BEDCB, we have

$$10(i_1 + i_2) + 50i_2 = 2 \Rightarrow 10i_1 + 60i_2 = 2 \quad \cdots \cdots \quad (2)$$

To find the currents $i_1$ and $i_2$, solve equations (1) and (2) using matrix method. The matrix is

$$\begin{pmatrix} 50 & 10 \\ 10 & 60 \end{pmatrix} \begin{pmatrix} i_1 \\ i_2 \end{pmatrix} = \begin{pmatrix} 6 \\ 2 \end{pmatrix}.$$

The augmented matrix is $\begin{pmatrix} 50 & 10 & | & 6 \\ 10 & 60 & | & 2 \end{pmatrix}$

$R_2 \to 5R_2 - R_1, \sim \begin{pmatrix} 50 & 10 & | & 6 \\ 0 & 290 & | & 4 \end{pmatrix}$

Here the coefficient matrix is in upper triangular form. By Gauss elimination method,

$$50i_1 + 10i_2 = 6$$
$$290i_2 = 4$$

By back substitution, we have

$$290i_2 = 4 \Rightarrow i_2 = 0.01379$$
$$50i_1 + 10i_2 = 6 \Rightarrow i_1 = 0.1172$$

$\therefore i_1 = 0.1172$ Amp, $i_2 = 0.01379$ Amp.

**Example:** Find the currents in the following circuit:

**Solution:** Using KCL, current at node Q

$$i_1 + i_2 = i_3 \Rightarrow i_1 + i_2 - i_3 = 0 \quad \cdots\cdots \quad (1)$$

Applying KVL to the loop PQTUP, we have

$$5i_1 + 10i_1 - 10i_2 + 4i_1 = -30 - 60 \Rightarrow 19i_1 - 10i_2 = -90 \quad \cdots\cdots \quad (2)$$

Applying KVL to the loop QRSTQ, we have

$$30i_3 + 10i_2 = 60 + 120 \Rightarrow i_2 + 3i_3 = 18 \quad \cdots\cdots \quad (3)$$

To find the currents $i_1$, $i_2$ and $i_3$, solve equations (1), (2) and (3) using matrix method. The matrix is

$$\begin{pmatrix} 1 & 1 & -1 \\ 19 & -10 & 0 \\ 0 & 1 & 3 \end{pmatrix} \begin{pmatrix} i_1 \\ i_2 \\ i_3 \end{pmatrix} = \begin{pmatrix} 0 \\ -90 \\ 18 \end{pmatrix}.$$

The augmented matrix is $\begin{pmatrix} 1 & 1 & -1 & | & 0 \\ 19 & -10 & 0 & | & -90 \\ 0 & 1 & 3 & | & 18 \end{pmatrix}$

$R_2 \to R_2 - 19R_1,\ R_3 \to \frac{R_3}{4},\ \sim \begin{pmatrix} 1 & 1 & -1 & | & 0 \\ 0 & -29 & 19 & | & -90 \\ 0 & 1 & 3 & | & 18 \end{pmatrix}$

$R_2 \to 29R_3 + R_2,\ \sim \begin{pmatrix} 1 & 1 & -1 & | & 0 \\ 0 & 0 & 106 & | & 432 \\ 0 & 1 & 3 & | & 18 \end{pmatrix}$

$R_2 \leftrightarrow R_3,\ \sim \begin{pmatrix} 1 & 1 & -1 & | & 0 \\ 0 & 1 & 3 & | & 18 \\ 0 & 0 & 106 & | & 432 \end{pmatrix}$

Here the coefficient matrix is in upper triangular form. By Gauss elimination method,

$$i_1 + i_2 - i_3 = 0$$
$$i_2 + 3i_3 = -18$$
$$106 i_3 = 432$$

By back substitution, we have

$$106 i_3 = 432 \Rightarrow i_3 = 4.0754$$
$$i_2 + 3i_3 = -18 \Rightarrow i_2 = 5.7738$$
$$i_1 + i_2 - i_3 = 0 \Rightarrow i_1 = -1.6984$$

$\therefore i_1 = -1.6984$ Amp, $i_2 = 5.7738$ Amp, $i_3 = 4.0754$ Amp.

**Example:** Find the currents in the following circuit:

**Solution:** Using KCL, current at node $B$

$$i_1 + i_2 = i_3 \Rightarrow i_1 + i_2 - i_3 = 0 \quad \cdots \cdots \quad (1)$$

Applying KVL to the loop ABEFA, we have

$$-4i_2 + 2i_1 = -12 + 10 \Rightarrow i_1 - 2i_2 = -1 \quad \cdots \cdots \quad (2)$$

Applying KVL to the loop BCDED, we have

$$8i_3 + 4i_2 = 12 \Rightarrow i_2 + 2i_3 = 3$$

To minimize the number of equations, substitute $i_3 = 1_1 + i_2$, then equation (1), we have $i_2 + 2(i_1 + i_2) = 3 \Rightarrow 2i_1 + 3i_2 = 3$ ...... (3)

To find the currents $i_1$ and $i_2$, solve equations (2) and (3) using matrix method. The matrix is

$$\begin{pmatrix} 1 & -2 \\ 2 & 3 \end{pmatrix} \begin{pmatrix} i_1 \\ i_2 \end{pmatrix} = \begin{pmatrix} -1 \\ 3 \end{pmatrix}.$$

The augmented matrix is $\begin{pmatrix} 1 & -2 & | & -1 \\ 2 & 3 & | & 3 \end{pmatrix}$

$R_2 \to R_2 - 2R_1, \sim \begin{pmatrix} 1 & -2 & | & -1 \\ 0 & 7 & | & 5 \end{pmatrix}$

Here the coefficient matrix is in upper triangular form. By Gauss elimination method,

$$i_1 - 2i_2 = -1$$
$$7i_2 = 5$$

By back substitution, we have

$$7i_2 = 5 \Rightarrow i_2 = 0.71$$
$$i_1 - 2i_2 = -1 \Rightarrow i_1 = 0.42$$
$$i_1 + i_2 - i_3 = 0 \Rightarrow i_3 = 1.13$$

$\therefore i_1 = 0.42$ Amp, $i_2 = 0.71$ Amp, $i_3 = 1.13$ Amp.

**Example:** Find the current in each cell considering the circuit given in the figure:

**Solution:** Using KCL, current at node $B$ is

$$i_1 + i_3 = i_2 \Rightarrow i_1 - i_2 + i_3 = 0 \quad \cdots\cdots \quad (1)$$

Applying KVL to the loop ABFGA, we have

$$i_3 - i_1 = -2 \Rightarrow i_1 - i_3 = 2 \quad \cdots\cdots \quad (2)$$

Applying KVL to the loop BCDFB, we have

$$-i_2 - i_3 = 4 \Rightarrow i_2 + i_3 = -4 \quad \cdots\cdots \quad (3)$$

To find the currents $i_1, i_2$ and $i_3$, solve equations (1), (2) and (3) using matrix method.

The matrix is

$$\begin{pmatrix} 1 & -1 & -1 \\ 1 & 0 & -1 \\ 0 & 1 & 1 \end{pmatrix} \begin{pmatrix} i_1 \\ i_2 \\ i_3 \end{pmatrix} = \begin{pmatrix} 0 \\ 2 \\ -4 \end{pmatrix}.$$

The augmented matrix is $\left(\begin{array}{ccc|c} 1 & -1 & 1 & 0 \\ 1 & 0 & -1 & 2 \\ 0 & 1 & 1 & -4 \end{array}\right)$

$R_2 \to R_2 - R_1, \sim \left(\begin{array}{ccc|c} 1 & -1 & 1 & 0 \\ 0 & 1 & -2 & 2 \\ 0 & 1 & 1 & -4 \end{array}\right)$

$R_3 \to R_3 - R_2, \sim \left(\begin{array}{ccc|c} 1 & -1 & 1 & 0 \\ 0 & 1 & -2 & 2 \\ 0 & 0 & 3 & -6 \end{array}\right)$

Here the coefficient matrix is in upper triangular form. By Gauss elimination method,

$$i_1 - i_2 + i_3 = 0$$
$$i_2 - 2i_3 = 2$$
$$3i_3 = -6$$

By back substitution, we have

$$3i_3 = -6 \Rightarrow i_3 = -2$$
$$i_2 - 2i_3 = 2 \Rightarrow i_2 = -2$$
$$i_1 - i_2 + i_3 = 0 \Rightarrow i_1 = 0$$

$\therefore i_1 = 0$ Amp, $i_2 = -2$ Amp, $i_3 = -2$ Amp.

## 1.8 Eigenvalues and Eigenvectors

Let $A = [a_{ij}]_{n \times n}$. A non-zero vector $X$ is said to be **eigenvector or characteristic vector or latent vector** if there exist a scalar $\lambda$ such that $AX = \lambda X \Rightarrow (A - \lambda I)X = 0$.

If $AX = \lambda X$ then $X$ is said to be eigenvector of $A$ corresponding to the eigenvalue

$\lambda$ of $A$.

$A - \lambda I$ is called characteristic matrix also $|A - \lambda I| = 0$ is called the characteristic equation.

**A Vector as a Linear Combination of Vectors:** A Vector $X$ which can be expressed in the form

$$X = k_1 X_1 + k_2 X_2 + \cdots + k_n X_n$$

is said to be a linear combination of the set of vectors $X_1, X_2, \cdots, X_n$. Here $k_1, k_2, \cdots, k_n$ are any scalars.

**Example:** $(2, -3)$ is a linear combination of the vectors $(2, 0)$ and $(0, 3)$ since

$$1(2, 0) + (-1)(0, 3) = (2, -3).$$

$(2, 7, -5)$ is a linear combination of the vectors $(2, 1, 0), (1, -1, 2)$ and $(1, 2, 1)$ since

$$(2)(2, 1, 0) + (-3)(1, -1, 2) + (1)(1, 2, 1) = (2, 7, -5).$$

**Linearly Dependent:** A set of vectors $X_1, X_2, \cdots, X_n$ is said to be *linearly dependent* if there exists scalars $k_1, k_2, \cdots, k_n$ not all zero such that

$$k_1 X_1 + k_2 X_2 + \cdots + k_n X_n = \mathbf{0}.$$

**Example:** Show that the vectors $(1, -2), (3, 1), (6, 4)$ are linearly dependent.
**Solution:** By taking $k_1 = 6, k_2 = -16$ and $k_3 = 7$, we have

$$k_1 X_1 + k_2 X_2 + k_3 X_3 = 6X_1 - 16X_2 + 7X_3$$
$$= 6(1, -2) - 16(3, 1) + 7(6, 4)$$
$$= (0, 0) = 0$$

Thus there exits three scalars $k_1 = 6, k_2 = -16$ and $k_3 = 7$, which are not all zero such that $k_1 X_1 + k_2 X_2 + k_3 X_3 = 0$.

Hence the vectors $X_1, X_2$ and $X_3$ are linearly dependent.

**Linearly Independent:** If the linear combination

$$k_1 X_1 + k_2 X_2 + \cdots + k_n X_n = 0$$

implies that all $k_i$'s are zero then the set of vectors $X_1, X_2, \cdots, X_n$ is *linearly independent*.

**Example:** Show that the vectors $(1,2,0), (0,3,1), (-1,0,1)$ are linearly independent.

**Solution:** Given $X_1 = (1,2,0), X_2 = (0,3,1)$ and $X_3 = (-1,0,1)$ are three vectors. Let $k_1, k_2, k_3$ be any three scalars satisfying the relation

$$k_1 X_1 + k_2 X_2 + k_3 X_3 = 0$$
$$\Rightarrow k_1(1,2,0) + k_2(0,3,1) + k_3(-1,0,1) = (0,0,0)$$
$$\Rightarrow (k_1, 2k_1, 0) + (0, 3k_2, k_2) + (-k_3, 0, k_3) = (0,0,0)$$
$$\Rightarrow (k_1 - k_3, 2k_1 + k_2, k_2 + k_3) = (0,0,0)$$

Obviously the relation is true if and only if $k_1 = 0, k_2 = 0$ and $k_3 = 0$. Hence the vectors $X_1, X_2$ and $X_3$ are linearly independent.

**Example:** Find the characteristic roots of the matrix $A = \begin{pmatrix} 2 & 2 & 1 \\ 1 & 3 & 1 \\ 1 & 2 & 2 \end{pmatrix}$.

**Solution:** Given $A = \begin{pmatrix} 2 & 2 & 1 \\ 1 & 3 & 1 \\ 1 & 2 & 2 \end{pmatrix}$.

The characteristic equation of $A$ is $|A - \lambda I| = 0$.

i.e., $\begin{vmatrix} 2-\lambda & 2 & 1 \\ 1 & 3-\lambda & 1 \\ 1 & 2 & 2-\lambda \end{vmatrix} = 0.$

$\Rightarrow (2-\lambda)[(3-\lambda)(2-\lambda) - 2] - 2[(2-\lambda) - 1] + 1[2 - (3-\lambda)] = 0$

$\Rightarrow (2-\lambda)[6 - 3\lambda - 2\lambda + \lambda^2 - 2] - 2[1 - \lambda] + 1[-1 + \lambda] = 0$

$\Rightarrow (2-\lambda)[\lambda^2 - 5\lambda + 4] + 2[\lambda - 1] + 1[\lambda - 1] = 0$

$\Rightarrow 2\lambda^2 - 10\lambda + 8 - \lambda^3 + 5\lambda^2 - 4\lambda + 2\lambda - 2 + \lambda - 1 = 0$

$\Rightarrow -\lambda^3 + 7\lambda^2 - 11\lambda + 5 = 0$

$\Rightarrow \lambda^3 - 7\lambda^2 + 11\lambda - 5 = 0$

$\Rightarrow \lambda = 1, 1, 5.$

Hence the characteristic roots are 1, 1, 5.

**Example:** The matrix $A = \begin{pmatrix} 1 & 1 & 3 \\ 1 & 5 & 1 \\ 3 & 1 & 1 \end{pmatrix}$ has an eigen vector $\begin{pmatrix} -1 \\ 0 \\ 1 \end{pmatrix}$, find the corresponding eigen value of $A$.

**Solution:** Let $\lambda$ be the eigen value of $A$ for which $X = \begin{pmatrix} -1 \\ 0 \\ 1 \end{pmatrix}$ is an eigen vector.

Then
$$(A - \lambda I)X = 0$$

$$\Rightarrow \begin{pmatrix} 1-\lambda & 1 & 3 \\ 1 & 5-\lambda & 1 \\ 3 & 1 & 1-\lambda \end{pmatrix} \begin{pmatrix} -1 \\ 0 \\ 1 \end{pmatrix} = \begin{pmatrix} 0 \\ 0 \\ 0 \end{pmatrix}$$

$$\Rightarrow \begin{pmatrix} -(1-\lambda) + 3 \\ -1 + 1 \\ -3 + 1 - \lambda \end{pmatrix} = \begin{pmatrix} 0 \\ 0 \\ 0 \end{pmatrix}$$

$\Rightarrow \lambda + 2 = 0$

$\Rightarrow \lambda = -2$

$\therefore$ Corresponding eigen value is $-2$.

**Example:** Find the eigenvalues and eigenvectors of the matrix $A = \begin{pmatrix} 8 & -6 & 2 \\ -6 & 7 & -4 \\ 2 & -4 & 3 \end{pmatrix}$.

(Dec-2020), (Jan-2020)

**Solution:** Let $\lambda$ be the eigenvalue of $A$.

The characteristic equation of $A$ is given by $|A - \lambda I| = 0$.

$\Rightarrow \lambda^3 - 18\lambda^2 + 45\lambda = 0.$

$\Rightarrow \lambda = 0, 3, 15.$

$\therefore$ Eigenvalues of $A$ are $\lambda = 0, 3, 15.$

## To find eigenvectors:

Let $X = \begin{pmatrix} x_1 \\ x_2 \\ x_3 \end{pmatrix}$ be the eigenvector of $A$ corresponding to $\lambda$.

Then
$$AX = \lambda X$$
$$\Rightarrow (A - \lambda I)X = 0$$
$$\Rightarrow \begin{pmatrix} 8-\lambda & -6 & 2 \\ -6 & 7-\lambda & -4 \\ 2 & -4 & 3-\lambda \end{pmatrix} \begin{pmatrix} x_1 \\ x_2 \\ x_3 \end{pmatrix} = \begin{pmatrix} 0 \\ 0 \\ 0 \end{pmatrix} \quad \cdots \cdots \quad (1)$$

**Case 1:** To find eigenvector when $\lambda = 0$.

Put $\lambda = 0$ in (1), we have
$$\begin{pmatrix} 8 & -6 & 2 \\ -6 & 7 & -4 \\ 2 & -4 & 3 \end{pmatrix} \begin{pmatrix} x_1 \\ x_2 \\ x_3 \end{pmatrix} = \begin{pmatrix} 0 \\ 0 \\ 0 \end{pmatrix}$$

This gives,
$$8x_1 - 6x_2 + 2x_3 = 0$$
$$-6x_1 + 7x_2 - 4x_3 = 0$$
$$2x_1 - 4x_2 + 3x_3 = 0$$

Solving first two equations by rules of cross multiplication, we have
$$\frac{x_1}{24-14} = \frac{x_2}{-12+32} = \frac{x_3}{56-36} = k_1 \quad (k_1 \neq 0)$$
$$\frac{x_1}{10} = \frac{x_2}{20} = \frac{x_3}{20} = k_1$$
$$\therefore x_1 = k_1, x_2 = 2k_1, x_3 = 2k_1.$$

The eigenvector corresponding to $\lambda = 0$ is $X_1 = \begin{pmatrix} k_1 \\ 2k_1 \\ 2k_1 \end{pmatrix}$.

When $k_1 = 1$, we have $X_1 = \begin{pmatrix} 1 \\ 2 \\ 2 \end{pmatrix}$.

**Case 2:** To find eigenvector when $\lambda = 3$.

Put $\lambda = 3$ in (1), we have

$$\begin{pmatrix} 5 & -6 & 2 \\ -6 & 4 & -4 \\ 2 & -4 & 0 \end{pmatrix} \begin{pmatrix} x_1 \\ x_2 \\ x_3 \end{pmatrix} = \begin{pmatrix} 0 \\ 0 \\ 0 \end{pmatrix}$$

This gives,

$$5x_1 - 6x_2 + 2x_3 = 0$$
$$-6x_1 + 4x_2 - 4x_3 = 0$$
$$2x_1 - 4x_2 = 0$$

Solving first two equations by cross multiplication, we have

$$\frac{x_1}{24-8} = \frac{x_2}{-12+20} = \frac{x_3}{20-36} = k_2 \quad (k_2 \neq 0)$$

$$\frac{x_1}{16} = \frac{x_2}{8} = \frac{x_3}{-16} = k_2$$

$$\therefore x_1 = 2k_2, x_2 = k_2, x_3 = -2k_2.$$

The eigenvector corresponding to $\lambda = 2$ is $X_2 = \begin{pmatrix} 2k_2 \\ k_2 \\ -2k_2 \end{pmatrix}$.

When $k_2 = 1$, the eigenvector is $X_2 = \begin{pmatrix} 2 \\ 1 \\ -2 \end{pmatrix}$.

**Case 3:** To find eigenvector when $\lambda = 15$.

Put $\lambda = 15$ in (1), we have

$$\begin{pmatrix} -7 & -6 & 2 \\ -6 & -8 & -4 \\ 2 & -4 & -12 \end{pmatrix} \begin{pmatrix} x_1 \\ x_2 \\ x_3 \end{pmatrix} = \begin{pmatrix} 0 \\ 0 \\ 0 \end{pmatrix}$$

This gives,

$$-7x_1 - 6x_2 + 2x_3 = 0$$

$$-6x_1 - 8x_2 - 4x_3 = 0$$
$$2x_1 - 4x_2 - 12x_3 = 0$$

Solving last two equations by cross multiplication, we have

$$\frac{x_1}{96-16} = \frac{x_2}{-8-72} = \frac{x_3}{24+16} = k_3 \quad (k_3 \neq 0)$$

$$\frac{x_1}{80} = \frac{x_2}{-80} = \frac{x_3}{40} = k_3$$

$$\therefore x_1 = 2k_3, x_2 = -2k_3, x_3 = k_3.$$

The eigenvector corresponding to $\lambda = 3$ is $X_3 = \begin{pmatrix} 2k_3 \\ -2k_3 \\ k_3 \end{pmatrix}$.

When $k_3 = 1$, the eigenvector is $X_3 = \begin{pmatrix} 2 \\ -2 \\ 1 \end{pmatrix}$.

**Example:** Find the latent roots and latent vectors of the matrix $A = \begin{pmatrix} 2 & 0 & 1 \\ 0 & 2 & 0 \\ 1 & 0 & 2 \end{pmatrix}$.

**Solution:** Let $\lambda$ be the eigenvalue of $A$.
The characteristic equation of $A$ is given by $|A - \lambda I| = 0$.

$$\Rightarrow \lambda^3 - 6\lambda^2 + 11\lambda - 6 = 0.$$

$$\Rightarrow \lambda = 1, 2, 3.$$

$\therefore$ Eigenvalues of $A$ are $\lambda = 1, 2, 3$.

**To find eigenvectors:**

Let $X = \begin{pmatrix} x_1 \\ x_2 \\ x_3 \end{pmatrix}$ be the eigenvector of $A$ corresponding to $\lambda$.

Then
$$AX = \lambda X$$
$$\Rightarrow (A - \lambda I)X = 0$$

$$\Rightarrow \begin{pmatrix} 2-\lambda & 0 & 1 \\ 0 & 2-\lambda & 0 \\ 1 & 0 & 2-\lambda \end{pmatrix} \begin{pmatrix} x_1 \\ x_2 \\ x_3 \end{pmatrix} = \begin{pmatrix} 0 \\ 0 \\ 0 \end{pmatrix} \quad \ldots \ldots \quad (1)$$

**Case 1:** To find eigenvector when $\lambda = 1$.

Put $\lambda = 1$ in (1), we have

$$\begin{pmatrix} 1 & 0 & 1 \\ 0 & 1 & 0 \\ 1 & 0 & 1 \end{pmatrix} \begin{pmatrix} x_1 \\ x_2 \\ x_3 \end{pmatrix} = \begin{pmatrix} 0 \\ 0 \\ 0 \end{pmatrix}$$

This gives,
$$x_1 + x_3 = 0$$
$$x_2 = 0$$

gives $x_1 = -x_3$. Let $x_1 = k_1$ then $x_3 = -k_1$

The eigenvector corresponding to $\lambda = 1$ is $X_1 = \begin{pmatrix} k_1 \\ 0 \\ -k_1 \end{pmatrix}$.

When $k_1 = 1$, we have $X_1 = \begin{pmatrix} 1 \\ 0 \\ -1 \end{pmatrix}$.

**Case 2:** To find eigenvector when $\lambda = 2$.

Put $\lambda = 2$ in (1), we have

$$\begin{pmatrix} 0 & 0 & 1 \\ 0 & 0 & 0 \\ 1 & 0 & 0 \end{pmatrix} \begin{pmatrix} x_1 \\ x_2 \\ x_3 \end{pmatrix} = \begin{pmatrix} 0 \\ 0 \\ 0 \end{pmatrix}$$

This gives,
$$x_3 = 0, \ x_1 = 0$$

Let $x_2 = k_2$.

The eigenvector corresponding to $\lambda = 2$ is $X_2 = \begin{pmatrix} 0 \\ k_2 \\ 0 \end{pmatrix}$.

When $k_2 = 1$, the eigenvector is $X_2 = \begin{pmatrix} 0 \\ 1 \\ 0 \end{pmatrix}$.

**Case 3:** To find eigenvector when $\lambda = 3$.

Put $\lambda = 3$ in (1), we have

$$\begin{pmatrix} -1 & 0 & 1 \\ 0 & -1 & 0 \\ 1 & 0 & -1 \end{pmatrix} \begin{pmatrix} x_1 \\ x_2 \\ x_3 \end{pmatrix} = \begin{pmatrix} 0 \\ 0 \\ 0 \end{pmatrix}$$

This gives,

$$-x_1 + x_3 = 0$$
$$-x_2 = 0$$
$$x_1 - x_3 = 0$$

Let $x_1 = k_1$. Then $x_3 = k_1$.

The eigenvector corresponding to $\lambda = 3$ is $X_3 = \begin{pmatrix} k_3 \\ 0 \\ k_3 \end{pmatrix}$.

When $k_3 = 1$, the eigenvector is $X_3 = \begin{pmatrix} 1 \\ 0 \\ 1 \end{pmatrix}$.

**Example:** Find the Eigenvalues and eigen vectors of the matrix $\begin{pmatrix} -1 & 2 & -2 \\ 1 & 2 & 1 \\ -1 & -1 & 0 \end{pmatrix}$.

(Jan-2020), (Dec-2021)

**Solution:** Let $\lambda$ be the eigenvalue of $A$.

The characteristic equation of $A$ is given by $|A - \lambda I| = 0$

$$\begin{vmatrix} -1-\lambda & 2 & -2 \\ 1 & 2-\lambda & 1 \\ -1 & -1 & -\lambda \end{vmatrix}$$
$$\Rightarrow \lambda^3 - \lambda^2 - 5\lambda + 5 = 0$$
$$\Rightarrow (\lambda - 1)(\lambda^2 - 5) = 0$$
$$\Rightarrow \lambda = 1, \pm\sqrt{5}.$$

$\therefore$ Eigenvalues of $A$ are $\lambda = 1, \sqrt{5}, -\sqrt{5}$.

**To find eigenvectors:**

Let $X = \begin{pmatrix} x_1 \\ x_2 \\ x_3 \end{pmatrix}$ be the eigenvector of $A$ corresponding to $\lambda$.

Then
$$AX = \lambda X$$
$$\Rightarrow (A - \lambda I)X = 0$$
$$\Rightarrow \begin{pmatrix} -1-\lambda & 2 & -2 \\ 1 & 2-\lambda & 1 \\ -1 & -1 & -\lambda \end{pmatrix} \begin{pmatrix} x_1 \\ x_2 \\ x_3 \end{pmatrix} = \begin{pmatrix} 0 \\ 0 \\ 0 \end{pmatrix} \quad \cdots\cdots \quad (1)$$

**Case 1:** To find eigenvector when $\lambda = 1$.

Put $\lambda = 1$ in (1), we have
$$\begin{pmatrix} -2 & 2 & -2 \\ 1 & 1 & 1 \\ -1 & -1 & -1 \end{pmatrix} \begin{pmatrix} x_1 \\ x_2 \\ x_3 \end{pmatrix} = \begin{pmatrix} 0 \\ 0 \\ 0 \end{pmatrix}$$

This gives,
$$-2x_1 + 2x_2 - 2x_3 = 0$$
$$x_1 + x_2 + x_3 = 0$$
$$-x_1 - x_2 - x_3 = 0$$

Solving first two equations by rules of cross multiplication, we have
$$\frac{x_1}{2+2} = \frac{x_2}{-2+2} = \frac{x_3}{-2-2} = k_1 \quad (k_1 \neq 0)$$

$$\frac{x_1}{1} = \frac{x_2}{0} = \frac{x_3}{-1} = k_1$$

$$\therefore x_1 = k_1, x_2 = 0, x_3 = -k_1.$$

The eigenvector corresponding to $\lambda = 0$ is $X_1 = \begin{pmatrix} k_1 \\ 0 \\ -k_1 \end{pmatrix}$.

When $k_1 = 1$, we have $X_1 = \begin{pmatrix} 1 \\ 0 \\ -1 \end{pmatrix}$.

**Case 2:** To find eigenvector when $\lambda = \sqrt{5}$.

Put $\lambda = \sqrt{5}$ in (1), we have

$$\begin{pmatrix} -1-\sqrt{5} & 2 & -2 \\ 1 & 2-\sqrt{5} & 1 \\ -1 & -1 & -\sqrt{5} \end{pmatrix} \begin{pmatrix} x_1 \\ x_2 \\ x_3 \end{pmatrix} = \begin{pmatrix} 0 \\ 0 \\ 0 \end{pmatrix}$$

This gives,

$$(-1-\sqrt{5})x_1 + 2x_2 - 2x_3 = 0$$

$$x_1 + (2-\sqrt{5})x_2 + x_3 = 0$$

$$-x_1 - x_2 - \sqrt{5}x_3 = 0$$

Solving last two equations by cross multiplication, we have

$$\frac{x_1}{6-2\sqrt{5}} = \frac{x_2}{-1+\sqrt{5}} = \frac{x_3}{-1+2-\sqrt{5}} = k_2 \quad (k_2 \neq 0)$$

$$\frac{x_1}{(\sqrt{5}-1)^2} = \frac{x_2}{\sqrt{5}-1} = \frac{x_3}{-(\sqrt{5}-1)} = k_2$$

$$\frac{x_1}{\sqrt{5}-1} = \frac{x_2}{1} = \frac{x_3}{-1} = k_2$$

$$\therefore x_1 = (\sqrt{5}-1)k_2, x_2 = k_2, x_3 = -k_2.$$

The eigenvector corresponding to $\lambda = \sqrt{5}$ is $X_2 = \begin{pmatrix} (\sqrt{5}-1)k_2 \\ k_2 \\ -k_2 \end{pmatrix}$.

When $k_2 = 1$, the eigenvector is $X_2 = \begin{pmatrix} \sqrt{5}-1 \\ 1 \\ -1 \end{pmatrix}$.

**Case 3:** To find eigenvector when $\lambda = -\sqrt{5}$.
Put $\lambda = -\sqrt{5}$ in (1), we have

$$\begin{pmatrix} -1+\sqrt{5} & 2 & -2 \\ 1 & 2+\sqrt{5} & 1 \\ -1 & -1 & \sqrt{5} \end{pmatrix} \begin{pmatrix} x_1 \\ x_2 \\ x_3 \end{pmatrix} = \begin{pmatrix} 0 \\ 0 \\ 0 \end{pmatrix}$$

This gives,

$$(-1+\sqrt{5})x_1 + 2x_2 - 2x_3 = 0$$
$$x_1 + (2+\sqrt{5})x_2 + x_3 = 0$$
$$-x_1 - x_2 + \sqrt{5}x_3 = 0$$

Solving last two equations by cross multiplication, we have

$$\frac{x_1}{6+2\sqrt{5}} = \frac{x_2}{-1-\sqrt{5}} = \frac{x_3}{-1+2+\sqrt{5}} = k_3 \quad (k_3 \neq 0)$$

$$\frac{x_1}{(\sqrt{5}+1)^2} = \frac{x_2}{-(\sqrt{5}+1)} = \frac{x_3}{\sqrt{5}+1} = k_3$$

$$\frac{x_1}{\sqrt{5}+1} = \frac{x_2}{-1} = \frac{x_3}{1} = k_3$$

$$\therefore x_1 = (\sqrt{5}+1)k_3, \, x_2 = -k_3, \, x_3 = k_3.$$

The eigenvector corresponding to $\lambda = -\sqrt{5}$ is $X_3 = \begin{pmatrix} (\sqrt{5}+1)k_3 \\ -k_3 \\ k_3 \end{pmatrix}$.

When $k_3 = 1$, the eigenvector is $X_3 = \begin{pmatrix} \sqrt{5}+1 \\ -1 \\ 1 \end{pmatrix}$.

**Example:** Find the eigen values and eigen vectors of the matrix $A = \begin{pmatrix} 1 & 1 & 3 \\ 1 & 5 & 1 \\ 3 & 1 & 1 \end{pmatrix}$.

(AU-2018)

**Solution:** Let $\lambda$ be the eigenvalue of $A$.
The characteristic equation of $A$ is given by

$$|A - \lambda I| = 0 \Rightarrow \begin{vmatrix} 1-\lambda & 1 & 3 \\ 1 & 5-\lambda & 1 \\ 3 & 1 & 1-\lambda \end{vmatrix} = 0 \Rightarrow \lambda^3 - 7\lambda^2 + 36 = 0.$$

$$\Rightarrow \lambda = -2, 3, 6.$$

$\therefore$ Eigenvalues of $A$ are $\lambda = -2, 3, 6$.

**To find eigenvectors:**

Let $X = \begin{pmatrix} x_1 \\ x_2 \\ x_3 \end{pmatrix}$ be the eigenvector of $A$ corresponding to $\lambda$.

Then
$$AX = \lambda X$$
$$\Rightarrow (A - \lambda I)X = 0$$
$$\Rightarrow \begin{pmatrix} 1-\lambda & 1 & 3 \\ 1 & 5-\lambda & 1 \\ 3 & 1 & 1-\lambda \end{pmatrix} \begin{pmatrix} x_1 \\ x_2 \\ x_3 \end{pmatrix} = \begin{pmatrix} 0 \\ 0 \\ 0 \end{pmatrix} \quad \cdots \cdots (1)$$

**Case 1:** To find eigenvector when $\lambda = -2$.

Put $\lambda = -2$ in (1), we have
$$\begin{pmatrix} 3 & 1 & 3 \\ 1 & 7 & 1 \\ 3 & 1 & 3 \end{pmatrix} \begin{pmatrix} x_1 \\ x_2 \\ x_3 \end{pmatrix} = \begin{pmatrix} 0 \\ 0 \\ 0 \end{pmatrix}$$

This gives,
$$3x_1 + x_2 + 3x_3 = 0$$
$$x_1 + 7x_2 + x_3 = 0$$
$$3x_1 + x_2 + 3x_3 = 0$$

Solving first two equations by cross multiplication, we have
$$\frac{x_1}{1-21} = \frac{x_2}{3-3} = \frac{x_3}{21-1}$$
$$\frac{x_1}{-20} = \frac{x_2}{0} = \frac{x_3}{20}$$
$$\therefore x_1 = -1, x_2 = 0, x_3 = 1.$$

The eigenvector corresponding to $\lambda = -2$ is $X_1 = \begin{pmatrix} -1 \\ 0 \\ 1 \end{pmatrix}$.

**Case 2:** To find eigenvector when $\lambda = 3$.

Put $\lambda = 3$ in (1), we have

$$\begin{pmatrix} -2 & 1 & 3 \\ 1 & 2 & 1 \\ 3 & 1 & -2 \end{pmatrix} \begin{pmatrix} x_1 \\ x_2 \\ x_3 \end{pmatrix} = \begin{pmatrix} 0 \\ 0 \\ 0 \end{pmatrix}$$

This gives,

$$-2x_1 + x_2 + 3x_3 = 0$$
$$x_1 + 2x_2 + x_3 = 0$$
$$3x_1 + x_2 - 2x_3 = 0$$

Solving first two equations by cross multiplication, we have

$$\frac{x_1}{1-6} = \frac{x_2}{3+2} = \frac{x_3}{-4-1}$$

$$\frac{x_1}{-5} = \frac{x_2}{5} = \frac{x_3}{-5}$$

$$\therefore x_1 = 1, x_2 = -1, x_3 = 1.$$

The eigenvector corresponding to $\lambda = 3$ is $X_2 = \begin{pmatrix} 1 \\ -1 \\ 1 \end{pmatrix}$.

**Case 3:** To find eigenvector when $\lambda = 6$.

Put $\lambda = 6$ in (1), we have

$$\begin{pmatrix} -5 & 1 & 3 \\ 1 & -1 & 1 \\ 3 & 1 & -5 \end{pmatrix} \begin{pmatrix} x_1 \\ x_2 \\ x_3 \end{pmatrix} = \begin{pmatrix} 0 \\ 0 \\ 0 \end{pmatrix}$$

This gives,

$$-5x_1 + x_2 + 3x_3 = 0$$
$$x_1 - x_2 + x_3 = 0$$
$$3x_1 + x_2 - 5x_3 = 0$$

Solving last two equations by cross multiplication, we have

$$\frac{x_1}{5-1} = \frac{x_2}{3+5} = \frac{x_3}{1+3}$$

$$\frac{x_1}{4} = \frac{x_2}{8} = \frac{x_3}{4}$$
$$\therefore x_1 = 1, x_2 = 2, x_3 = 1.$$

The eigenvector corresponding to $\lambda = 6$ is $X_3 = \begin{pmatrix} 1 \\ 2 \\ 1 \end{pmatrix}$.

**Example:** Find the Eigen values and Eigen vectors of $\begin{pmatrix} 3 & 1 & 4 \\ 0 & 2 & 6 \\ 0 & 0 & 5 \end{pmatrix}$.

**Solution:** Let $\lambda$ be the eigenvalue of $A$.
The characteristic equation of $A$ is given by

$$|A - \lambda I| = 0 \Rightarrow \begin{vmatrix} 3-\lambda & 1 & 4 \\ 0 & 2-\lambda & 6 \\ 0 & 0 & 5-\lambda \end{vmatrix} = 0 \Rightarrow (3-\lambda)(2-\lambda)(5-\lambda) = 0.$$

$$\Rightarrow \lambda = 2, 3, 5.$$

$\therefore$ Eigenvalues of $A$ are $\lambda = 2, 3, 5$.

**To find eigenvectors:**

Let $X = \begin{pmatrix} x_1 \\ x_2 \\ x_3 \end{pmatrix}$ be the eigenvector of $A$ corresponding to $\lambda$.

Then
$$AX = \lambda X$$
$$\Rightarrow (A - \lambda I)X = 0$$

$$\Rightarrow \begin{pmatrix} 3-\lambda & 1 & 4 \\ 0 & 2-\lambda & 6 \\ 0 & 0 & 5-\lambda \end{pmatrix} \begin{pmatrix} x_1 \\ x_2 \\ x_3 \end{pmatrix} = \begin{pmatrix} 0 \\ 0 \\ 0 \end{pmatrix} \quad \cdots \cdots \quad (1)$$

**Case 1:** To find eigenvector when $\lambda = 2$.

Put $\lambda = 2$ in (1), we have

$$\begin{pmatrix} 1 & 1 & 4 \\ 0 & 0 & 6 \\ 0 & 0 & 3 \end{pmatrix} \begin{pmatrix} x_1 \\ x_2 \\ x_3 \end{pmatrix} = \begin{pmatrix} 0 \\ 0 \\ 0 \end{pmatrix}$$

This gives,

$$x_1 + x_2 + 4x_3 = 0$$
$$6x_3 = 0$$
$$3x_3 = 0$$

$\therefore x_3 = 0, x_1 + x_2 = 0 \Rightarrow x_1 = -x_2$.

Let $x_2 = k_1$. Then $x_1 = k_1$.

The eigenvector corresponding to $\lambda = 2$ is $X_1 = \begin{pmatrix} k_1 \\ -k_1 \\ 0 \end{pmatrix} = k_1 \begin{pmatrix} 1 \\ -1 \\ 0 \end{pmatrix}$.

**Case 2:** To find eigenvector when $\lambda = 3$.

Put $\lambda = 3$ in (1), we have

$$\begin{pmatrix} 0 & 1 & 4 \\ 0 & -1 & 6 \\ 0 & 0 & 2 \end{pmatrix} \begin{pmatrix} x_1 \\ x_2 \\ x_3 \end{pmatrix} = \begin{pmatrix} 0 \\ 0 \\ 0 \end{pmatrix}$$

This gives,

$$x_2 + 4x_3 = 0$$
$$-x_2 + 6x_3 = 0$$
$$2x_3 = 0$$

$\therefore x_3 = 0, x_2 = 0$. Let $x_1 = k_2$.

The eigenvector corresponding to $\lambda = 3$ is $X_2 = \begin{pmatrix} k_2 \\ 0 \\ 0 \end{pmatrix} = k_2 \begin{pmatrix} 1 \\ 0 \\ 0 \end{pmatrix}$.

**Case 3:** To find eigenvector when $\lambda = 6$.

Put $\lambda = 6$ in (1), we have

$$\begin{pmatrix} -2 & 1 & 4 \\ 0 & -3 & 6 \\ 0 & 0 & 0 \end{pmatrix} \begin{pmatrix} x_1 \\ x_2 \\ x_3 \end{pmatrix} = \begin{pmatrix} 0 \\ 0 \\ 0 \end{pmatrix}$$

This gives,
$$-2x_1 + x_2 + 4x_3 = 0$$
$$-3x_2 + 6x_3 = 0$$

Solving the equations
$$\frac{x_1}{2+4} = \frac{x_2}{0+4} = \frac{x_3}{2} = k_3$$
$$\frac{x_1}{3} = \frac{x_2}{2} = \frac{x_3}{1} = k_3$$
$$\therefore x_1 = 3k_3, x_2 = 2k_3, x_3 = k_3.$$

The eigenvector corresponding to $\lambda = 6$ is $X_3 = \begin{pmatrix} 3k_3 \\ 2k_3 \\ k_3 \end{pmatrix} = k_3 \begin{pmatrix} 3 \\ 2 \\ 1 \end{pmatrix}$.

**Example:** Determine the eigen values and eigen vectors of the matrix $A = \begin{pmatrix} -2 & 2 & -3 \\ 2 & 1 & -6 \\ -1 & -2 & 0 \end{pmatrix}$. (AU-2021)

**Solution:** Let $\lambda$ be the eigenvalue of $A$.
The characteristic equation of $A$ is given by $|A - \lambda I| = 0$.
$$\Rightarrow \lambda^3 + \lambda^2 - 21\lambda - 45 = 0.$$
$$\Rightarrow \lambda = -3, -3, 5.$$
$\therefore$ Eigenvalues of $A$ are $\lambda = -3, -3, 5$.

**To find eigenvectors:**

Let $X = \begin{pmatrix} x_1 \\ x_2 \\ x_3 \end{pmatrix}$ be the eigenvector of $A$ corresponding to $\lambda$.

Then
$$AX = \lambda X$$
$$\Rightarrow (A - \lambda I)X = 0$$

$$\Rightarrow \begin{pmatrix} -2-\lambda & 2 & -3 \\ 2 & 1-\lambda & -6 \\ -1 & -2 & -\lambda \end{pmatrix} \begin{pmatrix} x_1 \\ x_2 \\ x_3 \end{pmatrix} = \begin{pmatrix} 0 \\ 0 \\ 0 \end{pmatrix} \quad \ldots\ldots \quad (1)$$

**Case 1:** To find eigenvector when $\lambda = -3$.
Put $\lambda = -3$ in (1), we have

$$\begin{pmatrix} 1 & 2 & -3 \\ 2 & 4 & -6 \\ -1 & -2 & 3 \end{pmatrix} \begin{pmatrix} x_1 \\ x_2 \\ x_3 \end{pmatrix} = \begin{pmatrix} 0 \\ 0 \\ 0 \end{pmatrix}$$

This gives,
$$x_1 + 2x_2 - 3x_3 = 0$$
$$2x_1 + 4x_2 - 6x_3 = 0$$
$$-x_1 - 2x_2 + 3x_3 = 0$$

The above of set of equations represents the single equation
$$x_1 + 2x_2 - 3x_3 = 0.$$

$\therefore \rho(A) = 1, n = 3$, arbitrary constants $= n - r = 3 - 1 = 2$.
Let $x_2 = k_1, x_3 = k_2$.
$\therefore x_1 = -2x_2 + 3x_3 = -2k_1 + 3k_2$, where $k_1$ and $k_2$ are arbitrary constants.
The eigenvector corresponding to $\lambda = -3$ is

$$X = \begin{pmatrix} -2k_1 + 3k_2 \\ k_1 \\ k_2 \end{pmatrix} = \begin{pmatrix} -2k_1 \\ k_1 \\ 0 \end{pmatrix} + \begin{pmatrix} 3k_2 \\ 0 \\ k_2 \end{pmatrix} = k_1 \begin{pmatrix} -2 \\ 1 \\ 0 \end{pmatrix} + k_2 \begin{pmatrix} 3 \\ 0 \\ 1 \end{pmatrix}.$$

When $k_1 = 0$ and $k_2 = 1$, the eigenvector corresponding to $\lambda = -3$ is $X_1 = \begin{pmatrix} 3 \\ 0 \\ 1 \end{pmatrix}$.

When $k_1 = 1$ and $k_2 = 0$, the eigenvector corresponding to $\lambda = -3$ is $X_2 = \begin{pmatrix} -2 \\ 1 \\ 0 \end{pmatrix}$.

**Case 2:** To find eigenvector when $\lambda = 5$.

Put $\lambda = 5$ in (1), we have

$$\begin{pmatrix} -7 & 2 & -3 \\ 2 & -4 & -6 \\ -1 & -2 & -5 \end{pmatrix} \begin{pmatrix} x_1 \\ x_2 \\ x_3 \end{pmatrix} = \begin{pmatrix} 0 \\ 0 \\ 0 \end{pmatrix}$$

This gives,

$$-7x_1 + 2x_2 - 3x_3 = 0$$
$$2x_1 - 4x_2 - 6x_3 = 0$$
$$-x_1 - 2x_2 - 5x_3 = 0$$

Solving last two equations by cross multiplication, we have

$$\frac{x_1}{20-12} = \frac{x_2}{6+10} = \frac{x_3}{-4-4} = k_3 \quad (k_3 \neq 0)$$

$$\frac{x_1}{8} = \frac{x_2}{10} = \frac{x_3}{-8} = k_3$$

$$\therefore x_1 = k_3, x_2 = 2k_3, x_3 = -k_3.$$

The eigenvector corresponding to $\lambda = 3$ is $X_3 = \begin{pmatrix} k_3 \\ 2k_3 \\ -k_3 \end{pmatrix}$.

When $k_3 = 1$, the eigenvector is $X_3 = \begin{pmatrix} 1 \\ 2 \\ -1 \end{pmatrix}$.

**Example:** Determine the eigen values and eigen vectors of the matrix $A = \begin{pmatrix} 2 & 2 & 1 \\ 1 & 3 & 1 \\ 1 & 2 & 2 \end{pmatrix}$. (AU-2021)

**Solution:** Let $\lambda$ be the eigenvalue of $A$.

The characteristic equation of $A$ is given by

$$|A - \lambda I| = 0 \Rightarrow \begin{pmatrix} 2-\lambda & 2 & 1 \\ 1 & 3-\lambda & 1 \\ 1 & 2 & 2-\lambda \end{pmatrix} = 0.$$

$\Rightarrow \lambda^3 - 7\lambda^2 + 11\lambda - 5 = 0.$

$\Rightarrow \lambda = 1, 1, 5.$

$\therefore$ Eigen values of $A$ are $\lambda = 1, 1, 5$.

**To find eigen vectors:**

Let $X = \begin{pmatrix} x_1 \\ x_2 \\ x_3 \end{pmatrix}$ be the eigenvector of $A$ corresponding to $\lambda$.

Then

$$AX = \lambda X$$

$$\Rightarrow (A - \lambda I)X = 0$$

$$\Rightarrow \begin{pmatrix} 2-\lambda & 2 & 1 \\ 1 & 3-\lambda & 1 \\ 1 & 2 & 2-\lambda \end{pmatrix} \begin{pmatrix} x_1 \\ x_2 \\ x_3 \end{pmatrix} = \begin{pmatrix} 0 \\ 0 \\ 0 \end{pmatrix}$$

$$\left.\begin{array}{r} (2-\lambda)x_1 + 2x_2 + x_3 = 0 \\ x_1 + (3-\lambda)x_2 + x_3 = 0 \\ x_1 + 2x_2 + (2-\lambda)x_3 = 0 \end{array}\right\} \quad \cdots\cdots \quad (I)$$

**Case (i)** If $\lambda = 5$, then the equations (I) become

$$-3x_1 + 2x_2 + x_3 = 0$$
$$x_1 - 2x_2 + x_3 = 0$$
$$-x_1 - 2x_2 - 3x_3 = 0$$

Solve the last two equations, by rule of cross multiplication we get

$$\frac{x_1}{6-2} = \frac{x_2}{1+3} = \frac{x_3}{2+2} = k_1 \quad (k_1 \neq 0)$$

$$\frac{x_1}{4} = \frac{x_2}{4} = \frac{x_3}{4} = k_1$$

$$\therefore x_1 = k_1, x_2 = k_1, x_3 = k_1.$$

The eigenvector corresponding to $\lambda = 5$ is

$$X = \begin{pmatrix} k_1 \\ k_1 \\ k_1 \end{pmatrix} = k_1 \begin{pmatrix} 1 \\ 1 \\ 1 \end{pmatrix}.$$

When $k_1 = 1$, the eigenvector corresponding to $\lambda = 5$ is $X_1 = \begin{pmatrix} 1 \\ 1 \\ 1 \end{pmatrix}$.

**Case (ii)** If $\lambda = 1$, then the equations (I) become

$$x_1 + 2x_2 + x_3 = 0$$
$$x_1 + 2x_2 + x_3 = 0$$
$$x_1 + 2x_2 + x_3 = 0$$

We have only one equation $x_1 + 2x_2 + x_3 = 0$ to solve for $x_1, x_2, x_3$. Assign arbitrary values for two variables and solve for the third.

Choose $x_3 = 0$, then $x_1 + 2x_2 = 0 \Rightarrow x_1 = -2x_2$.

Choose $x_2 = 1$. $\therefore x_1 = -2$, we get an eigen vector

$$X_2 = \begin{pmatrix} -2 \\ 1 \\ 0 \end{pmatrix}.$$

We shall find one more solution from $x_1 + 2x_2 + x_3 = 0$.

Choose $x_2 = 0$ then $x_1 + x_3 = 0 \Rightarrow x_3 = -x_1$.

Choose $x_1 = 1$. $\therefore x_3 = -1$.

$\therefore$ Another eigen vector corresponding to $\lambda = 1$

is
$$X_3 = \begin{pmatrix} 1 \\ 0 \\ -1 \end{pmatrix}.$$

Thus eigen values of $A$ are 5, 1, 1 and the corresponding eigen vectors are

$$X_1 = \begin{pmatrix} 1 \\ 1 \\ 1 \end{pmatrix}, \quad X_2 = \begin{pmatrix} -2 \\ 1 \\ 0 \end{pmatrix} \text{ and } X_3 = \begin{pmatrix} 1 \\ 0 \\ -1 \end{pmatrix}.$$

**Example:** Determine the eigenvalues and eigenvectors of the matrix $A = \begin{pmatrix} 6 & -2 & 2 \\ -2 & 3 & -1 \\ 2 & -1 & 3 \end{pmatrix}$.

**Solution:** Let $\lambda$ be the eigenvalue of $A$. **(Dec-2020), (Apr-2022)**

The characteristic equation of $A$ is given by $|A - \lambda I| = 0$.

$$\Rightarrow \lambda^3 - 12\lambda^2 + 36\lambda - 32 = 0.$$

$$\Rightarrow \lambda = 2, 2, 8.$$

$\therefore$ Eigenvalues of $A$ are $\lambda = 2, 2, 8$.

**To find eigenvectors:**

Let $X = \begin{pmatrix} x_1 \\ x_2 \\ x_3 \end{pmatrix}$ be the eigenvector of $A$ corresponding to $\lambda$.

Then

$$AX = \lambda X$$

$$\Rightarrow (A - \lambda I)X = 0$$

$$\Rightarrow \begin{pmatrix} 6-\lambda & -2 & 2 \\ -2 & 3-\lambda & -1 \\ 2 & -1 & 3-\lambda \end{pmatrix} \begin{pmatrix} x_1 \\ x_2 \\ x_3 \end{pmatrix} = \begin{pmatrix} 0 \\ 0 \\ 0 \end{pmatrix} \quad \ldots\ldots \text{(1)}$$

**Case 1:** To find eigenvector when $\lambda = 2$.

Put $\lambda = 2$ in (1), we have

$$\begin{pmatrix} 4 & -2 & 2 \\ -2 & 1 & -1 \\ 2 & -1 & 1 \end{pmatrix} \begin{pmatrix} x_1 \\ x_2 \\ x_3 \end{pmatrix} = \begin{pmatrix} 0 \\ 0 \\ 0 \end{pmatrix}$$

This gives,

$$4x_1 - 2x_2 + 2x_3 = 0$$

$$-2x_1 + x_2 - x_3 = 0$$
$$2x_1 - x_2 + x_3 = 0$$

The above of set of equations represents the single equation
$$2x_1 - x_2 + x_3 = 0.$$
$\therefore \rho(A) = 1, n = 3$, arbitrary constants $= n - r = 3 - 1 = 2$.

Let $x_2 = k_1, x_3 = k_2$.

$\therefore 2x_1 = x_2 - x_3 = k_1 - k_2 \Rightarrow x_1 = \frac{k_1 - k_2}{2}$, where $k_1$ and $k_2$ are arbitrary constants.

The eigenvector corresponding to $\lambda = 2$ is

$$X = \begin{pmatrix} x_1 \\ x_2 \\ x_3 \end{pmatrix} = \begin{pmatrix} \frac{k_1}{2} - \frac{k_2}{2} \\ k_1 \\ k_2 \end{pmatrix} = \begin{pmatrix} \frac{k_1}{2} \\ k_1 \\ 0 \end{pmatrix} + \begin{pmatrix} \frac{-k_2}{2} \\ 0 \\ k_2 \end{pmatrix} = \frac{k_1}{2}\begin{pmatrix} 1 \\ 2 \\ 0 \end{pmatrix} + \frac{k_2}{2}\begin{pmatrix} -1 \\ 0 \\ 2 \end{pmatrix}.$$

When $k_1 = 0$ and $k_2 = 2$, the eigen vector corresponding to $\lambda = 2$ is $X_1 = \begin{pmatrix} -1 \\ 0 \\ 2 \end{pmatrix}$.

When $k_1 = 2$ and $k_2 = 0$, the eigenvector corresponding to $\lambda = 2$ is $X_2 = \begin{pmatrix} 1 \\ 2 \\ 0 \end{pmatrix}$.

Case 2: To find eigenvector when $\lambda = 8$.

Put $\lambda = 8$ in (1), we have

$$\begin{pmatrix} -2 & -2 & 2 \\ -2 & -5 & -1 \\ 2 & -1 & -5 \end{pmatrix} \begin{pmatrix} x_1 \\ x_2 \\ x_3 \end{pmatrix} = \begin{pmatrix} 0 \\ 0 \\ 0 \end{pmatrix}$$

This gives,

$$-2x_1 - 2x_2 + 2x_3 = 0$$
$$-2x_1 - 5x_2 - x_3 = 0$$
$$2x_1 - x_2 - 5x_3 = 0$$

Solving last two equations by cross multiplication, we have

$$\frac{x_1}{25-1} = \frac{x_2}{-2-10} = \frac{x_3}{2+10} = k_3 \quad (k_3 \neq 0)$$

$$\frac{x_1}{24} = \frac{x_2}{-12} = \frac{x_3}{12} = k_3$$
$$\therefore x_1 = 2k_3, x_2 = -k_3, x_3 = k_3.$$

The eigenvector corresponding to $\lambda = 8$ is $X_3 = \begin{pmatrix} 2k_3 \\ -k_3 \\ k_3 \end{pmatrix}$.

When $k_3 = 1$, the eigenvector is $X_3 = \begin{pmatrix} 2 \\ -1 \\ 1 \end{pmatrix}$.

**Note:** The eigne values of a triangular or a diagonal matrix are just the diagonal elements of the matrix.

**Example:** Determine the eigenvalues and eigenvectors of the matrix $A = \begin{pmatrix} 3 & 0 & 0 \\ 5 & 7 & 0 \\ 2 & 6 & 1 \end{pmatrix}$.

**Solution:** Given $A = \begin{pmatrix} 3 & 0 & 0 \\ 5 & 7 & 0 \\ 2 & 6 & 1 \end{pmatrix}$.

Since $A$ is a lower triangular matrix, so the eigenvalues of $A$ are diagonal elements.

$\therefore$ Eigenvalues of $A$ are $\lambda = 1, 3, 7$.

**To find eigenvectors:**

Let $X = \begin{pmatrix} x_1 \\ x_2 \\ x_3 \end{pmatrix}$ be the eigenvector of $A$ corresponding to $\lambda$.

Then
$$AX = \lambda X$$
$$\Rightarrow (A - \lambda I)X = 0$$
$$\Rightarrow \begin{pmatrix} 3-\lambda & 0 & 0 \\ 5 & 7-\lambda & 0 \\ 2 & 6 & 1-\lambda \end{pmatrix} \begin{pmatrix} x_1 \\ x_2 \\ x_3 \end{pmatrix} = \begin{pmatrix} 0 \\ 0 \\ 0 \end{pmatrix} \quad \ldots\ldots \quad (1)$$

**Case 1:** To find eigenvector when $\lambda = 1$.

Put $\lambda = 1$ in (1), we have

$$\begin{pmatrix} 2 & 0 & 0 \\ 5 & 6 & 0 \\ 2 & 6 & 0 \end{pmatrix} \begin{pmatrix} x_1 \\ x_2 \\ x_3 \end{pmatrix} = \begin{pmatrix} 0 \\ 0 \\ 0 \end{pmatrix}$$

This gives,

$$2x_1 = 0$$
$$5x_1 + 6x_2 = 0$$
$$2x_1 + 6x_2 = 0$$
$$\therefore x_1 = 0 \text{ and } x_2 = 0.$$

Let $x_3 = k_1$.

The eigenvector corresponding to $\lambda = 1$ is $X_1 = \begin{pmatrix} 0 \\ 0 \\ k_1 \end{pmatrix}$, $k_1 \neq 0$ is any scalar.

When $k_1 = 1$, we have $X_1 = \begin{pmatrix} 0 \\ 0 \\ 1 \end{pmatrix}$.

Case 2: To find eigenvector when $\lambda = 3$.

Put $\lambda = 3$ in (1), we have

$$\begin{pmatrix} 0 & 0 & 0 \\ 5 & 4 & 0 \\ 2 & 6 & -2 \end{pmatrix} \begin{pmatrix} x_1 \\ x_2 \\ x_3 \end{pmatrix} = \begin{pmatrix} 0 \\ 0 \\ 0 \end{pmatrix}$$

This gives,

$$5x_1 + 4x_2 = 0$$
$$2x_1 + 6x_2 - 2x_3 = 0$$

Solving these equations by cross multiplication, we have

$$\frac{x_1}{-8} = \frac{x_2}{0+10} = \frac{x_3}{30-8} = k_2 \quad (k_2 \neq 0)$$

$$\frac{x_1}{-4} = \frac{x_2}{5} = \frac{x_3}{11} = k_2$$

$$\therefore x_1 = -4k_2, x_2 = 5k_2, x_3 = 11k_2.$$

The eigenvector corresponding to $\lambda = 3$ is $X_2 = \begin{pmatrix} -4k_2 \\ 5k_2 \\ 11k_2 \end{pmatrix}$.

When $k_2 = 1$, the eigenvector is $X_2 = \begin{pmatrix} -4 \\ 5 \\ 11 \end{pmatrix}$.

Case 3: To find eigenvector when $\lambda = 7$.

Put $\lambda = 7$ in (1), we have

$$\begin{pmatrix} -4 & 0 & 0 \\ 5 & 0 & 0 \\ 2 & 6 & -6 \end{pmatrix} \begin{pmatrix} x_1 \\ x_2 \\ x_3 \end{pmatrix} = \begin{pmatrix} 0 \\ 0 \\ 0 \end{pmatrix}$$

This gives,

$$-4x_1 = 0$$
$$5x_1 = 0$$
$$2x_1 + 6x_2 - 6x_3 = 0$$
$$\therefore x_1 = 0 \text{ and } x_2 = x_3.$$

Let $x_2 = k_3 \Rightarrow x_3 = k_3$.

The eigenvector corresponding to $\lambda = 7$ is $X_3 = \begin{pmatrix} 0 \\ k_3 \\ k_3 \end{pmatrix}$.

When $k_3 = 1$, the eigenvector is $X_3 = \begin{pmatrix} 0 \\ 1 \\ 1 \end{pmatrix}$.

## EXERCISE

1. Find the eigenvalues and the corresponding eigenvectors $A = \begin{pmatrix} -3 & -7 & -5 \\ 2 & 4 & 3 \\ 1 & 2 & 2 \end{pmatrix}$.

**Ans:** $\lambda = 1, 1, 1, X = \begin{pmatrix} -3 \\ 1 \\ 1 \end{pmatrix}$. (Dec-2020), (Aug-2022)

2. Find the eigenvalues and the corresponding eigen vectors of $\begin{pmatrix} 1 & 2 & -2 \\ 1 & 1 & 1 \\ 1 & 3 & -1 \end{pmatrix}$.

**Ans:** $\lambda = 1, 2, 3, \begin{pmatrix} 1 \\ 0 \\ 1 \end{pmatrix}, \begin{pmatrix} 1 \\ 0 \\ -1 \end{pmatrix}, \begin{pmatrix} 0 \\ 1 \\ 0 \end{pmatrix}$ (Dec-2020)

3. Find the eigenvalues and the corresponding eigen vectors of $\begin{pmatrix} 2 & -2 & 2 \\ 1 & 1 & 1 \\ 1 & 3 & -1 \end{pmatrix}$.

**Ans:** $\lambda = -2, 2, 2, \begin{pmatrix} 4 \\ 1 \\ -7 \end{pmatrix}, \begin{pmatrix} 0 \\ 1 \\ 1 \end{pmatrix}$.

4. Find the eigenvalues and eigenvectors of $\begin{pmatrix} 8 & -8 & 2 \\ 4 & -3 & -2 \\ 3 & -4 & -1 \end{pmatrix}$. (Jan-2020)

5. Find the latent values and latent roots of the matrix $A = \begin{pmatrix} 2 & 1 & 1 \\ 2 & 3 & 4 \\ -1 & -1 & 3 \end{pmatrix}$.

6. Find the Eigen values and Eigen vectors of the matrix $\begin{pmatrix} 0 & 1 & 1 \\ 1 & 0 & 1 \\ 1 & 1 & 0 \end{pmatrix}$.

7. Find the Eigen values and Eigen vectors of $A = \begin{pmatrix} 3 & -6 & 3 \\ 1 & 0 & -1 \\ 1 & 2 & -3 \end{pmatrix}$. (Apr-2022)

8. Find the Eigen values and Eigen vectors of $A = \begin{pmatrix} 2 & 1 & 1 \\ 1 & 2 & 1 \\ 0 & 0 & 1 \end{pmatrix}$. **(Apr-2022)**

9. Find the Eigen values and Eigen vectors of $A = \begin{pmatrix} 1 & 0 & 3 \\ 1 & 2 & 1 \\ 2 & 2 & 3 \end{pmatrix}$. **(Apr-2022)**

## 1.8.1 Properties of Eigenvalues and Eigenvectors

**Property:1** The sum of the eigenvalues of a square matrix $A$ is equal to the *trace* of $A$ and the product of the eigenvalues of $A$ is equal to the *determinant* of $A$.

**Proof:** Let $A = \begin{pmatrix} a_{11} & a_{12} & a_{13} \\ a_{21} & a_{22} & a_{23} \\ a_{31} & a_{32} & a_{33} \end{pmatrix}$. **(Dec-2020), (Sep-2021)**

The characteristic equation of $A$ is $|A - \lambda I| = 0$.

$$\Rightarrow \begin{vmatrix} a_{11} - \lambda & a_{12} & a_{13} \\ a_{21} & a_{22} - \lambda & a_{23} \\ a_{31} & a_{32} & a_{33} - \lambda \end{vmatrix} = 0.$$

$\Rightarrow (a_{11} - \lambda)\left[(a_{22} - \lambda)(a_{33} - \lambda) - a_{23}a_{32}\right] - a_{12}\left[a_{21}(a_{33} - \lambda) - a_{23}a_{31}\right]$
$+ a_{13}\left[a_{21}a_{32} - a_{31}(a_{22} - \lambda)\right] = 0$

$\Rightarrow (a_{11} - \lambda)\left[(a_{22}a_{33} - \lambda a_{22} - \lambda a_{33} + \lambda^2 - a_{23}a_{32}\right] - a_{12}\left[a_{21}a_{33} - \lambda a_{21} - a_{23}a_{31}\right]$
$+ a_{13}\left[a_{21}a_{32} - a_{31}a_{22} + \lambda a_{31}\right] = 0$

$\Rightarrow a_{11}a_{22}a_{33} - \lambda a_{11}a_{22} - \lambda a_{11}a_{33} + \lambda^2 a_{11} - a_{11}a_{23}a_{32} - \lambda a_{22}a_{33} + \lambda^2 a_{22} + \lambda^2 a_{33}$
$-\lambda^3 + \lambda a_{23}a_{32} - a_{12}a_{21}a_{33} + \lambda a_{12}a_{21} + a_{12}a_{23}a_{31} + a_{13}a_{21}a_{32} - a_{13}a_{31}a_{22} + \lambda a_{13}a_{31} = 0$

$\Rightarrow -\lambda^3 + \lambda^2(a_{11} + a_{22} + a_{33}) - \lambda(a_{11}a_{22} + a_{11}a_{33} + a_{22}a_{33} - a_{23}a_{32} - a_{12}a_{21} - a_{13}a_{31})$
$+ a_{11}a_{22}a_{33} - a_{11}a_{23}a_{32} - a_{12}a_{21}a_{33} + a_{12}a_{23}a_{31} + a_{13}a_{21}a_{32} - a_{13}a_{31}a_{22} = 0 \quad \cdots\cdots \quad (1)$

Also, if $\lambda_1, \lambda_2, \lambda_3$ are eigen values of $A$, then

$$|A - \lambda I| = (-1)^3(\lambda - \lambda_1)(\lambda - \lambda_2)(\lambda - \lambda_3)$$
$$= -(\lambda - \lambda_1)[\lambda^2 - \lambda\lambda_2 - \lambda\lambda_3 + \lambda_2\lambda_3]$$
$$= \lambda^2\lambda_1 - \lambda\lambda_1\lambda_2 - \lambda\lambda_1\lambda_3 + \lambda_1\lambda_2\lambda_3 - \lambda^3 + \lambda^2\lambda_2 + \lambda^2\lambda_3 - \lambda\lambda_2\lambda_3$$
$$= -\lambda^3 + \lambda^2(\lambda_1 + \lambda_2 + \lambda_3) - \lambda(\lambda_1\lambda_2 + \lambda_1\lambda_3 + \lambda_2\lambda_3) + \lambda_1\lambda_2\lambda_3 \quad \cdots\cdots \quad (2)$$

From (1) and (2), we have from the theory of equations

$$\lambda_1 + \lambda_2 + \lambda_3 = a_{11} + a_{22} + a_{33} = \text{trace of } A$$

and

$$\lambda_1\lambda_2\lambda_3 = a_{11}a_{22}a_{33} - a_{11}a_{23}a_{32} - a_{12}a_{21}a_{33} + a_{12}a_{23}a_{31} + a_{13}a_{21}a_{32} - a_{13}a_{31}a_{22} = |A|.$$

Alternatively, we have

$$|A - \lambda I| = (-1)^3(\lambda - \lambda_1)(\lambda - \lambda_2)(\lambda - \lambda_3)$$

Put $\lambda = 0$, then

$$|A| = -(-\lambda_1)(-\lambda_2)(-\lambda_3) = \lambda_1\lambda_2\lambda_3.$$
$$\lambda_1\lambda_2\lambda_3 = |A|.$$

**Property:2** If $\lambda = 0$ is an eigenvalue of $A$ then $A$ is singular matrix i.e., $|A| = 0$.

Proof: Let $\lambda = 0$ be the eigenvalue of $A$

$$|A - \lambda I| = 0$$
$$\Rightarrow |A| = 0. \qquad (\because \lambda = 0)$$

The matrix is singular.

**Property:3** If $X$ is an eigenvector of $A$ corresponding to the eigenvalue $\lambda$ then $kX$ is also eigenvector of $A$.

**Property:4** If $\lambda$ is an eigenvalue of a square matrix $A$ with $X$ as a corresponding eigenvector, then for any non-zero constant $k$, $k\lambda$ is an eigenvalue of the matrix $kA$ with $X$ as a corresponding eigenvector.

**Proof:** Since $\lambda$ is an eigenvalue of $A$ with $X$ as a corresponding eigenvector, we have

$$AX = \lambda X$$
$$\Rightarrow k(AX) = k(\lambda X)$$
$$\Rightarrow (kA)X = (k\lambda)X$$

$\therefore$ $k\lambda$ is the eigenvalue of $kA$.

**Property:5** If $\lambda$ is an eigenvalue of a square matrix $A$ with $X$ as a corresponding eigenvector, then for any positive integer $m$, $\lambda^m$ is an eigenvalue of $A^m$ with $X$ as a corresponding eigenvector. **(Dec-2020)**

Proof: If $\lambda$ is an eigenvalue of a square matrix $A$ with $X$ as a corresponding eigenvector, then we have $AX = \lambda X$. Consequently,

$$A^m X = A^{m-1}(AX) = A^{m-1}(\lambda X) = \lambda(A^{m-1}X)$$
$$= \lambda A^{m-2}(AX) = \lambda A^{m-2}(\lambda X) = \lambda^2(A^{m-2}X)$$
$$= \lambda^2 A^{m-3}(AX) = \lambda^2 A^{m-3}(\lambda X) = \lambda^3(A^{m-3}X)$$
$$\cdots\cdots$$
$$\cdots\cdots$$
$$= \lambda^{m-1}(AX) = \lambda^{m-1}(\lambda X) = \lambda^m X.$$

This shows that $\lambda^m$ is an eigenvalue of $A^m$ with $X$ as a corresponding eigenvector.

**Example:** Show that if $\lambda_1, \lambda_2, \cdots, \lambda_m$ are latent roots of a matrix $A$, then $A^3$ has latent roots $\lambda_1^3, \lambda_2^3, \cdots, \lambda_m^3$.

**Solution:** Let $\lambda$ be any eigen value of $A$. Then we can find a column matrix $X \neq 0$ such that

$$AX = \lambda X \quad \cdots\cdots \quad (2)$$

Now $A^2 X = A(AX) = A(\lambda X) = \lambda(AX) = \lambda(\lambda X)$

$$\Rightarrow A^2 X = \lambda^2 X \quad \cdots\cdots \quad (2)$$

and
$$A^3X = A(A^2X) = A(\lambda X) = A(\lambda^2 X)$$
$$= \lambda^2(AX) = \lambda^2(\lambda X)$$
$$\therefore A^3X = \lambda^3 X$$

$\therefore \lambda^3$ is an eigen value of $A^3$, by definition.

This is true for all eigen values of $A$.

$\therefore \lambda_1^3, \lambda_2^3, \cdots, \lambda_m^3$ are eigen values of $A^3$.

**Property:6** If $\lambda$ is an eigenvalue of a square matrix $A$ with $X$ as a corresponding eigenvector, then $\lambda^{-1}$ is an eigenvalue of $A^{-1}$ if $A^{-1}$ exist with the same eigenvector $X$. **(Apr-2017), (Nov-2017)**

**Proof:** If $\lambda$ is an eigenvalue of a square matrix $A$ with $X$ as a corresponding eigenvector, then we have

$$\Rightarrow AX = \lambda X$$
$$\Rightarrow A^{-1}(AX) = A^{-1}(\lambda X)$$
$$\Rightarrow (A^{-1}A)X = \lambda(A^{-1}X)$$
$$\Rightarrow X = \lambda A^{-1}X$$
$$\Rightarrow \lambda^{-1}X = A^{-1}X$$
$$\Rightarrow A^{-1}X = \lambda^{-1}X.$$

Therefore $\lambda^{-1}$ is an eigenvalue of $A^{-1}$.

**Property:7** The square matrix $A$ and its transpose $A^T$ have the same eigenvalues.

**Proof:** We note that

$$|A - \lambda I| = |(A - \lambda I)^T|, \text{ because for any square matrix } P, \text{ we have } |P| = |P^T|$$
$$= |A^T - (\lambda I)^T|, \text{ because } (A \pm B)^T = A^T \pm B^T$$
$$= |A^T - \lambda I|, \text{ because } (\lambda I)^T = \lambda I^T = \lambda I.$$

Form this, it follows that $|A^T - \lambda I| = 0$ if and only if $|A - \lambda I| = 0$. This means that $\lambda$ is an eigenvalue of $A^T$ if and only if $\lambda$ is an eigenvalue of $A$.

**Property:8** If $\lambda$ is an eigenvalue of a nonsingular matrix $A$ with $X$ as a corresponding eigenvector, then $\dfrac{|A|}{\lambda}$ is an eigenvalue of the adj$A$ with $X$ as a corresponding eigenvector. **(Apr-2017), (Apr-2018), (Dec-2020)**

**Proof:** We know that $A^{-1} = \dfrac{\text{adj } A}{|A|}$.

If $\lambda$ is an eigenvalue of a square matrix $A$ with $X$ as a corresponding eigenvector, then we have

$$\Rightarrow AX = \lambda X$$
$$\Rightarrow \text{adj } A(AX) = \text{adj } A(\lambda X)$$
$$\Rightarrow (\text{adj } A \cdot A)X = \lambda(\text{adj } A \cdot X)$$
$$\Rightarrow |A|X = \lambda(\text{adj } A \cdot X)$$
$$\Rightarrow \frac{|A|}{\lambda}X = \text{adj } A \cdot X$$
$$\Rightarrow \text{adj } A \cdot X = \frac{|A|}{\lambda}X.$$

This shows that $\dfrac{|A|}{\lambda}$ is an eigenvalue of adj $A$ with $X$ as a corresponding eigenvector.

**Property:9** If $\lambda$ is an eigenvalue of $A$ then the eigenvalue of $B = a_0 A^2 + a_1 A + a_2 I$ is $a_0 \lambda^2 + a_1 \lambda + a_2$. **(Dec-2020)**

**Proof:** If $X$ be the eigen vector corresponding to the eigen value $\lambda$, then

$$AX = \lambda X \qquad \cdots \cdots \quad (1)$$

Pre-multiply by $A$ on both sides

$$A(AX) = A(\lambda X)$$
$$\Rightarrow A^2 X = \lambda(AX)$$
$$\Rightarrow A^2 X = \lambda(\lambda X)$$
$$\Rightarrow A^2 X = \lambda^2 X$$

This shows that $\lambda^2$ is eigen value of $A^2$.

We have
$$B = a_0 A^2 + a_1 A + a_2 I$$
$$\Rightarrow BX = (a_0 A^2 + a_1 A + a_2 I)X = a_0 A^2 X + a_1 AX + a_2 X$$
$$= a_0 \lambda^2 X + a_1 \lambda X + a_2 X = (a_0 \lambda^2 + a_1 \lambda + a_2)X$$

This shows that $a_0 \lambda^2 + a_1 \lambda + a_2$ is an eigen value of $B$ and the corresponding eigen vector of $B$ is $X$.

**Property:10** If $A$ and $P$ are square matrices of the same order and $P$ is non-singular, then $A$ and $P^{-1}AP$ have the same eigenvalues.

**Proof:** For any $\lambda$, we have
$$|P^{-1}AP - \lambda I| = |P^{-1}AP - \lambda(P^{-1}P)|$$
$$= |P^{-1}AP - P^{-1}(\lambda P)|$$
$$= |P^{-1}(AP - \lambda P)|$$
$$= |P^{-1}(A - \lambda I)P|$$
$$= |P^{-1}||A - \lambda I||P|.$$

From this, it follows that $|P^{-1}AP - \lambda I| = 0$ if and only if $|A - \lambda I| = 0$. This means that $\lambda$ is an eigenvalue of $P^{-1}AP$ if and only if $\lambda$ is an eigenvalue of $A$. In other words, $A$ and $P^{-1}AP$ have the same eigenvalues.

**Property:11** If $A$ and $B$ are non-singular matrices of same order then $AB$ and $BA$ have the same eigenvalues. **(Nov-2017), (Apr-2018)**

**Proof:** $AB = IAB = (B^{-1}B)AB = B^{-1}(BA)B$.
By (10), $BA$ and $B^{-1}(BA)B$ have the same eigenvalues.
$\Rightarrow BA$ and $AB$ have the same eigenvalues.

**Property:12** The eigen values of $BA^{-1}$ and $A^{-1}B$ are same. **(Aug-2022)**

**Property:13** The eigenvectors corresponding to two distinct eigenvalues of a square matrix are linearly independent. **(Jan-2020), (Aug-2021), (Apr-2022)**

**Proof:** Let $\lambda_1$ and $\lambda_2$ be distinct eigenvalues of a square matrix. Let $X_1$ and $X_2$ be their corresponding eigenvectors so that

$$AX_1 = \lambda_1 X_1 \text{ and } AX_2 = \lambda_2 X_2.$$

Let us consider

$$k_1 X_1 + k_2 X_2 = 0 \quad \cdots \cdots \quad (1)$$

Multiply (1) by $A$ on both sides

$$A(k_1 X_1 + k_2 X_2) = 0$$
$$\Rightarrow Ak_1 X_1 + Ak_2 X_2 = 0$$
$$\Rightarrow k_1(AX_1) + k_2(AX_2) = 0$$
$$\Rightarrow k_1(\lambda_1 X_1) + k_2(\lambda_2 X_2) = 0 \quad \cdots \cdots \quad (2)$$

From (1), we get $k_2 X_2 = -k_1 X_1$.

Substitute in (2),

$$k_1(\lambda_1 X_1) - k_1 \lambda_2 X_1 = 0$$
$$\Rightarrow (\lambda_2 - \lambda_1) k_1 X_1 = 0$$
$$\Rightarrow k_1 = 0 \quad [\because \lambda_2 \neq \lambda_1, X_1 \neq 0]$$

From (1), $k_2 = 0$.

Therefore, the equation (1) holds only if $k_1 = 0$ and $k_2 = 0$.

$\therefore$ $X_1$ and $X_2$ are linearly independent.

**Property:14** The Eigen values of triangular matrix are its diagonal elements.

(Aug-2022)

**Example:** Find the sum and product of the eigen values of $\begin{pmatrix} 3 & 1 & 4 \\ 0 & 2 & 6 \\ 0 & 0 & 5 \end{pmatrix}$.

**Solution:** Sum of the eigen values=trace of the matrix= $3 + 2 + 5 = 10$

Product of the eigen values= $|A| = 3(10 - 0) - 0(5 - 0) + 0(6 - 8) = 30$.

**Example:** Two eigen values of the matrix $A = \begin{pmatrix} 2 & 2 & 1 \\ 1 & 3 & 1 \\ 1 & 2 & 2 \end{pmatrix}$ are equal to 1 each. Find the eigen values of $A^{-1}$.

**Solution:** Given two eigen values of $A$ are 1, 1. Let $\lambda$ be the third eigen value.

$\therefore 1 + 1 + \lambda =$ Trace of $A = 2 + 3 + 2 = 7 \Rightarrow \lambda = 5$

$\therefore$ The eigen values of $A$ are 1, 1, 5.

$\therefore$ The eigen values of $A^{-1}$ are $1, 1, \dfrac{1}{5}$.

**Example:** If the product of two eigen values of the matrix $A = \begin{pmatrix} 1 & 0 & 0 \\ 0 & 3 & -1 \\ 0 & -1 & 3 \end{pmatrix}$ is 2, find the third eigen value.

**Solution:** Let $\lambda_1, \lambda_2, \lambda_3$ be the eigen values of $A$.

Given that $\lambda_1 \lambda_2 = 2$.

We know that product of eigen values of $A$=determinant of $A$.

$\therefore \lambda_1 \lambda_2 \lambda_3 = |A| = 8$

$\Rightarrow 2\lambda_3 = 8$

$\Rightarrow \lambda_3 = 4$.

**Example:** If the eigen values of a $3 \times 3$ matrix $A$ are $-1, 1, 3$, find the eigen values of adj$A$.

**Solution:** The eigen values of $A$ are $-1, 1, 3$

$\therefore$ Eigen values of $A^{-1}$ are $-1, 1, \dfrac{1}{3}$

We know that $A^{-1} = \dfrac{\text{adj } A}{|A|} \Rightarrow \text{adj } A = |A|A^{-1}$.

But $|A|$=Product of the eigen values= $-1 \cdot 1 \cdot 3 = -3$

$\therefore$ Eigen values of adj $A$ are

$-3 \cdot -1, -3 \cdot 1, -3 \cdot \dfrac{1}{3} \Rightarrow 3, -3, -1$.

**Example:** For the matrix $A = \begin{pmatrix} 1 & 2 & -3 \\ 0 & 3 & -2 \\ 0 & 0 & -2 \end{pmatrix}$, find the eigenvalues of $3A^3 + 5A^2 - 6A + 2I$.

**Solution:** Given $A = \begin{pmatrix} 1 & 2 & -3 \\ 0 & 3 & -2 \\ 0 & 0 & -2 \end{pmatrix}$.

Since $A$ is an upper triangular matrix. Therefore eigenvalues of $A$ are $1, 3, -2$.

Since, if $\lambda$ is an eigenvalue of $A$, then the eigenvalue of $B = a_0 A^3 + a_1 A^2 + a_2 A + a_3 I$ is $a_0 \lambda^3 + a_1 \lambda^2 + a_2 \lambda + a_3$.

If $\lambda$ is an eigenvalue of $A$, then the eigenvalue of $B = 3A^3 + 5A^2 - 6A + 2I$ is $3\lambda^3 + 5\lambda^2 - 6\lambda + 2$.

$\lambda = 1$ for $A$, then eigenvalue of $B = 3(1)^3 + 5(1)^2 - 6(1) + 2 = 4$.

$\lambda = 3$ for $A$, then eigenvalue of $B = 3(3)^3 + 5(3)^2 - 6(3) + 2 = 110$.

$\lambda = -2$ for $A$, then eigenvalue of $B = 3(-2)^3 + 5(-2)^2 - 6(-2) + 2 = 10$.

$\therefore$ The eigenvalues of $B = 3A^3 + 5A^2 - 6A + 2I$ are $4, 110, 10$.

**Example:** If $2, 3, 5$ are the eigenvalues of matrix $A$, then find the eigenvalues of $2A^3 + 3A^2 + 5A + 3I$. **(June-2015)**

**Solution:** Given that the eigenvalues of $A$ are $2, 3, 5$.

By the property, if $\lambda$ is the eigenvalue of $A$, then the eigenvalue of $a_0 A^3 + a_1 A^2 + a_2 A + a_3 I$ is $a_0 \lambda^3 + a_1 \lambda^2 + a_2 \lambda + a_3$.

$\lambda = 2 \Rightarrow 2(2)^3 + 3(2)^2 + 5(2) + 3 = 41$

$\lambda = 3 \Rightarrow 2(3)^3 + 3(3)^2 + 5(3) + 3 = 99$

$\lambda = 5 \Rightarrow 2(5)^3 + 3(3)^2 + 5(3) + 3 = 353$.

Hence the eigenvalues of $2A^3 + 3A^2 + 5A + 3I$ are $41, 99$ and $353$.

**Example:** If the Eigen values of $A$ are $-1, 1, 3$ then find the Eigen values of (i) $\text{Adj} A$ (ii) $A - 3I$ (iii) $A^3$. **(Aug-2022)**

**Example:** If the Eigen values of $A$ are $1, 2, 3$ then find

(i) $\det A$ (ii) Trace of $A$ (iii) Eigen values of $A^3 - 5A + 6I$. (Aug-2022)

**Example:** Find the eigenvalues and eigenvectors of $A^{-1}$, where $A = \begin{pmatrix} -2 & 2 & -3 \\ 2 & 1 & -6 \\ -1 & -2 & 0 \end{pmatrix}$.

**Solution:** Let $\lambda$ be the eigenvalue of $A$.
The characteristic equation of $A$ is given by $|A - \lambda I| = 0$.

$$\Rightarrow \lambda^3 + \lambda^2 - 21\lambda - 45 = 0.$$

$$\Rightarrow \lambda = -3, -3, 5.$$

$\therefore$ Eigenvalues of $A$ are $\lambda = -3, -3, 5$.

**To find eigenvectors:**

Let $X = \begin{pmatrix} x_1 \\ x_2 \\ x_3 \end{pmatrix}$ be the eigenvector of $A$ corresponding to $\lambda$.

Then

$$AX = \lambda X$$

$$\Rightarrow (A - \lambda I)X = 0$$

$$\Rightarrow \begin{pmatrix} -2-\lambda & 2 & -3 \\ 2 & 1-\lambda & -6 \\ -1 & -2 & -\lambda \end{pmatrix} \begin{pmatrix} x_1 \\ x_2 \\ x_3 \end{pmatrix} = \begin{pmatrix} 0 \\ 0 \\ 0 \end{pmatrix} \quad \cdots \cdots \quad (1)$$

**Case 1:** To find eigenvector when $\lambda = -3$.
Put $\lambda = -3$ in (1), we have

$$\begin{pmatrix} 1 & 2 & -3 \\ 2 & 4 & -6 \\ -1 & -2 & 3 \end{pmatrix} \begin{pmatrix} x_1 \\ x_2 \\ x_3 \end{pmatrix} = \begin{pmatrix} 0 \\ 0 \\ 0 \end{pmatrix}$$

This gives,

$$x_1 + 2x_2 - 3x_3 = 0$$
$$2x_1 + 4x_2 - 6x_3 = 0$$
$$-x_1 - 2x_2 + 3x_3 = 0.$$

This is set of one equation with three unknowns. Hence it has infinite number of solutions.

$$\therefore x_1 + 2x_2 - 3x_3 = 0.$$

Let $x_2 = k_1$ and $x_3 = k_2$. Then

$$x_1 = -2x_2 + 3x_3 = -2k_1 + 3k_2.$$

Hence $X = \begin{pmatrix} x_1 \\ x_2 \\ x_3 \end{pmatrix} = \begin{pmatrix} -2k_1 + 3k_2 \\ k_1 \\ k_2 \end{pmatrix} = k_1 \begin{pmatrix} -2 \\ 1 \\ 0 \end{pmatrix} + k_2 \begin{pmatrix} 3 \\ 0 \\ 1 \end{pmatrix}.$

$\therefore$ One of the eigenvector corresponding to $\lambda = -3$ is $X_1 = \begin{pmatrix} -2 \\ 1 \\ 0 \end{pmatrix}$ and the other eigenvector for $\lambda = -3$ is $X_2 = \begin{pmatrix} 3 \\ 0 \\ 1 \end{pmatrix}$.

Case 2: To find eigenvector when $\lambda = 5$.

Put $\lambda = 5$ in (1), we have

$$\begin{pmatrix} -7 & 2 & -3 \\ 2 & -4 & -6 \\ -1 & -2 & -5 \end{pmatrix} \begin{pmatrix} x_1 \\ x_2 \\ x_3 \end{pmatrix} = \begin{pmatrix} 0 \\ 0 \\ 0 \end{pmatrix}$$

This gives,

$$-7x_1 + 2x_2 - 3x_3 = 0$$

$$2x_1 - 4x_2 - 6x_3 = 0$$

$$-x_1 - 2x_2 - 5x_3 = 0$$

Solving last two equations by cross multiplication, we have

$$\frac{x_1}{20-12} = \frac{x_2}{6+10} = \frac{x_3}{-4-4} = k_3 \quad (k_3 \neq 0)$$

$$\frac{x_1}{8} = \frac{x_2}{16} = \frac{x_3}{-8} = k_3$$

$$\therefore x_1 = k_3, x_2 = 2k_3, x_3 = -k_3.$$

The eigenvector corresponding to $\lambda = 5$ is $X_3 = \begin{pmatrix} k_3 \\ 2k_3 \\ -k_3 \end{pmatrix}$.

Put $k_3 = 1$, then the eigenvector for $\lambda = 5$ is $X_3 = \begin{pmatrix} 1 \\ 2 \\ -1 \end{pmatrix}$.

Since the eigenvalues of $A^{-1}$ are the reciprocal of eigenvalues of $A$. The eigenvalues of $A^{-1}$ are given by $\frac{-1}{3}, \frac{-1}{3}, \frac{1}{5}$ and the eigenvectors are $X_1 = \begin{pmatrix} -2 \\ 1 \\ 0 \end{pmatrix}$, $X_2 = \begin{pmatrix} 3 \\ 0 \\ 1 \end{pmatrix}$

and $X_3 = \begin{pmatrix} 1 \\ 2 \\ -1 \end{pmatrix}$.

## 1.8.2 Algebraic Multiplicity and Geometric Multiplicity

**Definition:** Algebraic multiplicity of an eigenvalue $\lambda$ is the order of the eigenvalue as a root of the characteristic equation (i.e., $\lambda$ is a double root then algebraic multiplicity is 2).

Geometric multiplicity of $\lambda$ is the number of linearly independent eigenvectors corresponding to $\lambda$.

$\therefore$ Geometric multiplicity $\leq$ Algebraic multiplicity

**Example:** Determine the algebraic multiplicity and geometric multiplicity of

$$\begin{pmatrix} 3 & 10 & 5 \\ -2 & -3 & -4 \\ 3 & 5 & 7 \end{pmatrix}.$$

**Solution:** Let $\lambda$ be the eigenvalue of $A$.
The characteristic equation of $A$ is given by $|A - \lambda I| = 0$.
$$\Rightarrow \lambda^3 - 7\lambda^2 + 16\lambda - 12 = 0.$$

$\Rightarrow \lambda = 2, 2, 3.$

$\therefore$ Eigenvalues of $A$ are $\lambda = 2, 2, 3$.

**To find eigenvectors:**

Let $X = \begin{pmatrix} x_1 \\ x_2 \\ x_3 \end{pmatrix}$ be the eigenvector of $A$ corresponding to $\lambda$.

Then

$$AX = \lambda X$$

$$\Rightarrow (A - \lambda I)X = 0$$

$$\Rightarrow \begin{pmatrix} 3-\lambda & 10 & 5 \\ -2 & -3-\lambda & -4 \\ 3 & 5 & 7-\lambda \end{pmatrix} \begin{pmatrix} x_1 \\ x_2 \\ x_3 \end{pmatrix} = \begin{pmatrix} 0 \\ 0 \\ 0 \end{pmatrix} \quad \ldots\ldots (1)$$

**Case 1:** To find eigenvector when $\lambda = 3$.

Put $\lambda = 3$ in (1), we have

$$\begin{pmatrix} 0 & 10 & 5 \\ -2 & -6 & -4 \\ 3 & 5 & 4 \end{pmatrix} \begin{pmatrix} x_1 \\ x_2 \\ x_3 \end{pmatrix} = \begin{pmatrix} 0 \\ 0 \\ 0 \end{pmatrix}$$

This gives,

$$10x_2 + 5x_3 = 0$$

$$-2x_1 - 6x_2 - 4x_3 = 0$$

$$3x_1 + 5x_2 + 4x_3 = 0$$

Solving last two equations by cross multiplication, we have

$$\frac{x_1}{-24+20} = \frac{x_2}{-12+8} = \frac{x_3}{-10+18} = k_1 \quad (k_1 \neq 0)$$

$$\frac{x_1}{-4} = \frac{x_2}{-4} = \frac{x_3}{8} = k_1$$

$\therefore x_1 = -k_1, x_2 = -k_1, x_3 = 2k_1$.

The eigenvector corresponding to $\lambda = 3$ is $X_1 = \begin{pmatrix} -k_1 \\ -k_1 \\ 2k_1 \end{pmatrix}$.

When $k_1 = 1$, $X_1 = \begin{pmatrix} -1 \\ -1 \\ 2 \end{pmatrix}$ is the eigenvector corresponding to $\lambda = 3$.

Case 2: To find eigenvector when $\lambda = 2$.

Put $\lambda = 2$ in (1), we have

$$\begin{pmatrix} 1 & 10 & 5 \\ -2 & -5 & -4 \\ 3 & 5 & 5 \end{pmatrix} \begin{pmatrix} x_1 \\ x_2 \\ x_3 \end{pmatrix} = \begin{pmatrix} 0 \\ 0 \\ 0 \end{pmatrix}$$

This gives,

$$x_1 + 10x_2 + 5x_3 = 0$$
$$-2x_1 - 5x_2 - 4x_3 = 0$$
$$3x_1 + 5x_2 + 5x_3 = 0$$

Solving first two equations by cross multiplication, we have

$$\frac{x_1}{-40+25} = \frac{x_2}{-10+4} = \frac{x_3}{-5+20} = k_1 \ (k_2 \neq 0)$$

$$\frac{x_1}{-15} = \frac{x_2}{-6} = \frac{x_3}{15} = k_2$$

$$\therefore x_1 = -5k_2, x_2 = -2k_2, x_3 = 5k_2.$$

The eigenvector corresponding to $\lambda = 2$ is $X_2 = \begin{pmatrix} -5k_2 \\ -2k_2 \\ 5k_2 \end{pmatrix}$.

When $k_2 = 1$, $X_2 = \begin{pmatrix} -5 \\ -2 \\ 5 \end{pmatrix}$ is the eigenvector corresponding to $\lambda = 2$.

Therefore eigenvalues of $A$ are 2, 2, 3.

$\therefore$ Algebraic multiplicity of 2 is 2.

There is only one eigenvector $X_2$ corresponding to the repeated eigenvalue $\lambda = 2$.

The geometric multiplicity of the eigenvalue $\lambda = 2$ is one.

**Example:** Determine the algebraic multiplicity and geometric multiplicity of

$$\begin{pmatrix} 1 & 2 & 2 \\ 0 & 2 & 1 \\ -1 & 2 & 2 \end{pmatrix}.$$

**Solution:** Let $\lambda$ be the eigenvalue of $A$.

The characteristic equation of $A$ is given by $|A - \lambda I| = 0$.

$$\Rightarrow \lambda^3 - 5\lambda^2 + 8\lambda - 4 = 0.$$

$$\Rightarrow \lambda = 1, 2, 2.$$

$\therefore$ Eigenvalues of $A$ are $\lambda = 1, 2, 2$.

**To find eigenvectors:**

Let $X = \begin{pmatrix} x_1 \\ x_2 \\ x_3 \end{pmatrix}$ be the eigenvector of $A$ corresponding to $\lambda$.

Then

$$AX = \lambda X$$

$$\Rightarrow (A - \lambda I)X = 0$$

$$\Rightarrow \begin{pmatrix} 1-\lambda & 2 & 2 \\ 0 & 2-\lambda & 1 \\ -1 & 2 & 2-\lambda \end{pmatrix} \begin{pmatrix} x_1 \\ x_2 \\ x_3 \end{pmatrix} = \begin{pmatrix} 0 \\ 0 \\ 0 \end{pmatrix} \quad \cdots\cdots \quad (1)$$

**Case 1:** To find eigenvector when $\lambda = 1$.

Put $\lambda = 1$ in (1), we have

$$\begin{pmatrix} 0 & 2 & 2 \\ 0 & 1 & 1 \\ -1 & 2 & 1 \end{pmatrix} \begin{pmatrix} x_1 \\ x_2 \\ x_3 \end{pmatrix} = \begin{pmatrix} 0 \\ 0 \\ 0 \end{pmatrix}$$

This gives,

$$2x_2 + 2x_3 = 0$$

$$x_2 + x_3 = 0$$
$$-x_1 + 2x_2 + x_3 = 0$$
$$\therefore x_2 = -x_3$$
$$x_1 = 2x_2 + x_3 = -2x_3 + x_3 = -x_3$$
$$\Rightarrow x_1 = x_2 = -x_3.$$

The eigenvector corresponding to $\lambda = 1$ is $X_1 = \begin{pmatrix} 1 \\ 1 \\ -1 \end{pmatrix}$.

Case 2: To find eigenvector when $\lambda = 2$.

Put $\lambda = 2$ in (1), we have

$$\begin{pmatrix} -1 & 2 & 2 \\ 0 & 0 & 1 \\ -1 & 2 & 0 \end{pmatrix} \begin{pmatrix} x_1 \\ x_2 \\ x_3 \end{pmatrix} = \begin{pmatrix} 0 \\ 0 \\ 0 \end{pmatrix}$$

This gives,
$$-x_1 + 2x_2 + 2x_3 = 0$$
$$x_3 = 0$$
$$-x_1 + 2x_2 = 0$$
$$\therefore x_3 = 0, x_1 = 2x_2.$$
$$\therefore x_1 = 2, x_2 = 1, x_3 = 0.$$

The eigenvector corresponding to $\lambda = 2$ is $X_2 = \begin{pmatrix} 2 \\ 1 \\ 0 \end{pmatrix}$.

Therefore eigenvalues of $A$ are 1, 2, 2.

$\therefore$ Algebraic multiplicity of 2 is 2.

There is only one eigenvector $X_2$ corresponding to the repeated eigenvalue $\lambda = 2$. The geometric multiplicity of the eigenvalue $\lambda = 2$ is one.

### 1.8.3 Applications of Eigenvalues and Eigenvectors

**Example:** Find the natural frequencies and normal modes of a vibrating system

$m'' + kx = 0$ for mass $m = \begin{pmatrix} 2 & 0 \\ 0 & 1 \end{pmatrix}$ and stiffness $k = \begin{pmatrix} 9 & -3 \\ -3 & 3 \end{pmatrix}$. **(Apr-2017)**

**Solution:** The eigenvalue equation for $m'' + kx = 0$ is

$$\begin{bmatrix} k - \lambda m \end{bmatrix} \overline{x} = 0$$

i.e., $\begin{bmatrix} 9 - 2\lambda & -3 \\ -3 & 3 - \lambda \end{bmatrix} \begin{bmatrix} x_1 \\ x_2 \end{bmatrix} = \begin{bmatrix} 0 \\ 0 \end{bmatrix}$ ...... (1)

The characteristic equation is $|k - \lambda m| = 0$.

$$\Rightarrow 2\lambda^2 - 15\lambda + 18 = 0.$$
$$\Rightarrow (\lambda - 6)(2\lambda - 3) = 0.$$
$$\Rightarrow \lambda = 6, 3/2.$$

$\therefore \lambda = 6$ and $\lambda = 3/2$ are natural frequencies.

**Mode shapes at $\lambda = 6$:**

Put $\lambda = 6$ in (1), we have

$$\begin{pmatrix} -3 & -3 \\ -3 & -3 \end{pmatrix} \begin{pmatrix} x_1 \\ x_2 \end{pmatrix} = \begin{pmatrix} 0 \\ 0 \end{pmatrix}$$

This gives,

$$x_1 + x_2 = 0 \Rightarrow x_1 = -x_2$$

Let $x_2 = k$, we get $x_1 = -k$.

The mode shape at $\lambda = 6$ is $\begin{pmatrix} -1 \\ 1 \end{pmatrix}$.

**Mode shapes at $\lambda = 3/2$:**

Put $\lambda = 3/2$ in (1), we have

$$\begin{pmatrix} 6 & -3 \\ -3 & 3/2 \end{pmatrix} \begin{pmatrix} x_1 \\ x_2 \end{pmatrix} = \begin{pmatrix} 0 \\ 0 \end{pmatrix}$$

This gives,

$$6x_1 - 3x_2 = 0 \Rightarrow x_1 = \frac{1}{2} x_2$$

Let $x_2 = 2k$, we get $x_1 = k$.

The mode shape at $\lambda = 3/2$ is $\begin{pmatrix} 1 \\ 2 \end{pmatrix}$.

**Example:** Determine the natural frequencies and normal modes of vibrating system for which $m = \begin{pmatrix} 2 & 0 \\ 0 & 4 \end{pmatrix}$ and stiffness $k = \begin{pmatrix} 6 & -2 \\ -2 & 9 \end{pmatrix}$.

**Example:** Determine the natural frequencies and normal modes of vibrating system for which mass $m = \begin{pmatrix} 1 & 0 \\ 0 & 2 \end{pmatrix}$ and stiffness $k = \begin{pmatrix} 2 & 1 \\ 1 & 3 \end{pmatrix}$. **(Apr-2017), (Apr-2018)**

Ans: $\lambda = 1, 5/2$, $\begin{pmatrix} -1 \\ 1 \end{pmatrix}, \begin{pmatrix} 2 \\ 1 \end{pmatrix}$.

**Example:** Find the natural frequencies and corresponding vibrating shapes of the system $m\ddot{x} + kx = 0$ for which $m = \begin{pmatrix} 1 & 0 & 0 \\ 0 & 2 & 0 \\ 0 & 0 & 1 \end{pmatrix}$ and $k = \begin{pmatrix} 1 & -2 & 1 \\ -2 & 4 & -2 \\ 1 & -2 & 1 \end{pmatrix}$.

Ans: $\lambda = 0, 0, 4$, $\begin{pmatrix} 2 \\ 1 \\ 0 \end{pmatrix}, \begin{pmatrix} -1 \\ 0 \\ 1 \end{pmatrix}, \begin{pmatrix} 1 \\ -1 \\ 1 \end{pmatrix}$.

# Chapter 2

# Cayley-Hamilton Theorem and Quadratic forms

## 2.1 Cayley-Hamilton Theorem

**Statement:** Every Square matrix satisfies its own characteristic equation.

Cayley-Hamilton theorem has two important uses

(1) to find the inverse of a nonsingular matrix $A$ and

(2) to find higher integral powers of $A$.

**Example:** Find the characteristics equation of the matrix $A = \begin{pmatrix} 1 & 3 & 7 \\ 4 & 2 & 3 \\ 1 & 2 & 1 \end{pmatrix}$.

Show that the equation is satisfied by $A$ and hence obtain the inverse of the given matrix.

**Solution:** The characteristic equation of $A$ is

$$|A - \lambda I| = 0 \Rightarrow \begin{vmatrix} 1-\lambda & 3 & 7 \\ 4 & 2-\lambda & 3 \\ 1 & 2 & 1-\lambda \end{vmatrix} = 0 \Rightarrow \lambda^3 - 4\lambda^2 - 20\lambda - 35 = 0.$$

If $A$ satisfies it,

$$A^3 - 4A^2 - 20A - 35I = \mathbf{0}.$$

Now,
$$A^2 = \begin{pmatrix} 20 & 23 & 23 \\ 15 & 22 & 37 \\ 10 & 9 & 14 \end{pmatrix}, \quad A^3 = \begin{pmatrix} 135 & 152 & 232 \\ 140 & 163 & 208 \\ 60 & 76 & 111 \end{pmatrix}.$$

$\therefore A^3 - 4A^2 - 20A - 35I$

$$= \begin{pmatrix} 135 & 152 & 232 \\ 140 & 163 & 208 \\ 60 & 76 & 111 \end{pmatrix} - 4\begin{pmatrix} 20 & 23 & 23 \\ 15 & 22 & 37 \\ 10 & 9 & 14 \end{pmatrix} - 20\begin{pmatrix} 1 & 3 & 7 \\ 4 & 2 & 3 \\ 1 & 2 & 1 \end{pmatrix} - 35\begin{pmatrix} 1 & 0 & 0 \\ 0 & 1 & 0 \\ 0 & 0 & 1 \end{pmatrix}$$

$$= \begin{pmatrix} 0 & 0 & 0 \\ 0 & 0 & 0 \\ 0 & 0 & 0 \end{pmatrix} = \mathbf{0}.$$

Hence $A$ satisfies the characteristic equation.

**To find $A^{-1}$:**

To obtain inverse of given matrix $A$ we proceed as follows:

Since $A^3 - 4A^2 - 20A - 35I = 0$.

Multiply both sides by $A^{-1}$, we have

$$A^2 - 4A - 20I - 35A^{-1} = \mathbf{0}$$

$$\Rightarrow 35A^{-1} = A^2 - 4A - 20I$$

$$= \begin{pmatrix} 20 & 23 & 23 \\ 15 & 22 & 37 \\ 10 & 9 & 14 \end{pmatrix} - 4\begin{pmatrix} 1 & 3 & 7 \\ 4 & 2 & 3 \\ 1 & 2 & 1 \end{pmatrix} - 20\begin{pmatrix} 1 & 0 & 0 \\ 0 & 1 & 0 \\ 0 & 0 & 1 \end{pmatrix}$$

$$= \begin{pmatrix} -4 & 11 & -5 \\ -1 & -6 & 25 \\ 6 & 1 & -10 \end{pmatrix}$$

$$\therefore A^{-1} = \frac{1}{35}\begin{pmatrix} -4 & 11 & -5 \\ -1 & -6 & 25 \\ 6 & 1 & -10 \end{pmatrix}$$

**Example:** Find the characteristics equation of the matrix $A = \begin{pmatrix} 1 & 1 & 3 \\ 1 & 3 & -3 \\ -2 & -4 & -4 \end{pmatrix}$ and hence find its inverse.

**Solution:** The characteristic equation of $A$ is

$$|A - \lambda I| = 0 \Rightarrow \begin{vmatrix} 1-\lambda & 1 & 3 \\ 1 & 3-\lambda & -3 \\ -2 & -4 & -4-\lambda \end{vmatrix} = 0 \Rightarrow \lambda^3 - 20\lambda + 8 = 0.$$

By Calyley-Hamilton theorem, we have

$$A^3 - 20A + 8I = \mathbf{0}.$$

Now,

$$A^2 = \begin{pmatrix} -4 & -8 & -12 \\ 10 & 22 & 6 \\ 2 & 2 & 22 \end{pmatrix}.$$

**To find $A^{-1}$:**

To obtain inverse of given matrix $A$ we proceed as follows:
Since $A^3 - 20A + 8I = \mathbf{0}$.

Multiply both sides by $A^{-1}$, we have

$$A^2 - 20I + 8A^{-1} = 0$$

$$\Rightarrow 8A^{-1} = 20I - A^2$$

$$= 20 \begin{pmatrix} 1 & 0 & 0 \\ 0 & 1 & 0 \\ 0 & 0 & 1 \end{pmatrix} - \begin{pmatrix} -4 & -8 & -12 \\ 10 & 22 & 6 \\ 2 & 2 & 22 \end{pmatrix}$$

$$= \begin{pmatrix} 24 & 8 & 12 \\ -10 & -2 & -6 \\ -2 & -2 & -2 \end{pmatrix}$$

$$\therefore A^{-1} = \frac{1}{8} \begin{pmatrix} 24 & 8 & 12 \\ -10 & -2 & -6 \\ -2 & -2 & -2 \end{pmatrix}$$

**Example:** Using Cayley-Hamilton theorem, find the inverse of $\begin{pmatrix} 5 & 3 \\ 3 & 2 \end{pmatrix}$.

**Solution:** The characteristic equation of $A$ is

$$|A - \lambda I| = 0 \Rightarrow \begin{vmatrix} 5 - \lambda & 3 \\ 3 & 2 - \lambda \end{vmatrix} = 0 \Rightarrow \lambda^2 - 7\lambda + 1 = 0.$$

By Cayley-Hamilton theorem, we have,

$$A^2 - 7A + I = 0.$$

**To find $A^{-1}$:**

To obtain inverse of given matrix $A$ we proceed as follows:

Since $A^2 - 7A + I = 0$.

Multiply both sides by $A^{-1}$, we have

$$A - 7I + A^{-1} = \mathbf{0}$$
$$\Rightarrow A^{-1} = -A + 7I$$
$$= -\begin{pmatrix} 5 & 3 \\ 3 & 2 \end{pmatrix} + 7\begin{pmatrix} 1 & 0 \\ 0 & 1 \end{pmatrix}$$
$$= \begin{pmatrix} 2 & -3 \\ -3 & 5 \end{pmatrix}$$

**Example:** Using Cayley-Hamilton theorem, find the inverse of $\begin{pmatrix} 1 & 1 & 2 \\ 0 & -2 & 0 \\ 0 & 0 & 3 \end{pmatrix}$.

**Solution:** The characteristic equation of $A$ is

$$|A - \lambda I| = 0 \Rightarrow \begin{vmatrix} 1-\lambda & 1 & 2 \\ 0 & -2-\lambda & 0 \\ 0 & 0 & 3-\lambda \end{vmatrix} = 0 \Rightarrow \lambda^3 - 2\lambda^2 - 5\lambda + 6 = 0.$$

By Cayley-Hamilton theorem, we have,
$$A^3 - 2A^2 - 5A + 6I = \mathbf{0}.$$

Now,
$$A^2 = \begin{pmatrix} 1 & -1 & 8 \\ 0 & 4 & 0 \\ 0 & 0 & 9 \end{pmatrix}.$$

**To find $A^{-1}$:**

To obtain inverse of given matrix $A$ we proceed as follows:

Since $A^3 - 2A^2 - 5A + 6I = \mathbf{0}.$

Multiply both sides by $A^{-1}$, we have

$$A^2 - 2A - 5I + 6A^{-1} = 0$$

$$\Rightarrow 6A^{-1} = -A^2 + 2A + 5I$$

$$= -\begin{pmatrix} 1 & -1 & 8 \\ 0 & 4 & 0 \\ 0 & 0 & 9 \end{pmatrix} + 2\begin{pmatrix} 1 & 1 & 2 \\ 0 & -2 & 0 \\ 0 & 0 & 3 \end{pmatrix} + 5\begin{pmatrix} 1 & 0 & 0 \\ 0 & 1 & 0 \\ 0 & 0 & 1 \end{pmatrix}$$

$$= \begin{pmatrix} 6 & 3 & -4 \\ 0 & -3 & 0 \\ 0 & 0 & 2 \end{pmatrix}$$

$$\therefore A^{-1} = \frac{1}{6}\begin{pmatrix} 6 & 3 & -4 \\ 0 & -3 & 0 \\ 0 & 0 & 2 \end{pmatrix}$$

**Example:** Using Cayley-Hamilton theorem, find the inverse of the matrix $\begin{pmatrix} 7 & -1 & 3 \\ 6 & 1 & 4 \\ 2 & 4 & 8 \end{pmatrix}$.

**Solution:** The characteristic equation of $A$ is

$$|A - \lambda I| = 0 \Rightarrow \begin{vmatrix} 7-\lambda & -1 & 3 \\ 6 & 1-\lambda & 4 \\ 2 & 4 & 8-\lambda \end{vmatrix} = 0 \Rightarrow \lambda^3 - 16\lambda^2 + 55\lambda - 50 = 0.$$

By Cayley-Hamilton theorem, we have,

$$A^3 - 16A^2 + 55A - 50I = \mathbf{0}.$$

Now,

$$A^2 = \begin{pmatrix} 49 & 4 & 41 \\ 56 & 11 & 54 \\ 54 & 54 & 86 \end{pmatrix}.$$

**To find $A^{-1}$:**
To obtain inverse of given matrix $A$ we proceed as follows:

Since $A^3 - 16A^2 + 55A - 50I = 0$.

Multiply both sides by $A^{-1}$, we have

$$A^2 - 16A + 55I - 50A^{-1} = 0$$

$$\Rightarrow 50A^{-1} = A^2 - 16A + 55I$$

$$= \begin{pmatrix} 49 & 4 & 41 \\ 56 & 11 & 54 \\ 54 & 54 & 86 \end{pmatrix} - 16 \begin{pmatrix} 7 & -1 & 3 \\ 6 & 1 & 4 \\ 2 & 4 & 8 \end{pmatrix} + 55 \begin{pmatrix} 1 & 0 & 0 \\ 0 & 1 & 0 \\ 0 & 0 & 1 \end{pmatrix}$$

$$= \begin{pmatrix} -8 & 20 & -7 \\ -40 & 50 & -10 \\ 22 & -39 & 13 \end{pmatrix}$$

$$\therefore A^{-1} = \frac{1}{50} \begin{pmatrix} -8 & 20 & -7 \\ -40 & 50 & -10 \\ 22 & -39 & 13 \end{pmatrix}$$

**Example:** Using Cayley-Hamilton theorem, find the inverse of the matrix $\begin{pmatrix} 1 & 1 & 3 \\ 1 & 3 & -3 \\ 2 & -4 & -4 \end{pmatrix}$.

**Solution:** The characteristic equation of $A$ is

$$|A - \lambda I| = 0 \Rightarrow \begin{vmatrix} 1-\lambda & 1 & 3 \\ 1 & 3-\lambda & -3 \\ 2 & -4 & -4-\lambda \end{vmatrix} = 0 \Rightarrow \lambda^3 - 32\lambda + 56 = 0.$$

By Cayley-Hamilton theorem, we have,

$$A^3 - 32A + 56I = 0.$$

Now,

$$A^2 = \begin{pmatrix} 8 & -8 & -12 \\ -2 & 22 & 6 \\ -10 & 6 & 34 \end{pmatrix}.$$

**To find $A^{-1}$:**

To obtain inverse of given matrix $A$ we proceed as follows:

Since $A^3 - 32A + 56I = \mathbf{0}$.

Multiply both sides by $A^{-1}$, we have

$$A^2 - 32I + 56A^{-1} = \mathbf{0}$$

$$\Rightarrow 56A^{-1} = 32I - A^2$$

$$= 32\begin{pmatrix} 1 & 0 & 0 \\ 0 & 1 & 0 \\ 0 & 0 & 1 \end{pmatrix} - \begin{pmatrix} 8 & -8 & -12 \\ -2 & 22 & 6 \\ -10 & 6 & 34 \end{pmatrix} = \begin{pmatrix} 24 & 8 & 12 \\ 2 & 0 & -6 \\ 10 & -6 & -2 \end{pmatrix}$$

$$\therefore A^{-1} = \frac{1}{56}\begin{pmatrix} 24 & 8 & 12 \\ 2 & 0 & -6 \\ 10 & -6 & -2 \end{pmatrix}$$

**Example:** Using Cayley-Hamilton, find $A^8$, if $A = \begin{pmatrix} 1 & 2 \\ 2 & -1 \end{pmatrix}$.

**Solution:** The characteristic equation of $A$ is

$$|A - \lambda I| = 0 \Rightarrow \begin{vmatrix} 1-\lambda & 2 \\ 2 & -1-\lambda \end{vmatrix} = 0 \Rightarrow \lambda^2 - 5 = 0. \quad \cdots \cdots \quad (1)$$

By Cayley-Hamilton theorem, $A$ satisfies (1), i.e., $A^2 - 5I = \mathbf{0}$.

**To find $A^8$:** We have

$$A^2 - 5I = \mathbf{0}$$

$$\Rightarrow A^2 = 5I$$

$$\Rightarrow A^8 = (5I)^4$$

$$= 625I$$

$$= \begin{pmatrix} 625 & 0 \\ 0 & 625 \end{pmatrix}$$

**Example:** Find the characteristic equation of the matrix $A = \begin{pmatrix} 2 & 1 & 1 \\ 0 & 1 & 0 \\ 1 & 1 & 2 \end{pmatrix}$ and hence compute $A^{-1}$. Also find the matrix represented by $A^8 - 5A^7 + 7A^6 - 3A^5 +$

$A^4 - 5A^3 + 8A^2 - 2A + I$.

**Solution:** The characteristic equation of $A$ is

$$|A - \lambda I| = 0 \Rightarrow \begin{vmatrix} 2-\lambda & 1 & 1 \\ 0 & 1-\lambda & 0 \\ 1 & 1 & 2-\lambda \end{vmatrix} = 0 \Rightarrow \lambda^3 - 5\lambda^2 + 7\lambda - 3 = 0. \quad \cdots \cdots \quad (1)$$

By Cayley-Hamilton theorem, the matrix $A$ satisfy the characteristic equation (1).

$$\text{i.e., } A^3 - 5A^2 + 7A - 3I = \mathbf{0}.$$

**To find $A^{-1}$:** Now, $A^3 - 5A^2 + 7A - 3I = 0$.

But $A^2 = \begin{pmatrix} 5 & 4 & 4 \\ 0 & 1 & 0 \\ 4 & 4 & 5 \end{pmatrix}$. Multiply both sides by $A^{-1}$, we have

$A^2 - 5A + 7I - 3A^{-1} = 0$

$\Rightarrow 3A^{-1} = A^2 - 5A + 7I$

$$= \begin{pmatrix} 5 & 4 & 4 \\ 0 & 1 & 0 \\ 4 & 4 & 5 \end{pmatrix} - 5 \begin{pmatrix} 2 & 1 & 1 \\ 0 & 1 & 0 \\ 1 & 1 & 2 \end{pmatrix} + 7 \begin{pmatrix} 1 & 0 & 0 \\ 0 & 1 & 0 \\ 0 & 0 & 1 \end{pmatrix} = \begin{pmatrix} 2 & -1 & -1 \\ 0 & 3 & 0 \\ -1 & -1 & 2 \end{pmatrix}$$

$$\therefore A^{-1} = \frac{1}{3} \begin{pmatrix} 2 & -1 & -1 \\ 0 & 3 & 0 \\ -1 & -1 & 2 \end{pmatrix}.$$

We can write the given expression as

$$A^8 - 5A^7 + 7A^6 - 3A^5 + A^4 - 5A^3 + 8A^2 - 2A + I$$
$$= A^5(A^3 - 5A^2 + 7A - 3I) + A(A^3 - 5A^2 + 8A - 2I) + I$$
$$= A^5(0) + A(A^3 - 5A^2 + 7A - 3I) + A^2 + A + I$$
$$= 0 + A(0) + A^2 + A + I$$
$$= A^2 + A + I$$

$$= \begin{pmatrix} 5 & 4 & 4 \\ 0 & 1 & 0 \\ 4 & 4 & 5 \end{pmatrix} + \begin{pmatrix} 2 & 1 & 1 \\ 0 & 1 & 0 \\ 1 & 1 & 2 \end{pmatrix} + \begin{pmatrix} 1 & 0 & 0 \\ 0 & 1 & 0 \\ 0 & 0 & 1 \end{pmatrix} = \begin{pmatrix} 8 & 5 & 5 \\ 0 & 3 & 0 \\ 5 & 5 & 8 \end{pmatrix}.$$

**Alter:**

$$A^8 - 5A^7 + 7A^6 - 3A^5 + A^4 - 5A^3 + 8A^2 - 2A + I$$
$$= (A^3 - 5A^2 + 7A - 3I)(A^5 + A) + A^2 + A + I$$
$$= (0)(A^5 + A) + A^2 + A + I$$
$$= A^2 + A + I$$

$$= \begin{pmatrix} 5 & 4 & 4 \\ 0 & 1 & 0 \\ 4 & 4 & 5 \end{pmatrix} + \begin{pmatrix} 2 & 1 & 1 \\ 0 & 1 & 0 \\ 1 & 1 & 2 \end{pmatrix} + \begin{pmatrix} 1 & 0 & 0 \\ 0 & 1 & 0 \\ 0 & 0 & 1 \end{pmatrix} = \begin{pmatrix} 8 & 5 & 5 \\ 0 & 3 & 0 \\ 5 & 5 & 8 \end{pmatrix}.$$

**Example:** If $A = \begin{pmatrix} 2 & -1 & 1 \\ -1 & 2 & -1 \\ 1 & -1 & 2 \end{pmatrix}$ then find $A^{-1}$ by using Cayley-Hamilton theorem. Also express $A^6 - 6A^5 + 9A^4 - 2A^3 - 12A^2 + 23A - 9I$ as a linear polynomial in $A$.

**Solution:** The characteristic equation of $A$ is

$$|A - \lambda I| = 0 \Rightarrow \begin{vmatrix} 2-\lambda & -1 & 1 \\ -1 & 2-\lambda & -1 \\ 1 & -1 & 2-\lambda \end{vmatrix} = 0 \Rightarrow \lambda^3 - 6\lambda^2 + 9\lambda - 4 = 0.$$

By Cayley-Hamilton theorem, we have,
$$A^3 - 6A^2 + 9A - 4I = 0.$$
Now,
$$A^2 = \begin{pmatrix} 6 & -5 & 5 \\ -5 & 6 & -5 \\ 5 & -5 & 6 \end{pmatrix}.$$

**To find $A^{-1}$:**

To obtain inverse of given matrix $A$ we proceed as follows:

Since $A^3 - 6A^2 + 9A - 4I = \mathbf{0}$.

Multiply both sides by $A^{-1}$, we have

$$A^2 - 6A + 9I - 4A^{-1} = \mathbf{0}$$

$$\Rightarrow 4A^{-1} = A^2 - 6A + 9I$$

$$= \begin{pmatrix} 6 & -5 & 5 \\ -5 & 6 & -5 \\ 5 & -5 & 6 \end{pmatrix} - 6\begin{pmatrix} 2 & -1 & 1 \\ -1 & 2 & -1 \\ 1 & -1 & 2 \end{pmatrix} + 9\begin{pmatrix} 1 & 0 & 0 \\ 0 & 1 & 0 \\ 0 & 0 & 1 \end{pmatrix}$$

$$= \begin{pmatrix} 3 & 1 & -1 \\ 1 & 3 & 1 \\ -1 & 1 & 3 \end{pmatrix}$$

$$\therefore A^{-1} = \frac{1}{4}\begin{pmatrix} 3 & 1 & -1 \\ 1 & 3 & 1 \\ -1 & 1 & 3 \end{pmatrix}$$

We can write the given expression as

$$A^6 - 6A^5 + 9A^4 - 2A^3 - 12A^2 + 23A - 9I$$
$$= A^3(A^3 - 6A^2 + 9A - 4I) + 2A^3 - 12A^2 + 23A - 9I$$
$$= 2A^3 - 12A^2 + 23A - 9I$$
$$= 2(A^3 - 6A^2 + 9A - 4I) + 5A - I$$
$$= 5A - I$$

$$= 5\begin{pmatrix} 2 & -1 & 1 \\ -1 & 2 & -1 \\ 1 & -1 & 2 \end{pmatrix} - \begin{pmatrix} 1 & 0 & 0 \\ 0 & 1 & 0 \\ 0 & 0 & 1 \end{pmatrix} = \begin{pmatrix} 9 & -5 & 5 \\ -5 & 9 & -5 \\ 5 & -5 & 0 \end{pmatrix}.$$

**Alter:**

$$A^6 - 6A^5 + 9A^4 - 2A^3 - 12A^2 + 23A - 9I$$
$$= (A^3 - 6A^2 + 9A - 4I)(A^3 + 2I) + 5A - I$$
$$= (0)(A^3 + 2I) + 5A - I$$
$$= 5A - I$$

$$= 5\begin{pmatrix} 2 & -1 & 1 \\ -1 & 2 & -1 \\ 1 & -1 & 2 \end{pmatrix} - \begin{pmatrix} 1 & 0 & 0 \\ 0 & 1 & 0 \\ 0 & 0 & 1 \end{pmatrix} = \begin{pmatrix} 9 & -5 & 5 \\ -5 & 9 & -5 \\ 5 & -5 & 0 \end{pmatrix}.$$

**Example:** Using Cayley-Hamilton theorem find $A^{-1}$ and $A^4$ if $A = \begin{pmatrix} 1 & 0 & 3 \\ 2 & 1 & -1 \\ 1 & -1 & 1 \end{pmatrix}$.

(Apr-2017), (Mar-2022)

**Solution:** The characteristic equation of $A$ is

$$|A - \lambda I| = 0 \Rightarrow \begin{vmatrix} 1-\lambda & 0 & 3 \\ 2 & 1-\lambda & -1 \\ 1 & -1 & 1-\lambda \end{vmatrix} = 0 \Rightarrow \lambda^3 - 3\lambda^2 - \lambda + 9 = 0. \quad \cdots \cdots (1)$$

By Cayley-Hamilton theorem, the matrix $A$ satisfy the characteristic equation (1).

i.e., $A^3 - 3A^2 - A + 9I = \mathbf{0}$.

Now

$$A^2 = \begin{pmatrix} 4 & -3 & 6 \\ 3 & 2 & 4 \\ 0 & -2 & 5 \end{pmatrix}.$$

**To find $A^{-1}$:** Now, $A^3 - 3A^2 - A + 9I = \mathbf{0}$.

Multiply both sides by $A^{-1}$, we have

$$A^2 - 3A - I + 9A^{-1} = 0$$
$$\Rightarrow 9A^{-1} = -A^2 + 3A + I$$

$$= -\begin{pmatrix} 4 & -3 & 6 \\ 3 & 2 & 4 \\ 0 & -2 & 5 \end{pmatrix} + 3\begin{pmatrix} 1 & 0 & 3 \\ 2 & 1 & -1 \\ 1 & -1 & 1 \end{pmatrix} + \begin{pmatrix} 1 & 0 & 0 \\ 0 & 1 & 0 \\ 0 & 0 & 1 \end{pmatrix}$$

$$= \begin{pmatrix} 0 & 3 & 3 \\ 3 & 2 & -7 \\ 3 & -1 & -1 \end{pmatrix}$$

$$\therefore A^{-1} = \frac{1}{9}\begin{pmatrix} 0 & 3 & 3 \\ 3 & 2 & -7 \\ 3 & -1 & -1 \end{pmatrix}.$$

**Example:** Using Cayley-Hamilton theorem, find $A^{-2}$, where $A = \begin{pmatrix} 1 & 2 & 0 \\ 2 & -1 & 0 \\ 0 & 0 & -1 \end{pmatrix}$.

**Solution:** The characteristic equation of $A$ is

$$|A - \lambda I| = 0 \Rightarrow \begin{vmatrix} 1-\lambda & 2 & 0 \\ 2 & -1-\lambda & 0 \\ 0 & 0 & -1-\lambda \end{vmatrix} = 0 \Rightarrow \lambda^3 + \lambda^2 - 5\lambda - 5 = 0.$$

By Cayley-Hamilton theorem, we have,

$$A^3 + A^2 - 5A - 5I = \mathbf{0}.$$

Now,
$$A^2 = \begin{pmatrix} 5 & 0 & 0 \\ 0 & 5 & 0 \\ 0 & 0 & 1 \end{pmatrix}.$$

**To find $A^{-1}$:**

To obtain inverse of given matrix $A$ we proceed as follows:

Since $A^3 + A^2 - 5A - 5I = \mathbf{0}$.

Multiply both sides by $A^{-1}$, we have

$$A^2 + A - 5I - 5A^{-1} = \mathbf{0}$$

$$\Rightarrow 5A^{-1} = A^2 + A - 5I$$

$$= \begin{pmatrix} 5 & 0 & 0 \\ 0 & 5 & 0 \\ 0 & 0 & 1 \end{pmatrix} + \begin{pmatrix} 1 & 2 & 0 \\ 2 & -1 & 0 \\ 0 & 0 & -1 \end{pmatrix} - 5\begin{pmatrix} 1 & 0 & 0 \\ 0 & 1 & 0 \\ 0 & 0 & 1 \end{pmatrix}$$

$$= \begin{pmatrix} 1 & 2 & 0 \\ 2 & -1 & 0 \\ 0 & 0 & -5 \end{pmatrix}$$

$$\therefore A^{-1} = \frac{1}{5}\begin{pmatrix} 1 & 2 & 0 \\ 2 & -1 & 0 \\ 0 & 0 & -5 \end{pmatrix}$$

**To find $A^{-2}$:**

To obtain $A^{-2}$ of given matrix $A$ we proceed as follows:

Since $A^3 + A^2 - 5A - 5I = \mathbf{0}$.

Multiply both sides by $A^{-2}$, we have

$$A + I - 5A^{-1} - 5A^{-2} = 0$$

$$\Rightarrow 5A^{-2} = A + I - 5A^{-1}$$

$$= \begin{pmatrix} 1 & 2 & 0 \\ 2 & -1 & 0 \\ 0 & 0 & -1 \end{pmatrix} + \begin{pmatrix} 1 & 0 & 0 \\ 0 & 1 & 0 \\ 0 & 0 & 1 \end{pmatrix} - \begin{pmatrix} 1 & 2 & 0 \\ 2 & -1 & 0 \\ 0 & 0 & -5 \end{pmatrix}$$

$$= \begin{pmatrix} 1 & 0 & 0 \\ 0 & 1 & 0 \\ 0 & 0 & 5 \end{pmatrix}$$

$$\therefore A^{-2} = \frac{1}{5} \begin{pmatrix} 1 & 0 & 0 \\ 0 & 1 & 0 \\ 0 & 0 & 5 \end{pmatrix}$$

**Example:** Verify Cayley-Hamilton theorem for the matrix $A = \begin{pmatrix} 2 & -1 & 1 \\ -1 & 2 & -1 \\ 1 & -1 & 2 \end{pmatrix}$ and find its inverse.

**Solution:** The characteristic equation of $A$ is

$$|A - \lambda I| = 0 \Rightarrow \begin{vmatrix} 2-\lambda & -1 & 1 \\ -1 & 2-\lambda & -1 \\ 1 & -1 & 2-\lambda \end{vmatrix} = 0 \Rightarrow \lambda^3 - 6\lambda^2 + 9\lambda - 4 = 0.$$

To verify Cayley-Hamilton theorem, we have to show that

$$A^3 - 6A^2 + 9A - 4I = 0. \quad \cdots \cdots \quad (1)$$

Now,

$$A^2 = \begin{pmatrix} 6 & -5 & 5 \\ -5 & 6 & -5 \\ 5 & -5 & 6 \end{pmatrix}, \quad A^3 = \begin{pmatrix} 22 & -22 & -21 \\ -21 & 22 & -21 \\ 21 & -21 & 22 \end{pmatrix}.$$

$\therefore A^3 - 6A^2 + 9A - 4I$

$= \begin{pmatrix} 22 & -22 & -21 \\ -21 & 22 & -21 \\ 21 & -21 & 22 \end{pmatrix} - 6 \begin{pmatrix} 6 & -5 & 5 \\ -5 & 6 & -5 \\ 5 & -5 & 6 \end{pmatrix} + 9 \begin{pmatrix} 2 & -1 & 1 \\ -1 & 2 & -1 \\ 1 & -1 & 2 \end{pmatrix} - 4 \begin{pmatrix} 1 & 0 & 0 \\ 0 & 1 & 0 \\ 0 & 0 & 1 \end{pmatrix}$

$= \begin{pmatrix} 0 & 0 & 0 \\ 0 & 0 & 0 \\ 0 & 0 & 0 \end{pmatrix} = \mathbf{0}.$

Hence Cayley-Hamilton theorem is verified.

**To find $A^{-1}$:** Now, $A^3 - 6A^2 + 9A - 4I = \mathbf{0}$.

Multiply both sides by $A^{-1}$, we have

$$A^2 - 6A + 9I - 4A^{-1} = \mathbf{0}$$

$$\Rightarrow 4A^{-1} = A^2 - 6A + 9I$$

$= \begin{pmatrix} 6 & -5 & 5 \\ -5 & 6 & -5 \\ 5 & -5 & 6 \end{pmatrix} - 6 \begin{pmatrix} 2 & -1 & 1 \\ -1 & 2 & -1 \\ 1 & -1 & 2 \end{pmatrix} + 9 \begin{pmatrix} 1 & 0 & 0 \\ 0 & 1 & 0 \\ 0 & 0 & 1 \end{pmatrix}$

$= \begin{pmatrix} 3 & 1 & -1 \\ 1 & 3 & 1 \\ -1 & 1 & 3 \end{pmatrix}$

$\therefore A^{-1} = \dfrac{1}{4} \begin{pmatrix} 3 & 1 & -1 \\ 1 & 3 & 1 \\ -1 & 1 & 3 \end{pmatrix}$

**Example:** Verify Cayley-Hamilton theorem for the matrix $A = \begin{pmatrix} 7 & 2 & -2 \\ -6 & -1 & 2 \\ 6 & 2 & -1 \end{pmatrix}$.

and find $A^{-1}$.

**Solution:** The characteristic equation of $A$ is

$$|A - \lambda I| = 0 \Rightarrow \begin{vmatrix} 7-\lambda & 2 & -2 \\ -6 & -1-\lambda & 2 \\ 6 & 2 & -1-\lambda \end{vmatrix} = 0 \Rightarrow \lambda^3 - 5\lambda^2 + 7\lambda - 3 = 0.$$

To verify Cayley-Hamilton theorem, we have to show that

$$A^3 - 5A^2 + 7A - 3I = \mathbf{0}. \quad \cdots \cdots \quad (1)$$

Now,

$$A^2 = \begin{pmatrix} 25 & 8 & -8 \\ -24 & -7 & 8 \\ 24 & 8 & -7 \end{pmatrix}, \quad A^3 = \begin{pmatrix} 79 & 26 & -26 \\ -78 & 25 & 26 \\ 78 & 26 & -25 \end{pmatrix}.$$

$\therefore A^3 - 5A^2 + 7A - 3I$

$$= \begin{pmatrix} 79 & 26 & -26 \\ -78 & 25 & 26 \\ 78 & 26 & -25 \end{pmatrix} - 5\begin{pmatrix} 25 & 8 & -8 \\ -24 & -7 & 8 \\ 24 & 8 & -7 \end{pmatrix} + 7\begin{pmatrix} 7 & 2 & -2 \\ -6 & -1 & 2 \\ 6 & 2 & -1 \end{pmatrix} - 3\begin{pmatrix} 1 & 0 & 0 \\ 0 & 1 & 0 \\ 0 & 0 & 1 \end{pmatrix}$$

$$= \begin{pmatrix} 0 & 0 & 0 \\ 0 & 0 & 0 \\ 0 & 0 & 0 \end{pmatrix} = \mathbf{0}.$$

Hence Cayley-Hamilton theorem is verified.

**To find $A^{-1}$:** Now, $A^3 - 5A^2 + 7A - 3I = \mathbf{0}.$

Multiply both sides by $A^{-1}$, we have

$$A^2 - 5A + 7I - 3A^{-1} = 0$$

$$\Rightarrow 3A^{-1} = A^2 - 5A + 7I$$

$$= \begin{pmatrix} 25 & 8 & -8 \\ -24 & -7 & 8 \\ 24 & 8 & -7 \end{pmatrix} - 5 \begin{pmatrix} 7 & 2 & -2 \\ -6 & -1 & 2 \\ 6 & 2 & -1 \end{pmatrix} + 7 \begin{pmatrix} 1 & 0 & 0 \\ 0 & 1 & 0 \\ 0 & 0 & 1 \end{pmatrix}$$

$$= \begin{pmatrix} 3 & -2 & 2 \\ 6 & 5 & -2 \\ -6 & -2 & 5 \end{pmatrix}$$

$$\therefore A^{-1} = \frac{1}{3} \begin{pmatrix} 3 & -2 & 2 \\ 6 & 5 & -2 \\ -6 & -2 & 5 \end{pmatrix}$$

**Example:** Verify Cayley-Hamilton theorem and hence find its inverse of the matrix

$$A = \begin{pmatrix} 1 & 0 & 1 \\ 2 & 1 & -1 \\ 1 & -1 & 1 \end{pmatrix}.$$

**Solution:** The characteristic equation of $A$ is

$$|A - \lambda I| = 0 \Rightarrow \begin{vmatrix} 1-\lambda & 0 & 1 \\ 2 & 1-\lambda & -1 \\ 1 & -1 & 1-\lambda \end{vmatrix} = 0 \Rightarrow \lambda^3 - 3\lambda^2 + \lambda + 3 = 0. \quad \cdots \cdots \quad (1)$$

To verify Cayley-Hamilton theorem, we have to show that

$$A^3 - 3A^2 + A + 3I = 0.$$

Now,

$$A^2 = \begin{pmatrix} 2 & -1 & 2 \\ 3 & 2 & 0 \\ 0 & -2 & 3 \end{pmatrix}, \quad A^3 = \begin{pmatrix} 2 & -3 & 5 \\ 7 & 2 & 1 \\ -1 & -5 & 5 \end{pmatrix}.$$

$\therefore A^3 - 3A^2 + A + 3I$

$$= \begin{pmatrix} 2 & -3 & 5 \\ 7 & 2 & 1 \\ -1 & -5 & 5 \end{pmatrix} - 3\begin{pmatrix} 2 & -1 & 2 \\ 3 & 2 & 0 \\ 0 & -2 & 3 \end{pmatrix} + \begin{pmatrix} 1 & 0 & 1 \\ 2 & 1 & -1 \\ 1 & -1 & 1 \end{pmatrix} + 3\begin{pmatrix} 1 & 0 & 0 \\ 0 & 1 & 0 \\ 0 & 0 & 1 \end{pmatrix}$$

$$= \begin{pmatrix} 0 & 0 & 0 \\ 0 & 0 & 0 \\ 0 & 0 & 0 \end{pmatrix} = \mathbf{0}.$$

Hence Cayley-Hamilton theorem is verified.

**To find** $A^{-1}$: Now, $A^3 - 3A^2 + A + 3I = \mathbf{0}$.

Multiply both sides by $A^{-1}$, we have

$$A^2 - 3A + I + 3A^{-1} = 0$$

$$\Rightarrow 3A^{-1} = -A^2 + 3A - I$$

$$= -\begin{pmatrix} 2 & -1 & 2 \\ 3 & 2 & 0 \\ 0 & -2 & 3 \end{pmatrix} + 3\begin{pmatrix} 1 & 0 & 1 \\ 2 & 1 & -1 \\ 1 & -1 & 1 \end{pmatrix} - \begin{pmatrix} 1 & 0 & 0 \\ 0 & 1 & 0 \\ 0 & 0 & 1 \end{pmatrix}$$

$$= \begin{pmatrix} 0 & 1 & 1 \\ 3 & 0 & -3 \\ 3 & -1 & -1 \end{pmatrix}$$

$$\therefore A^{-1} = \frac{1}{3}\begin{pmatrix} 0 & 1 & 1 \\ 3 & 0 & -3 \\ 3 & -1 & -1 \end{pmatrix}$$

**Example:** Verify Cayley-Hamilton theorem and find $A^{-1}$, for $A = \begin{pmatrix} -1 & -2 & 0 \\ 1 & 0 & 2 \\ 2 & 3 & 4 \end{pmatrix}$.

**Solution:** The characteristic equation of $A$ is

$$|A - \lambda I| = 0 \Rightarrow \begin{vmatrix} -1-\lambda & -2 & 0 \\ 1 & 0-\lambda & 2 \\ 2 & 3 & 4-\lambda \end{vmatrix} = 0 \Rightarrow \lambda^3 - 3\lambda^2 - 8\lambda - 6 = 0. \quad \cdots\cdots \quad (1)$$

To verify Cayley-Hamilton theorem, we have to show that
$$A^3 - 3A^2 - 8A - 6I = 0.$$

Now,
$$A^2 = \begin{pmatrix} -1 & 2 & -4 \\ 3 & 4 & 8 \\ 9 & 8 & 22 \end{pmatrix}, \quad A^3 = \begin{pmatrix} -5 & -10 & -12 \\ 17 & 18 & 40 \\ 43 & 48 & 104 \end{pmatrix}.$$

$\therefore A^3 - 3A^2 - 8A - 6I$

$$= \begin{pmatrix} -5 & -10 & -12 \\ 17 & 18 & 40 \\ 43 & 48 & 104 \end{pmatrix} - 3\begin{pmatrix} -1 & 2 & -4 \\ 3 & 4 & 8 \\ 9 & 8 & 22 \end{pmatrix} - 8\begin{pmatrix} -1 & -2 & 0 \\ 1 & 0 & 2 \\ 2 & 3 & 4 \end{pmatrix} - 6\begin{pmatrix} 1 & 0 & 0 \\ 0 & 1 & 0 \\ 0 & 0 & 1 \end{pmatrix}$$

$$= \begin{pmatrix} 0 & 0 & 0 \\ 0 & 0 & 0 \\ 0 & 0 & 0 \end{pmatrix} = 0$$

Hence Cayley-Hamilton theorem is verified.

**To find $A^{-1}$:** Now, $A^3 - 3A^2 - 8A - 6I = 0$.

Multiply both sides by $A^{-1}$, we have

$$A^2 - 3A - 8I - 6A^{-1} = 0$$

$$\Rightarrow 6A^{-1} = A^2 - 3A - 8I$$

$$= \begin{pmatrix} -1 & 2 & -4 \\ 3 & 4 & 8 \\ 9 & 8 & 22 \end{pmatrix} - 3\begin{pmatrix} -1 & -2 & 0 \\ 1 & 0 & 2 \\ 2 & 3 & 4 \end{pmatrix} - 8\begin{pmatrix} 1 & 0 & 0 \\ 0 & 1 & 0 \\ 0 & 0 & 1 \end{pmatrix}$$

$$= \begin{pmatrix} -6 & 8 & -4 \\ 0 & -4 & 2 \\ 3 & -1 & 2 \end{pmatrix}$$

$$\therefore A^{-1} = \frac{1}{6}\begin{pmatrix} -6 & 8 & -4 \\ 0 & -4 & 2 \\ 3 & -1 & 2 \end{pmatrix}$$

**Example:** Verify Cayley-Hamilton theorem for the matrix $A = \begin{pmatrix} 1 & 2 & -1 \\ 2 & 1 & -2 \\ 2 & -2 & 1 \end{pmatrix}$
and hence find $A^{-1}$ and $A^4$.  **Apr-2017, Apr-2018, Dec-2020**

**Solution:** The characteristic equation of $A$ is

$$|A - \lambda I| = 0 \Rightarrow \begin{vmatrix} 1-\lambda & 2 & -1 \\ 2 & 1-\lambda & -2 \\ 2 & -2 & 1-\lambda \end{vmatrix} = 0 \Rightarrow \lambda^3 - 3\lambda^2 - 3\lambda + 9 = 0.$$

To verify Cayley-Hamilton theorem, we have to show that

$$A^3 - 3A^2 - 3A + 9I = \mathbf{0}.$$

Now,

$$A^2 = \begin{pmatrix} 3 & 6 & -6 \\ 0 & 9 & -6 \\ 0 & 0 & 3 \end{pmatrix}, \quad A^3 = \begin{pmatrix} 3 & 24 & -21 \\ 6 & 21 & -24 \\ 6 & -6 & 3 \end{pmatrix}.$$

$\therefore A^3 - 3A^2 - 3A + 9I$

$= \begin{pmatrix} 3 & 24 & -21 \\ 6 & 21 & -24 \\ 6 & -6 & 3 \end{pmatrix} - 3\begin{pmatrix} 3 & 6 & -6 \\ 0 & 9 & -6 \\ 0 & 0 & 3 \end{pmatrix} - 3\begin{pmatrix} 1 & 2 & -1 \\ 2 & 1 & -2 \\ 2 & -2 & 1 \end{pmatrix} + 9\begin{pmatrix} 1 & 0 & 0 \\ 0 & 1 & 0 \\ 0 & 0 & 1 \end{pmatrix}$

$= \begin{pmatrix} 0 & 0 & 0 \\ 0 & 0 & 0 \\ 0 & 0 & 0 \end{pmatrix} = \mathbf{0}$

Hence Cayley-Hamilton theorem is verified.

**To find $A^{-1}$:** Now, $A^3 - 3A^2 - 3A + 9I = \mathbf{0}$.
Multiply both sides by $A^{-1}$, we have

$$A^2 - 3A - 3I + 9A^{-1} = 0$$
$$\Rightarrow 9A^{-1} = -A^2 + 3A + 3I$$
$$= -\begin{pmatrix} 3 & 6 & -6 \\ 0 & 9 & -6 \\ 0 & 0 & 3 \end{pmatrix} + 3\begin{pmatrix} 1 & 2 & -1 \\ 2 & 1 & -2 \\ 2 & -2 & 1 \end{pmatrix} + 3\begin{pmatrix} 1 & 0 & 0 \\ 0 & 1 & 0 \\ 0 & 0 & 1 \end{pmatrix}$$
$$= \begin{pmatrix} 3 & 0 & 3 \\ 6 & -3 & 0 \\ 6 & -6 & 3 \end{pmatrix}$$
$$\therefore A^{-1} = \frac{1}{9}\begin{pmatrix} 3 & 0 & 3 \\ 6 & -3 & 0 \\ 6 & -6 & 3 \end{pmatrix}$$
$$= \frac{1}{3}\begin{pmatrix} 1 & 0 & 1 \\ 2 & -1 & 0 \\ 2 & -2 & 1 \end{pmatrix}.$$

**To find $A^4$:** Now, $A^3 - 3A^2 - 3A + 9I = 0$.

Multiply both sides by $A$, we have

$$A^4 - 3A^3 - 3A^2 + 9A = 0$$
$$\Rightarrow A^4 = 3A^3 + 3A^2 - 9A$$
$$= 3\begin{pmatrix} 3 & 24 & -21 \\ 6 & 21 & -24 \\ 6 & -6 & 3 \end{pmatrix} + 3\begin{pmatrix} 3 & 6 & -6 \\ 0 & 9 & -6 \\ 0 & 0 & 3 \end{pmatrix} - 9\begin{pmatrix} 1 & 2 & -1 \\ 2 & 1 & -2 \\ 2 & -2 & 1 \end{pmatrix}$$
$$= \begin{pmatrix} 9 & 72 & -72 \\ 0 & 81 & -72 \\ 0 & 0 & 9 \end{pmatrix}.$$

**Example:** Verify Cayley-Hamilton theorem for the matrix $A = \begin{pmatrix} 1 & 4 \\ 2 & 3 \end{pmatrix}$ and hence find $A^{-1}$ and find $B = A^5 - 4A^4 - 7A^3 + 11A^2 - A - 10I$ as a linear polynomial in $A$.

**Solution:** The characteristic equation of $A$ is

$$|A - \lambda I| = 0 \Rightarrow \begin{vmatrix} 1-\lambda & 4 \\ 2 & 3-\lambda \end{vmatrix} = 0 \Rightarrow \lambda^2 - 4\lambda - 5 = 0.$$

To verify Cayley-Hamilton theorem, we have to show that

i.e., $A^2 - 4A - 5I = \mathbf{0}.$

Now, $A^2 = \begin{pmatrix} 9 & 16 \\ 8 & 17 \end{pmatrix}.$

$\therefore A^2 - 4A - 5I = \begin{pmatrix} 9 & 16 \\ 8 & 17 \end{pmatrix} - 4\begin{pmatrix} 1 & 4 \\ 2 & 3 \end{pmatrix} - 5\begin{pmatrix} 1 & 0 \\ 0 & 1 \end{pmatrix} = \begin{pmatrix} 0 & 0 \\ 0 & 0 \end{pmatrix} = \mathbf{0}.$

Hence Cayley-Hamilton theorem is verified.

**To find $A^{-1}$:** Now, $A^2 - 4A - 5I = \mathbf{0}.$

Multiply both sides by $A^{-1}$, we have

$$A - 4I - 5A^{-1} = \mathbf{0}$$

$$\Rightarrow 5A^{-1} = A - 4I$$

$$= \begin{pmatrix} 1 & 4 \\ 2 & 3 \end{pmatrix} - 4\begin{pmatrix} 1 & 0 \\ 0 & 1 \end{pmatrix}$$

$$= \begin{pmatrix} -3 & 4 \\ 2 & -1 \end{pmatrix}$$

$$\therefore A^{-1} = \frac{1}{5}\begin{pmatrix} -3 & 4 \\ 2 & -1 \end{pmatrix}.$$

We can write the given expression as

$$B = A^5 - 4A^4 - 7A^3 + 11A^2 - A - 10I$$
$$= A^3(A^2 - 4A - 5I) - 2A^3 + 11A^2 - A - 10I$$
$$= A^3(0) - 2A^3 + 11A^2 - A - 10I$$
$$= -2A(A^2 - 4A - 5I) + 3A^2 - 11A - 10I$$
$$= 3A^2 - 11A - 10I$$
$$= 3(A^2 - 4A - 5I) + A + 5I$$
$$= A + 5I$$
$$= \begin{pmatrix} 6 & 4 \\ 2 & 8 \end{pmatrix}.$$

**Alter:**

$$A^5 - 4A^4 - 7A^3 + 11A^2 - A - 10I = (A^2 - 4A - 5I)(A^3 - 2A + 5I) + A + 5I$$
$$= (0)(A^3 - 2A + 5I) + A + 5I$$
$$= A + 5I$$
$$= \begin{pmatrix} 6 & 4 \\ 2 & 8 \end{pmatrix}.$$

**Example:** Verify Cayley-Hamilton theorem for the matrix $A = \begin{pmatrix} 1 & 6 & 1 \\ 1 & 2 & 0 \\ 0 & 0 & 3 \end{pmatrix}$ and hence find $A^4 - 6A^3 + 5A^2$.

**Solution:** The characteristic equation of $A$ is

$$|A - \lambda I| = 0 \Rightarrow \begin{vmatrix} 1-\lambda & 6 & 1 \\ 1 & 2-\lambda & 0 \\ 0 & 0 & 3-\lambda \end{vmatrix} = 0 \Rightarrow \lambda^3 - 6\lambda^2 + 5\lambda + 12 = 0.$$

To verify Cayley-Hamilton theorem, we have to show that

i.e., $A^3 - 6A^2 + 5A + 12I = \mathbf{0}.$

Now,
$$A^2 = \begin{pmatrix} 7 & 18 & 4 \\ 3 & 10 & 1 \\ 0 & 0 & 9 \end{pmatrix}, \quad A^3 = \begin{pmatrix} 25 & 78 & 19 \\ 13 & 38 & 6 \\ 0 & 0 & 27 \end{pmatrix}.$$

$\therefore A^3 - 6A^2 + 5A + 12I$

$$= \begin{pmatrix} 25 & 78 & 19 \\ 13 & 38 & 6 \\ 0 & 0 & 27 \end{pmatrix} - 6\begin{pmatrix} 7 & 18 & 4 \\ 3 & 10 & 1 \\ 0 & 0 & 9 \end{pmatrix} + 5\begin{pmatrix} 1 & 6 & 1 \\ 1 & 2 & 0 \\ 0 & 0 & 3 \end{pmatrix} + 12\begin{pmatrix} 1 & 0 & 0 \\ 0 & 1 & 0 \\ 0 & 0 & 1 \end{pmatrix}$$

$$= \begin{pmatrix} 0 & 0 & 0 \\ 0 & 0 & 0 \\ 0 & 0 & 0 \end{pmatrix} = \mathbf{0}.$$

Hence Cayley-Hamilton theorem is verified.

Now,
$$A^4 - 6A^3 + 5A^2 = A(A^3 - 6A^2 + 5A)$$
$$= A(-12I) \quad (\because A^3 - 6A^2 + 5A + 12I = \mathbf{0})$$
$$= -12A$$
$$= \begin{pmatrix} -12 & -72 & -12 \\ -12 & -24 & 0 \\ 0 & 0 & -36 \end{pmatrix}$$

**Example:** If $A = \begin{pmatrix} 1 & 0 & 0 \\ 1 & 0 & 1 \\ 0 & 1 & 0 \end{pmatrix}$, then show that $A^n = A^{n-2} + A^2 - I$ for $n \geq 3$.

Hence find $A^{50}$.

**Solution:** Given $A = \begin{pmatrix} 1 & 0 & 0 \\ 1 & 0 & 1 \\ 0 & 1 & 0 \end{pmatrix}$.

The characteristic equation of $A$ is

$$|A - \lambda I| = 0 \Rightarrow \begin{vmatrix} 1-\lambda & 0 & 0 \\ 1 & -\lambda & 1 \\ 0 & 1 & -\lambda \end{vmatrix} = 0 \Rightarrow \lambda^3 - \lambda^2 + \lambda + 1 = 0 \quad \cdots \cdots \quad (1)$$

By Cayley-Hamilton theorem $A$ satisfies (1).

$$A^3 - A^2 + A + I = 0 \Rightarrow A^3 - A^2 = A - I \quad \cdots \cdots \quad (2)$$

Multiplying (2) by $A, A^2, \cdots, A^{n-3}$, we get the equations

$$A^4 - A^3 = A^2 - A$$
$$A^5 - A^4 = A^3 - A^2$$
$$A^6 - A^5 = A^4 - A^3$$
$$\vdots \qquad \vdots$$
$$A^n - A^{n-1} = A^{n-2} - A^{n-3}$$

Adding (2) and all these equations, we get

$$A^n - A^2 = A^{n-2} - I$$
$$\Rightarrow A^n = A^{n-2} + (A^2 - I) \text{ for all } n \geq 3 \quad \cdots \cdots \quad (i)$$
$$\therefore A^{n-2} = A^{n-4} + A^2 - I$$
$$A^{n-4} = A^{n-6} + A^2 - I$$
$$\vdots \qquad \vdots$$
$$A^n = (A^{n-3} + A^2 - I) + (A^2 + I)$$
$$= A^{n-4} + 2(A^2 - I) \quad \cdots \cdots \quad (ii)$$
$$= A^{n-6} + A^2 - I + 2(A^2 - I)$$
$$= A^{n-6} + 3(A^2 - I) \quad \cdots \cdots \quad (iii)$$
$$= A^{n-8} + 4(A^2 - I) \quad \cdots \cdots \quad (iv)$$

If $n$ is even, then
$$A^n = A^{n-(n-2)} + \left(\frac{n-2}{2}\right)(A^2 - I).$$

Observe the coefficients of $A^2 - I$ in (i), (ii), (iii), $\cdots$ and index of $A$, we see $\frac{2}{2} = 1$ in (i), $\frac{4}{2} = 2$ in (ii), $\frac{3}{2} = 3$ in (iii), $\frac{8}{2} = 4$ in (iv) and so on $\frac{n-2}{2}$ in the last one.

$$\therefore A^n = A^2 + \left(\frac{n-2}{2}\right)A^2 - \left(\frac{n-2}{2}\right)I$$
$$= \frac{n}{2}A^2 - \left(\frac{n-2}{2}\right)I$$

Putting $n = 50$, we get
$$A^{50} = 25A^2 - 24I.$$

But $A^2 = \begin{pmatrix} 1 & 0 & 0 \\ 1 & 1 & 0 \\ 1 & 0 & 1 \end{pmatrix}$.

$\therefore A^{50} = 25A^2 - 24I$

$$= 25\begin{pmatrix} 1 & 0 & 0 \\ 1 & 1 & 0 \\ 1 & 0 & 1 \end{pmatrix} - 24\begin{pmatrix} 1 & 0 & 0 \\ 0 & 1 & 0 \\ 0 & 0 & 1 \end{pmatrix} = \begin{pmatrix} 1 & 0 & 0 \\ 25 & 1 & 0 \\ 25 & 0 & 1 \end{pmatrix}$$

**EXERCISE**

1. Verify Cayley-Hamilton theorem for the matrix $A = \begin{pmatrix} 1 & 2 & 3 \\ 2 & 4 & 5 \\ 3 & 5 & 6 \end{pmatrix}$. Hence find $A^{-1}$.

Apr-2018

**Solution:** The characteristic equation is $\lambda^3 - 11\lambda^2 - 4\lambda + 1 = 0$.

$$A^{-1} = \begin{pmatrix} 1 & -3 & 2 \\ -3 & 3 & -1 \\ 2 & -1 & 0 \end{pmatrix}.$$

2. Verify Cayley-Hamilton theorem for $A = \begin{pmatrix} 10 & 1 & 1 \\ 1 & 10 & -1 \\ 1 & -2 & 10 \end{pmatrix}$ and find $A^{-1}$.

**Solution:** The characteristic equation is $\lambda^3 - 30\lambda^2 + 296\lambda - 957 = 0$.

$A^{-1} = \dfrac{1}{957} \begin{pmatrix} 98 & -12 & -11 \\ -11 & 99 & 11 \\ -12 & 21 & 99 \end{pmatrix}$ 

Dec-2020

3. Verify Cayley-Hamilton theorem for $A = \begin{pmatrix} 3 & 1 & 2 \\ 2 & -3 & 1 \\ 1 & 2 & 1 \end{pmatrix}$ and find $A^{-1}$.

Dec-2020

4. Verify Cayley-Hamilton theorem and find $A^{-1}$ if $A = \begin{pmatrix} 3 & 1 & 1 \\ -1 & 5 & -1 \\ 1 & -1 & 5 \end{pmatrix}$.

**Solution:** The characteristic equation is $\lambda^3 - 13\lambda^2 + 54\lambda - 72 = 0$.

$A^{-1} = \dfrac{1}{72} \begin{pmatrix} 24 & -6 & -6 \\ 4 & 14 & 2 \\ -4 & 4 & 16 \end{pmatrix}$.   (Dec-2021), (Nov-2021), (Aug-2022)

5. Verify Cayley-Hamilton theorem for the matrix $A = \begin{pmatrix} 8 & -8 & -2 \\ 4 & -3 & -2 \\ 3 & -4 & 1 \end{pmatrix}$ also find $A^{-1}$ and $A^4$. (Jan-2020), (July-2021), (Sep-2021), (Mar-2022), (Aug-2022)

**Solution:** The characteristic equation is $\lambda^3 - 6\lambda^2 - \lambda + 22 = 0$.

$A^{-1} = \dfrac{1}{22} \begin{pmatrix} 9 & 0 & -22 \\ 10 & -4 & -24 \\ 7 & -8 & -10 \end{pmatrix}$ and $A^4 = \begin{pmatrix} 1146 & -1904 & 1226 \\ 322 & -639 & 476 \\ 359 & -544 & 407 \end{pmatrix}$.

6. Verify Cayley-Hamilton theorem for $A = \begin{pmatrix} -1 & 2 & -2 \\ 1 & -2 & 1 \\ -1 & -1 & 0 \end{pmatrix}$. Also find $A^{-1}$ and $A^4$. (Jan-2020)

7. Verify Cayley-Hamilton theorem for $A = \begin{pmatrix} 2 & -1 & 0 \\ 3 & 1 & -1 \\ 2 & 0 & 3 \end{pmatrix}$ and hence find $A^{-1}$.

8. Verify Cayley-Hamilton theorem for $A = \begin{pmatrix} 4 & 1 & 1 \\ 1 & 4 & 1 \\ 1 & 1 & 4 \end{pmatrix}$ and hence find $A^{-1}$.

(Apr-2022)

9. Verify Cayley-Hamilton theorem for $A = \begin{pmatrix} 1 & 1 & 1 \\ 1 & 6 & 1 \\ 2 & 3 & 1 \end{pmatrix}$ and hence find $A^{-1}$.

(Apr-2022)

10. Verify Cayley-Hamilton theorem for $A = \begin{pmatrix} 1 & 1 & 2 \\ 1 & 2 & 1 \\ 2 & 1 & 1 \end{pmatrix}$ and hence find $A^{-1}$.

(Apr-2022)

11. Verify Cayley-Hamilton theorem for $A = \begin{pmatrix} 1 & -3 & 2 \\ 6 & 3 & 1 \\ 1 & 3 & 1 \end{pmatrix}$ and hence find $A^{-1}$.

(Apr-2022)

12. Verify Cayley-Hamilton theorem and hence find $A^{-1}$ and $A^4$ if $A = \begin{pmatrix} 2 & 2 & 1 \\ 1 & 3 & 1 \\ 1 & 2 & 2 \end{pmatrix}$.

(Aug-2022)

13. Verify Cayley-Hamilton theorem for $A = \begin{pmatrix} 1 & -2 & 4 \\ 0 & -1 & 2 \\ 2 & 0 & 3 \end{pmatrix}$ and hence find $A^{-1}$.

(Aug-2022)

## 2.2 Diagonalization of a Matrix

**Definition:** A square matrix $A$ is said to be diagonalizable if there exists a non-singular square matrix $P$ such that $P^{-1}AP = D$, where $D$ is a diagonal matrix. In other words, $A$ is diagonalizable if $A$ is similar to diagonal matrix.

**Similar matrix:** Two matrices $A$ and $B$ are said to be similar if there exists a non-singular matrix $P$ such that $B = P^{-1}AP$.

**Note:** An $n \times n$ matrix is diagonalizable if and only if it possess $n$ linearly independent eigenvectors.

**Modal Matrix:** The matrix defined by $P^{-1}AP = D$ which diagonalized the square matrix $A$ is called the modal matrix.

**Diagonalization of $A$:** Given a square matrix $A$, the process of finding the diagonal matrix $D$ (through the construction of the matrix $P$) is called ***diagonalization of*** $A$. The matrix $D$ is known as the ***spectral matrix of*** $A$.

**Procedure to find the diagonal matrix:**

1. Find the characteristic equation of $A$. i.e., $|A - \lambda I| = 0$.
2. Solve the characteristic equation, find the eigenvalues of $A$.
3. Find linearly independent eigenvectors for each of the eigenvalue $\lambda$.
4. Form the modal matrix $P$, whose columns are the eigenvectors of $A$.
    i.e., $P = \begin{bmatrix} X_1 & X_2 & X_3 \end{bmatrix}$.
5. Find $P^{-1}$.
6. Required diagonal matrix $D = P^{-1}AP$, whose diagonal elements are the eigenvalues.

**Calculation of powers of a matrix:** Let $A$ be the given matrix of order 3.

Consider
$$D = P^{-1}AP$$
$$\Rightarrow D^2 = (P^{-1}AP)(P^{-1}AP)$$
$$= P^{-1}A(PP^{-1})AP$$
$$= P^{-1}AIAP$$
$$= P^{-1}A^2P.$$

In general,
$$D^n = P^{-1}A^nP$$
$$\Rightarrow PD^nP^{-1} = P(P^{-1}A^nP)P^{-1}$$
$$= (PP^{-1})A^n(PP^{-1})$$
$$= IA^nI$$
$$= A^n$$
$$\therefore A^n = PD^nP^{-1}$$
$$= P \begin{pmatrix} \lambda_1^n & 0 & 0 \\ 0 & \lambda_2^n & 0 \\ 0 & 0 & \lambda_3^n \end{pmatrix} P^{-1}$$

where $\lambda_1, \lambda_2, \lambda_3$ are eigenvalues of $A$.

**Example:** Show that the linear transformation $T = \begin{pmatrix} \cos\theta & \sin\theta \\ -\sin\theta & \cos\theta \end{pmatrix}$, where $\theta = \dfrac{1}{2}\tan^{-1}\dfrac{2h}{a-b}$, changes the matrix $A = \begin{pmatrix} a & h \\ h & b \end{pmatrix}$ to the diagonal form $D = T^{-1}AT$.

**Solution:** Given $T = \begin{pmatrix} \cos\theta & \sin\theta \\ -\sin\theta & \cos\theta \end{pmatrix}$

$$\therefore T^{-1} = \begin{pmatrix} \cos\theta & -\sin\theta \\ \sin\theta & \cos\theta \end{pmatrix}.$$

Now

$$T^{-1}AT = \begin{pmatrix} \cos\theta & -\sin\theta \\ \sin\theta & \cos\theta \end{pmatrix} \begin{pmatrix} a & h \\ h & b \end{pmatrix} \begin{pmatrix} \cos\theta & \sin\theta \\ -\sin\theta & \cos\theta \end{pmatrix}$$

$$= \begin{pmatrix} a\cos\theta - h\sin\theta & h\cos\theta - b\sin\theta \\ a\sin\theta + h\cos\theta & h\sin\theta + b\cos\theta \end{pmatrix} \begin{pmatrix} \cos\theta & -\sin\theta \\ \sin\theta & \cos\theta \end{pmatrix}$$

$$= \begin{pmatrix} a\cos^2\theta - 2h\sin\theta\cos\theta + b\sin^2\theta & (a-b)\sin\theta\cos\theta - h\sin^2\theta + h\cos^2\theta \\ (a-b)\sin\theta\cos\theta + h\cos^2\theta - h\sin^2\theta & a\sin^2\theta + 2h\sin\theta\cos\theta + b\cos^2\theta \end{pmatrix}$$

$$= \begin{pmatrix} a\cos^2\theta - h\sin 2\theta + b\sin^2\theta & (a-b)\sin\theta\cos\theta + h\cos 2\theta \\ (a-b)\sin\theta\cos\theta + h\cos 2\theta & a\sin^2\theta + h\sin 2\theta + b\cos^2\theta \end{pmatrix}$$

$$= \begin{pmatrix} d_1 & 0 \\ 0 & d_2 \end{pmatrix} \text{ being diagonal matrix}$$

$\therefore (a-b)\sin\theta\cos\theta + h\cos 2\theta = 0$

$\Rightarrow \dfrac{a-b}{2}\sin 2\theta + h\cos 2\theta = 0$

$\Rightarrow \dfrac{a-b}{2}\sin 2\theta = -h\cos 2\theta$

$\Rightarrow \tan 2\theta = \dfrac{2h}{b-a}$

$\Rightarrow \theta = \dfrac{1}{2}\tan^{-1}\dfrac{2h}{b-a}.$

**Example:** Reduce the matrix $A = \begin{pmatrix} -19 & 7 \\ -42 & 16 \end{pmatrix}$ to the diagonal form.

**Solution:** Let $\lambda$ be the eigenvalue of $A$.

The characteristic equation of $A$ is given by $|A - \lambda I| = 0$

$$\begin{vmatrix} -19-\lambda & 7 \\ -42 & 16-\lambda \end{vmatrix} = 0$$

$\Rightarrow \lambda^2 - 3\lambda - 10 = 0$

$\Rightarrow \lambda = -5, 2.$

$\therefore$ Eigenvalues of $A$ are $\lambda = -5, 2.$

**To find eigenvectors:**

Let $X = \begin{pmatrix} x_1 \\ x_2 \end{pmatrix}$ be the eigenvector of $A$ corresponding to $\lambda$.

Then
$$AX = \lambda X$$
$$\Rightarrow (A - \lambda I)X = 0$$
$$\Rightarrow \begin{pmatrix} -19-\lambda & 7 \\ -42 & 16-\lambda \end{pmatrix} \begin{pmatrix} x_1 \\ x_2 \end{pmatrix} = \begin{pmatrix} 0 \\ 0 \end{pmatrix} \quad \ldots\ldots \quad (1)$$

**Case 1:** To find eigenvector when $\lambda = -5$.

Put $\lambda = -5$ in (1), we have
$$\begin{pmatrix} -21 & 7 \\ -42 & 14 \end{pmatrix} \begin{pmatrix} x_1 \\ x_2 \end{pmatrix} = \begin{pmatrix} 0 \\ 0 \end{pmatrix}$$

This gives,
$$-21x_1 + 7x_2 = 0$$
$$-42x_1 + 14x_2 = 0$$

The above two equations represent the single equation $-3x_1 + x_2 = 0$.

Put $x_1 = k_1$, then $x_2 = 3k_1$.

The eigenvector corresponding to $\lambda = 5$ is $X_1 = \begin{pmatrix} k_1 \\ 3k_1 \end{pmatrix}$.

When $k_1 = 1$, we have $X_1 = \begin{pmatrix} 1 \\ 3 \end{pmatrix}$.

**Case 2:** To find eigenvector when $\lambda = -2$.

Put $\lambda = -2$ in (1), we have
$$\begin{pmatrix} -14 & 7 \\ -42 & 21 \end{pmatrix} \begin{pmatrix} x_1 \\ x_2 \end{pmatrix} = \begin{pmatrix} 0 \\ 0 \end{pmatrix}$$

This gives,
$$-14x_1 + 7x_2 = 0$$
$$-42x_1 + 21x_2 = 0$$

The above two equations represent the single equation $-2x_1 + x_2 = 0$.

Put $x_1 = k_2$, then $x_2 = 2k_2$.

The eigenvector corresponding to $\lambda = -2$ is $X_2 = \begin{pmatrix} k_2 \\ 2k_2 \end{pmatrix}$.

When $k_2 = 1$, we have $X_2 = \begin{pmatrix} 1 \\ 2 \end{pmatrix}$.

$\therefore$ The modal matrix is given by $P = \begin{pmatrix} 1 & 1 \\ 2 & 3 \end{pmatrix}$.

$\therefore P^{-1} = \begin{pmatrix} 3 & -1 \\ -2 & 1 \end{pmatrix}$.

$$P^{-1}AP = \begin{pmatrix} 3 & -1 \\ -2 & 1 \end{pmatrix} \begin{pmatrix} -19 & 7 \\ -42 & 16 \end{pmatrix} \begin{pmatrix} 1 & 1 \\ 2 & 3 \end{pmatrix}$$

$$= \begin{pmatrix} -5 & 0 \\ 0 & 2 \end{pmatrix} = D$$

**Example:** Reduce the matrix $A = \begin{pmatrix} -1 & 2 & -2 \\ 1 & 2 & 1 \\ -1 & -1 & 0 \end{pmatrix}$ to the diagonal form.

**Solution:** Let $\lambda$ be the eigenvalue of $A$.

The characteristic equation of $A$ is given by $|A - \lambda I| = 0$

$$\begin{vmatrix} -1-\lambda & 2 & -2 \\ 1 & 2-\lambda & 1 \\ -1 & -1 & -\lambda \end{vmatrix}$$

$\Rightarrow \lambda^3 - \lambda^2 - 5\lambda + 5 = 0$

$\Rightarrow (\lambda - 1)(\lambda^2 - 5) = 0$

$\Rightarrow \lambda = 1, \pm\sqrt{5}$.

$\therefore$ Eigenvalues of $A$ are $\lambda = 1, \sqrt{5}, -\sqrt{5}$.

**To find eigenvectors:**

Let $X = \begin{pmatrix} x_1 \\ x_2 \\ x_3 \end{pmatrix}$ be the eigenvector of $A$ corresponding to $\lambda$.

Then

$$AX = \lambda X$$
$$\Rightarrow (A - \lambda I)X = 0$$

$$\Rightarrow \begin{pmatrix} -1-\lambda & 2 & -2 \\ 1 & 2-\lambda & 1 \\ -1 & -1 & -\lambda \end{pmatrix} \begin{pmatrix} x_1 \\ x_2 \\ x_3 \end{pmatrix} = \begin{pmatrix} 0 \\ 0 \\ 0 \end{pmatrix} \quad \ldots\ldots \quad (1)$$

**Case 1:** To find eigenvector when $\lambda = 1$.
Put $\lambda = 1$ in (1), we have

$$\begin{pmatrix} -2 & 2 & -2 \\ 1 & 1 & 1 \\ -1 & -1 & -1 \end{pmatrix} \begin{pmatrix} x_1 \\ x_2 \\ x_3 \end{pmatrix} = \begin{pmatrix} 0 \\ 0 \\ 0 \end{pmatrix}$$

This gives,

$$-2x_1 + 2x_2 - 2x_3 = 0$$
$$x_1 + x_2 + x_3 = 0$$
$$-x_1 - x_2 - x_3 = 0$$

Solving first two equations by rules of cross multiplication, we have

$$\frac{x_1}{2+2} = \frac{x_2}{-2+2} = \frac{x_3}{-2-2} = k_1 \quad (k_1 \neq 0)$$

$$\frac{x_1}{1} = \frac{x_2}{0} = \frac{x_3}{-1} = k_1$$

$$\therefore x_1 = k_1, x_2 = 0, x_3 = -k_1.$$

The eigenvector corresponding to $\lambda = 0$ is $X_1 = \begin{pmatrix} k_1 \\ 0 \\ -k_1 \end{pmatrix}$.

When $k_1 = 1$, we have $X_1 = \begin{pmatrix} 1 \\ 0 \\ -1 \end{pmatrix}$.

**Case 2:** To find eigenvector when $\lambda = \sqrt{5}$.

Put $\lambda = \sqrt{5}$ in (1), we have

$$\begin{pmatrix} -1-\sqrt{5} & 2 & -2 \\ 1 & 2-\sqrt{5} & 1 \\ -1 & -1 & -\sqrt{5} \end{pmatrix} \begin{pmatrix} x_1 \\ x_2 \\ x_3 \end{pmatrix} = \begin{pmatrix} 0 \\ 0 \\ 0 \end{pmatrix}$$

This gives,

$$(-1-\sqrt{5})x_1 + 2x_2 - 2x_3 = 0$$
$$x_1 + (2-\sqrt{5})x_2 + x_3 = 0$$
$$-x_1 - x_2 - \sqrt{5}x_3 = 0$$

Solving last two equations by cross multiplication, we have

$$\frac{x_1}{6-2\sqrt{5}} = \frac{x_2}{-1+\sqrt{5}} = \frac{x_3}{-1+2-\sqrt{5}} = k_2 \quad (k_2 \neq 0)$$

$$\frac{x_1}{(\sqrt{5}-1)^2} = \frac{x_2}{\sqrt{5}-1} = \frac{x_3}{-(\sqrt{5}-1)} = k_2$$

$$\frac{x_1}{\sqrt{5}-1} = \frac{x_2}{1} = \frac{x_3}{-1} = k_2$$

$$\therefore x_1 = (\sqrt{5}-1)k_2, x_2 = k_2, x_3 = -k_2.$$

The eigenvector corresponding to $\lambda = \sqrt{5}$ is $X_2 = \begin{pmatrix} (\sqrt{5}-1)k_2 \\ k_2 \\ -k_2 \end{pmatrix}$.

When $k_2 = 1$, the eigenvector is $X_2 = \begin{pmatrix} \sqrt{5}-1 \\ 1 \\ -1 \end{pmatrix}$.

**Case 3:** To find eigenvector when $\lambda = -\sqrt{5}$.

Put $\lambda = -\sqrt{5}$ in (1), we have

$$\begin{pmatrix} -1+\sqrt{5} & 2 & -2 \\ 1 & 2+\sqrt{5} & 1 \\ -1 & -1 & \sqrt{5} \end{pmatrix} \begin{pmatrix} x_1 \\ x_2 \\ x_3 \end{pmatrix} = \begin{pmatrix} 0 \\ 0 \\ 0 \end{pmatrix}$$

This gives,

$$(-1+\sqrt{5})x_1 + 2x_2 - 2x_3 = 0$$
$$x_1 + (2+\sqrt{5})x_2 + x_3 = 0$$

$$-x_1 - x_2 + \sqrt{5}x_3 = 0$$

Solving last two equations by cross multiplication, we have

$$\frac{x_1}{6+2\sqrt{5}} = \frac{x_2}{-1-\sqrt{5}} = \frac{x_3}{-1+2+\sqrt{5}} = k_3 \quad (k_3 \neq 0)$$

$$\frac{x_1}{(\sqrt{5}+1)^2} = \frac{x_2}{-(\sqrt{5}+1)} = \frac{x_3}{\sqrt{5}+1} = k_3$$

$$\frac{x_1}{\sqrt{5}+1} = \frac{x_2}{-1} = \frac{x_3}{1} = k_3$$

$$\therefore x_1 = (\sqrt{5}+1)k_3, \; x_2 = -k_3, \; x_3 = k_3.$$

The eigenvector corresponding to $\lambda = -\sqrt{5}$ is $X_2 = \begin{pmatrix} (\sqrt{5}+1)k_2 \\ -k_2 \\ k_2 \end{pmatrix}$.

When $k_2 = 1$, the eigenvector is $X_2 = \begin{pmatrix} \sqrt{5}+1 \\ -1 \\ 1 \end{pmatrix}$.

$\therefore$ The modal matrix is given by

$$P = \begin{pmatrix} 1 & \sqrt{5}-1 & \sqrt{5}+1 \\ 0 & 1 & -1 \\ 1 & -1 & 1 \end{pmatrix}$$

and

$$D = P^{-1}AP = \begin{pmatrix} \lambda_1 & 0 & 0 \\ 0 & \lambda_2 & 0 \\ 0 & 0 & \lambda_3 \end{pmatrix} = \begin{pmatrix} 1 & 0 & 0 \\ 0 & \sqrt{5} & 0 \\ 0 & 0 & -\sqrt{5} \end{pmatrix}.$$

**Example:** Diagonalize the matrix $A = \begin{pmatrix} 1 & 0 & -1 \\ 1 & 2 & 1 \\ 2 & 2 & 3 \end{pmatrix}$ and find $A^4$ using the modal matrix.

(Nov-2017), (Nov-2021)

**Solution:** Let $\lambda$ be the eigenvalue of $A$.

The characteristic equation of $A$ is given by $|A - \lambda I| = 0$.

$$\Rightarrow \lambda^3 - 6\lambda^2 + 11\lambda - 6 = 0.$$

$$\Rightarrow \lambda = 1, 2, 3.$$

∴ Eigenvalues of $A$ are $\lambda = 1, 2, 3$ which are real and distinct. Hence, $A$ is diagonalizable.

**To find eigenvectors:**

Let $X = \begin{pmatrix} x_1 \\ x_2 \\ x_3 \end{pmatrix}$ be the eigenvector of $A$ corresponding to $\lambda$.

Then
$$AX = \lambda X$$
$$\Rightarrow (A - \lambda I)X = 0$$
$$\Rightarrow \begin{pmatrix} 1-\lambda & 0 & -1 \\ 1 & 2-\lambda & 1 \\ 2 & 2 & 3-\lambda \end{pmatrix} \begin{pmatrix} x_1 \\ x_2 \\ x_3 \end{pmatrix} = \begin{pmatrix} 0 \\ 0 \\ 0 \end{pmatrix} \quad \cdots\cdots \quad (1)$$

**Case 1:** To find eigenvector when $\lambda = 1$.

Put $\lambda = 1$ in (1), we have

$$\begin{pmatrix} 0 & 0 & -1 \\ 1 & 1 & 1 \\ 2 & 2 & 2 \end{pmatrix} \begin{pmatrix} x_1 \\ x_2 \\ x_3 \end{pmatrix} = \begin{pmatrix} 0 \\ 0 \\ 0 \end{pmatrix}$$

This gives,
$$-x_3 = 0$$
$$x_1 + x_2 + x_3 = 0$$
$$2x_1 + 2x_2 + 2x_3 = 0$$
$$\therefore x_3 = 0$$
$$x_1 + x_2 = 0 \Rightarrow x_1 = -x_2$$
$$\therefore x_1 = 1, x_2 = -1, x_3 = 0.$$

The eigenvector corresponding to $\lambda = 1$ is $X_1 = \begin{pmatrix} 1 \\ -1 \\ 0 \end{pmatrix}$.

**Case 2:** To find eigenvector when $\lambda = 2$.

Put $\lambda = 2$ in (1), we have

$$\begin{pmatrix} -1 & 0 & -1 \\ 1 & 0 & 1 \\ 2 & 2 & 1 \end{pmatrix} \begin{pmatrix} x_1 \\ x_2 \\ x_3 \end{pmatrix} = \begin{pmatrix} 0 \\ 0 \\ 0 \end{pmatrix}$$

This gives,

$$-x_1 - x_3 = 0$$
$$x_1 + x_3 = 0$$
$$2x_1 + 2x_2 + x_3 = 0$$

Solving last two equations by cross multiplication, we have

$$\frac{x_1}{0-2} = \frac{x_2}{2-1} = \frac{x_3}{2-0}$$

$$\frac{x_1}{-2} = \frac{x_2}{1} = \frac{x_3}{2}$$

$$\therefore x_1 = -2, x_2 = 1, x_3 = 2.$$

The eigenvector corresponding to $\lambda = 2$ is $X_2 = \begin{pmatrix} -2 \\ 1 \\ 2 \end{pmatrix}$.

**Case 3:** To find eigenvector when $\lambda = 3$.

Put $\lambda = 3$ in (1), we have

$$\begin{pmatrix} -2 & 0 & -1 \\ 1 & -1 & 1 \\ 2 & 2 & 0 \end{pmatrix} \begin{pmatrix} x_1 \\ x_2 \\ x_3 \end{pmatrix} = \begin{pmatrix} 0 \\ 0 \\ 0 \end{pmatrix}$$

This gives,

$$-2x_1 - x_3 = 0$$
$$x_1 - x_2 + x_3 = 0$$
$$2x_1 + 2x_2 = 0$$

Solving last two equations by cross multiplication, we have

$$\frac{x_1}{0-2} = \frac{x_2}{2-0} = \frac{x_3}{2+2}$$

$$\frac{x_1}{-2} = \frac{x_2}{2} = \frac{x_3}{4}$$

$$\therefore x_1 = -1, x_2 = 1, x_3 = 2.$$

The eigenvector corresponding to $\lambda = 3$ is $X_3 = \begin{pmatrix} -1 \\ 1 \\ 2 \end{pmatrix}$.

**To find Modal Matrix:**

Now, the modal matrix $P = \begin{bmatrix} X_1 & X_2 & X_3 \end{bmatrix}$

$$= \begin{pmatrix} 1 & -2 & -1 \\ -1 & 1 & 1 \\ 0 & 2 & 2 \end{pmatrix}.$$

**To obtain $P^{-1}$:** $|P| = 1(2-2) + 2(-2-0) - 1(-2-0) = -4 + 2 = -2 \neq 0$.

$$\therefore P^{-1} = \frac{-1}{2} \begin{pmatrix} 0 & 2 & -1 \\ 2 & 2 & 0 \\ -2 & -2 & -1 \end{pmatrix}.$$

**Diagonalization:**

$D = P^{-1}AP$

$$= \frac{-1}{2} \begin{pmatrix} 0 & 2 & -1 \\ 2 & 2 & 0 \\ -2 & -2 & -1 \end{pmatrix} \begin{pmatrix} 1 & 0 & -1 \\ 1 & 2 & 1 \\ 2 & 2 & 3 \end{pmatrix} \begin{pmatrix} 1 & -2 & -1 \\ -1 & 1 & 1 \\ 0 & 2 & 2 \end{pmatrix}$$

$$= \begin{pmatrix} 1 & 0 & 0 \\ 0 & 2 & 0 \\ 0 & 0 & 3 \end{pmatrix}$$

whose principle diagonal elements are the eigenvalues of given matrix $A$.

**To find $A^4$:** We know that

$$A^4 = PD^4P^{-1}$$

$$= \begin{pmatrix} 1 & -2 & -1 \\ -1 & 1 & 1 \\ 0 & 2 & 2 \end{pmatrix} \begin{pmatrix} 1^4 & 0 & 0 \\ 0 & 2^4 & 0 \\ 0 & 0 & 3^4 \end{pmatrix} \frac{-1}{2} \begin{pmatrix} 0 & 2 & -1 \\ 2 & 2 & 0 \\ -2 & -2 & -1 \end{pmatrix}$$

$$= \begin{pmatrix} -49 & -50 & -40 \\ 65 & 66 & 40 \\ 130 & 130 & 81 \end{pmatrix}$$

**Example:** Diagonalize the matrix $A = \begin{pmatrix} 3 & -1 & 1 \\ -1 & 5 & -1 \\ 1 & -1 & 3 \end{pmatrix}$ and find $A^4$ using the modal matrix.

**(Apr-2018), (Aug-2022)**

**Solution:** Let $\lambda$ be the eigenvalue of $A$.

The characteristic equation of $A$ is given by $|A - \lambda I| = 0$.

$$\Rightarrow \lambda^3 - 11\lambda^2 + 36\lambda - 36 = 0.$$

$$\Rightarrow \lambda = 2, 3, 6.$$

$\therefore$ Eigenvalues of $A$ are $\lambda = 2, 3, 6$ which are real and distinct. Hence, $A$ is diagonalizable.

**To find eigenvectors:**

Let $X = \begin{pmatrix} x_1 \\ x_2 \\ x_3 \end{pmatrix}$ be the eigenvector of $A$ corresponding to $\lambda$.

Then

$$AX = \lambda X$$

$$\Rightarrow (A - \lambda I)X = 0$$

$$\Rightarrow \begin{pmatrix} 3-\lambda & -1 & 1 \\ -1 & 5-\lambda & -1 \\ 1 & -1 & 3-\lambda \end{pmatrix} \begin{pmatrix} x_1 \\ x_2 \\ x_3 \end{pmatrix} = \begin{pmatrix} 0 \\ 0 \\ 0 \end{pmatrix} \quad \cdots \cdots \quad (1)$$

**Case 1:** To find eigenvector when $\lambda = 2$.

Put $\lambda = 2$ in (1), we have

$$\begin{pmatrix} 1 & -1 & 1 \\ -1 & 3 & -1 \\ 1 & -1 & 1 \end{pmatrix} \begin{pmatrix} x_1 \\ x_2 \\ x_3 \end{pmatrix} = \begin{pmatrix} 0 \\ 0 \\ 0 \end{pmatrix}$$

This gives,
$$x_1 - x_2 + x_3 = 0$$
$$-x_1 + 3x_2 - x_3 = 0$$
$$x_1 - x_2 + x_3 = 0$$

Solving first two equations by cross multiplication, we have

$$\frac{x_1}{1-3} = \frac{x_2}{-1+1} = \frac{x_3}{3-1}$$

$$\frac{x_1}{-2} = \frac{x_2}{0} = \frac{x_3}{2}$$

$$\therefore x_1 = -1, x_2 = 0, x_3 = 1.$$

The eigenvector corresponding to $\lambda = 2$ is $X_1 = \begin{pmatrix} -1 \\ 0 \\ 1 \end{pmatrix}$.

**Case 2:** To find eigenvector when $\lambda = 3$.

Put $\lambda = 3$ in (1), we have

$$\begin{pmatrix} 0 & -1 & 1 \\ -1 & 2 & -1 \\ 1 & -1 & 0 \end{pmatrix} \begin{pmatrix} x_1 \\ x_2 \\ x_3 \end{pmatrix} = \begin{pmatrix} 0 \\ 0 \\ 0 \end{pmatrix}$$

This gives,
$$-x_2 + x_3 = 0$$
$$-x_1 + 2x_2 - x_3 = 0$$
$$x_1 - x_2 = 0$$
$$\therefore x_1 = x_2 = x_3 = 1.$$

The eigenvector corresponding to $\lambda = 3$ is $X_2 = \begin{pmatrix} 1 \\ 1 \\ 1 \end{pmatrix}$.

**Case 3:** To find eigenvector when $\lambda = 6$.

Put $\lambda = 6$ in (1), we have

$$\begin{pmatrix} -3 & -1 & 1 \\ -1 & -1 & -1 \\ 1 & -1 & -3 \end{pmatrix} \begin{pmatrix} x_1 \\ x_2 \\ x_3 \end{pmatrix} = \begin{pmatrix} 0 \\ 0 \\ 0 \end{pmatrix}$$

This gives,

$$-3x_1 - x_2 - x_3 = 0$$
$$-x_1 - x_2 - x_3 = 0$$
$$x_1 - x_2 - 3x_3 = 0$$

Solving last two equations by cross multiplication, we have

$$\frac{x_1}{3-1} = \frac{x_2}{-1-3} = \frac{x_3}{1+1}$$

$$\frac{x_1}{2} = \frac{x_2}{-4} = \frac{x_3}{2}$$

$$\therefore x_1 = 1, x_2 = -2, x_3 = 1.$$

The eigenvector corresponding to $\lambda = 6$ is $X_3 = \begin{pmatrix} 1 \\ -2 \\ 1 \end{pmatrix}$.

**To find Modal Matrix:**

Now, the modal matrix $P = \begin{bmatrix} X_1 & X_2 & X_3 \end{bmatrix}$

$$= \begin{pmatrix} -1 & 1 & 1 \\ 0 & 1 & -2 \\ 1 & 1 & 1 \end{pmatrix}.$$

**To obtain $P^{-1}$:** $|P| = -1(1+2) - 1(0+2) + 1(0-1) = -3 - 2 - 1 = -6 \neq 0$.

$$\therefore P^{-1} = \frac{-1}{6} \begin{pmatrix} 3 & 0 & -3 \\ -2 & -2 & -2 \\ -1 & 2 & -1 \end{pmatrix}.$$

## Diagonalization:

$$D = P^{-1}AP$$

$$= \frac{-1}{6}\begin{pmatrix} 3 & 0 & -3 \\ -2 & -2 & -2 \\ -1 & 2 & -1 \end{pmatrix}\begin{pmatrix} 3 & -1 & 1 \\ -1 & 5 & -1 \\ 1 & -1 & 3 \end{pmatrix}\begin{pmatrix} -1 & 1 & 1 \\ 0 & 1 & -2 \\ 1 & 1 & 1 \end{pmatrix}$$

$$= \begin{pmatrix} 2 & 0 & 0 \\ 0 & 3 & 0 \\ 0 & 0 & 6 \end{pmatrix}$$

whose principle diagonal elements are the eigenvalues of given matrix $A$.

## To find $A^4$:

We know that

$$A^4 = PD^4P^{-1}$$

$$= \begin{pmatrix} -1 & 1 & 1 \\ 0 & 1 & -2 \\ 1 & 1 & 1 \end{pmatrix}\begin{pmatrix} 2^4 & 0 & 0 \\ 0 & 3^4 & 0 \\ 0 & 0 & 6^4 \end{pmatrix}\frac{-1}{2}\begin{pmatrix} 3 & 0 & -3 \\ -2 & -2 & -2 \\ -1 & 2 & -1 \end{pmatrix}$$

$$= \begin{pmatrix} 251 & -405 & 235 \\ -405 & 891 & -405 \\ 235 & -405 & 251 \end{pmatrix}$$

**Example:** Find the matrix $P$ which transform the matrix $A = \begin{pmatrix} 1 & 1 & 3 \\ 1 & 5 & 1 \\ 3 & 1 & 1 \end{pmatrix}$ to the diagonal form. Hence find $A^4$.

**Solution:** Let $\lambda$ be the eigenvalue of $A$.

The characteristic equation of $A$ is given by $|A - \lambda I| = 0$.

$$\Rightarrow \lambda^3 - 7\lambda^2 + 36 = 0.$$

$$\Rightarrow \lambda = -2, 3, 6.$$

$\therefore$ Eigenvalues of $A$ are $\lambda = -2, 3, 6$ which are real and distinct. Hence, $A$ is diagonalizable.

**To find eigenvectors:**

Let $X = \begin{pmatrix} x_1 \\ x_2 \\ x_3 \end{pmatrix}$ be the eigenvector of $A$ corresponding to $\lambda$.

Then
$$AX = \lambda X$$
$$\Rightarrow (A - \lambda I)X = 0$$

$$\Rightarrow \begin{pmatrix} 1-\lambda & 1 & 3 \\ 1 & 5-\lambda & 1 \\ 3 & 1 & 1-\lambda \end{pmatrix} \begin{pmatrix} x_1 \\ x_2 \\ x_3 \end{pmatrix} = \begin{pmatrix} 0 \\ 0 \\ 0 \end{pmatrix} \quad \cdots\cdots \quad (1)$$

**Case 1:** To find eigenvector when $\lambda = -2$.

Put $\lambda = -2$ in (1), we have

$$\begin{pmatrix} 3 & 1 & 3 \\ 1 & 7 & 1 \\ 3 & 1 & 3 \end{pmatrix} \begin{pmatrix} x_1 \\ x_2 \\ x_3 \end{pmatrix} = \begin{pmatrix} 0 \\ 0 \\ 0 \end{pmatrix}$$

This gives,
$$3x_1 + x_2 + 3x_3 = 0$$
$$x_1 + 7x_2 + x_3 = 0$$
$$3x_1 + x_2 + 3x_3 = 0$$

Solving first two equations by cross multiplication, we have

$$\frac{x_1}{1-21} = \frac{x_2}{3-3} = \frac{x_3}{21-1}$$

$$\frac{x_1}{-20} = \frac{x_2}{0} = \frac{x_3}{20}$$

$$\therefore x_1 = -1, x_2 = 0, x_3 = 1.$$

The eigenvector corresponding to $\lambda = -2$ is $X_1 = \begin{pmatrix} -1 \\ 0 \\ 1 \end{pmatrix}$.

**Case 2:** To find eigenvector when $\lambda = 3$.

Put $\lambda = 3$ in (1), we have

$$\begin{pmatrix} -2 & 1 & 3 \\ 1 & 2 & 1 \\ 3 & 1 & -2 \end{pmatrix} \begin{pmatrix} x_1 \\ x_2 \\ x_3 \end{pmatrix} = \begin{pmatrix} 0 \\ 0 \\ 0 \end{pmatrix}$$

This gives,

$$-2x_1 + x_2 + 3x_3 = 0$$

$$x_1 + 2x_2 + x_3 = 0$$

$$3x_1 + x_2 - 2x_3 = 0$$

Solving first two equations by cross multiplication, we have

$$\frac{x_1}{1-6} = \frac{x_2}{3+2} = \frac{x_3}{-4-1}$$

$$\frac{x_1}{-5} = \frac{x_2}{5} = \frac{x_3}{-5}$$

$$\therefore x_1 = 1, x_2 = -1, x_3 = 1.$$

The eigenvector corresponding to $\lambda = 3$ is $X_2 = \begin{pmatrix} 1 \\ -1 \\ 1 \end{pmatrix}$.

**Case 3:** To find eigenvector when $\lambda = 6$.

Put $\lambda = 6$ in (1), we have

$$\begin{pmatrix} -5 & 1 & 3 \\ 1 & -1 & 1 \\ 3 & 1 & -5 \end{pmatrix} \begin{pmatrix} x_1 \\ x_2 \\ x_3 \end{pmatrix} = \begin{pmatrix} 0 \\ 0 \\ 0 \end{pmatrix}$$

This gives,

$$-5x_1 + x_2 + 3x_3 = 0$$

$$x_1 - x_2 + x_3 = 0$$

$$3x_1 + x_2 - 5x_3 = 0$$

Solving last two equations by cross multiplication, we have

$$\frac{x_1}{5-1} = \frac{x_2}{3+5} = \frac{x_3}{1+3}$$

$$\frac{x_1}{4} = \frac{x_2}{8} = \frac{x_3}{4}$$

$$\therefore x_1 = 1, x_2 = 2, x_3 = 1.$$

The eigenvector corresponding to $\lambda = 6$ is $X_3 = \begin{pmatrix} 1 \\ 2 \\ 1 \end{pmatrix}$.

**To find Modal Matrix:**

Now, the modal matrix $P = \begin{bmatrix} X_1 & X_2 & X_3 \end{bmatrix}$

$$= \begin{pmatrix} -1 & 1 & 1 \\ 0 & -1 & 2 \\ 1 & 1 & 1 \end{pmatrix}.$$

**To obtain $P^{-1}$:** $|P| = -1(-1-2) - 1(0-2) + 1(0+1) = 3 + 2 + 1 = 6 \neq 0$.

$$\therefore P^{-1} = \frac{1}{6} \begin{pmatrix} -3 & 0 & 3 \\ 2 & -2 & 2 \\ 1 & 2 & 1 \end{pmatrix}.$$

**Diagonalization:**

$$D = P^{-1}AP$$

$$= \frac{1}{6} \begin{pmatrix} -3 & 0 & 3 \\ 2 & -2 & 2 \\ 1 & 2 & 1 \end{pmatrix} \begin{pmatrix} 1 & 1 & 3 \\ 1 & 5 & 1 \\ 3 & 1 & 1 \end{pmatrix} \begin{pmatrix} -1 & 1 & 1 \\ 0 & -1 & 2 \\ 1 & 1 & 1 \end{pmatrix}$$

$$= \begin{pmatrix} -2 & 0 & 0 \\ 0 & 3 & 0 \\ 0 & 0 & 6 \end{pmatrix}$$

whose principle diagonal elements are the eigenvalues of given matrix $A$.

**To find $A^4$:**

We know that
$$A^4 = PD^4P^{-1}$$

$$= \begin{pmatrix} -1 & 1 & 1 \\ 0 & -1 & 2 \\ 1 & 1 & 1 \end{pmatrix} \begin{pmatrix} (-2)^4 & 0 & 0 \\ 0 & 3^4 & 0 \\ 0 & 0 & 6^4 \end{pmatrix} \frac{1}{6} \begin{pmatrix} -3 & 0 & 3 \\ 2 & -2 & 2 \\ 1 & 2 & 1 \end{pmatrix}$$

$$= \begin{pmatrix} 251 & 485 & 235 \\ 485 & 1051 & 485 \\ 235 & 485 & 251 \end{pmatrix}$$

**Example:** Diagonalize the matrix $A = \begin{pmatrix} 1 & 1 & 1 \\ 1 & 1 & 1 \\ 1 & 1 & 1 \end{pmatrix}$ and find $A^4$ using the modal matrix.

**Solution:** The characteristic equation of $A$ is $|A - \lambda I| = 0$.

$$\Rightarrow \begin{vmatrix} 1-\lambda & 1 & 1 \\ 1 & 1-\lambda & 1 \\ 1 & 1 & 1-\lambda \end{vmatrix} = 0$$

$$\Rightarrow \lambda^3 - 3\lambda^2 = 0$$

$$\Rightarrow \lambda = 0, 0, 3.$$

The eigenvalues of $A$ are 0, 0, 3.

**To find eigenvectors:**

Let $X = \begin{pmatrix} x_1 \\ x_2 \\ x_3 \end{pmatrix}$ be the eigenvector of $A$ corresponding to $\lambda$.

Then
$$AX = \lambda X$$
$$\Rightarrow (A - \lambda I)X = 0$$

$$\Rightarrow \begin{pmatrix} 1-\lambda & 1 & 1 \\ 1 & 1-\lambda & 1 \\ 1 & 1 & 1-\lambda \end{pmatrix} \begin{pmatrix} x_1 \\ x_2 \\ x_3 \end{pmatrix} = \begin{pmatrix} 0 \\ 0 \\ 0 \end{pmatrix} \quad \cdots \cdots \quad (1)$$

**Case 1:** To find eigenvector when $\lambda = 0$.

Put $\lambda = 0$ in (1), we have

$$\begin{pmatrix} 1 & 1 & 1 \\ 1 & 1 & 1 \\ 1 & 1 & 1 \end{pmatrix} \begin{pmatrix} x_1 \\ x_2 \\ x_3 \end{pmatrix} = \begin{pmatrix} 0 \\ 0 \\ 0 \end{pmatrix}$$

This gives,

$$x_1 + x_2 + x_3 = 0.$$

Let $x_2 = k_1, x_3 = k_2$ then $x_1 = -k_1 - k_2$.

$$X = \begin{pmatrix} -k_1 - k_2 \\ k_1 \\ k_2 \end{pmatrix} = k_1 \begin{pmatrix} -1 \\ 1 \\ 0 \end{pmatrix} + k_2 \begin{pmatrix} -1 \\ 0 \\ 1 \end{pmatrix}.$$

Let $k_1 = 1, k_2 = 0$, then the eigenvector is $X_1 = \begin{pmatrix} -1 \\ 1 \\ 0 \end{pmatrix}$ and letting $k_2 = 1, k_1 = 0$,

then the eigenvector is $X_2 = \begin{pmatrix} -1 \\ 0 \\ 1 \end{pmatrix}$.

**Case 2:** To find eigenvector when $\lambda = 3$.

Put $\lambda = 3$ in (1), we have

$$\begin{pmatrix} -2 & 1 & 1 \\ 1 & -2 & 1 \\ 1 & 1 & -2 \end{pmatrix} \begin{pmatrix} x_1 \\ x_2 \\ x_3 \end{pmatrix} = \begin{pmatrix} 0 \\ 0 \\ 0 \end{pmatrix}$$

This gives,

$$-2x_1 + x_2 + x_3 = 0$$
$$x_1 - 2x_2 + x_3 = 0$$
$$x_1 + x_2 - 2x_3 = 0$$

Solving first two equations by cross multiplication, we have

$$\frac{x_1}{1+2} = \frac{x_2}{1=2} = \frac{x_3}{4-1}$$

$$\frac{x_1}{3} = \frac{x_2}{3} = \frac{x_3}{3}$$
$$\therefore x_1 = 1, x_2 = 1, x_3 = 1.$$

The eigenvector corresponding to $\lambda = 3$ is $X_3 = \begin{pmatrix} 1 \\ 1 \\ 1 \end{pmatrix}$.

The three linearly independent eigenvectors are $X_1, X_2, X_3$ and the matrix $A$ is diagonalizable.

**To find Modal Matrix:**

Now, the modal matrix $P = \begin{bmatrix} X_1 & X_2 & X_3 \end{bmatrix}$
$$= \begin{pmatrix} -1 & -1 & 1 \\ 1 & 0 & 1 \\ 0 & 1 & 1 \end{pmatrix}$$

**To obtain $P^{-1}$:** $P = -1(0-1) + (1-0) + 1(1-0) = 3.$

$$\therefore P^{-1} = \frac{1}{3} \begin{pmatrix} -1 & 2 & -1 \\ -1 & -1 & 2 \\ 1 & 1 & 1 \end{pmatrix}.$$

**Diagonalization:**

$$D = P^{-1}AP = \frac{1}{3} \begin{pmatrix} -1 & 2 & -1 \\ -1 & -1 & 2 \\ 1 & 1 & 1 \end{pmatrix} \begin{pmatrix} 1 & 1 & 1 \\ 1 & 1 & 1 \\ 1 & 1 & 1 \end{pmatrix} \begin{pmatrix} -1 & -1 & 1 \\ 1 & 0 & 1 \\ 0 & 1 & 1 \end{pmatrix} = \begin{pmatrix} 0 & 0 & 0 \\ 0 & 0 & 0 \\ 0 & 0 & 3 \end{pmatrix}$$

whose principle diagonal elements are the eigenvalues of given matrix $A$.

**To find $A^4$:**

We know that

$$A^4 = PD^4P^{-1}$$

$$= \begin{pmatrix} -1 & -1 & 1 \\ 1 & 0 & 1 \\ 0 & 1 & 1 \end{pmatrix} \begin{pmatrix} 0^4 & 0 & 0 \\ 0 & 0^4 & 0 \\ 0 & 0 & 3^4 \end{pmatrix} \frac{1}{3} \begin{pmatrix} -1 & 2 & -1 \\ -1 & -1 & 2 \\ 1 & 1 & 1 \end{pmatrix}$$

$$= \begin{pmatrix} 27 & 27 & 27 \\ 27 & 27 & 27 \\ 27 & 27 & 27 \end{pmatrix}.$$

**Example:** Is the matrix $\begin{pmatrix} 3 & 10 & 5 \\ -2 & -3 & -4 \\ 3 & 5 & 7 \end{pmatrix}$ diagonalizable?

**Solution:** Let $\lambda$ be the eigenvalue of $A$.
The characteristic equation of $A$ is given by $|A - \lambda I| = 0$.

$$\Rightarrow \lambda^3 - 7\lambda^2 + 16\lambda - 12 = 0.$$

$$\Rightarrow \lambda = 2, 2, 3.$$

$\therefore$ Eigenvalues of $A$ are $\lambda = 2, 2, 3$ which are real but two of them are equal.

So if we find two linearly independent eigenvectors corresponding to $\lambda = 2$, then only the matrix $A$ is diagonalizable.

**To find eigenvectors:**

Let $X = \begin{pmatrix} x_1 \\ x_2 \\ x_3 \end{pmatrix}$ be the eigenvector of $A$ corresponding to $\lambda$.

Then

$$AX = \lambda X$$

$$\Rightarrow (A - \lambda I)X = 0$$

$$\Rightarrow \begin{pmatrix} 3-\lambda & 10 & 5 \\ -2 & -3-\lambda & -4 \\ 3 & 5 & 7-\lambda \end{pmatrix} \begin{pmatrix} x_1 \\ x_2 \\ x_3 \end{pmatrix} = \begin{pmatrix} 0 \\ 0 \\ 0 \end{pmatrix} \quad \cdots \cdots \quad (1)$$

**Case 1:** To find eigenvector when $\lambda = 3$.

Put $\lambda = 3$ in (1), we have

$$\begin{pmatrix} 0 & 10 & 5 \\ -2 & -6 & -4 \\ 3 & 5 & 4 \end{pmatrix} \begin{pmatrix} x_1 \\ x_2 \\ x_3 \end{pmatrix} = \begin{pmatrix} 0 \\ 0 \\ 0 \end{pmatrix}$$

This gives,

$$10x_2 + 5x_3 = 0$$
$$-2x_1 - 6x_2 - 4x_3 = 0$$
$$3x_1 + 5x_2 + 4x_3 = 0$$

Solving first two equations by cross multiplication, we have

$$\frac{x_1}{-40+30} = \frac{x_2}{-10-0} = \frac{x_3}{0+20}$$

$$\frac{x_1}{-10} = \frac{x_2}{-10} = \frac{x_3}{20}$$

$$\therefore x_1 = -1, x_2 = -1, x_3 = 2.$$

The eigenvector corresponding to $\lambda = 3$ is $X_2 = \begin{pmatrix} -1 \\ -1 \\ 2 \end{pmatrix}$.

**Case 2:** To find eigenvector when $\lambda = 2$.

Put $\lambda = 2$ in (1), we have

$$\begin{pmatrix} 1 & 10 & 5 \\ -2 & -5 & -4 \\ 3 & 5 & 5 \end{pmatrix} \begin{pmatrix} x_1 \\ x_2 \\ x_3 \end{pmatrix} = \begin{pmatrix} 0 \\ 0 \\ 0 \end{pmatrix}$$

This gives,

$$x_1 + 10x_2 + 5x_3 = 0$$
$$-2x_1 - 5x_2 - 4x_3 = 0$$
$$3x_1 + 5x_2 + 5x_3 = 0$$

Solving first two equations by cross multiplication, we have

$$\frac{x_1}{1-3} = \frac{x_2}{-1+1} = \frac{x_3}{3-1}$$

$$\frac{x_1}{-2} = \frac{x_2}{0} = \frac{x_3}{2}$$

$$\therefore x_1 = -1, x_2 = 0, x_3 = 1.$$

The eigenvector corresponding to $\lambda = 2$ is $X_1 = \begin{pmatrix} -1 \\ 0 \\ 1 \end{pmatrix}$.

There is only one linearly independent eigenvector corresponding to repeated eigenvalue $\lambda = 2$. Since the total number of independent eigenvector is two. Therefore the matrix is not diagonalizable.

**EXERCISE**

1. Diagonalize the matrix $\begin{pmatrix} 1 & 2 & 3 \\ 0 & 3 & 1 \\ 0 & 0 & 1 \end{pmatrix}$ if possible.  (Apr-2018), (Mar-2022)

**Solution:** Eigenvalues are $1, 1, 3$. For $\lambda = 1$, the eigen vector is $\begin{pmatrix} 1 \\ 0 \\ 0 \end{pmatrix}$.

The algebraic multiplicity of eigenvalue $\lambda = 1$ is 2 and geometric multiplicity of the eigenvalue $\lambda = 1$ is 1. Since algebraic multiplicity$\neq$Geometric multiplicity. The matrix cannot be diagolizable.

2. Diagonalize the matrix $\begin{pmatrix} 1 & 2 \\ 2 & 1 \end{pmatrix}$  (Dec-2021)

3. Diagonalize the matrix $\begin{pmatrix} 4 & 6 \\ 6 & -1 \end{pmatrix}$  (Aug-2022)

4. Diagonalize the matrix $\begin{pmatrix} 5 & 4 \\ 1 & 2 \end{pmatrix}$  (Aug-2022)

5. Diagonalize the matrix $\begin{pmatrix} 3 & 1 \\ 1 & 3 \end{pmatrix}$  (Aug-2022)

6. Diagonalize the matrix $\begin{pmatrix} -1 & 2 \\ 2 & -1 \end{pmatrix}$ Hence find $A^6$.

**Ans:** $\lambda = 1, -3$. $A^6 = \begin{pmatrix} 365 & -364 \\ -364 & 365 \end{pmatrix}$

7. Find the matrix $P$ that reduces the matrix $\begin{pmatrix} 2 & 2 & 0 \\ 2 & 5 & 0 \\ 0 & 0 & 3 \end{pmatrix}$ to the diagonal form. Hence find $A^4$.

**Ans:** $\lambda = 1, 2, 3$. $A^4 = \begin{pmatrix} 260 & 518 & 0 \\ 518 & 1037 & 0 \\ 0 & 0 & 81 \end{pmatrix}$.

8. Diagonalize the matrix $\begin{pmatrix} 1 & 1 & 1 \\ 0 & 2 & 1 \\ -4 & 4 & 3 \end{pmatrix}$. Hence find $A^5$. **(Jan-2020)**

**Ans:** $\lambda = 1, 2, 3$. $A^5 = \begin{pmatrix} -359 & 391 & 211 \\ -360 & 392 & 211 \\ -484 & 484 & 243 \end{pmatrix}$.

8. Diagonalize the matrix $\begin{pmatrix} 3 & 2 & 2 \\ 1 & 2 & 1 \\ -2 & -2 & -1 \end{pmatrix}$. Hence find $A^4$. **(Jan-2020)**

**Ans:** $\lambda = 1, 1, 2$. $P = \begin{pmatrix} -1 & -1 & -2 \\ 0 & 1 & -1 \\ 1 & 0 & 2 \end{pmatrix}$.

10. Diagonalize the matrix $\begin{pmatrix} 1 & 1 & 3 \\ 1 & 5 & 1 \\ 3 & 1 & 1 \end{pmatrix}$. Hence find $A^4$. **(Jan-2020)**

**Ans:** $\lambda = -2, 3, 6$, $P = \begin{pmatrix} -1 & 1 & 1 \\ 0 & -1 & 2 \\ 1 & 1 & 1 \end{pmatrix}$, $A^4 = \begin{pmatrix} 251 & 485 & 485 \\ 485 & 1051 & 2 \\ 235 & 485 & 251 \end{pmatrix}$.

11. Diagonalize the matrix $\begin{pmatrix} 2 & 0 & 4 \\ 0 & 6 & 0 \\ 4 & 0 & 2 \end{pmatrix}$. Hence find $A^4$. (July-2021)

12. Diagonalize the matrix $\begin{pmatrix} 6 & -2 & 2 \\ -2 & 3 & -1 \\ 2 & -1 & 3 \end{pmatrix}$. Hence find $A^4$. (Jan-2020)

**Ans:** $\lambda = 2, 2, 8$.

13. Diagonalize the matrix $\begin{pmatrix} 1 & 1 & 3 \\ 1 & 5 & 1 \\ 3 & 1 & 1 \end{pmatrix}$. (Mar-2022)

14. Diagonalize the matrix $\begin{pmatrix} 8 & -6 & 2 \\ -6 & 7 & -4 \\ 2 & -4 & 3 \end{pmatrix}$. (Aug-2022)

## 2.3 Quadratic Forms

**Real Matrix:** A matrix $A = [a_{ij}]$ is said to be a real matrix if every element $a_{ij}$ of $A$ is real.

A real square matrix $A = [a_{ij}]$ is said to be

(i). *symmetric* if $A = A^T$, i.e., $a_{ij} = a_{ji}$

(ii). *skew-symmetric* if $A = -A^T$, i.e., $a_{ij} = -a_{ji}$

(iii). *orthogonal* if $AA^T = A^TA = I$, i.e., $A^T = A^{-1}$.

**Hermitian matrix:** A square matrix $A$ is called Hermitian if $\overline{a_{ji}} = a_{ij}$.
If $A$ is Hermitian then $A^* = \overline{A}^T = A$.

**Skew Hermitian matrix:** A square matrix $A$ is said to be Skew Hermitian if $\overline{a_{ji}} = -a_{ij}$.
If $A$ is Skew Hermitian then $A^* = -A$.

**Unitary matrix:** A square matrix $A$ is said to be unitary if $AA^* = A^*A = I$.

**Properties of Orthogonal matrices:**

1). If $A$ is an orthogonal matrix, then $A^T$ is also orthogonal.

2). If $A$ is orthogonal, then $A^{-1}$ is also orthogonal.

3). If $A$ and $B$ are orthogonal matrices of same order then $AB$ is orthogonal.

4). The determinant of an orthogonal matrix is $\pm 1$.

$AA^T = I \Rightarrow |A||A^T| = 1 \Rightarrow |A|^2 = 1 \Rightarrow |A| = \pm 1$.

5). If $\lambda_1, \lambda_2, \cdots, \lambda_n$ are eigenvalues of $A$, then $\frac{1}{\lambda_1}, \frac{1}{\lambda_2}, \cdots, \frac{1}{\lambda_n}$ are also eigenvalues of $A$.

We know that, $\frac{1}{\lambda_1}, \frac{1}{\lambda_2}, \cdots, \frac{1}{\lambda_n}$ are eigenvalues of $A^{-1}$, but $A^{-1} = A^T$ from orthogonal matrix. The eigenvalues of $A^T$ =The eigenvalues of $A$.

Therefore $\frac{1}{\lambda_1}, \frac{1}{\lambda_2}, \cdots, \frac{1}{\lambda_n}$ are also eigenvalues of $A$.

**Properties of unitary matrices:**

1). If $A$ is an unitary matrix, then $A^T$ is also unitary.

2). If $A$ is unitary, then $A^{-1}$ is also unitary.

3). The product of two unitary matrices is unitary.

4). The modulus of an eigenvalue of an unitary matrix is unity.

**Quadratic Form:** A homogeneous expression of the second degree in any number of variables is called a quadratic form.

**Examples:**

1. $2x^2 + 4xy + 5y^2$ is quadratic form in two variables $x$ and $y$.

2. $x^2 - 4y^2 + 2xy + 6z^2 - 4xz + 6yz$ is a quadratic form in three variables $x, y$ and $z$.

3. $x_1^2 + x_2^2 + 10x_1x_2$ is quadratic form in two variables $x_1$ and $x_2$.

**Theorem:** Every quadratic form in variables $x_1, x_2, \cdots, x_n$ can be express in the form $X^T A X$ where $X = \begin{pmatrix} x_1 \\ x_2 \\ \vdots \\ x_n \end{pmatrix}$, and $A$ is symmetric matrix of order $n$.

**Proof:** Let $A = [a_{ij}] = \begin{pmatrix} a_{11} & a_{12} & \cdots & a_{1n} \\ a_{21} & a_{22} & \cdots & a_{2n} \\ \cdots & \cdots & \cdots & \cdots \\ a_{n1} & a_{n1} & \cdots & a_{nn} \end{pmatrix}$

$X = \begin{pmatrix} x_1 \\ x_2 \\ \vdots \\ x_n \end{pmatrix}$ and $X^T = \begin{pmatrix} x_1 & x_2 & \cdots & x_n \end{pmatrix}$.

Consider

$X^T A X = \begin{pmatrix} x_1 & x_2 & \cdots & x_n \end{pmatrix} \begin{pmatrix} a_{11} & a_{12} & \cdots & a_{1n} \\ a_{21} & a_{22} & \cdots & a_{2n} \\ \cdots & \cdots & \cdots & \cdots \\ a_{n1} & a_{n1} & \cdots & a_{nn} \end{pmatrix} \begin{pmatrix} x_1 \\ x_2 \\ \vdots \\ x_n \end{pmatrix}$

$= \begin{pmatrix} a_{11}x_1 + a_{21}x_2 + \cdots + a_{n1}x_n + \cdots + a_{1n}x_1 + a_{2n}x_2 + \cdots + a_{nn}x_n \end{pmatrix} \begin{pmatrix} x_1 \\ x_2 \\ \vdots \\ x_n \end{pmatrix}$

$= a_{11}x_1^2 + a_{21}x_2 x_1 + \cdots + a_{n1}x_n x_1 + a_{21}x_1 x_2 + a_{22}x_2^2 + \cdots + a_{n2}x_n x_2$

$\cdots \quad \cdots \quad \cdots$

$+ a_{n1}x_n x_1 + a_{n2}x_n x_2 + \cdots + a_{nn}x_n^2$

$= \sum_{i=1}^{n} \sum_{j=1}^{n} a_{ij} x_i x_j$

Hence any quadratic form can be express as $Q = X^T A X$.

### 2.3.1 General Form of a Quadratic Form

The general quadratic form in $n$ variables $x_1, x_2, \cdots, x_n$ is defined as

$$Q = \sum_{i=1}^{n} \sum_{j=1}^{n} a_{ij} x_i x_j$$

where $a_{ij}$'s are constants. If $a_{ij}$'s are real, then the quadratic form is known as real quadratic form.

## 2.3.2 Matrix Form of a Quadratic Form

The general quadratic form
$$\sum_{i=1}^{n}\sum_{j=1}^{n} a_{ij}x_ix_j$$

where $a_{ij} = a_{ji}$ can always be written as $X^TAX$ where $X = \begin{pmatrix} x_1 \\ x_2 \\ \vdots \\ x_n \end{pmatrix}$, $X^T = \begin{pmatrix} x_1 & x_2 & \cdots & x_n \end{pmatrix}$, $A = [a_{ij}]$.

The symmetric matrix

$$A = [a_{ij}] = \begin{pmatrix} a_{11} & a_{12} & \cdots & a_{1n} \\ a_{21} & a_{22} & \cdots & a_{2n} \\ \cdots & \cdots & \cdots & \cdots \\ a_{n1} & a_{n1} & \cdots & a_{nn} \end{pmatrix}$$

is called the matrix of the quadratic form.

$$\text{i.e., } X^TAX = \sum_{i=1}^{n}\sum_{j=1}^{n} a_{ij}x_ix_j.$$

**Note:**
1. The symmetric matrix $A$ is called the matrix of the quadratic form
2. Given a quadratic form we can find its matrix and conversely.

## 2.3.3 Rank of the Quadratic form

1. The rank $r$ of the matrix $A$ is called the rank of the quadratic form $X^TAX$.
2. If the rank of $A$ is $r < n$ (order of $A$) or $|A| = 0$ or $A$ is singular then the quadratic

form is singular, otherwise non-singular.

**Example:** Find the symmetric matrix corresponding to the quadratic form $x_1^2 + 6x_1x_2 + 5x_2^2$.

**Solution:** The given quadratic form can be written as
$$x_1 \cdot x_1 + 3x_1 \cdot x_2 + 3x_2 \cdot x_1 + 5x_2 \cdot x_2.$$
Let $A$ be the symmetric matrix of this quadratic form. Then $A = \begin{pmatrix} 1 & 3 \\ 3 & 5 \end{pmatrix}$.

Let $X = \begin{pmatrix} x_1 \\ x_2 \end{pmatrix}$ and $X^T = \begin{pmatrix} x_1 & x_2 \end{pmatrix}$.

We have $X^T A X = \begin{pmatrix} x_1 & x_2 \end{pmatrix} \begin{pmatrix} 1 & 3 \\ 3 & 5 \end{pmatrix} \begin{pmatrix} x_1 \\ x_2 \end{pmatrix} = x_1^2 + 6x_1x_2 + 5x_2^2$.

**Example:** Write the matrix relating the quadratic form $ax^2 + 2hxy + by^2$.

**Solution:** The given quadratic form can be written as
$$\begin{pmatrix} x & y \end{pmatrix} \begin{pmatrix} a & h \\ h & b \end{pmatrix} \begin{pmatrix} x \\ y \end{pmatrix}.$$

The corresponding matrix is $\begin{pmatrix} a & h \\ h & b \end{pmatrix}$.

**Example:** Find the quadratic form corresponding to $\begin{pmatrix} a & h & g \\ h & b & f \\ g & f & c \end{pmatrix}$.

**Solution:** Let $X = \begin{pmatrix} x \\ y \\ z \end{pmatrix}$ then $X^T = \begin{pmatrix} x & y & z \end{pmatrix}$.

Then $X^T A X$ is the quadratic form corresponding to the matrix $A$.

Hence
$$X^T AX = \begin{pmatrix} x & y & z \end{pmatrix} \begin{pmatrix} a & h & g \\ h & b & f \\ g & f & c \end{pmatrix} \begin{pmatrix} x \\ y \\ z \end{pmatrix}$$

$$= \begin{pmatrix} x & y & z \end{pmatrix} \begin{pmatrix} ax + hy + gz \\ hx + by + fz \\ gx + fy + cz \end{pmatrix}$$

$$= ax^2 + by^2 + cz^2 + 2hxy + 2fyz + 2gzx.$$

**Example:** Find the quadratic form relating to the matrix $\text{diag}[\lambda_1, \lambda_2, \cdots, \lambda_n]$.

**Solution:** Let $X = \begin{pmatrix} x_1 \\ x_2 \\ \vdots \\ x_n \end{pmatrix}$ so that $X^T = \begin{pmatrix} x_1 & x_2 & \cdots & x_n \end{pmatrix}$.

Let $A = \text{diag}[\lambda_1, \lambda_2, \cdots, \lambda_n] = \begin{pmatrix} \lambda_1 & 0 & \cdots & 0 \\ 0 & \lambda_2 & \cdots & 0 \\ \cdots & \cdots & \cdots & \cdots \\ 0 & 0 & 0 & \lambda_n \end{pmatrix}$.

The required quadratic form is

$$X^T AX = \begin{pmatrix} x_1 & x_2 & \cdots & x_n \end{pmatrix} \begin{pmatrix} \lambda_1 & 0 & \cdots & 0 \\ 0 & \lambda_2 & \cdots & 0 \\ \cdots & \cdots & \cdots & \cdots \\ 0 & 0 & 0 & \lambda_n \end{pmatrix} \begin{pmatrix} x_1 \\ x_2 \\ \vdots \\ x_n \end{pmatrix}$$

$$= \lambda_1 x_1^2 + \lambda_2 x_2^2 + \cdots + \lambda_n x_n^2$$

**Example:** Write the quadratic form corresponding to the symmetric matrices:(**Apr-2017**)

$$\begin{pmatrix} 1 & 0 & 4 \\ 0 & -2 & -1 \\ 4 & -1 & 3 \end{pmatrix}, \begin{pmatrix} 0 & 5/1 & 3 \\ 5/2 & 7 & 1 \\ 3 & 1 & 2 \end{pmatrix}, \begin{pmatrix} 2 & -3 & 5 \\ -3 & 2 & -2 \\ 5 & -2 & 2 \end{pmatrix}, \begin{pmatrix} 3 & 5 & 0 \\ 5 & 5 & 4 \\ 0 & 4 & 7 \end{pmatrix}.$$

## 2.3.4 Linear Transformation of a Quadratic Form

Let $X^T A X$ be a quadratic form in $n$ variables $x_1, x_2, \cdots, x_n$ and $X = PY$ be the linear transformation, that transforms the variable set $X = \begin{pmatrix} x_1 & x_2 & \cdots & x_n \end{pmatrix}^T$ to a new variable set $Y = \begin{pmatrix} y_1 & y_2 & \cdots & y_n \end{pmatrix}^T$. Then the non-singular matrix $P$ is known as matrix of the transformation.

Apply the linear transformation $X = PY$ to the quadratic form $X^T A X$, we have

$$X^T A X = (PY)^T A (PY)$$
$$= Y^T P^T A P Y$$
$$= Y^T (P^T A P) Y$$
$$= Y^T B Y \quad \text{(where } B = P^T A P\text{)}$$
$$\therefore \ X^T A X = Y^T B Y.$$

Here $Y^T B Y$ is also a quadratic form in $n$ variables $y_1, y_2, \cdots, y_n$.
Thus $Y^T B Y$ is the linear transform of the quadratic form $X^T A X$ under the linear transformation $X = PY$.

**Note:** Here $B$ is symmetric matrix.

## 2.3.5 Orthogonal Transformation

If $P$ is an orthogonal matrix then the transformation $X = PY$ is called the orthogonal transformation.

## 2.3.6 Canonical Form or Normal Form of a Quadratic Form

Let $X^T A X$ be a quadratic form in $n$ variables. Then there exists a real non-singular linear transformation $X = PY$ which transforms $X^T A X$ to another quadratic form of the type $Y^T D Y = \lambda_1 y_1^2 + \lambda_2 y_2^2 + \cdots + \lambda_n y_n^n$, then $Y^T D Y$ is called the canonical form $X^T A X$. Here $D = \text{diag}(\lambda_1, \lambda_2, \cdots, \lambda_n)$.

**Theorem:** Any quadratic form can be reduced to canonical form by means of a non-singular transformation.

**Note:** 1. This form is also known as sum of the squares of quadratic form.
2. If $\rho(A) = r$, then the quadratic form $X^T A X$ will contains only $r$ terms.

### 2.3.7 Rank, Index and Signature of the Quadratic Form

**Rank of Q.F:** The number of non-zero terms in the canonical form of a quadratic form is called the rank of the quadratic form and it is denoted by $r$.

**Index of Q. F:** The number of positive terms in the canonical form is called the index of the quadratic form and it is denoted by $s$.

**Signature of Q. F:** The excess number of positive terms over the number of negative terms in the canonical form i.e., $s - (s - r) = 2s - r$ is called the signature of the quadratic form. In other words, signature of the quadratic form is defined as the difference between the number of positive terms and negative terms in the canonical form.

### 2.3.8 Nature of the Quadratic Form

The quadratic form $X^T A X$ in $n$ variables is said to be

**Positive definite** if $r = n$ and $s = n$ or if all eigenvalues of $A$ are positive

**Negative definite** if $r = n$ and $s = 0$ or if all eigenvalues of $A$ are negative

**Positive semi definite** if $r < n$ and $s = r$ or if atleast one eigenvalue is zero and remaining are positive

**Negative semi definite** if $r < n$ and $s = 0$ or if atleast one eigenvalue is zero and remaining are negative

**Indefinite** in all other cases.

**Example:** Identify the nature of the quadratic form $x_1^2 + 4x_2^2 + x_3^2 - 4x_1x_2 + 2x_1x_3 - 4x_2x_3$.

**Solution:** Given quadratic form can be written in the matrix form as $A = \begin{pmatrix} 1 & -2 & 1 \\ -2 & 4 & -2 \\ 1 & -2 & 1 \end{pmatrix}$.

The characteristic equation of $A$ is given by $|A - \lambda I| = 0$.

$$\Rightarrow \begin{vmatrix} 1-\lambda & -2 & 1 \\ -2 & 4-\lambda & -2 \\ 1 & -2 & 1-\lambda \end{vmatrix}.$$

$\Rightarrow \lambda^3 - 6\lambda^2 = 0$

$\Rightarrow \lambda = 0, 0, 6$.

Eigenvalues are $\lambda = 0, 0, 6$ which are positive and two values are zero.

$\therefore$ The quadratic form is positive semi definite.

**Example:** Discuss the nature of the quadric form $x^2 + 4xy + 6xz - y^2 + 2yz + 4z^2$.

**Solution:** Given quadratic form can be written in the matrix form as $A = \begin{pmatrix} 1 & 2 & 3 \\ 2 & -1 & 1 \\ 3 & 1 & 4 \end{pmatrix}$.

The characteristic equation of $A$ is given by $|A - \lambda I| = 0$.

$$\Rightarrow \begin{vmatrix} 1-\lambda & 2 & 3 \\ 2 & -1-\lambda & 1 \\ 3 & 1 & 4-\lambda \end{vmatrix}.$$

$\Rightarrow \lambda^3 - 4\lambda^2 - 15\lambda = 0$

$\Rightarrow \lambda = 0, 2 + \sqrt{19}, 2 - \sqrt{19}$.

$\therefore$ The quadratic form is indefinite.

## 2.4 Method of Reduction of Quadratic Form to Canonical Form

A given quadratic form (Q.F) can be reduced to a canonical form (C.F) by the following methods:

1. Diagonalization (Reduction to canonical form by linear transformation)

2. Orthogonalization (Reduction to canonical form by orthogonal transformation)
3. Lagrange's reduction.

### 2.4.1 Diagonalization

Let $X^T A X$ be a quadratic form where $A$ is the matrix of the quadratic form. Let $X = PY$ be the non-singular linear transformation. Then, we have

$$\begin{aligned} X^T A X &= (PY)^T A(PY) \\ &= (Y^T P^T) A(PY) \\ &= Y^T (P^T A P) Y \\ &= Y^T D Y \quad \text{where } P^T A P = D \end{aligned}$$

Here $Y^T D Y$ is also a quadratic form in $n$ variables $y_1, y_2, \cdots, y_n$ and known as canonical form.

**Working Rule to reduce Q.F to C. F by Diagonalization**

**Step 1:** Write the symmetric matrix of the given Q. F

**Step 2:** Write the matrix $A$ in the following relation: $A_{n \times n} = I_n A I_n$

**Step 3:** Reduce the matrix $A$ on the left hand side to a diagonal matrix

(i) by applying elementary row operations on the left identity and on $A$ on left hand side

(ii) by applying elementary column operations on the right identity and on $A$ on left hand side

**Step 4:** By these operations, $A = IAI$ will be reduced to the form $D = P^T A P$, where $D$ is the diagonal matrix with the elements $d_1, d_2, \cdots, d_n$ and $P$ is the matrix used in the linear transformation.

∴ The canonical form is given by

$$Y^T DY = \begin{pmatrix} y_1 & y_2 & \cdots & y_n \end{pmatrix} \begin{pmatrix} d_1 & 0 & \cdots & 0 \\ 0 & d_2 & \cdots & 0 \\ \cdots & \cdots & \cdots & \cdots \\ 0 & 0 & \cdots & d_n \end{pmatrix} \begin{pmatrix} y_1 \\ y_2 \\ \vdots \\ y_n \end{pmatrix}$$

$$= d_1 y_1^2 + d_2 y_2^2 + \cdots + d_n y_n^2$$

Here some of the coefficients $d_1, d_2, \cdots, d_n$ may be zero.

**Example:** Reduce the quadratic form $10x^2 + 2y^2 + 5z^2 + 6yz - 10zx - 4xy$ into a sum of squares and find the matrix of the transformation.

(Nov-2017), (Apr-2018), (Jan-2020)

**Solution:** The given quadratic form can be written as $X^T A X$, where $X^T = \begin{bmatrix} x & y & z \end{bmatrix}$ and the symmetric matrix

$$A = \begin{pmatrix} 10 & -2 & -5 \\ -2 & 2 & 3 \\ -5 & 3 & 5 \end{pmatrix}.$$

Now we reduce $A$ to diagonal matrix by using elementary row and column operations.

We know that $A = I_3 A I_3$

$$\begin{pmatrix} 10 & -2 & -5 \\ -2 & 2 & 3 \\ -5 & 3 & 5 \end{pmatrix} = \begin{pmatrix} 1 & 0 & 0 \\ 0 & 1 & 0 \\ 0 & 0 & 1 \end{pmatrix} A \begin{pmatrix} 1 & 0 & 0 \\ 0 & 1 & 0 \\ 0 & 0 & 1 \end{pmatrix}$$

$R_2 \to 5R_2 + R_1$, $R_3 \to 2R_3 + R_1$, we have

$$\begin{pmatrix} 10 & -2 & -5 \\ 0 & 8 & 10 \\ 0 & 4 & 5 \end{pmatrix} = \begin{pmatrix} 1 & 0 & 0 \\ 1 & 5 & 0 \\ 1 & 0 & 2 \end{pmatrix} A \begin{pmatrix} 1 & 0 & 0 \\ 0 & 1 & 0 \\ 0 & 0 & 1 \end{pmatrix}$$

$C_2 \to 5C_2 + C_1$, $C_3 \to 2C_3 + C_1$, we have

$$\begin{pmatrix} 10 & 0 & 0 \\ 0 & 40 & 20 \\ 0 & 20 & 10 \end{pmatrix} = \begin{pmatrix} 1 & 0 & 0 \\ 1 & 5 & 0 \\ 1 & 0 & 2 \end{pmatrix} A \begin{pmatrix} 1 & 1 & 1 \\ 0 & 5 & 0 \\ 0 & 0 & 2 \end{pmatrix}$$

$R_3 \to 2R_3 - R_2$, we have
$$\begin{pmatrix} 10 & 0 & 0 \\ 0 & 40 & 20 \\ 0 & 0 & 0 \end{pmatrix} = \begin{pmatrix} 1 & 0 & 0 \\ 1 & 5 & 0 \\ 1 & -5 & 4 \end{pmatrix} A \begin{pmatrix} 1 & 1 & 1 \\ 0 & 5 & 0 \\ 0 & 0 & 2 \end{pmatrix}$$

$C_3 \to 2C_3 - C_2$, we have
$$\begin{pmatrix} 10 & 0 & 0 \\ 0 & 40 & 0 \\ 0 & 0 & 0 \end{pmatrix} = \begin{pmatrix} 1 & 0 & 0 \\ 1 & 5 & 0 \\ 1 & -5 & 4 \end{pmatrix} A \begin{pmatrix} 1 & 1 & 1 \\ 0 & 5 & -5 \\ 0 & 0 & 4 \end{pmatrix}$$

$$\Rightarrow D = P^T A P$$

where $D$=a diagonal matrix=$\begin{pmatrix} 10 & 0 & 0 \\ 0 & 40 & 0 \\ 0 & 0 & 0 \end{pmatrix}$ and $P = \begin{pmatrix} 1 & 1 & 1 \\ 0 & 5 & -5 \\ 0 & 0 & 4 \end{pmatrix}$.

The linear transformation is $X = PY$

i.e., $\begin{pmatrix} x \\ y \\ z \end{pmatrix} = \begin{pmatrix} 1 & 1 & 1 \\ 0 & 5 & -5 \\ 0 & 0 & 4 \end{pmatrix} \begin{pmatrix} y_1 \\ y_2 \\ y_3 \end{pmatrix}$

i.e., $x = y_1 + y_2 + y_3$

$y = 5y_2 - 5y_3$

$z = 4y_3$.

$\therefore$ The canonical form of the given quadratic form is

$$Y^T D Y = \begin{pmatrix} y_1 & y_2 & y_3 \end{pmatrix} \begin{pmatrix} 10 & 0 & 0 \\ 0 & 40 & 0 \\ 0 & 0 & 0 \end{pmatrix} \begin{pmatrix} y_1 \\ y_2 \\ y_3 \end{pmatrix}$$

$$= 10y_1^2 + 40y_2^2.$$

Here,

$r$=Rank of $A$=2 (No. of non-zero terms in its canonical form)

$s$=Index Q.F=2 (No. of positive terms in its canonical form)

Signature=$2s - r = 2(2) - 2 = 2$.

Since $r < n$ and $s = r$, the given quadratic form is positive semi definite.

**Example:** Reduce the following quadratic form to canonical form by diagonalization $6x^2+3y^2+3z^2-4xy-2yz+4zx$. **(Apr-2017)**

**Solution:** The given quadratic form can be written as $X^T A X$, where $X^T = \begin{bmatrix} x & y & z \end{bmatrix}$ and the symmetric matrix

$$A = \begin{pmatrix} 6 & -2 & 2 \\ -2 & 3 & -1 \\ 2 & -1 & 3 \end{pmatrix}.$$

Now we reduce $A$ to diagonal matrix by using elementary row and column operations.

We know that $A = I_3 A I_3$

$$\begin{pmatrix} 6 & -2 & 2 \\ -2 & 3 & -1 \\ 2 & -1 & 3 \end{pmatrix} = \begin{pmatrix} 1 & 0 & 0 \\ 0 & 1 & 0 \\ 0 & 0 & 1 \end{pmatrix} A \begin{pmatrix} 1 & 0 & 0 \\ 0 & 1 & 0 \\ 0 & 0 & 1 \end{pmatrix}$$

$R_2 \to 3R_2 + R_1,\; R_3 \to 3R_3 - R_1$, we have

$$\begin{pmatrix} 6 & -2 & 2 \\ 0 & 7 & -1 \\ 0 & -1 & 7 \end{pmatrix} = \begin{pmatrix} 1 & 0 & 0 \\ 1 & 3 & 0 \\ -1 & 0 & 3 \end{pmatrix} A \begin{pmatrix} 1 & 0 & 0 \\ 0 & 1 & 0 \\ 0 & 0 & 1 \end{pmatrix}$$

$C_2 \to 3C_2 + C_1,\; C_3 \to 3C_3 - C_1$, we have

$$\begin{pmatrix} 6 & 0 & 0 \\ 0 & 21 & -3 \\ 0 & -3 & 21 \end{pmatrix} = \begin{pmatrix} 1 & 0 & 0 \\ 1 & 3 & 0 \\ -1 & 0 & 3 \end{pmatrix} A \begin{pmatrix} 1 & 1 & -1 \\ 0 & 3 & 0 \\ 0 & 0 & 3 \end{pmatrix}$$

$R_3 \to 7R_3 + R_2$, we have

$$\begin{pmatrix} 6 & 0 & 0 \\ 0 & 21 & -3 \\ 0 & 0 & 144 \end{pmatrix} = \begin{pmatrix} 1 & 0 & 0 \\ 1 & 3 & 0 \\ -6 & 3 & 21 \end{pmatrix} A \begin{pmatrix} 1 & 1 & -1 \\ 0 & 3 & 0 \\ 0 & 0 & 3 \end{pmatrix}$$

$C_3 \to 7C_3 + C_2$, we have

$$\begin{pmatrix} 6 & 0 & 0 \\ 0 & 21 & 0 \\ 0 & 0 & 1008 \end{pmatrix} = \begin{pmatrix} 1 & 0 & 0 \\ 1 & 3 & 0 \\ -6 & 3 & 21 \end{pmatrix} A \begin{pmatrix} 1 & 1 & -6 \\ 0 & 3 & 3 \\ 0 & 0 & 21 \end{pmatrix}$$

$$\Rightarrow D = P^T AP$$

where $D$=a diagonal matrix= $\begin{pmatrix} 6 & 0 & 0 \\ 0 & 21 & 0 \\ 0 & 0 & 1008 \end{pmatrix}$ and $P = \begin{pmatrix} 1 & 1 & -6 \\ 0 & 3 & 3 \\ 0 & 0 & 21 \end{pmatrix}$.

∴ The canonical form of the given quadratic form is

$$Y^T DY = \begin{pmatrix} y_1 & y_2 & y_3 \end{pmatrix} \begin{pmatrix} 6 & 0 & 0 \\ 0 & 21 & 0 \\ 0 & 0 & 1008 \end{pmatrix} \begin{pmatrix} y_1 \\ y_2 \\ y_3 \end{pmatrix}$$

$$= 6y_1^2 + 21y_2^2 + 1008y_3^2.$$

Here,

$r$=Rank of $A$=3 (No. of non-zero terms in its canonical form)

$s$=Index Q.F=3 (No. of positive terms in its canonical form)

Signature=$2s - r = 2(3) - 3 = 3$.

Since $r = n$ and $s = n$, the given quadratic form is positive definite.

**Example:** Reduce the Q. F $3x^2 + 3y^2 + 3z^2 + 4xy + 8yz + 8xz$ into a canonical form by a linear transformatio and find its nature, rank, index and signature.

**Solution:** The given quadratic form can be written as $X^T AX$, where $X^T = \begin{bmatrix} x & y & z \end{bmatrix}$ and the symmetric matrix

$$A = \begin{pmatrix} 3 & 2 & 4 \\ 2 & 3 & 4 \\ 4 & 4 & 3 \end{pmatrix}.$$

Now we reduce $A$ to diagonal matrix by using elementary row and column operations.

We know that $A = I_3 A I_3$

$$\begin{pmatrix} 3 & 2 & 4 \\ 2 & 3 & 4 \\ 4 & 4 & 3 \end{pmatrix} = \begin{pmatrix} 1 & 0 & 0 \\ 0 & 1 & 0 \\ 0 & 0 & 1 \end{pmatrix} A \begin{pmatrix} 1 & 0 & 0 \\ 0 & 1 & 0 \\ 0 & 0 & 1 \end{pmatrix}$$

$R_2 \to 3R_2 - 2R_1$, $R_3 \to 3R_3 - 4R_1$, we have

$$\begin{pmatrix} 3 & 2 & 4 \\ 0 & 5 & 4 \\ 0 & 4 & -7 \end{pmatrix} = \begin{pmatrix} 1 & 0 & 0 \\ -2 & 3 & 0 \\ -4 & 0 & 3 \end{pmatrix} A \begin{pmatrix} 1 & 0 & 0 \\ 0 & 1 & 0 \\ 0 & 0 & 1 \end{pmatrix}$$

$C_2 \to 3C_2 - 2C_1$, $C_3 \to 3C_3 - 4C_1$, we have

$$\begin{pmatrix} 3 & 0 & 0 \\ 0 & 15 & 12 \\ 0 & 12 & -21 \end{pmatrix} = \begin{pmatrix} 1 & 0 & 0 \\ -2 & 3 & 0 \\ -4 & 0 & 3 \end{pmatrix} A \begin{pmatrix} 1 & -2 & -4 \\ 0 & 3 & 0 \\ 0 & 0 & 3 \end{pmatrix}$$

$R_3 \to 15R_3 - 12R_2$, we have

$$\begin{pmatrix} 3 & 0 & 0 \\ 0 & 15 & 12 \\ 0 & 0 & -459 \end{pmatrix} = \begin{pmatrix} 1 & 0 & 0 \\ -2 & 3 & 0 \\ -36 & -36 & 45 \end{pmatrix} A \begin{pmatrix} 1 & -2 & -4 \\ 0 & 3 & 0 \\ 0 & 0 & 3 \end{pmatrix}$$

$C_3 \to 15C_3 - 12C_2$, we have

$$\begin{pmatrix} 3 & 0 & 0 \\ 0 & 15 & 0 \\ 0 & 0 & -6885 \end{pmatrix} = \begin{pmatrix} 1 & 0 & 0 \\ -2 & 3 & 0 \\ -36 & -36 & 45 \end{pmatrix} A \begin{pmatrix} 1 & -2 & -36 \\ 0 & 3 & -36 \\ 0 & 0 & 45 \end{pmatrix}$$

$$\Rightarrow D = P^T A P$$

where $D$=a diagonal matrix= $\begin{pmatrix} 3 & 0 & 0 \\ 0 & 15 & 0 \\ 0 & 0 & -6885 \end{pmatrix}$ and $P = \begin{pmatrix} 1 & -2 & -36 \\ 0 & 3 & -36 \\ 0 & 0 & 45 \end{pmatrix}$.

The linear transformation is $X = PY$

i.e., $\begin{pmatrix} x \\ y \\ z \end{pmatrix} = \begin{pmatrix} 1 & -2 & -36 \\ 0 & 3 & -36 \\ 0 & 0 & 45 \end{pmatrix} \begin{pmatrix} y_1 \\ y_2 \\ y_3 \end{pmatrix}$

i.e., $x = y_1 - 2y_2 - 36y_3$

$y = 3y_2 - 36y_3$

$z = 45y_3$.

∴ The canonical form of the given quadratic form is

$$Y^T DY = \begin{pmatrix} y_1 & y_2 & y_3 \end{pmatrix} \begin{pmatrix} 3 & 0 & 0 \\ 0 & 15 & 0 \\ 0 & 0 & -6885 \end{pmatrix} \begin{pmatrix} y_1 \\ y_2 \\ y_3 \end{pmatrix}$$

$$= 3y_1^2 + 15y_2^2 - 6885y_3^2.$$

Here,

$r$=Rank of $A$=3 (No. of non-zero terms in its canonical form)

$s$=Index Q.F=2 (No. of positive terms in its canonical form)

Signature=$2s - r = 2(2) - 3 = 1$.

Since $r = n$ and $s < n$, the given quadratic from is indefinite.

**Example:** Find Rank index and signature of quadratic form using diagonalization method $7x^2 + 6y^2 + 5z^2 - 4xy - 4yz$. **(Jan-2020)**

**Solution:** The given quadratic form can be written as $X^T AX$, where $X^T = \begin{bmatrix} x & y & z \end{bmatrix}$ and the symmetric matrix

$$A = \begin{pmatrix} 7 & -2 & 0 \\ -2 & 6 & -2 \\ 0 & -2 & 5 \end{pmatrix}.$$

Now we reduce $A$ to diagonal matrix by using elementary row and column operations.

We know that $A = I_3 A I_3$

$$\begin{pmatrix} 7 & -2 & 0 \\ -2 & 6 & -2 \\ 0 & -2 & 5 \end{pmatrix} = \begin{pmatrix} 1 & 0 & 0 \\ 0 & 1 & 0 \\ 0 & 0 & 1 \end{pmatrix} A \begin{pmatrix} 1 & 0 & 0 \\ 0 & 1 & 0 \\ 0 & 0 & 1 \end{pmatrix}$$

$R_2 \to (7/2)R_2 + R_1$, we have

$$\begin{pmatrix} 7 & -2 & 0 \\ 0 & 19 & -7 \\ 0 & -2 & 5 \end{pmatrix} = \begin{pmatrix} 1 & 0 & 0 \\ 0 & 7/2 & 0 \\ 0 & 0 & 1 \end{pmatrix} A \begin{pmatrix} 1 & 0 & 0 \\ 0 & 1 & 0 \\ 0 & 0 & 1 \end{pmatrix}$$

$C_2 \to (7/2)C_2 + C_1$, we have

$$\begin{pmatrix} 7 & 0 & 0 \\ 0 & 133/2 & -7 \\ 0 & -7 & 5 \end{pmatrix} = \begin{pmatrix} 1 & 0 & 0 \\ 1 & 7/2 & 0 \\ 0 & 0 & 1 \end{pmatrix} A \begin{pmatrix} 1 & 1 & 0 \\ 0 & 7/2 & 0 \\ 0 & 0 & 1 \end{pmatrix}$$

$R_2 \to 2R_2$, we have

$$\begin{pmatrix} 7 & 0 & 0 \\ 0 & 133 & -14 \\ 0 & -7 & 5 \end{pmatrix} = \begin{pmatrix} 1 & 0 & 0 \\ 2 & 7 & 0 \\ 0 & 0 & 1 \end{pmatrix} A \begin{pmatrix} 1 & 1 & 0 \\ 0 & 7/2 & 0 \\ 0 & 0 & 1 \end{pmatrix}$$

$C_2 \to 2C_2$, we have

$$\begin{pmatrix} 7 & 0 & 0 \\ 0 & 266 & -14 \\ 0 & -14 & 5 \end{pmatrix} = \begin{pmatrix} 1 & 0 & 0 \\ 2 & 7 & 0 \\ 0 & 0 & 1 \end{pmatrix} A \begin{pmatrix} 1 & 2 & 0 \\ 0 & 7 & 0 \\ 0 & 0 & 1 \end{pmatrix}$$

$C_3 \to 19C_3 + C_2$, we have

$$\begin{pmatrix} 7 & 0 & 0 \\ 0 & 266 & 0 \\ 0 & -14 & 81 \end{pmatrix} = \begin{pmatrix} 1 & 0 & 0 \\ 2 & 7 & 0 \\ 0 & 0 & 1 \end{pmatrix} A \begin{pmatrix} 1 & 2 & 2 \\ 0 & 7 & 7 \\ 0 & 0 & 19 \end{pmatrix}$$

$R_3 \to 19R_3 + R_2$, we have

$$\begin{pmatrix} 7 & 0 & 0 \\ 0 & 266 & 0 \\ 0 & 0 & 1539 \end{pmatrix} = \begin{pmatrix} 1 & 0 & 0 \\ 2 & 7 & 0 \\ 2 & 7 & 10 \end{pmatrix} A \begin{pmatrix} 1 & 2 & 2 \\ 0 & 7 & 7 \\ 0 & 0 & 19 \end{pmatrix}$$

$$\Rightarrow D = P^T A P$$

where $D$ = a diagonal matrix = $\begin{pmatrix} 7 & 0 & 0 \\ 0 & 266 & 0 \\ 0 & 0 & 1539 \end{pmatrix}$ and $P = \begin{pmatrix} 1 & 2 & 2 \\ 0 & 7 & 7 \\ 0 & 0 & 19 \end{pmatrix}$.

∴ The canonical form of the given quadratic form is

$$Y^T D Y = \begin{pmatrix} y_1 & y_2 & y_3 \end{pmatrix} \begin{pmatrix} 7 & 0 & 0 \\ 0 & 266 & 0 \\ 0 & 0 & 1539 \end{pmatrix} \begin{pmatrix} y_1 \\ y_2 \\ y_3 \end{pmatrix}$$

$$= 7y_1^2 + 266y_2^2 + 1539y_3^2.$$

Here,

$r$ = Rank of $A$ = 3 (No. of non-zero terms in its canonical form) s=Index Q.F=3 (No. of positive terms in its canonical form)

Signature=$2s - r = 2(3) - 3 = 3$.

Since $r = n$ and $s = n$, the given quadratic form is positive definite.

**Example:** Reduce the quadratic form $x^2 + y^2 + 2z^2 - 2xy + 4zx + 4yz$ in to canonical form using diagonalization method hence find rank, index and signature.

(Jan-2020)

**Solution:** The given quadratic form can be written as $X^T A X$, where $X^T = \begin{bmatrix} x & y & z \end{bmatrix}$ and the symmetric matrix

$$A = \begin{pmatrix} 1 & -1 & 2 \\ -1 & 1 & 2 \\ 2 & 2 & 2 \end{pmatrix}.$$

Now we reduce $A$ to diagonal matrix by using elementary row and column operations.

We know that $A = I_3 A I_3$

$$\begin{pmatrix} 1 & -1 & 2 \\ -1 & 1 & 2 \\ 2 & 2 & 2 \end{pmatrix} = \begin{pmatrix} 1 & 0 & 0 \\ 0 & 1 & 0 \\ 0 & 0 & 1 \end{pmatrix} A \begin{pmatrix} 1 & 0 & 0 \\ 0 & 1 & 0 \\ 0 & 0 & 1 \end{pmatrix}$$

$R_2 \to R_2 + R_1$, $R_3 \to R_3 - 2R_1$ we have

$$\begin{pmatrix} 1 & -1 & 2 \\ 0 & 0 & 4 \\ 0 & 4 & -2 \end{pmatrix} = \begin{pmatrix} 1 & 0 & 0 \\ 1 & 1 & 0 \\ -2 & 0 & 1 \end{pmatrix} A \begin{pmatrix} 1 & 0 & 0 \\ 0 & 1 & 0 \\ 0 & 0 & 1 \end{pmatrix}$$

$C_2 \to C_2 + C_1$, $C_3 \to C_3 - 2C_1$ we have

$$\begin{pmatrix} 1 & 0 & 0 \\ 0 & 0 & 4 \\ 0 & 4 & -2 \end{pmatrix} = \begin{pmatrix} 1 & 0 & 0 \\ 1 & 1 & 0 \\ -2 & 0 & 1 \end{pmatrix} A \begin{pmatrix} 1 & 1 & -2 \\ 0 & 1 & 0 \\ 0 & 0 & 1 \end{pmatrix}$$

$R_3 \to R_2 + 2R_3$ we have

$$\begin{pmatrix} 1 & 0 & 0 \\ 0 & 8 & 0 \\ 0 & 4 & -2 \end{pmatrix} = \begin{pmatrix} 1 & 0 & 0 \\ -3 & 1 & 2 \\ -2 & 0 & 1 \end{pmatrix} A \begin{pmatrix} 1 & 1 & -2 \\ 0 & 1 & 0 \\ 0 & 0 & 1 \end{pmatrix}$$

$C_2 \to C_2 + 2C_3$ we have

$$\begin{pmatrix} 1 & 0 & 0 \\ 0 & 8 & 0 \\ 0 & 0 & -2 \end{pmatrix} = \begin{pmatrix} 1 & 0 & 0 \\ -3 & 1 & 2 \\ -2 & 0 & 1 \end{pmatrix} A \begin{pmatrix} 1 & -3 & -2 \\ 0 & 1 & 0 \\ 0 & 2 & 1 \end{pmatrix}$$

$$\Rightarrow D = P^T A P$$

where $D$=a diagonal matrix=$\begin{pmatrix} 1 & 0 & 0 \\ 0 & 8 & 0 \\ 0 & 0 & -2 \end{pmatrix}$ and $P = \begin{pmatrix} 1 & -3 & -2 \\ 0 & 1 & 0 \\ 0 & 2 & 1 \end{pmatrix}$.

∴ The canonical form of the given quadratic form is

$$Y^T DY = \begin{pmatrix} y_1 & y_2 & y_3 \end{pmatrix} \begin{pmatrix} 1 & 0 & 0 \\ 0 & 8 & 0 \\ 0 & 0 & -2 \end{pmatrix} \begin{pmatrix} y_1 \\ y_2 \\ y_3 \end{pmatrix}$$

$$= y_1^2 + 8y_2^2 - 2y_3^2.$$

Here,

$r$=Rank of $A$=3 (No. of non-zero terms in its canonical form)

$s$=Index Q.F=2 (No. of positive terms in its canonical form)

Signature=$2s - r = 2(2) - 3 = 1$.

Since $r = n$ and $s \neq n$, the given quadratic form is indefinite.

The matrix of the transformation is

$$X = PY \Rightarrow \begin{pmatrix} x \\ y \\ z \end{pmatrix} = \begin{pmatrix} 1 & -3 & -2 \\ 0 & 1 & 0 \\ 0 & 2 & 1 \end{pmatrix} \begin{pmatrix} y_1 \\ y_2 \\ y_3 \end{pmatrix}.$$

**Example:** Find the nature, rank, index and signature of the quadratic from by reduce in canonical form $2x^2 + y^2 - 3z^2 + 12xy - 4xz - 8yz$. **(Sep-2021)**

**Solution:** The given quadratic form can be written as $X^T AX$, where $X^T = \begin{bmatrix} x & y & z \end{bmatrix}$ and the symmetric matrix

$$A = \begin{pmatrix} 2 & 6 & -2 \\ 6 & 1 & -4 \\ -2 & -4 & -3 \end{pmatrix}.$$

Now we reduce $A$ to diagonal matrix by using elementary row and column operations.

We know that $A = I_3 A I_3$

$$\begin{pmatrix} 2 & 6 & -2 \\ 6 & 1 & -4 \\ -2 & -4 & -3 \end{pmatrix} = \begin{pmatrix} 1 & 0 & 0 \\ 0 & 1 & 0 \\ 0 & 0 & 1 \end{pmatrix} A \begin{pmatrix} 1 & 0 & 0 \\ 0 & 1 & 0 \\ 0 & 0 & 1 \end{pmatrix}$$

$R_2 \to R_2 - 3R_1$, $R_3 \to R_3 + R_2$ we have

$$\begin{pmatrix} 2 & 6 & -2 \\ 6 & -17 & 2 \\ 0 & 2 & -5 \end{pmatrix} = \begin{pmatrix} 1 & 0 & 0 \\ -3 & 1 & 0 \\ 1 & 0 & 1 \end{pmatrix} A \begin{pmatrix} 1 & 0 & 0 \\ 0 & 1 & 0 \\ 0 & 0 & 1 \end{pmatrix}$$

$C_2 \to C_2 - 3C_1$, $C_3 \to C_3 + C_2$ we have

$$\begin{pmatrix} 2 & 0 & 0 \\ 0 & -17 & 2 \\ 0 & 2 & -5 \end{pmatrix} = \begin{pmatrix} 1 & 0 & 0 \\ -3 & 1 & 0 \\ 1 & 0 & 1 \end{pmatrix} A \begin{pmatrix} 1 & -3 & 1 \\ 0 & 1 & 0 \\ 0 & 0 & 1 \end{pmatrix}$$

$R_3 \to 17R_3 + R_2$, we have

$$\begin{pmatrix} 2 & 0 & 0 \\ 0 & -17 & 2 \\ 0 & 0 & -81 \end{pmatrix} = \begin{pmatrix} 1 & 0 & 0 \\ -3 & 1 & 0 \\ 11 & 2 & 17 \end{pmatrix} A \begin{pmatrix} 1 & -3 & 1 \\ 0 & 1 & 0 \\ 0 & 0 & 1 \end{pmatrix}$$

$C_3 \to 17C_3 + C_2$, we have

$$\begin{pmatrix} 2 & 0 & 0 \\ 0 & -17 & 0 \\ 0 & 0 & -1377 \end{pmatrix} = \begin{pmatrix} 1 & 0 & 0 \\ -3 & 1 & 0 \\ 11 & 2 & 17 \end{pmatrix} A \begin{pmatrix} 1 & -3 & 11 \\ 0 & 1 & 2 \\ 0 & 0 & 17 \end{pmatrix}$$

$$\Rightarrow D = P^T A P$$

where $D$ = a diagonal matrix = $\begin{pmatrix} 2 & 0 & 0 \\ 0 & -17 & 0 \\ 0 & 0 & -1377 \end{pmatrix}$ and $P = \begin{pmatrix} 1 & -3 & 11 \\ 0 & 1 & 2 \\ 0 & 0 & 17 \end{pmatrix}$.

$\therefore$ The canonical form of the given quadratic form is

$$Y^T DY = \begin{pmatrix} y_1 & y_2 & y_3 \end{pmatrix} \begin{pmatrix} 2 & 0 & 0 \\ 0 & -17 & 0 \\ 0 & 0 & -1377 \end{pmatrix} \begin{pmatrix} y_1 \\ y_2 \\ y_3 \end{pmatrix}$$

$$= 2y_1^2 - 17y_2^2 - 1377y_3^2.$$

Here,

$r$=Rank of $A$=3 (No. of non-zero terms in its canonical form)

$s$=Index Q.F=1 (No. of positive terms in its canonical form)

Signature=$2s - r = 2(1) - 3 = -1$.

The given quadratic form is indefinite.

The matrix of the transformation is

$$X = PY \Rightarrow \begin{pmatrix} x \\ y \\ z \end{pmatrix} = \begin{pmatrix} 1 & -3 & 11 \\ 0 & 1 & 2 \\ 0 & 0 & 17 \end{pmatrix} \begin{pmatrix} y_1 \\ y_2 \\ y_3 \end{pmatrix}.$$

**Example:** Reduce the quadratic form $x^2 + 3y^2 + 3z^2 - 2yz$ into canonical form and find its rank, index and signature of the quadratic form. **(Nov-2017), (Mar-2022)**

**Solution:** Symmetric matrix $A = \begin{pmatrix} 1 & 0 & 0 \\ 0 & 3 & -1 \\ 0 & -1 & 3 \end{pmatrix}.$

Diagonal matrix $D = \begin{pmatrix} 1 & 0 & 0 \\ 0 & 3 & 0 \\ 0 & 0 & 24 \end{pmatrix}$ and matrix of transformation $P = \begin{pmatrix} 1 & 0 & 0 \\ 0 & 1 & 1 \\ 0 & 0 & 3 \end{pmatrix}$

The required canonical form is $Y^T DY = y_1^2 + 3y_2^2 + 24y_3^2$.

Here $n = 3$, rank$(r) = 3$, index $s = 3$ and signature=3.

$r = n$ and $s = n$. Nature of the Q.F is positive definite.

**EXERCISE**

1. Reduce $4x^2 + 3y^2 + z^2 - 8xy - 6yz + 4zx$ to a canonical form using a linear transformation. Indicate the nature, rank, index and signature of the quadratic form.

**Ans:** $Y^T DY = 4y_1^2 - y_2^2 + 4y_3^2$. Rank=3, index=2, signature=1. Nature is indefinite.

2. Reduce the quadratic form $3x^2 - 2y^2 + z^2 + 12yz + 8zx - 4xy$ to the canonical form using diagonilazation method hence find the rank index signature.

**(Jan-2020), (Aug-2021)**

3. Find the nature, rank, index and signature of the quadratic from by reduce in to canonical form $x^2 + y^2 + 2z^2 + 2xy - 4xz + 4yz$. **(Sep-2021)**

4. Reduce the quadratic form $x^2 + 3y^2 + 3z^2 + 4t^2 + 4xy - 2xz + 6xt + 4yt + 2yz$ to the canonical form and hence find the nature, index, rank and signature of the quadratic form.

5. Reduce the quadratic form $x^2 - 2y^2 + 3z^2 - 4yz + 6zx$ in to canonical form using diagonalization method hence find rank, index and signature. **(Jan-2020)**

6. Reduce the quadratic form $3x^2 + 5y^2 - 3z^2 - 2yz + 2zx - 2xy$ to the canonical form using diagonalization method hence find rank, index and signature. **(Mar-2022)**

7. Reduce the quadratic form $x^2 + 4y^2 + z^2 + 4xy + 6yz + 2zx$ to canonical form. Also find signature and rank of the quadratic form.

8. Reduce the quadratic form $x^2 + y^2 + 2z^2 - 2xy + 4xz + 4yz$ to the canonical form and find the rank, index and signature.

9. Reduce the quadratic form $8x^2 + 7y^2 + 3z^2 - 12xy - 8yz + 4xz$ to the canonical form hence find the rank, index and signature.

10. Reduce the quadratic form $6x_1^2 + 3x_2^2 + 14x_3^2 + 4x_1x_2 + 18x_3x_1 + 4x_3x_2$ to canonical form. Also find signature and rank of the quadratic form.

## 2.4.2 Lagrange's Reduction Method

Let $X^T AX$ be the Q. F in $n$ variables.

1. Take the common terms from the product terms of the given Q. F.
2. Make prefect squares suitable by regrouping the terms.
3. The resulting relation gives the required C. F.

**Example:** Using Lagrange's method, reduce the quadratic form $x_1^2 + 4x_2^2 + x_3^2 - 4x_1x_2 + 2x_3x_1 - 4x_2x_3$ to canonical form. Also, find the nature, rank, index and signature of the quadratic form. **(July-2021)**

**Solution:** The given Q. F. is

$$x_1^2 + 4x_2^2 + x_3^2 - 4x_1x_2 + 2x_3x_1 - 4x_2x_3$$
$$= x_1^2 + (-2x_2)^2 + x_3^2 + 2(x_1)(-2x_2) + 2(x_3)(x_1) + 2(-2x_2)(x_3)$$
$$= (x_1 - 2x_2 + x_3)^2$$
$$= y_1^2$$

where $y_1 = x_1 - 2x_2 + x_3$.

The required Canonical form is $y_1^2$.

Number of variables in Q. F=$n$=3

Rank of the Q. F=$r$=1

Index of the Q. F=$s$=1

Signature=$2s - r$=2-1=1

Since $r < n$ and $s = r$.

Therefore, the nature of the given Q. F is positive semi definite.

**Example:** Using Lagrange's method, reduce the quadratic form $2x_1^2 + 7x_2^2 + 5x_3^2 - 8x_1x_2 - 10x_2x_3 + 4x_1x_3$ to sum of squares form. Hence, find the nature, rank, index and signature of the quadratic form.

**Solution:** The given Q. F. is

$$2x_1^2 + 7x_2^2 + 5x_3^2 - 8x_1x_2 - 10x_2x_3 + 4x_1x_3$$
$$= 2\left[x_1^2 + 4x_2^2 + x_3^2 - 4x_1x_2 - 4x_2x_3 + 2x_3x_1\right] - x_2^2 + 3x_3^2 - 2x_2x_3$$
$$= 2(x_1 - 2x_2 + x_3)^2 - (x_2^2 - 3x_3^2 + 2x_2x_3)$$
$$= 2(x_1 - 2x_2 + x_3)^2 - (x_2^2 + x_3^2 + 2x_2x_3 - 4x_3^2)$$
$$= 2(x_1 - 2x_2 + x_3)^2 - (x_2 + x_3)^2 + 4x_3^2$$
$$= 2y_1^2 - y_2^2 + 4y_3^2$$

where $y_1 = x_1 - 2x_2 + x_3$

$y_2 = x_2 + x_3$

$y_3 = x_3$.

Thus, the Canonical form is $2y_1^2 - y_2^2 + 4y_3^2$.

Number of variables in Q. F=$n$=3

Rank of the Q. F=$r$=3

Index of the Q. F=$s$=2

Signature=$2s - r$=2(2)-1=3

Since $r = n$ and $s < n$.

Therefore, the nature of the given Q. F is indefinite.

**Example:** Reduce the quadratic form $x_1^2 + 2x_2^2 - 7x_3^2 - 4x_1x_2 + 8x_1x_3$ by Lagrange's method of reduction to canonical form.

**Solution:** The given Q. F. is

$x_1^2 + 2x_2^2 - 7x_3^2 - 4x_1x_2 + 8x_1x_3$

$= x_1^2 - 2x_1(2x_2 - 4x_3) - 7x_3^2 + 2x_2^2$

$= \left[x_1^2 - 2x_1(2x_2 - 4x_3) + (2x_2 - 4x_3)^2 - (2x_2 - 4x_3)^2\right] - 7x_3^2 + 2x_2^2$

$= \left[x_1 - (2x_2 - 4x_3)\right]^2 - (2x_2 - 4x_3)^2 - 7x_3^2 + 2x_2^2$

$= (x_1 - 2x_2 + 4x_3)^2 - 4x_2^2 - 16x_3^2 + 16x_2x_3 - 7x_3^2 + 2x_2^2$

$= (x_1 - 2x_2 + 4x_3)^2 - 2x_2^2 - 23x_3^2 + 16x_2x_3$

$= (x_1 - 2x_2 + 4x_3)^2 - 2\left[x_2^2 - 8x_2x_3\right] - 23x_3^2$

$= (x_1 - 2x_2 + 4x_3)^2 - 2\left[x_2^2 - 2 \cdot x_2 \cdot 4x_3 + 4^2x_3^2 - 4^2x_3^2\right] - 23x_3^2$

$= (x_1 - 2x_2 + 4x_3)^2 - 2\left[x_2^2 - 2 \cdot x_2 \cdot 4x_3 + 4^2x_3^2\right] + 9x_3^2$

$= (x_1 - 2x_2 + 4x_3)^2 - 2(x_2 - 4x_3)^2 + 9x_3^2$

$= y_1^2 - 2y_2^2 + 9y_3^2$

where $y_1 = x_1 - 2x_2 + 4x_3$

$y_2 = x_2 - 4x_3$

$y_3 = x_3$.

Thus, the Canonical form is $y_1^2 - 2y_2^2 + 9y_3^2$.

Number of variables in Q. F=n=3

Rank of the Q. F=r=3

Index of the Q. F=s=2

Signature=$2s - r$=2(2)-1=3

Since $r = n$ and $s < n$.

Therefore, the nature of the given Q. F is indefinite.

**Example:** By Lagrange's reduction, transform the quadratic form $X^T A X$ to sum of squares form for $A = \begin{pmatrix} 1 & 2 & 4 \\ 2 & 6 & -2 \\ 4 & -2 & 18 \end{pmatrix}$.

**Solution:** The given quadratic form $X^T A X$ for the given matrix $A$ is

$$X^T A X = \begin{pmatrix} x & y & z \end{pmatrix} \begin{pmatrix} 1 & 2 & 4 \\ 2 & 6 & -2 \\ 4 & -2 & 18 \end{pmatrix} \begin{pmatrix} x \\ y \\ z \end{pmatrix}$$

$$= \begin{pmatrix} x & y & z \end{pmatrix} \begin{pmatrix} x + 2y + 4z \\ 2x + 6y - 2z \\ 4x - 2y + 18z \end{pmatrix}$$

$$= x^2 + 2xy + 4zx + 2xy + 6y^2 - 2yz + 4zx - 2yz + 18z^2$$

$$= x^2 + 6y^2 + 18z^2 + 4xy - 4yz + 8zx$$

$$= \left[ x^2 + 4y^2 + 16z^2 + 4xy - 4yz + 8zx \right] + 2y^2 + 2z^2$$

$$= \left[ x^2 + (2y)^2 + (4z)^2 + 2(x)(2y) + 2(2y)(4z) + 2(4z)(x) \right] + 2y^2 + 2z^2 - 20yz$$

$$= (x + 2y + 4z)^2 + 2\left[ y^2 + z^2 - 10yz \right]$$

$$= (x + 2y + 4z)^2 + 2\left[ y^2 + (-5z)^2 + 2(y)(-5z) \right] - 48z^2$$

$$= (x + 2y + 4z)^2 + 2(y - 5z)^2 - 48z^2$$

$$= y_1^2 + 2y_2^2 - 48y_3^2$$

where $y_1 = x + 2y + 4z$

$y_2 = y - 5z$

$y_3 = z$.

Thus, the Canonical form is $y_1^2 + 2y_2^2 - 48y_3^2$.

Number of variables in Q. F=$n$=3

Rank of the Q. F=$r$=3

Index of the Q. F=$s$=2

Signature=$2s - r$=2(2)-1=3

Since $r = n$ and $s < n$.

Therefore, the nature of the given Q. F is indefinite.

## 2.4.3 Orthogonal Transformation

In this process we can reduce the quadratic form to the canonical form by orthogonal transformation $X = PY$, where $P$ is an orthogonal matrix.

To find the orthogonal transformation $X = PY$, where $P$ is an orthogonal matrix, we have to proceed as follows:

1). First write the symmetric matrix $A$ of the given quadratic form.

2). Find the eigen values of the matrix $A$.

3). Find the eigen vectors of the matrix $A$ corresponding to the eigen values such that eigen vectors must be pairwise orthogonal.

Then find the normalized eigen vectors of $A$ and form the required orthogonal matrix $P$.

$$\text{i.e., } P = \begin{pmatrix} \frac{X_1}{\|X_1\|} & \frac{X_2}{\|X_2\|} & \frac{X_3}{\|X_3\|} \end{pmatrix}.$$

**Note: 1.** Normalized eigen vector of $X$ is $\frac{X}{\|X\|}$.

**Ex:** If $X = \begin{pmatrix} 1 & 2 & 3 \end{pmatrix}^T$, $\|X\| = \sqrt{1+4+9} = \sqrt{14}$.

Then its normalized eigen vector is $\frac{X}{\|X\|} = \begin{pmatrix} \frac{1}{\sqrt{14}} \\ \frac{2}{\sqrt{14}} \\ \frac{3}{\sqrt{14}} \end{pmatrix}$.

2. If the eigen values of a matrix are distinct then its corresponding eigen vectors are always pairwise orthogonal.

3. If the eigen values of a matrix $A$ are not distinct then we have to find the corresponding linear independent eigen vectors in such away that they must pairwise orthogonal.

Any two vectors $X_1 = \begin{pmatrix} x_1 \\ x_2 \\ x_3 \end{pmatrix}$ and $X_2 = \begin{pmatrix} a \\ b \\ c \end{pmatrix}$ are said to be pairwise orthogonal, if their linear combination i.e., $ax_1 + bx_2 + cx_3 = 0$.

**Example:** Find the index, signature and nature of the quadratic from $x_1^2 + 2x_2^2 -$

$3x_3^2$.

**Solution:** The quadratic form is in canonical form in 3 variables.

$$\therefore \lambda_1 = 1,\ \lambda_2 = 2,\ \lambda_3 = -3.$$

There are 2 positive eigen values.

$$\therefore \text{Index} = 2$$

One negative eigen value.

$$\therefore \text{Signature} = 2 - 1 = 1$$

Rank=3(i.e., the total number of +ve and -ve eigen values).

**Note:** Its nature is indefinite.

**Example:** Reduce the quadratic form $3x^2 + 5y^2 + 3z^2 - 2xy - 2yz + 2zx$ to the canonical form by orthogonal transformation and hence find the rank, index, signature and nature of the quadratic form. **(Apr-2017), (Apr-2018), (Apr-2022)**

**Solution:** The given quadratic form can be written in the matrix form as $X^T A X$ where $X^T = \begin{bmatrix} x & y & z \end{bmatrix}$ and the symmetric matrix

$$A = \begin{pmatrix} 3 & -1 & 1 \\ -1 & 5 & -1 \\ 1 & -1 & 3 \end{pmatrix}.$$

The characteristic equation of $A$ is $|A - \lambda I| = 0$.

$$\Rightarrow \begin{vmatrix} 3-\lambda & -1 & 1 \\ -1 & 5-\lambda & -1 \\ 1 & -1 & 3-\lambda \end{vmatrix} = 0.$$

$$\Rightarrow \lambda^3 - 11\lambda^2 + 36\lambda - 36 = 0$$

$$\Rightarrow \lambda = 2, 3, 6.$$

Thus the eigenvalues of $A$ are $\lambda = 2, 3, 6$, which are distinct.

**To find eigenvectors:** Let $X = \begin{pmatrix} x_1 \\ x_2 \\ x_3 \end{pmatrix}$ be the eigenvector of $A$ corresponding to the eigenvalue $\lambda$. Then

$$(A - \lambda I)X = 0$$

$$\Rightarrow \begin{bmatrix} 3-\lambda & -1 & 1 \\ -1 & 5-\lambda & -1 \\ 1 & -1 & 3-\lambda \end{bmatrix} \begin{pmatrix} x_1 \\ x_2 \\ x_3 \end{pmatrix} = \begin{pmatrix} 0 \\ 0 \\ 0 \end{pmatrix} \quad \ldots \ldots \quad (1)$$

**Case:1** Put $\lambda = 2$ in (1), we get

$$\Rightarrow \begin{bmatrix} 1 & -1 & 1 \\ -1 & 3 & -1 \\ 1 & -1 & 1 \end{bmatrix} \begin{pmatrix} x_1 \\ x_2 \\ x_3 \end{pmatrix} = \begin{pmatrix} 0 \\ 0 \\ 0 \end{pmatrix}$$

This gives,

$$x_1 - x_2 + x_3 = 0$$

$$-x_1 + 3x_2 - x_3 = 0$$

$$x_1 - x_2 + x_3 = 0.$$

Solving last two equations by cross multiplication, we have

$$\frac{x_1}{3-1} = \frac{x_2}{-1+1} = \frac{x_3}{1-3}$$

$$\Rightarrow \frac{x_1}{2} = \frac{x_2}{0} = \frac{x_3}{-2}$$

$$\Rightarrow \frac{x_1}{1} = \frac{x_2}{0} = \frac{x_3}{-1}$$

$$\Rightarrow x_1 = 1, \ x_2 = 0, \ x_3 = -1$$

$\therefore$ The eigenvector corresponding to $\lambda = 2$ is $X_1 = \begin{pmatrix} 1 \\ 0 \\ -1 \end{pmatrix}$

and $\|X_1\| = \sqrt{1^2 + 0 + (-1)^2} = \sqrt{2}$.

Normalized eigenvector is $\begin{pmatrix} \frac{1}{\sqrt{2}} \\ 0 \\ \frac{-1}{\sqrt{2}} \end{pmatrix}$.

Similarly, corresponding to $\lambda = 3$, the eigenvector is $X_2 = \begin{pmatrix} 1 \\ 1 \\ 1 \end{pmatrix}$.

Normalized eigenvector is $\begin{pmatrix} \frac{1}{\sqrt{3}} \\ \frac{1}{\sqrt{3}} \\ \frac{1}{\sqrt{3}} \end{pmatrix}$.

Eigenvector corresponding to $\lambda = 6$ is $X_3 = \begin{pmatrix} 1 \\ -2 \\ 1 \end{pmatrix}$.

Normalized eigenvector is $\begin{pmatrix} \frac{1}{\sqrt{6}} \\ \frac{-2}{\sqrt{6}} \\ \frac{1}{\sqrt{6}} \end{pmatrix}$.

Here the eigenvectors $X_1, X_2, X_3$ are mutually orthogonal.

Modal matrix $= \begin{pmatrix} X_1 & X_2 & X_3 \end{pmatrix} = \begin{pmatrix} 1 & 1 & 1 \\ 0 & 1 & -2 \\ -1 & 1 & 1 \end{pmatrix}$.

Normalized modal matrix $P = \begin{pmatrix} \frac{1}{\sqrt{2}} & \frac{1}{\sqrt{3}} & \frac{1}{\sqrt{6}} \\ 0 & \frac{1}{\sqrt{3}} & \frac{-2}{\sqrt{6}} \\ \frac{-1}{\sqrt{2}} & \frac{1}{\sqrt{3}} & \frac{1}{\sqrt{6}} \end{pmatrix}$.

This is an orthogonal matrix.

We can verify that $PP^T = P^T P = I$.

Diagonalized matrix

$$D = P^{-1}AP = P^T AP = \begin{pmatrix} \frac{1}{\sqrt{2}} & 0 & \frac{-1}{\sqrt{2}} \\ \frac{1}{\sqrt{3}} & \frac{1}{\sqrt{3}} & \frac{1}{\sqrt{3}} \\ \frac{1}{\sqrt{6}} & \frac{-2}{\sqrt{6}} & \frac{1}{\sqrt{6}} \end{pmatrix} \begin{pmatrix} 3 & -1 & 1 \\ -1 & 5 & -1 \\ 1 & -1 & 3 \end{pmatrix} \begin{pmatrix} \frac{1}{\sqrt{2}} & \frac{1}{\sqrt{3}} & \frac{1}{\sqrt{6}} \\ 0 & \frac{1}{\sqrt{3}} & \frac{-2}{\sqrt{6}} \\ \frac{-1}{\sqrt{2}} & \frac{1}{\sqrt{3}} & \frac{1}{\sqrt{6}} \end{pmatrix}$$

$$= \begin{pmatrix} 2 & 0 & 0 \\ 0 & 3 & 0 \\ 0 & 0 & 6 \end{pmatrix}$$

Hence the required canonical form is

$$Y^T DY = \begin{pmatrix} y_1 & y_2 & y_3 \end{pmatrix} \begin{pmatrix} 2 & 0 & 0 \\ 0 & 3 & 0 \\ 0 & 0 & 6 \end{pmatrix} \begin{pmatrix} y_1 \\ y_2 \\ y_3 \end{pmatrix} = 2y_1^2 + 3y_2^2 + 6y_3^2.$$

Number of variables in Q. F=n=3

Rank of the Q. F=r=3

Index of the Q. F=s=3

Signature of the Q.F=$2s - r = 2(3) - 3 = 3$

Since $r = n$ and $s = n$, the nature of the given Q. F is positive definite.

**Example:** Reduce the quadratic form $6x_1^2 + 3x_2^2 + 3x_3^2 - 4x_1x_2 - 2x_2x_3 + 4x_1x_3$ to canonical form and hence state nature, rank, index and signature of the quadratic form. **(Apr-2017), (Apr-2022)**

Determine the diagonal matrix orthogonally similar to the following symmetric matrix $A = \begin{pmatrix} 6 & -2 & 2 \\ -2 & 3 & -1 \\ 2 & -1 & 3 \end{pmatrix}$. **(Apr-2018)**

Find the eigen vectors of the matrix $A = \begin{pmatrix} 6 & -2 & 2 \\ -2 & 3 & -1 \\ 2 & -1 & 3 \end{pmatrix}$ and hence reduce $6x^2 + 3y^2 + 3z^2 - 2yz + 4zx - 4xy$ to a "sum of squares". Also write the nature of the matrix.

**Solution:** The given quadratic form is $6x_1^2 + 3x_2^2 + 3x_3^2 - 4x_1x_2 - 2x_2x_3 + 4x_1x_3$.

The matrix of the quadratic form is $A = \begin{pmatrix} 6 & -2 & 2 \\ -2 & 3 & -1 \\ 2 & -1 & 3 \end{pmatrix}$.

The characteristic equation of $A$ is $|A - \lambda I| = 0$.

$$\Rightarrow \begin{vmatrix} 6-\lambda & -2 & 2 \\ -2 & 3-\lambda & -1 \\ 2 & -1 & 3-\lambda \end{vmatrix} = 0.$$

$$\Rightarrow \lambda^3 - 12\lambda^2 + 36\lambda - 32 = 0$$

$$\Rightarrow \lambda = 2, 2, 8.$$

Thus the eigenvalues of $A$ are $\lambda = 2, 2, 8$ which are not distinct. Now the eigenvectors $X$ corresponding to the eigenvalue $\lambda$ are obtained by solving the system of equations

$$(A - \lambda I)X = 0 \quad \cdots\cdots \quad (1)$$

For $\lambda = 2$, the system (1) can be written as

$$\begin{pmatrix} 4 & -2 & 2 \\ -2 & 1 & -1 \\ 2 & -1 & 1 \end{pmatrix} \begin{pmatrix} x_1 \\ x_2 \\ x_3 \end{pmatrix} = \begin{pmatrix} 0 \\ 0 \\ 0 \end{pmatrix}$$

$$\Rightarrow 4x_1 - 2x_2 + 2x_3 = 0 \quad \cdots\cdots \quad (2)$$

Let us take $x_1 = 1, x_2 = 2, x_3 = 0$ [$\because$ Take any set of values satisfying the equation]

i.e., $X_1 = \begin{pmatrix} 1 \\ 2 \\ 0 \end{pmatrix}$ is the linearly independent eigenvector of $A$ corresponding to the eigenvalue $\lambda = 2$.

Now we have to find the another linearly independent eigenvector $X_2$ of $A$ corresponding to the same eigenvalue such that $X_1$ and $X_2$ are mutually orthogonal.

Let $X_2 = \begin{pmatrix} a \\ b \\ c \end{pmatrix}$ be the another linearly independent eigenvector of $A$ corresponding to $\lambda = 2$ and is orthogonal to $X_1$ so that

$$a + 2b + 0c = 0 \text{ and}$$
$$4a - 2b + 2c = 0 \quad [\because X_2 \text{ satisfies equation (2)}]$$

Solving these equations by cross multiplication, we have

$$\frac{a}{4-0} = \frac{b}{-2} = \frac{c}{-2-8}$$

$$\Rightarrow \frac{a}{4} = \frac{b}{-2} = \frac{c}{-10}$$

$$\Rightarrow \frac{a}{2} = \frac{b}{-1} = \frac{c}{-5}$$

i.e., $a = 2, b = -1, c = -5$

Therefore, $X_2 = \begin{pmatrix} 2 \\ -1 \\ -5 \end{pmatrix}$ is the another linearly independent eigenvector of $A$ corresponding to $\lambda = 2$ and is orthogonal to $X_1$.

For $\lambda = 8$, the system (1) can be written as
$$\begin{pmatrix} -2 & -2 & 2 \\ -2 & -5 & -1 \\ 2 & -1 & -5 \end{pmatrix} \begin{pmatrix} x_1 \\ x_2 \\ x_3 \end{pmatrix} = \begin{pmatrix} 0 \\ 0 \\ 0 \end{pmatrix}$$

$$x_1 + x_2 - x_3 = 0$$
$$2x_1 + 5x_2 + x_3 = 0$$
$$2x_1 - x_2 - 5x_3 = 0$$

Solving these equations by cross multiplication, we have

$$\frac{x_1}{1+5} = \frac{x_2}{-2-1} = \frac{x_3}{5-2}$$
$$\frac{x_1}{6} = \frac{x_2}{-3} = \frac{x_3}{3}$$
$$\frac{x_1}{2} = \frac{x_2}{-1} = \frac{x_3}{1}$$

i.e., $x_1 = -2, x_2 = 1, x_3 = -1$.

Therefore, $X_3 = \begin{pmatrix} -2 \\ 1 \\ -1 \end{pmatrix}$ is the linearly independent eigenvector of $A$ corresponding to $\lambda = 8$ which is orthogonal to both $X_1$ and $X_2$.

i.e., The eigenvectors $X_1, X_2, X_3$ are mutually orthogonal.

Now $\|X_1\| = \sqrt{1+4+0} = \sqrt{5}$
$\|X_2\| = \sqrt{4+1+25} = \sqrt{30}$
$\|X_3\| = \sqrt{4+1+1} = \sqrt{6}$.

Let $P = \begin{pmatrix} \frac{X_1}{\|X_1\|} & \frac{X_2}{\|X_2\|} & \frac{X_3}{\|X_3\|} \end{pmatrix} = \begin{pmatrix} \frac{1}{\sqrt{5}} & \frac{2}{\sqrt{30}} & \frac{-2}{\sqrt{6}} \\ \frac{2}{\sqrt{5}} & \frac{-1}{\sqrt{30}} & \frac{1}{\sqrt{6}} \\ 0 & \frac{-5}{\sqrt{30}} & \frac{-1}{\sqrt{6}} \end{pmatrix}$ which is called the orthogonal matrix of $A$.

We can verify that $PP^T = P^TP = I$ and $P^TAP = \text{diag}(2,2,4)$.

Take $X = PY$.

$$X^TAX = Y^TDY = Y^T \begin{pmatrix} 2 & 0 & 0 \\ 0 & 2 & 0 \\ 0 & 0 & 8 \end{pmatrix} Y = 2y_1^2 + 2y_2^2 + 8y_3^2.$$

This is the required canonical form.

Number of variables in Q. F=n=3

Rank of the Q. F=r=3

Index of the Q. F=s=3

Signature of the Q.F=$2s - r = 2(3) - 3 = 3$

Since $r = n$ and $s = n$, the nature of the given Q. F is positive definite.

**Example:** Reduce the quadratic form $2x^2 + 2y^2 + 2z^2 - 2xy - 2yz - 2zx$ in to canonical form by an orthogonal transformation and hence find rank, index, signature and nature of the quadratic form.  **(Apr-2017), (Jan-2020)**

**Solution:** The given quadratic form is $2x^2 + 2y^2 + 2z^2 - 2xy - 2yz - 2zx$.

The matrix of the quadratic form is $A = \begin{pmatrix} 2 & -1 & -1 \\ -1 & 2 & -1 \\ -1 & -1 & 2 \end{pmatrix}$.

The characteristic equation of $A$ is $|A - \lambda I| = 0$.

$$\Rightarrow \begin{vmatrix} 2-\lambda & -1 & -1 \\ -1 & 2-\lambda & -1 \\ -1 & -1 & 2-\lambda \end{vmatrix} = 0.$$

$\Rightarrow \lambda^3 - 6\lambda^2 + 9\lambda = 0$

$\Rightarrow \lambda = 0, 3, 3$.

Thus the eigenvalues of $A$ are $\lambda = 0, 3, 3$ which are not distinct. Now the eigenvectors $X$ corresponding to the eigenvalue $\lambda$ are obtained by solving the system of equations

$$(A - \lambda I)X = 0 \quad \cdots\cdots \quad (1)$$

For $\lambda = 3$, the system (1) can be written as
$$\begin{pmatrix} -1 & -1 & -1 \\ -1 & -1 & -1 \\ -1 & -1 & -1 \end{pmatrix} \begin{pmatrix} x_1 \\ x_2 \\ x_3 \end{pmatrix} = \begin{pmatrix} 0 \\ 0 \\ 0 \end{pmatrix}$$

$$\Rightarrow -x_1 - x_2 - x_3 = 0 \quad \cdots \cdots \quad (2)$$

Let us take $x_1 = -1, x_2 = 1, x_3 = 0$ [$\because$ Take any set of values satisfying the equation]

i.e., $X_1 = \begin{pmatrix} -1 \\ 1 \\ 0 \end{pmatrix}$ is the linearly independent eigenvector of $A$ corresponding to the eigenvalue $\lambda = 3$.

Now we have to find the another linearly independent eigenvector $X_2$ of $A$ corresponding to the same eigenvalue such that $X_1$ and $X_2$ are mutually orthogonal.

Let $X_2 = \begin{pmatrix} a \\ b \\ c \end{pmatrix}$ be the another linearly independent eigenvector of $A$ corresponding to $\lambda = 3$ and is orthogonal to $X_1$ so that

$$-a + b + 0c = 0 \text{ and}$$

$$-a - b - c = 0 \quad [\because X_2 \text{ satisfies equation (2)}]$$

Solving these equations by cross multiplication, we have

$$\frac{a}{-1-0} = \frac{b}{0-1} = \frac{c}{1+1}$$

$$\Rightarrow \frac{a}{-1} = \frac{b}{-1} = \frac{c}{2}$$

$$\Rightarrow \frac{a}{1} = \frac{b}{1} = \frac{c}{-2}$$

i.e., $a = 1, b = 1, c = -2$

Therefore, $X_2 = \begin{pmatrix} 1 \\ 1 \\ -2 \end{pmatrix}$ is the another linearly independent eigenvector of $A$ corresponding to $\lambda = 3$ and is orthogonal to $X_1$.

For $\lambda = 0$, the system (1) can be written as
$$\begin{pmatrix} 2 & -1 & -1 \\ -1 & 2 & -1 \\ -1 & -1 & 2 \end{pmatrix} \begin{pmatrix} x_1 \\ x_2 \\ x_3 \end{pmatrix} = \begin{pmatrix} 0 \\ 0 \\ 0 \end{pmatrix}$$

$$2x_1 - x_2 - x_3 = 0$$
$$-x_1 + 2x_2 - x_3 = 0$$
$$-x_1 - x_2 + 2x_3 = 0$$

Solving these equations by cross multiplication, we have
$$\frac{x_1}{1+2} = \frac{x_2}{1+2} = \frac{x_3}{4-1}$$
$$\frac{x_1}{3} = \frac{x_2}{3} = \frac{x_3}{3}$$
$$\frac{x_1}{1} = \frac{x_2}{1} = \frac{x_3}{1}$$

i.e., $x_1 = 1, x_2 = 1, x_3 = 1$.

Therefore, $X_3 = \begin{pmatrix} 1 \\ 1 \\ 1 \end{pmatrix}$ is the linearly independent eigenvector of $A$ corresponding to $\lambda = 0$ which is orthogonal to both $X_1$ and $X_2$.

i.e., The eigenvectors $X_1, X_2, X_3$ are mutually orthogonal.

Now $\|X_1\| = \sqrt{1+1+0} = \sqrt{2}$
$\|X_2\| = \sqrt{1+1+4} = \sqrt{6}$
$\|X_3\| = \sqrt{1+1+1} = \sqrt{3}$.

Let $P = \begin{pmatrix} \frac{X_1}{\|X_1\|} & \frac{X_2}{\|X_2\|} & \frac{X_3}{\|X_3\|} \end{pmatrix} = \begin{pmatrix} \frac{-1}{\sqrt{2}} & \frac{1}{\sqrt{6}} & \frac{1}{\sqrt{3}} \\ \frac{1}{\sqrt{2}} & \frac{1}{\sqrt{6}} & \frac{1}{\sqrt{3}} \\ 0 & \frac{-2}{\sqrt{6}} & \frac{1}{\sqrt{3}} \end{pmatrix}$ which is called the orthogonal matrix of $A$.

We can verify that $PP^T = P^T P = I$ and $P^T A P = \text{diag}(3, 3, 0)$.

Take $X = PY$.

$$X^T A X = Y^T D Y = Y^T \begin{pmatrix} 3 & 0 & 0 \\ 0 & 3 & 0 \\ 0 & 0 & 0 \end{pmatrix} Y = 3y_1^2 + 3y_2^2 + 0y_3^2 = 3y_1^2 + 3y_2^2.$$

This is the required canonical form.

Number of variables in Q. F=$n = 3$

Rank of the Q. F=$r = 2$

Index of the Q. F=$s = 2$

Signature of the Q.F=$2s - r = 2(2) - 2 = 2$

Since $r < n$ and $s = r$, the nature of the given Q. F is positive semidefinite.

**Example:** Reduce the quadratic form $2x^2 + 2y^2 + 2z^2 - 2xy + 2zx - 2yz$ to canonical form by an orthogonal transformation and hence find rank, index, signature and nature of the quadratic form. **(Dec-2020)**

**Solution:** The given quadratic form is $2x^2 + 2y^2 + 2z^2 - 2xy + 2zx - 2yz$.

The matrix of the quadratic form is $A = \begin{pmatrix} 2 & -1 & 1 \\ -1 & 2 & -1 \\ 1 & -1 & 2 \end{pmatrix}$.

The characteristic equation of $A$ is $|A - \lambda I| = 0$.

$$\Rightarrow \begin{vmatrix} 2-\lambda & -1 & 1 \\ -1 & 2-\lambda & -1 \\ 1 & -1 & 2-\lambda \end{vmatrix} = 0.$$

$\Rightarrow \lambda^3 - 6\lambda^2 + 9\lambda - 4 = 0$

$\Rightarrow \lambda = 1, 1, 4$.

Thus the eigenvalues of $A$ are $\lambda = 1, 1, 4$ which are not distinct. Now the eigenvectors $X$ corresponding to the eigenvalue $\lambda$ are obtained by solving the system of equations

$$(A - \lambda I)X = 0 \quad \cdots \cdots \quad (1)$$

For $\lambda = 1$, the system (1) can be written as

$$\begin{pmatrix} 1 & -1 & 1 \\ -1 & 1 & -1 \\ 1 & -1 & 1 \end{pmatrix} \begin{pmatrix} x_1 \\ x_2 \\ x_3 \end{pmatrix} = \begin{pmatrix} 0 \\ 0 \\ 0 \end{pmatrix}$$

$$\Rightarrow x_1 - x_2 + x_3 = 0 \quad \cdots \cdots \quad (2)$$

Let us take $x_1 = 1, x_2 = 1, x_3 = 0$ [$\because$ Take any set of values satisfying the equation]

i.e., $X_1 = \begin{pmatrix} 1 \\ 1 \\ 0 \end{pmatrix}$ is the linearly independent eigenvector of $A$ corresponding to the eigenvalue $\lambda = 1$.

Now we have to find the another linearly independent eigenvector $X_2$ of $A$ corresponding to the same eigenvalue such that $X_1$ and $X_2$ are mutually orthogonal.

Let $X_2 = \begin{pmatrix} a \\ b \\ c \end{pmatrix}$ be the another linearly independent eigenvector of $A$ corresponding to $\lambda = 1$ and is orthogonal to $X_1$ so that

$$a - b + c = 0 \text{ and}$$
$$a + b + 0c = 0 \quad [\because X_2 \text{ satisfies equation } (2)]$$

Solving these equations by cross multiplication, we have

$$\frac{a}{0-1} = \frac{b}{1-0} = \frac{c}{1+1}$$

$$\Rightarrow \frac{a}{-1} = \frac{b}{1} = \frac{c}{2}$$

$$\Rightarrow \frac{a}{1} = \frac{b}{-1} = \frac{c}{-2}$$

i.e., $a = 1, b = -1, c = -2$

Therefore, $X_2 = \begin{pmatrix} 1 \\ -1 \\ -2 \end{pmatrix}$ is the another linearly independent eigenvector of $A$ corresponding to $\lambda = 1$ and is orthogonal to $X_1$.

For $\lambda = 4$, the system (1) can be written as

$$\begin{pmatrix} -2 & -1 & 1 \\ -1 & -2 & -1 \\ 1 & -1 & -2 \end{pmatrix} \begin{pmatrix} x_1 \\ x_2 \\ x_3 \end{pmatrix} = \begin{pmatrix} 0 \\ 0 \\ 0 \end{pmatrix}$$

$$-2x_1 - x_2 + x_3 = 0$$
$$-x_1 - 2x_2 - x_3 = 0$$
$$x_1 - x_2 - 2x_3 = 0$$

Solving these equations by cross multiplication, we have

$$\frac{x_1}{1+2} = \frac{x_2}{-1-2} = \frac{x_3}{4-1}$$
$$\frac{x_1}{3} = \frac{x_2}{-3} = \frac{x_3}{3}$$
$$\frac{x_1}{1} = \frac{x_2}{-1} = \frac{x_3}{1}$$

i.e., $x_1 = 1, x_2 = -1, x_3 = 1$.

Therefore, $X_3 = \begin{pmatrix} 1 \\ -1 \\ 1 \end{pmatrix}$ is the linearly independent eigenvector of $A$ corresponding to $\lambda = 4$ which is orthogonal to both $X_1$ and $X_2$.

i.e., The eigenvectors $X_1, X_2, X_3$ are mutually orthogonal.

Now $\|X_1\| = \sqrt{1+1+0} = \sqrt{2}$
$\|X_2\| = \sqrt{1+1+4} = \sqrt{6}$
$\|X_3\| = \sqrt{1+1+1} = \sqrt{3}$.

Let $P = \begin{pmatrix} \frac{X_1}{\|X_1\|} & \frac{X_2}{\|X_2\|} & \frac{X_3}{\|X_3\|} \end{pmatrix} = \begin{pmatrix} \frac{1}{\sqrt{2}} & \frac{1}{\sqrt{6}} & \frac{1}{\sqrt{3}} \\ \frac{1}{\sqrt{2}} & \frac{-1}{\sqrt{6}} & \frac{-1}{\sqrt{3}} \\ 0 & \frac{-2}{\sqrt{6}} & \frac{1}{\sqrt{3}} \end{pmatrix}$ which is called the orthogonal matrix of $A$.

We can verify that $PP^T = P^T P = I$ and $P^T A P = \text{diag}(1, 1, 4)$.

Take $X = PY$.

$$X^T AX = Y^T DY = Y^T \begin{pmatrix} 1 & 0 & 0 \\ 0 & 1 & 0 \\ 0 & 0 & 4 \end{pmatrix} Y = 1y_1^2 + 1y_2^2 + 4y_3^2 = y_1^2 + y_2^2 + 4y_3^2.$$

This is the required canonical form.

Number of variables in Q. F=$n = 3$

Rank of the Q. F=$r = 3$

Index of the Q. F=$s = 3$

Signature of the Q.F=$2s - r = 2(3) - 3 = 3$

Since $r = n$ and $s = n$, the nature of the given Q. F is positive definite.

**Example:** Reduce the quadratic form $3x_1^2 + 3x_2^2 + 3x_3^2 + 2x_1x_2 + 2x_1x_3 - 2x_2x_3$ into a sum of squares form by an orthogonal transformation and indicate the matrix of the transformation. **(Apr-2022)**

**Solution:** Given $3x_1^2 + 3x_2^2 + 3x_3^2 + 2x_1x_2 + 2x_1x_3 - 2x_2x_3$.

The matrix of the quadratic form is $A = \begin{pmatrix} 3 & 1 & 1 \\ 1 & 3 & -1 \\ 1 & -1 & 3 \end{pmatrix}$.

The characteristic equation of $A$ is $|A - \lambda I| = 0$

$$\Rightarrow \lambda^3 - S_1 \lambda^2 + S_2 \lambda - S_3 = 0$$

where $S_1$ =Sum of the diagonal elements of $A$

$$= 3 + 3 + 3 = 9$$

$S_2$ =Sum of minors of elements of diagonal of $A$

$$= \begin{vmatrix} 3 & -1 \\ -1 & 3 \end{vmatrix} + \begin{vmatrix} 3 & 1 \\ 1 & 3 \end{vmatrix} + \begin{vmatrix} 3 & 1 \\ 1 & 3 \end{vmatrix}$$

$$= (9 - 1) + (9 - 1) + (9 - 1) = 24$$

$S_3 = |A| = 3(9 - 1) - 1(3 + 1) + 1(-1 - 3) = 16.$

∴ The characteristic equation is $\lambda^3 - 9\lambda^2 + 24\lambda - 16 = 0$

$$\Rightarrow \lambda = 1, 4, 4$$

∴ The eigen values are $\lambda = 1, 4, 4$.

**To find eigen vectors:**

Let $X = \begin{bmatrix} x \\ y \\ z \end{bmatrix}$ be an eigen vector corresponding $\lambda$.

Then

$$(A - \lambda I)X = 0 \Rightarrow \begin{bmatrix} 3-\lambda & 1 & 1 \\ 1 & 3-\lambda & -1 \\ 1 & -1 & 3-\lambda \end{bmatrix} \begin{bmatrix} x \\ y \\ z \end{bmatrix} = \begin{bmatrix} 0 \\ 0 \\ 0 \end{bmatrix}$$

$$\left. \begin{array}{r} (3-\lambda)x + y + z = 0 \\ x + (3-\lambda)y - z = 0 \\ x - y + (3-\lambda)z = 0 \end{array} \right\} \quad \cdots\cdots \quad (I)$$

**Case (i)** If $\lambda = 1$ then equations (I) become

$$2x + y + z = 0$$
$$x + 2y - z = 0$$
$$x - y + 2z = 0$$

Choosing the first two equations, we have

$$\frac{x}{-1-2} = \frac{y}{1+2} = \frac{z}{4-1}$$

$$\Rightarrow \frac{x}{-3} = \frac{y}{3} = \frac{z}{3}$$

$$\Rightarrow \frac{x}{-1} = \frac{y}{1} = \frac{z}{1}$$

Choosing $x_1 = -1, x_2 = 1, x_3 = 1$, we get an eigen vector

$$X_1 = \begin{bmatrix} -1 \\ 1 \\ 1 \end{bmatrix}.$$

**Case (ii)** If $\lambda = 4$, then the equations (I) become

$$-x + y + z = 0 \Rightarrow x - y - z = 0$$
$$x - y - z = 0$$
$$x - y - z = 0$$

We get only one equation $x - y - z = 0$ ...... (1)

To solve for $x, y, z$, we can assign arbitrary values for two of the variables and we shall find 2 orthogonal vectors.

Put $z = 0, y = 1$, then $x = 1$, we get an eigen vector

$$X_2 = \begin{bmatrix} 1 \\ 1 \\ 0 \end{bmatrix}.$$

Let $X_3 = \begin{bmatrix} a \\ b \\ c \end{bmatrix}$ be orthogonal to $X_2$. Then $a + b = 0 \Rightarrow b = -a$ and $X_3$ should satisfy (1). $\therefore a - b - c = 0$.

Choosing $a = 1$, we get $b = -1$ and $c = 2$.

$$\therefore X_3 = \begin{bmatrix} 1 \\ -1 \\ 2 \end{bmatrix}.$$

Thus the eigen values are $\lambda = 1, 4, 4$, and the corresponding eigen vector are

$$X_1 = \begin{bmatrix} -1 \\ 1 \\ 1 \end{bmatrix}, \quad X_2 = \begin{bmatrix} 1 \\ 1 \\ 0 \end{bmatrix}, \quad X_3 = \begin{bmatrix} 1 \\ -1 \\ 2 \end{bmatrix}.$$

The normalised eigen vectors are

$$\begin{bmatrix} -\frac{1}{\sqrt{3}} \\ \frac{1}{\sqrt{3}} \\ \frac{1}{\sqrt{3}} \end{bmatrix}, \quad \begin{bmatrix} \frac{1}{\sqrt{2}} \\ \frac{1}{\sqrt{2}} \\ 0 \end{bmatrix}, \quad \begin{bmatrix} \frac{1}{\sqrt{6}} \\ \frac{-1}{\sqrt{6}} \\ \frac{2}{\sqrt{6}} \end{bmatrix}.$$

So, the normalised modal matrix

$$N = \begin{pmatrix} -\frac{1}{\sqrt{3}} & \frac{1}{\sqrt{2}} & \frac{1}{\sqrt{6}} \\ \frac{1}{\sqrt{3}} & \frac{1}{\sqrt{2}} & \frac{-1}{\sqrt{6}} \\ -\frac{1}{\sqrt{3}} & 0 & \frac{2}{\sqrt{6}} \end{pmatrix}$$

$$N^T A N = \begin{pmatrix} -\frac{1}{\sqrt{3}} & \frac{1}{\sqrt{3}} & \frac{1}{\sqrt{3}} \\ \frac{1}{\sqrt{2}} & \frac{1}{\sqrt{2}} & 0 \\ \frac{1}{\sqrt{6}} & -\frac{1}{\sqrt{6}} & \frac{2}{\sqrt{6}} \end{pmatrix} \begin{pmatrix} 3 & 1 & 1 \\ 1 & 3 & -1 \\ 1 & -1 & 3 \end{pmatrix} \begin{pmatrix} -\frac{1}{\sqrt{3}} & \frac{1}{\sqrt{2}} & \frac{1}{\sqrt{6}} \\ \frac{1}{\sqrt{3}} & \frac{1}{\sqrt{2}} & \frac{-1}{\sqrt{6}} \\ -\frac{1}{\sqrt{3}} & 0 & \frac{2}{\sqrt{6}} \end{pmatrix}$$

$$= \begin{pmatrix} 1 & 0 & 0 \\ 0 & 4 & 0 \\ 0 & 0 & 4 \end{pmatrix}, \text{ which is a diagonal matrix.}$$

The orthogonal transformation $X = NY$, where $\begin{bmatrix} y_1 \\ y_2 \\ y_3 \end{bmatrix}$ reduces the given quadratic form to

$$Y^T D Y = \begin{bmatrix} y_1 & y_2 & y_3 \end{bmatrix} \begin{bmatrix} 1 & 0 & 0 \\ 0 & 4 & 0 \\ 0 & 0 & 4 \end{bmatrix} \begin{bmatrix} y_1 \\ y_2 \\ y_3 \end{bmatrix} = y_1^2 + 4y_2^2 + 4y_3^2$$

which is required canonical form.

Since all the eigen values of $A$ are positive, the given quadratic form is positive definite.

**Example:** Reduce the quadratic form $2x_1x_2 + 2x_1x_3 - 2x_2x_3$ into a sum of squares form by an orthogonal reduction and discuss its nature. Also find the modal matrix.

**Solution:** Given quadratic form is $2x_1x_2 + 2x_1x_3 - 2x_2x_3$.

The matrix of the quadratic form is $A = \begin{pmatrix} 0 & 1 & 1 \\ 1 & 0 & -1 \\ 1 & -1 & 0 \end{pmatrix}$.

The characteristic equation of $A$ is $|A - \lambda I| = 0$

$$\Rightarrow \lambda^3 - S_1 \lambda^2 + S_2 \lambda - S_3 = 0$$

where $S_1$ =Sum of the diagonal elements of $A$

$$= 0 + 0 + 0 = 0$$

$S_2$ =Sum of minors of elements of diagonal of $A$

$$= \begin{vmatrix} 0 & -1 \\ -1 & 0 \end{vmatrix} + \begin{vmatrix} 0 & 1 \\ 1 & 0 \end{vmatrix} + \begin{vmatrix} 0 & 1 \\ 1 & 0 \end{vmatrix}$$

$$= -1 - 1 - 1 = -3$$

$$S_3 = |A| = 0(0-1) - 1(0+1) + 1(-1-0) = -2.$$

$\therefore$ The characteristic equation is $\lambda^3 - 3\lambda + 2 = 0$

$$\Rightarrow \lambda = -2, 1, 1$$

$\therefore$ The eigen values are $\lambda = -2, 1, 1$.

**To find eigen vectors:**

Let $X = \begin{bmatrix} x \\ y \\ z \end{bmatrix}$ be an eigen vector corresponding $\lambda$.

Then

$$(A - \lambda I)X = 0 \Rightarrow \begin{bmatrix} -\lambda & 1 & 1 \\ 1 & -\lambda & -1 \\ 1 & -1 & -\lambda \end{bmatrix} \begin{bmatrix} x \\ y \\ z \end{bmatrix} = \begin{bmatrix} 0 \\ 0 \\ 0 \end{bmatrix}$$

$$\left. \begin{array}{r} -\lambda x + y + z = 0 \\ x - \lambda y - z = 0 \\ x - y - \lambda z = 0 \end{array} \right\} \quad \ldots \ldots \text{ (I)}$$

**Case (i)** If $\lambda = -2$ then equations (I) become

$$2x + y + z = 0$$

$$x + 2y - z = 0$$

$$x - y + 2z = 0$$

Choosing the first two equations, we have

$$\frac{x}{-1-2} = \frac{y}{1+2} = \frac{z}{4-1}$$

$$\Rightarrow \frac{x}{-3} = \frac{y}{3} = \frac{z}{3}$$

$$\Rightarrow \frac{x}{-1} = \frac{y}{1} = \frac{z}{1}$$

Choosing $x_1 = -1, x_2 = 1, x_3 = 1$, we get an eigen vector

$$X_1 = \begin{bmatrix} -1 \\ 1 \\ 1 \end{bmatrix}.$$

**Case (ii)** If $\lambda = 4$, then the equations (I) become

$$-x + y + z = 0 \Rightarrow x - y - z = 0$$
$$x - y - z = 0$$
$$x - y - z = 0$$

We get only one equation $x - y - z = 0$ ...... (1)

To solve for $x, y, z$, we can assign arbitrary values for two of the variables and we shall find 2 orthogonal vectors.

Put $z = 0, y = 1$, then $x = 1$, we get an eigen vector

$$X_2 = \begin{bmatrix} 1 \\ 1 \\ 0 \end{bmatrix}.$$

Let $X_3 = \begin{bmatrix} a \\ b \\ c \end{bmatrix}$ be orthogonal to $X_2$. Then $a + b = 0 \Rightarrow b = -a$ and $X_3$ should satisfy (1). $\therefore a - b - c = 0$.

Choosing $a = 1$, we get $b = -1$ and $c = 2$.

$$\therefore X_3 = \begin{bmatrix} 1 \\ -1 \\ 2 \end{bmatrix}.$$

Thus the eigen values are $\lambda = -2, 1, 1$, and the corresponding eigen vector are

$$X_1 = \begin{bmatrix} -1 \\ 1 \\ 1 \end{bmatrix}, \quad X_2 = \begin{bmatrix} 1 \\ 1 \\ 0 \end{bmatrix}, \quad X_3 = \begin{bmatrix} 1 \\ -1 \\ 2 \end{bmatrix}.$$

The normalised eigen vectors are

$$\begin{bmatrix} -\frac{1}{\sqrt{3}} \\ \frac{1}{\sqrt{3}} \\ \frac{1}{\sqrt{3}} \end{bmatrix}, \quad \begin{bmatrix} \frac{1}{\sqrt{2}} \\ \frac{1}{\sqrt{2}} \\ 0 \end{bmatrix}, \quad \begin{bmatrix} \frac{1}{\sqrt{6}} \\ \frac{-1}{\sqrt{6}} \\ \frac{2}{\sqrt{6}} \end{bmatrix}.$$

So, the normalised modal matrix

$$N = \begin{pmatrix} -\frac{1}{\sqrt{3}} & \frac{1}{\sqrt{2}} & \frac{1}{\sqrt{6}} \\ \frac{1}{\sqrt{3}} & \frac{1}{\sqrt{2}} & \frac{-1}{\sqrt{6}} \\ -\frac{1}{\sqrt{3}} & 0 & \frac{2}{\sqrt{6}} \end{pmatrix}$$

$$N^T A N = \begin{pmatrix} -\frac{1}{\sqrt{3}} & \frac{1}{\sqrt{3}} & \frac{1}{\sqrt{3}} \\ \frac{1}{\sqrt{2}} & \frac{1}{\sqrt{2}} & 0 \\ \frac{1}{\sqrt{6}} & -\frac{1}{\sqrt{6}} & \frac{2}{\sqrt{6}} \end{pmatrix} \begin{pmatrix} 0 & 1 & 1 \\ 1 & 0 & -1 \\ 1 & -1 & 0 \end{pmatrix} \begin{pmatrix} -\frac{1}{\sqrt{3}} & \frac{1}{\sqrt{2}} & \frac{1}{\sqrt{6}} \\ \frac{1}{\sqrt{3}} & \frac{1}{\sqrt{2}} & \frac{-1}{\sqrt{6}} \\ -\frac{1}{\sqrt{3}} & 0 & \frac{2}{\sqrt{6}} \end{pmatrix}$$

$$= \begin{pmatrix} -2 & 0 & 0 \\ 0 & 1 & 0 \\ 0 & 0 & 1 \end{pmatrix}, \text{ which is a diagonal matrix.}$$

The orthogonal transformation $X = NY$, where $\begin{bmatrix} y_1 \\ y_2 \\ y_3 \end{bmatrix}$ reduces the given quadratic form to

$$Y^T DY = \begin{bmatrix} y_1 & y_2 & y_3 \end{bmatrix} \begin{bmatrix} -2 & 0 & 0 \\ 0 & 1 & 0 \\ 0 & 0 & 1 \end{bmatrix} \begin{bmatrix} y_1 \\ y_2 \\ y_3 \end{bmatrix} = -2y_1^2 + y_2^2 + y_3^2$$

which is required canonical form.

Since some of the eigen values of $A$ are positive and others are negative, the given quadratic form is indefinite.

**Example:** Reduce the quadratic form $x_1^2 + 3x_2^2 + 3x_3^2 - 2x_2 x_3$ into a canonical form.

**Solution:** Given $x_1^2 + 3x_2^2 + 3x_3^2 - 2x_2 x_3$.

The matrix of the quadratic form is $A = \begin{pmatrix} 1 & 0 & 0 \\ 0 & 3 & -1 \\ 0 & -1 & 3 \end{pmatrix}$.

The characteristic equation of $A$ is $|A - \lambda I| = 0$

$$\Rightarrow \lambda^3 - S_1 \lambda^2 + S_2 \lambda - S_3 = 0$$

where $S_1$ =Sum of the diagonal elements of $A$

$$= 3 + 3 + 3 = 9$$

$S_2$ =Sum of minors of elements of diagonal of $A$

$$= \begin{vmatrix} 3 & -1 \\ -1 & 3 \end{vmatrix} + \begin{vmatrix} 1 & 0 \\ 0 & 3 \end{vmatrix} + \begin{vmatrix} 1 & 0 \\ 0 & 3 \end{vmatrix}$$
$$= (9-1) + (3-0) + (3-0) = 14$$

$S_3 = |A| = 1(9-1) = 8.$

$\therefore$ The characteristic equation is $\lambda^3 - 9\lambda^2 + 14\lambda - 8 = 0$

$$\Rightarrow \lambda = 1, 2, 4$$

$\therefore$ The eigen values are $\lambda = 1, 2, 4$.

**To find eigen vectors:**

Let $X = \begin{bmatrix} x \\ y \\ z \end{bmatrix}$ be an eigen vector corresponding $\lambda$.

Then

$$(A - \lambda I)X = 0 \Rightarrow \begin{bmatrix} 1-\lambda & 0 & 0 \\ 0 & 3-\lambda & -1 \\ 0 & -1 & 3-\lambda \end{bmatrix} \begin{bmatrix} x \\ y \\ z \end{bmatrix} = \begin{bmatrix} 0 \\ 0 \\ 0 \end{bmatrix}$$

$$\left. \begin{array}{l} (1-\lambda)x + 0y + 0z = 0 \\ 0x + (3-\lambda)y + (-1)z = 0 \\ 0x + (-1)y + (3-\lambda)z = 0 \end{array} \right\} \quad \cdots \cdots \quad (I)$$

**Case (i)** If $\lambda = 1$ then equations (I) become

$$0x + 0y + z = 0 \Rightarrow z = 0$$
$$0x + 2y - z = 0 \Rightarrow 2y = z$$
$$0x - y + 2z = 0 \Rightarrow 2y = z$$

$\therefore y = 0.$

$x$ can be any real value, we shall take $x = 1$.

∴ An eigen vector is

$$X_1 = \begin{bmatrix} 1 \\ 0 \\ 0 \end{bmatrix}.$$

**Case (ii)** If $\lambda = 2$, then the equations (I) become

$$-x = 0 \Rightarrow x = 0$$
$$0x + y - z \Rightarrow y = z$$
$$0x - y + z \Rightarrow y = z$$

Take $y = 1$. $\therefore z = 1$.

∴ An eigen vector is

$$X_2 = \begin{bmatrix} 0 \\ 1 \\ 1 \end{bmatrix}.$$

**Case (iii)** If $\lambda = 4$, then the equations (I) become

$$-3x + 0y + 0z = 0 \Rightarrow x = 0$$
$$0x - y - z \Rightarrow y = -z$$
$$0x - y - z \Rightarrow y = -z$$

Take $z = 1$. $\therefore y = -1$.

∴ An eigen vector is

$$X_2 = \begin{bmatrix} 0 \\ -1 \\ 1 \end{bmatrix}.$$

Thus the eigen values are $\lambda = 1, 2, 4$, and the corresponding eigen vector are

$$X_1 = \begin{bmatrix} 1 \\ 0 \\ 0 \end{bmatrix}, \quad X_2 = \begin{bmatrix} 0 \\ 1 \\ 1 \end{bmatrix}, \quad X_3 = \begin{bmatrix} 0 \\ -1 \\ 1 \end{bmatrix}.$$

The normalised eigen vectors are

$$\begin{bmatrix} 1 \\ 0 \\ 0 \end{bmatrix}, \begin{bmatrix} 0 \\ \frac{1}{\sqrt{2}} \\ \frac{1}{\sqrt{2}} \end{bmatrix}, \begin{bmatrix} 0 \\ \frac{-1}{\sqrt{2}} \\ \frac{1}{\sqrt{2}} \end{bmatrix}.$$

So, the normalised modal matrix

$$N = \begin{pmatrix} 1 & 0 & 0 \\ 0 & \frac{1}{\sqrt{2}} & \frac{-1}{\sqrt{2}} \\ 0 & \frac{1}{\sqrt{2}} & \frac{1}{\sqrt{2}} \end{pmatrix}$$

$$\therefore N^T A N = \begin{pmatrix} 1 & 0 & 0 \\ 0 & \frac{1}{\sqrt{2}} & \frac{1}{\sqrt{2}} \\ 0 & \frac{-1}{\sqrt{2}} & \frac{1}{\sqrt{2}} \end{pmatrix} \begin{pmatrix} 1 & 0 & 0 \\ 0 & 3 & -1 \\ 0 & -1 & 3 \end{pmatrix} \begin{pmatrix} 1 & 0 & 0 \\ 0 & \frac{1}{\sqrt{2}} & \frac{-1}{\sqrt{2}} \\ 0 & \frac{1}{\sqrt{2}} & \frac{1}{\sqrt{2}} \end{pmatrix}$$

$$= \begin{pmatrix} 1 & 0 & 0 \\ 0 & 2 & 0 \\ 0 & 0 & 4 \end{pmatrix}, \text{ which is a diagonal matrix.}$$

The orthogonal transformation $X = NY$, where $\begin{bmatrix} y_1 \\ y_2 \\ y_3 \end{bmatrix}$ reduces the given quadratic form to

$$Y^T DY = \begin{bmatrix} y_1 & y_2 & y_3 \end{bmatrix} \begin{bmatrix} 1 & 0 & 0 \\ 0 & 2 & 0 \\ 0 & 0 & 4 \end{bmatrix} \begin{bmatrix} y_1 \\ y_2 \\ y_3 \end{bmatrix} = y_1^2 + 2y_2^2 + 4y_3^2$$

which is required canonical form.

Since all the eigen values of $A$ are positive, the given quadratic form is positive definite.

**Example:** Reduce the quadratic form $x^2 + 5y^2 + z^2 + 2xy + 2yz + 6zx$ to canonical form through orthogonal transformation.

**Solution:** Given $x^2 + 5y^2 + z^2 + 2xy + 2yz + 6zx$.

The matrix of the quadratic form is $A = \begin{pmatrix} 1 & 1 & 3 \\ 1 & 5 & 1 \\ 3 & 1 & 1 \end{pmatrix}$.

The characteristic equation of $A$ is $|A - \lambda I| = 0$

$$\Rightarrow \lambda^3 - S_1 \lambda^2 + S_2 \lambda - S_3 = 0$$

where $S_1$ = Sum of the diagonal elements of $A$

$$= 1 + 5 + 1 = 7$$

$S_2$ = Sum of minors of elements of diagonal of $A$

$$= \begin{vmatrix} 5 & 1 \\ 1 & 1 \end{vmatrix} + \begin{vmatrix} 1 & 1 \\ 1 & 5 \end{vmatrix} + \begin{vmatrix} 1 & 3 \\ 3 & 1 \end{vmatrix}$$

$$= (5-1) + (5-1) + (1-9) = 4 + 4 - 8 = 0$$

$S_3 = |A| = 1(5-1) - (1-3) + 3(1-15) = 4 + 2 - 42 = -36.$

$\therefore$ The characteristic equation is $\lambda^3 - 7\lambda^2 - 36 = 0$

$$\Rightarrow \lambda = -2, 3, 6$$

$\therefore$ The eigen values are $\lambda = -2, 3, 6$.

**To find eigen vectors:**

Let $X = \begin{bmatrix} x_1 \\ x_2 \\ x_3 \end{bmatrix}$ be an eigen vector corresponding $\lambda$.

Then

$$(A - \lambda I)X = 0 \Rightarrow \begin{bmatrix} 1-\lambda & 1 & 3 \\ 1 & 5-\lambda & 1 \\ 3 & 1 & 1-\lambda \end{bmatrix} \begin{bmatrix} x_1 \\ x_2 \\ x_3 \end{bmatrix} = \begin{bmatrix} 0 \\ 0 \\ 0 \end{bmatrix}$$

$$\left. \begin{array}{l} (1-\lambda)x_1 + x_2 + 3x_3 = 0 \\ x_1 + (5-\lambda)x_2 + x_3 = 0 \\ 3x_1 + x_2 + (1-\lambda)x_3 = 0 \end{array} \right\} \quad \cdots \cdots \text{ (I)}$$

**Case (i)** If $\lambda = -2$ then equations (I) become

$$3x_1 + x_2 + 3x_3 = 0$$

$$x_1 + 7x_2 + x_3 = 0$$

$$3x_1 + x_2 + 3x_3 = 0$$

From first two equations, we get

$$\frac{x_1}{1-21} = \frac{x_2}{3-3} = \frac{x_3}{21-1}$$

$$\Rightarrow \frac{x_1}{-20} = \frac{x_2}{0} = \frac{x_3}{20}$$

$$\Rightarrow \frac{x_1}{1} = \frac{x_2}{0} = \frac{x_3}{-1}$$

Choosing $x_1 = 1, x_2 = 0, x_3 = -1$, we get an eigen vector is

$$X_1 = \begin{bmatrix} 1 \\ 0 \\ -1 \end{bmatrix}.$$

**Case (ii)** If $\lambda = 3$ then equations (I) become

$$-2x_1 + x_2 + 3x_3 = 0$$

$$x_1 + 2x_2 + x_3 = 0$$

$$3x_1 + x_2 - 2x_3 = 0$$

From first two equations, we get

$$\frac{x_1}{1-6} = \frac{x_2}{3+2} = \frac{x_3}{-4-1}$$

$$\Rightarrow \frac{x_1}{-5} = \frac{x_2}{5} = \frac{x_3}{-5}$$

$$\Rightarrow \frac{x_1}{1} = \frac{x_2}{-1} = \frac{x_3}{1}$$

Choosing $x_1 = 1, x_2 = -1, x_3 = 1$, we get an eigen vector is

$$X_2 = \begin{bmatrix} 1 \\ -1 \\ 1 \end{bmatrix}.$$

**Case (iii)** If $\lambda = 6$ then equations (I) become

$$-5x_1 + x_2 + 3x_3 = 0$$

$$x_1 - x_2 + x_3 = 0$$

$$3x_1 + x_2 - 5x_3 = 0$$

From first two equations, we get

$$\frac{x_1}{1+3} = \frac{x_2}{3+5} = \frac{x_3}{5-1}$$

$$\Rightarrow \frac{x_1}{4} = \frac{x_2}{8} = \frac{x_3}{4}$$

$$\Rightarrow \frac{x_1}{1} = \frac{x_2}{2} = \frac{x_3}{1}$$

Choosing $x_1 = 1, x_2 = 2, x_3 = 1$, we get an eigen vector is

$$X_3 = \begin{bmatrix} 1 \\ 0 \\ -1 \end{bmatrix}.$$

Thus the eigen values are $\lambda = 1, 2, 4$, and the corresponding eigen vector are

$$X_1 = \begin{bmatrix} 1 \\ 0 \\ 0 \end{bmatrix}, \quad X_2 = \begin{bmatrix} 0 \\ 1 \\ 1 \end{bmatrix}, \quad X_3 = \begin{bmatrix} 0 \\ -1 \\ 1 \end{bmatrix}.$$

Since the eigen values are different, the eigen vectors are mutually orthogonal.

The normalised eigen vectors are

$$\begin{bmatrix} \frac{1}{\sqrt{2}} \\ 0 \\ \frac{-1}{\sqrt{2}} \end{bmatrix}, \quad \begin{bmatrix} \frac{1}{\sqrt{3}} \\ \frac{-1}{\sqrt{3}} \\ \frac{1}{\sqrt{3}} \end{bmatrix}, \quad \begin{bmatrix} \frac{1}{\sqrt{6}} \\ \frac{2}{\sqrt{6}} \\ \frac{1}{\sqrt{6}} \end{bmatrix}.$$

So, the normalised modal matrix

$$N = \begin{pmatrix} \frac{1}{\sqrt{2}} & \frac{1}{\sqrt{3}} & \frac{1}{\sqrt{6}} \\ 0 & \frac{-1}{\sqrt{3}} & \frac{2}{\sqrt{6}} \\ \frac{-1}{\sqrt{2}} & \frac{1}{\sqrt{3}} & \frac{1}{\sqrt{6}} \end{pmatrix}$$

$$\therefore N^T A N = \begin{pmatrix} \frac{1}{\sqrt{2}} & 0 & \frac{-1}{\sqrt{2}} \\ \frac{1}{\sqrt{3}} & \frac{-1}{\sqrt{3}} & \frac{1}{\sqrt{3}} \\ \frac{1}{\sqrt{6}} & \frac{2}{\sqrt{6}} & \frac{1}{\sqrt{6}} \end{pmatrix} \begin{pmatrix} 1 & 1 & 3 \\ 1 & 5 & 1 \\ 3 & 1 & 1 \end{pmatrix} \begin{pmatrix} \frac{1}{\sqrt{2}} & \frac{1}{\sqrt{3}} & \frac{1}{\sqrt{6}} \\ 0 & \frac{-1}{\sqrt{3}} & \frac{2}{\sqrt{6}} \\ \frac{-1}{\sqrt{2}} & \frac{1}{\sqrt{3}} & \frac{1}{\sqrt{6}} \end{pmatrix}$$

$$= \begin{pmatrix} -2 & 0 & 0 \\ 0 & 3 & 0 \\ 0 & 0 & 6 \end{pmatrix}, \text{ which is a diagonal matrix.}$$

The orthogonal transformation $X = NY$, where $\begin{bmatrix} y_1 \\ y_2 \\ y_3 \end{bmatrix}$ reduces the given quadratic form to

$$Y^T D Y = \begin{bmatrix} y_1 & y_2 & y_3 \end{bmatrix} \begin{bmatrix} -2 & 0 & 0 \\ 0 & 3 & 0 \\ 0 & 0 & 6 \end{bmatrix} \begin{bmatrix} y_1 \\ y_2 \\ y_3 \end{bmatrix} = -2y_1^2 + 3y_2^2 + 6y_3^2$$

which is required canonical form.

Since some of the eigen values of $A$ are positive and some are negative, the given quadratic form is indefinite.

**Example:** Reduce the quadratic form $8x^2 + 7y^2 + 3z^2 - 12xy - 8yz + 4zx$ to canonical form by an orthogonal transformation and give the matrix of transformation. Also state nature.

**Solution:** Given quadratic form is $8x^2 + 7y^2 + 3z^2 - 12xy - 8yz + 4zx$.

The matrix of the quadratic form is $A = \begin{pmatrix} 8 & -6 & 2 \\ -6 & 7 & -4 \\ 2 & -4 & 3 \end{pmatrix}$.

The characteristic equation of $A$ is $|A - \lambda I| = 0$

$$\Rightarrow \lambda^3 - S_1 \lambda^2 + S_2 \lambda - S_3 = 0$$

where $S_1$ =Sum of the diagonal elements of $A$

$= 8 + 7 + 3 = 18$

$S_2$ =Sum of minors of elements of diagonal of $A$

$$= \begin{vmatrix} 7 & -4 \\ -4 & 3 \end{vmatrix} + \begin{vmatrix} 8 & -6 \\ -6 & 7 \end{vmatrix} + \begin{vmatrix} 8 & 2 \\ 2 & 3 \end{vmatrix}$$
$$= (21-16) + (56-36) + (24-4) = 5 + 20 + 20 = 45.$$
$$S_3 = |A| = 8(21-16) + 6(-18+8) + 2(24-14) = 40 - 60 + 20 = 0.$$

∴ The characteristic equation is $\lambda^3 - 18\lambda^2 + 45 = 0$
$$\Rightarrow \lambda = 0, 3, 15$$

∴ The eigen values are $\lambda = 0, 3, 15$.

**To find eigen vectors:**

Let $X = \begin{bmatrix} x_1 \\ x_2 \\ x_3 \end{bmatrix}$ be an eigen vector corresponding $\lambda$.

Then

$$(A - \lambda I)X = 0 \Rightarrow \begin{bmatrix} 8-\lambda & -6 & 2 \\ -6 & 7-\lambda & -4 \\ 2 & -4 & 3-\lambda \end{bmatrix} \begin{bmatrix} x_1 \\ x_2 \\ x_3 \end{bmatrix} = \begin{bmatrix} 0 \\ 0 \\ 0 \end{bmatrix}$$

$$\left. \begin{array}{r} (8-\lambda)x_1 - 6x_2 + 2x_3 = 0 \\ -6x_1 + (7-\lambda)x_2 - 4x_3 = 0 \\ 2x_1 - 4x_2 + (3-\lambda)x_3 = 0 \end{array} \right\} \quad \cdots \cdots \text{ (I)}$$

**Case (i)** If $\lambda = 0$ then equations (I) become

$$8x_1 - 6x_2 + 2x_3 = 0 \Rightarrow 4x_1 - 3x_2 + x_3 = 0$$
$$-6x_1 + 7x_2 - 4x_3 = 0$$
$$2x_1 - 4x_2 + 3x_3 = 0$$

From the first and third equations we get
$$\frac{x_1}{-9+4} = \frac{x_2}{2-12} = \frac{x_3}{-16+6}$$
$$\Rightarrow \frac{x_1}{-5} = \frac{x_2}{-10} = \frac{x_3}{-10}$$
$$\Rightarrow \frac{x_1}{1} = \frac{x_2}{2} = \frac{x_3}{2}$$

Choosing $x_1 = 1, x_2 = 2, x_3 = 2$, we get an eigen vector is

$$X_1 = \begin{bmatrix} 1 \\ 2 \\ 2 \end{bmatrix}.$$

Similarly, we can find for $\lambda = 3$, $X_2 = \begin{bmatrix} 2 \\ 1 \\ -2 \end{bmatrix}$ and for $\lambda = 15$, $X_3 = \begin{bmatrix} 2 \\ -2 \\ 1 \end{bmatrix}$.

Thus the eigen values are $\lambda = 0, 3, 15$, and the corresponding eigen vector are

$$X_1 = \begin{bmatrix} 1 \\ 2 \\ 2 \end{bmatrix}, \quad X_2 = \begin{bmatrix} 2 \\ -2 \\ 1 \end{bmatrix}, \quad X_3 = \begin{bmatrix} 2 \\ 1 \\ -2 \end{bmatrix}.$$

Since the eigen values are different, the eigen vectors are mutually orthogonal. The normalised eigen vectors are

$$\begin{bmatrix} \frac{1}{3} \\ \frac{2}{3} \\ \frac{2}{3} \end{bmatrix}, \begin{bmatrix} \frac{2}{3} \\ \frac{1}{3} \\ \frac{-2}{3} \end{bmatrix}, \begin{bmatrix} \frac{2}{3} \\ \frac{-2}{3} \\ \frac{2}{3} \end{bmatrix}.$$

So, the normalised modal matrix

$$N = \begin{pmatrix} \frac{1}{3} & \frac{2}{3} & \frac{2}{3} \\ \frac{2}{3} & \frac{1}{3} & \frac{-2}{3} \\ \frac{2}{3} & \frac{-2}{3} & \frac{1}{3} \end{pmatrix} = \frac{1}{3} \begin{pmatrix} 1 & 2 & -2 \\ 2 & 1 & -2 \\ 2 & -2 & 1 \end{pmatrix}$$

$$\therefore N^T A N = \frac{1}{3} \begin{pmatrix} 1 & 2 & 2 \\ 2 & 1 & -2 \\ -2 & -2 & 1 \end{pmatrix} \begin{pmatrix} 8 & -6 & 2 \\ -6 & 7 & -4 \\ 2 & -4 & 3 \end{pmatrix} \frac{1}{3} \begin{pmatrix} 1 & 2 & -2 \\ 2 & 1 & -2 \\ 2 & -2 & 1 \end{pmatrix}$$

$$= \begin{pmatrix} 0 & 0 & 0 \\ 0 & 3 & 0 \\ 0 & 0 & 15 \end{pmatrix}, \text{ which is a diagonal matrix.}$$

The orthogonal transformation $X = NY$, where $\begin{bmatrix} y_1 \\ y_2 \\ y_3 \end{bmatrix}$ reduces the given quadratic form to

$$Y^T DY = \begin{bmatrix} y_1 & y_2 & y_3 \end{bmatrix} \begin{bmatrix} 0 & 0 & 0 \\ 0 & 3 & 0 \\ 0 & 0 & 15 \end{bmatrix} \begin{bmatrix} y_1 \\ y_2 \\ y_3 \end{bmatrix} = 0y_1^2 + 3y_2^2 + 15y_3^2$$

which is required canonical form.
The transformation is

$$\begin{pmatrix} x \\ y \\ z \end{pmatrix} = \frac{1}{3} \begin{pmatrix} 1 & 2 & -2 \\ 2 & 1 & -2 \\ 2 & -2 & 1 \end{pmatrix} \begin{pmatrix} y_1 \\ y_2 \\ y_3 \end{pmatrix}$$

$$x = \frac{1}{3}(y_1 + 2y_2 - 2y_3)$$
$$y = \frac{1}{3}(2y_1 + y_2 - 2y_3)$$
$$z = \frac{1}{3}(2y_1 - 2y_2 + y_3)$$

**EXERCISE**

1. Reduce the quadratic form $3x^2 + 2y^2 + 3z^2 - 2xy - 2yz$ to the normal form by orthogonal transformation. **(Mar-2022), (Apr-2022)**

**Ans:** $\lambda = 1, 3, 4$.

3. Reduce the quadratic form $3x^2 - 2y^2 - z^2 - 4xy + 12yz + 8xz$ in to canonical form by orthogonal transformation, find rank, index and signature.

**Ans:** $3, 6, -9$. Indefinite. **(Jan-2020)**

5. Find the nature of the quadratic form $6x^2 + 3y^2 + 3z^2 - 2xy - 4yz - 4xz$ to the canonical form. **(Nov-2021)**

6. Find the transformation which will transform $2xy + 2zx + 2xy$ in to a sum of

squares hence find the rank, index, signature.  (Jan-2020), (Nov-2021)
**Ans:** $\lambda^3 - 3\lambda - 2 = 0, \lambda = 2, -1, -1$.

## 2.5 Complex Matrices

**Definition:** A matrix with at least one element as complex number is called a complex matrix.

**Conjugate of a Matrix:** If all the elements of a matrix $A$ are replaced by their conjugate complex numbers (i.e., in the place of complex numbers only), then the resulting matrix is called as the conjugate of the given matrix. It is denoted by $\overline{A}$.

In the other words, if $A = [a_{ij}]$ and $\overline{a_{ij}}$ are complex conjugates of $a_{ij}$, then $\overline{A} = [\overline{a_{ij}}]$ is called conjugate of $A$.

**Example:** $A = \begin{bmatrix} 2+4i & -6 \\ 6-2i & 5+i \end{bmatrix}$ is a complex matrix.

The conjugate of $A$ is $\overline{A} = \begin{bmatrix} \overline{2+4i} & \overline{-6} \\ \overline{6-2i} & \overline{5+i} \end{bmatrix} = \begin{bmatrix} 2-4i & -6 \\ 6+2i & 5-i \end{bmatrix}$.

**Note:** If $\overline{A}$ and $\overline{B}$ be the conjugate matrices of $A$ and $B$ respectively, then

(i) $\overline{(\overline{A})} = A$

(ii) $\overline{(A+B)} = \overline{A} + \overline{B}$

(iii) $\overline{(kA)} = \overline{k} \cdot \overline{A}$, where $k$ being a complex number

(iv) $\overline{(AB)} = \overline{A} \cdot \overline{B}$

**Transpose of Conjugate of a Matrix:** The transpose of a conjugate of a matrix $A$ is called the transpose conjugate of a matrix $A$. It is denoted by $A^\theta$ or $A^*$.

i.e., $A^\theta$ or $A^* = (\overline{A})' = \overline{(A')}$

**Example:** If $A = \begin{bmatrix} 1-i & 1+i & 2-i \\ 2 & 3-2i & 1+2i \end{bmatrix}$, then

$A^* = \overline{A}^T = \begin{bmatrix} 1+i & 1-i & 2+i \\ 2 & 3+2i & 1-2i \end{bmatrix}^T = \begin{bmatrix} 1+i & 2 \\ 1-i & 3+2i \\ 2+i & 1-2i \end{bmatrix}$.

**Note:** If $A^*$ and $B^*$ be the transposed conjugates matrices of $A$ and $B$ respectively, then

(i) $(A^*)^* = A$

(ii) $(A+B)^* = A^* + B^*$

(iii) $(kA)^* = \bar{k} \cdot A^*$, where $k$ being a complex number

(iv) $(AB)^* = B^* \cdot A^*$.

## 2.5.1 Hermitian Matrix

**Definition:** A complex square matrix $A$ is said to be a Hermitian matrix if $A^* = A$
i.e., $a_{ij} = \overline{a_{ji}}$, for all $i, j$.

If $i = j$, we have $a_{ii} = \overline{a_{ii}}$

i.e., $a_{ii}$ real for all values of $i$. Therefore, the diagonal elements of a Hermitian matrix are real.

**Example:** Show that $A = \begin{bmatrix} 2 & 3+4i \\ 3-4i & -5 \end{bmatrix}$ is a Hermitian matrix.

Solution: Given $A = \begin{bmatrix} 2 & 3+4i \\ 3-4i & -5 \end{bmatrix}$.

$$\therefore A^* = (\overline{A})^T = \begin{bmatrix} \overline{2} & \overline{3+4i} \\ \overline{3-4i} & \overline{-5} \end{bmatrix}^T$$

$$= \begin{bmatrix} 2 & 3-4i \\ 3+4i & -5 \end{bmatrix}^T = \begin{bmatrix} 2 & 3+4i \\ 3-4i & -5 \end{bmatrix} = A.$$

Hence $A$ is a Hermitian matrix.

**Example:** Show that $A = \begin{bmatrix} -1 & 3-i \\ 3+i & 2 \end{bmatrix}$ is a Hermitian matrix.

Solution: Given $A = \begin{bmatrix} -1 & 3-i \\ 3+i & 2 \end{bmatrix}$.

$$\therefore A^* = (\overline{A})^T = \begin{bmatrix} \overline{-1} & \overline{3-i} \\ \overline{3+i} & \overline{2} \end{bmatrix}^T$$

$$= \begin{bmatrix} -1 & 3+i \\ 3-i & 2 \end{bmatrix}^T = \begin{bmatrix} -1 & 3-i \\ 3+i & 2 \end{bmatrix} = A.$$

Hence $A$ is a Hermitian matrix.

**Note:** The elements of the leading diagonal of a Hermitian matrix are real, while every other element is the complex conjugate of the element in the transposed position.

**Example:** Show that $\begin{pmatrix} 3 & 7-4i & -2+5i \\ 7+4i & -2 & 3+i \\ -2-5i & 3-i & 4 \end{pmatrix}$ is a Hermitian matrix.

**Solution:** Since the diagonal elements are real and the elements equidistant from the main diagonal are conjugates.

$\therefore$ Given matrix is a Hermitian matrix.

## 2.5.2 Skew-Hermitian Matrix

**Definition:** A complex matrix $A$ is said to be a Skew-Hermitian matrix if $A^* = -A$

i.e., $a_{ij} = -\overline{a_{ji}}$, forall $i, j$

If $i = j$, we have $a_{ii} = -\overline{a_{ii}}$

If $a_{ii} = \alpha + i\beta$, then $\overline{a_{ii}} = \alpha - i\beta$.

$\therefore \alpha + i\beta = -(\alpha - i\beta) \Rightarrow 2\alpha = 0 \Rightarrow \alpha = 0$

$\therefore a_{ii} = i\beta$, which is purely imaginary if $\beta \neq 0$ and 0 if $\beta = 0$.

$\therefore$ The diagonal elements of a Skew-Hermitian matrix are all purely imaginary or 0 and the elements equidistant from the main diagonal are conjugates with opposite sign.

**Theorem:** Show that every square matrix can be uniquely expressed as the sum of a Hermitian and a Skew-Hermitian matrix.

**Proof:** Let $A$ be a complex square matrix and $A^*$ denote the conjugate transpose of $A$.

Since $A$ is square complex matrix, we can write $A$ as

$$A = \frac{1}{2}(A + A^*) + \frac{1}{2}(A - A^*) = P + Q \text{ (say)} \quad \cdots \cdots \quad (1)$$

Now

$$P^* = \left[\frac{1}{2}(A+A^*)\right]^* = \frac{1}{2}(A+A^*)^* = \frac{1}{2}\left(A^* + (A^*)^*\right) = \frac{1}{2}\left(A^* + A\right) = P \quad \ldots \ldots \quad (2)$$

and

$$Q^* = \left[\frac{1}{2}(A-A^*)\right]^* = \frac{1}{2}(A-A^*)^* = \frac{1}{2}\left(A^* - (A^*)^*\right) = \frac{1}{2}\left(A^* - A\right) = -Q \quad \ldots \ldots \quad (3)$$

∴ From (2) and (3)

$$P^* = P \Rightarrow P \text{ is a Hermitian}$$
$$Q^* = -Q \Rightarrow Q \text{ is a Skew-Hermitian}$$

∴ $A$ is expressed as the sum of a Hermitian and a Skew-Hermitian matrix.

**To prove uniqueness:**

Let $A = R + S$ be another such representation of $A$, where $R$ is a Hermitian and $S$ is a Skew-Hermitian.

Then we have to prove that $P = R$ and $Q = S$.

$R$ is Hermitian $\Rightarrow R^* = R$ and $S$ is a Skew-Hermitian $\Rightarrow S^* = -S$.

∴ $A^* = (R+S)^* = R^* + S^* = R - S$.

Now $A = R+S$, $A^* = R - S \Rightarrow R = \frac{1}{2}(A+A^*) = P$

and $S = \frac{1}{2}(A - A^*) = Q$.

∴ $P = R$ and $Q = S$.

Hence the representation is unique.

**Theorem:** Prove that every Hermitian matrix can be written as $A + iB$, where $A$ is a real and symmetric and $B$ is a real and skew-symmetric.

**Proof:** Let $C$ be a Hermitian matrix, then

$$C^* = C \quad \ldots \ldots \quad (1)$$

Let us consider

$$A = \frac{1}{2}(C + \overline{C}) \text{ and } B = \frac{1}{2i}(C - \overline{C}).$$

Then obviously both $A$ and $B$ are real matrices.

We may write
$$C = \frac{1}{2}(C + \overline{C}) + i\left[\frac{1}{2i}(C - \overline{C})\right]$$
$$= A + iB$$

Thus $C$ is expressed as the matrix $A + iB$.

Now, we prove that $A$ is symmetric and $B$ is skew-symmetric.

$$A^T = \frac{1}{2}(C + \overline{C})^T = \frac{1}{2}\left(C^T + (\overline{C})^T\right)$$
$$= \frac{1}{2}(C^T + C^*)$$
$$= \frac{1}{2}\left[(C^*)^T + C\right] \quad \text{by (1)}$$
$$= \frac{1}{2}\left[((\overline{C})^T)^T + C\right]$$
$$= \frac{1}{2}\left[\overline{C} + C\right] = A$$
$$\Rightarrow A^T = A$$

Hence, $A$ is symmetric.

Now
$$B^T = \frac{1}{2i}(C - \overline{C})^T = \frac{1}{2i}\left(C^T - (\overline{C})^T\right)$$
$$= \frac{1}{2i}(C^T - C^*)$$
$$= \frac{1}{2i}\left[(C^*)^T - C\right] \quad \text{by (1)}$$
$$= \frac{1}{2i}\left[((\overline{C})^T)^T - C\right]$$
$$= \frac{1}{2i}\left[\overline{C} - C\right]$$
$$= \frac{-1}{2i}\left[C - \overline{C}\right] = -B$$
$$\Rightarrow B^T = -B$$

Thus, $B$ is skew-symmetric.

Hence the result.

**Theorem:** Show that every square complex matrix can be expressed uniquely as

$P + iQ$, where $P$ and $Q$ are Hermitian matrices.

**Proof:** Let $A$ be any square complex matrix. It can rewritten as

$$A = \frac{1}{2}(A + A^*) + i\left[\frac{1}{2i}(A - A^*)\right].$$

Put $P = \frac{1}{2}(A + A^*)$ and $Q = \frac{1}{2i}(A - A^*)$ then $A = P + iQ$.

Now we have show that $P$ and $Q$ are Hermitian matrices.

$$P^* = \left[\frac{1}{2}(A + A^*)\right]^* = \frac{1}{2}(A + A^*)^* = \frac{1}{2}\left(A^* + (A^*)^*\right) = \frac{1}{2}\left(A^* + A\right) = P$$

Thus $P$ is Hermitian matrix.

$$Q^* = \left[\frac{1}{2i}(A - A^*)\right]^* = \frac{1}{2i}(A - A^*)^* = -\frac{1}{2i}\left(A^* - (A^*)^*\right) = -\frac{1}{2i}\left(A^* - A\right) = Q$$

Thus $Q$ is Hermitian matrix.

Hence $A$ is expressed as $P + iQ$, where $P$ and $Q$ are Hermitian matrices.

**To prove uniqueness:**

We shall now prove the uniqueness of the expression $A = P + iQ$.

If possible let $A = R + iS$ be another such representation of $A$, where $R$ and $S$ are Hermitian matrices.

Then we have to prove that $P = R$ and $Q = S$.

$R$ is Hermitian $\Rightarrow R^* = R$ and $S$ is a Hermitian $\Rightarrow S^* = S$.

$\therefore A^* = (R + iS)^* = R^* + (iS)^* = R^* - iS^* = R - iS.$

Now $A = R + iS$, $A^* = R - iS \Rightarrow R = \frac{1}{2}(A + A^*) = P$

and $S = \frac{1}{2i}(A - A^*) = Q.$

$\therefore P = R$ and $Q = S$.

Hence $A = P + iQ$ is the unique expression, where $P$ and $Q$ are Hermitian matrices.

**Theorem:** The eigen values of a Hermitian matrix are real.

**Proof:** Let $A$ be a Hermitian matrix and $\lambda$ be the eigen value of $A$.

Then there exists a eigen vector $X$ corresponding eigen value $\lambda$ such that

$$AX = \lambda X \quad \cdots \cdots \quad (1)$$

Pre-multiplying both sides of (1) by $X^*$, we have

$$X^*(AX) = X^*(\lambda X)$$
$$\Rightarrow X^*AX = \lambda(X^*X) \quad \cdots \cdots \quad (2)$$

Taking transposed conjugate on both sides of (2), we have

$$(X^*AX)^* = (X^*\lambda X)^*$$
$$\Rightarrow X^*A^*(X^*)^* = \bar{\lambda}X^*(X^*)^*$$
$$\Rightarrow X^*AX = \bar{\lambda}X^*X \quad \cdots \cdots \quad (3)$$

$$\left[\because A \text{ is Hermitian, so that } A^* = A\right]$$

From (2) and (3), we have

$$\lambda X^*X = \bar{\lambda}X^*X$$
$$\Rightarrow (\lambda - \bar{\lambda})X^*X = 0$$
$$\Rightarrow \lambda - \bar{\lambda} = 0 \quad \left[\because X \neq 0 \Rightarrow X^*X \neq 0\right]$$
$$\Rightarrow \lambda = \bar{\lambda} \Rightarrow \lambda \text{ is real.}$$

Hence the eigen values of a Hermitian matrix are real.

**Theorem:** The eigen values of a Skew-Hermitian matrix are purely imaginary or zero.

**Proof:** Let $A$ be a Skew-Hermitian matrix and $\lambda$ be the eigen value of $A$.

Then there exists an eigen vector $X$ corresponding to the eigen value $\lambda$ such that

$$AX = \lambda X$$
$$\Rightarrow i(AX) = i(\lambda X)$$
$$\Rightarrow (iA)X = (i\lambda)X$$

This shows that $i\lambda$ is the eigen value of the Hermitian matrix $iA$.
Since the eigen values of a Hermitian matrix are real

$\Rightarrow i\lambda$ is purely real

$\Rightarrow \overline{i\lambda} = i\lambda$

$\Rightarrow -i\overline{\lambda} = i\lambda$

$\Rightarrow (\lambda + \overline{\lambda})i = 0 \Rightarrow \lambda + \overline{\lambda} = 0$

This shows that $\lambda$ is either 0 or purely imaginary.

Hence the eigen values of a Skew-Hermitian matrix are either purely imaginary or zero.

**Theorem:** If $A$ is a Hermitian matrix, then $iA$ is a Skew-Hermitian Matrix.

**Proof:** Given $A$ is a Hermitian matrix

$\Rightarrow A^* = A$

We have
$$(iA)^* = \overline{(iA)}^T$$
$$= (\overline{i} \cdot \overline{A})^T = -i\overline{A}^T = -i(\overline{A})^T = -iA^* = -iA$$
$$\Rightarrow (iA)^* = -(iA)$$

Thus, $iA$ is a Skew-Hermitian matrix.

**Example:** Show that $A = \begin{bmatrix} 2i & 1+i \\ -(1-i) & 0 \end{bmatrix}$ is a Skew-Hermitian matrix.

**Solution:** Given $A = \begin{bmatrix} 2i & 1+i \\ -(1-i) & 0 \end{bmatrix}$.

Since the diagonal elements are purely imaginary or zero and $1+i$ and $-(1-i)$ are conjugate with opposite sign, $A$ is Skew-Hermitian.

**Example:** If $A = \begin{pmatrix} -1 & 2+i & 5-3i \\ 2-i & 7 & 5i \\ 5+3i & -5i & 2 \end{pmatrix}$, then show that $A$ is Hermitian and $iA$ is Skew-Hermitian.

**Solution:** Given $A = \begin{bmatrix} -1 & 2+i & 5-3i \\ 2-i & 7 & 5i \\ 5+3i & -5i & 2 \end{bmatrix}$.

Now $\overline{A} = \begin{bmatrix} -1 & 2-i & 5+3i \\ 2+i & 7 & -5i \\ 5-3i & 5i & 2 \end{bmatrix}$.

$\therefore A^* = (\overline{A})^T = \begin{bmatrix} -1 & 2-i & 5+3i \\ 2+i & 7 & -5i \\ 5-3i & 5i & 2 \end{bmatrix}^T = \begin{bmatrix} -1 & 2+i & 5-3i \\ 2-i & 7 & 5i \\ 5+3i & -5i & 2 \end{bmatrix} = A$

$\therefore A^* = A \Rightarrow A$ is a Hermitian matrix.

Let $B = iA = \begin{bmatrix} -i & -1+2i & 3+5i \\ 1+2i & 7i & -5 \\ -3+5i & 5 & 2i \end{bmatrix}$.

$\overline{B} = \begin{bmatrix} i & -1-2i & 3-5i \\ 1-2i & -7i & -5 \\ -3-5i & 5 & -2i \end{bmatrix}$.

$\therefore B^* = (\overline{B})^T = \begin{bmatrix} i & 1-2i & -3-5i \\ -1-2i & -7i & 5 \\ 3-5i & -5 & -2i \end{bmatrix}$

$= -\begin{bmatrix} -i & -1+2i & 3+5i \\ 1+2i & 7i & -5 \\ -3+5i & 5 & 2i \end{bmatrix} = -A$

$\therefore B^* = -B \Rightarrow B$, i.e., $iA$ is a Skew-Hermitian matrix.

**Example:** If $A = \begin{bmatrix} 2+i & 3 & -1+3i \\ -5 & i & 4-2i \end{bmatrix}$, then show that $AA^*$ is a Hermitian matrix.

**Solution:** Given $A = \begin{bmatrix} 2+i & 3 & -1+3i \\ -5 & i & 4-2i \end{bmatrix}$.

$$\therefore A^* = \left(\overline{A}\right)^T = \begin{bmatrix} 2-i & 3 & -1-3i \\ -5 & -i & 4+2i \end{bmatrix}^T = \begin{bmatrix} 2-i & -5 \\ 3 & -i \\ -1-3i & 4+2i \end{bmatrix}.$$

Therefore,

$$AA^* = \begin{bmatrix} 2+i & 3 & -1+3i \\ -5 & i & 4-2i \end{bmatrix} \begin{bmatrix} 2-i & -5 \\ 3 & -i \\ -1-3i & 4+2i \end{bmatrix}$$

$$= \begin{bmatrix} (2+i)(2-i)+9+(-1+3i)(-1-3i) & -5(2+i)-3i+(-1+3i)(4+2i) \\ -5(2-i)+3i+(4-2i)(-1-3i) & 25-i^2+(4-2i)(4+2i) \end{bmatrix}$$

$$= \begin{bmatrix} 4-i^2+9+1+-9i^2 & -10-5i-3i-10+10i \\ -10+5i+3i-10-10i & 25-i^2+16-4i^2 \end{bmatrix}$$

$$= \begin{bmatrix} 24 & -20+2i \\ -20-2i & 46 \end{bmatrix}$$

$$\left(AA^*\right)^T = \begin{bmatrix} 24 & -20-2i \\ -20+2i & 46 \end{bmatrix}$$

$$= \overline{(AA^*)}$$

This shows that $AA^*$ is a Hermitian matrix, since the diagonal elements are real and elements equidistant from main diagonal are conjugates.

**Example:** Express the matrix $A = \begin{bmatrix} 1+i & -i & 2-3i \\ 2 & 1+2i & 3+i \\ -1+i & 3 & 1-2i \end{bmatrix}$ as the sum of a Hermitian and a Skew-Hermitian.

**Solution:** Given $A = \begin{bmatrix} 1+i & -i & 2-3i \\ 2 & 1+2i & 3+i \\ -1+i & 3 & 1-2i \end{bmatrix}$.

Now $\overline{A} = \begin{bmatrix} 1-i & i & 2+3i \\ 2 & 1-2i & 3-i \\ -1-i & 3 & 1+2i \end{bmatrix}$.

$A^* = (\overline{A})^T = \begin{bmatrix} 1-i & 2 & -1-i \\ i & 1-2i & 3 \\ 2+3i & 3-i & 1+2i \end{bmatrix}$.

Let $P = \dfrac{1}{2}(A + A^*)$ and $Q = \dfrac{1}{2}(A - A^*)$.

i.e., $P = \dfrac{1}{2}(A + A^*) = \dfrac{1}{2}\begin{bmatrix} 2 & 2-i & 1-4i \\ 2+i & 2 & 6+i \\ 1+4i & 6-i & 2 \end{bmatrix}$.

This is a Hermitian matrix.

$Q = \dfrac{1}{2}(A - A^*) = \dfrac{1}{2}\begin{bmatrix} 2i & -i-2 & 3-2i \\ 2-i & 4i & i \\ -3-2i & i & -4i \end{bmatrix}$.

This is a Skew-Hermitian matrix.

$\therefore P + Q = \dfrac{1}{2}\begin{bmatrix} 2 & 2-i & 1-4i \\ 2+i & 2 & 6+i \\ 1+4i & 6-i & 2 \end{bmatrix} + \dfrac{1}{2}\begin{bmatrix} 2i & -i-2 & 3-2i \\ 2-i & 4i & i \\ -3-2i & i & -4i \end{bmatrix}$

$= \dfrac{1}{2}\begin{bmatrix} 2+2i & -2i & 4-6i \\ 4 & 2+4i & 6+2i \\ -2+2i & 6 & 2-4i \end{bmatrix}$

$= \begin{bmatrix} 1+i & -i & 2-3i \\ 2 & 1+2i & 3+i \\ -1+i & 3 & 1-2i \end{bmatrix}$

Thus $A = P + Q$, where $P$ is a Hermitian and $Q$ is a Skew-Hermitian.

**Example:** Express the matrix $A = \begin{bmatrix} 2i & 2+i & 1-i \\ -2+i & -i & 3i \\ -1-i & 3i & 0 \end{bmatrix}$ as $P + iQ$ where $P$

is real and skew-symmetric and $Q$ is real and symmetric.

**Solution:** Given $A = \begin{bmatrix} 2i & 2+i & 1-i \\ -2+i & -i & 3i \\ -1-i & 3i & 0 \end{bmatrix}$.

Now $\overline{A} = \begin{bmatrix} -2i & 2-i & 1+i \\ -2-i & i & -3i \\ -1+i & -3i & 0 \end{bmatrix}$.

$A + \overline{A} = \begin{bmatrix} 0 & 4 & 2 \\ -4 & 0 & 0 \\ -2 & 0 & 0 \end{bmatrix}$

$A - \overline{A} = \begin{bmatrix} 4i & 2i & -2i \\ 2i & -2i & 6i \\ -2i & 6i & 0 \end{bmatrix}$.

$$P = \frac{1}{2}(A + A^*) = \begin{bmatrix} 0 & 2 & 1 \\ -2 & 0 & 0 \\ -1 & 0 & 0 \end{bmatrix}.$$

$$Q = \frac{1}{2i}(A - A^*) = \begin{bmatrix} 2 & 1 & -1 \\ 1 & -1 & 3 \\ -1 & 3 & 0 \end{bmatrix}.$$

Here $P^T = -P$ and $Q^T = Q$.

$\therefore$ $P$ is a Skew-symmetric matrix and $Q$ is a symmetric.

$$\therefore P + iQ = \begin{bmatrix} 0 & 2 & 1 \\ -2 & 0 & 0 \\ -1 & 0 & 0 \end{bmatrix} + i \begin{bmatrix} 2 & 1 & -1 \\ 1 & -1 & 3 \\ -1 & 3 & 0 \end{bmatrix}$$

$$= \begin{bmatrix} 2i & 2+i & 1-i \\ -2+i & -i & 3i \\ -1-i & 3i & 0 \end{bmatrix}$$

Thus $A = P + iQ$, where $P$ is a skew-symmetric and $Q$ is a symmetric.

### 2.5.3  Unitary Matrix

**Definition:** A complex square matrix is said to be unitary if $AA^* = A^*A = I$.
From the definition it is obvious that $A^*$ is the inverse of $A$.
$$\therefore A^* = A^{-1}.$$

**Theorem:** The eigen values of unitary matrix is of unit modulus.

**Proof:** Let $A$ be a unitary matrix. Then

$$AA^8 = A^*A = I \quad \cdots\cdots \quad (1)$$

Let $\lambda$ be the eigen value of $A$ and $X$ be the corresponding eigen vector of $A$. Then

$$AX = \lambda X \quad \cdots\cdots \quad (2)$$

Taking transposed conjugate on both sides of (2), we have

$$(AX)^* = (\lambda X)^*$$
$$\Rightarrow X^*A^* = \overline{\lambda}X^* \quad \cdots\cdots \quad (3)$$

Multiplying (3) and (2)

$$X^*A^*(AX) = \overline{\lambda}X^*(\lambda X)$$
$$\Rightarrow X^*(A^*A)X = \lambda\overline{\lambda}(X^*X)$$
$$\Rightarrow X^*IX = \lambda\overline{\lambda}(X^*X) \quad [\text{From } (1)]$$
$$\Rightarrow X^*X = |\lambda|^2 X^*X$$
$$\Rightarrow X^*X(1 - |\lambda|^2) = 0$$
$$\Rightarrow 1 - |\lambda|^2 = 0 \quad [\because X \neq 0 \text{ and } X^*X \neq 0]$$
$$\Rightarrow |\lambda|^2 = 1$$
$$\Rightarrow |\lambda| = 1$$

**Theorem:** Product of unitary matrices is unitary.

**Proof:** Let $A$ and $B$ be two unitary matrices.

So that $AA^* = A^*A = I$ and $BB^* = B^*B = I$ ...... (1)

Now we have to prove that $(AB)(AB)^* = (AB)^*(AB) = I$.

$$(AB)(AB)^* = (AB)(B^*A^*)$$
$$= A(BB^*)A^* = AIA^* = AA^* = I \quad [\because \text{From (1)}]$$
$$\Rightarrow (AB)(AB)^* = I$$

Similarly $(AB)^*(AB) = (B^*A^*)(AB)$
$$= B^*(A^*A)B$$
$$= B^*IB \quad [\because \text{From (1)}]$$
$$= B^*B = I$$
$$\Rightarrow (AB)^*(AB) = I$$
$$\therefore (AB)(AB)^* = (AB)^*(AB) = I$$

Thus $AB$ is a unitary matrix.

**Theorem:** Inverse of a unitary is unitary.

**Proof:** Let $A$ be a unitary matrix, then
$$AA^* = A^*A = I$$
$$\Rightarrow (AA^*)^{-1} = (A^*A)^{-1} = (I)^{-1}$$
$$\Rightarrow (A^*)^{-1}A^{-1} = A^{-1}(A^*)^{-1} = I$$
$$\Rightarrow (A^{-1})^*A^{-1} = A^{-1}(A^{-1})^* = I$$

Hence $A^{-1}$ is unitary.

**Theorem:** Transpose of a unitary matrix is unitary.

**Proof:** Let $A$ be a unitary matrix, then
$$AA^* = A^*A = I$$
$$\Rightarrow (AA^*)^T = (A^*A)^T = (I)^T$$
$$\Rightarrow (A^*)^T A^T = A^T (A^*)^T = I$$
$$\Rightarrow (A^T)^* A^T = A^T (A^T)^* = I$$

Hence $A^T$ is unitary.

**Example:** Prove that $\dfrac{1}{2}\begin{bmatrix} 1+i & -1+i \\ 1+i & 1-i \end{bmatrix}$ is unitary and hence find $A^{-1}$.

**Solution:** Let $A = \dfrac{1}{2}\begin{bmatrix} 1+i & -1+i \\ 1+i & 1-i \end{bmatrix}$.

$$\overline{A} = \dfrac{1}{2}\begin{bmatrix} 1-i & -1-i \\ 1-i & 1+i \end{bmatrix}.$$

Hence $A^* = \left(\overline{A}\right)^T = \dfrac{1}{2}\begin{bmatrix} 1-i & -1-i \\ 1-i & 1+i \end{bmatrix}^T = \dfrac{1}{2}\begin{bmatrix} 1-i & 1-i \\ -1-i & 1+i \end{bmatrix}.$

$\therefore AA^* = \dfrac{1}{4}\begin{bmatrix} 1+i & -1+i \\ 1+i & 1-i \end{bmatrix}\begin{bmatrix} 1-i & 1-i \\ -1-i & 1+i \end{bmatrix}$

$= \dfrac{1}{4}\begin{bmatrix} (1+i)(1-i)+(-1+i)(-1-i) & (1+i)(1-i)+(-1+i)(1+i) \\ (1+i)(1-i)+(1-i)(-1-i) & (1+i)(1-i)+(1-i)(1+i) \end{bmatrix}$

$= \dfrac{1}{4}\begin{bmatrix} 1+1+1+1 & 1+1-1-1 \\ 1+1-1-1 & 1+1+1+1 \end{bmatrix}$

$= \dfrac{1}{4}\begin{bmatrix} 4 & 0 \\ 0 & 4 \end{bmatrix}$

$= \begin{bmatrix} 1 & 0 \\ 0 & 1 \end{bmatrix} = I$

Similarly, $A^*A = I$.

Hence $AA^* = A^*A = I$, i.e., $A$ is unitary.

Also $A^{-1} = A^* = \dfrac{1}{2}\begin{bmatrix} 1-i & 1-i \\ -1-i & 1+i \end{bmatrix}.$

**Example:** Show that the matrix $A = \dfrac{1}{\sqrt{3}}\begin{bmatrix} 1 & 1+i \\ 1-i & -1 \end{bmatrix}$ is unitary.

**Solution:** Given $A = \dfrac{1}{\sqrt{3}}\begin{bmatrix} 1 & 1+i \\ 1-i & -1 \end{bmatrix}.$

$$\overline{A} = \frac{1}{\sqrt{3}}\begin{bmatrix} 1 & 1-i \\ 1+i & -1 \end{bmatrix}.$$

and $A^* = (\overline{A})^T = \frac{1}{\sqrt{3}}\begin{bmatrix} 1 & 1+i \\ 1-i & -1 \end{bmatrix}.$

$$\therefore AA^* = \frac{1}{3}\begin{bmatrix} 1 & 1+i \\ 1-i & -1 \end{bmatrix}\begin{bmatrix} 1 & 1+i \\ 1-i & -1 \end{bmatrix}$$

$$= \frac{1}{3}\begin{bmatrix} 1+1+1 & 1+i-1-i \\ 1-i-1+i & 1+1+1 \end{bmatrix}$$

$$= \frac{1}{3}\begin{bmatrix} 3 & 0 \\ 0 & 3 \end{bmatrix}$$

$$= \begin{bmatrix} 1 & 0 \\ 0 & 1 \end{bmatrix} = I$$

This shows that $A^* = A^{-1}$. Therefore, $A$ is a unitary matrix.

**Example:** Show that the matrix $A = \frac{1}{3}\begin{bmatrix} 2+i & 2i \\ 2i & 2-i \end{bmatrix}.$

**Solution:** Given $A = \frac{1}{3}\begin{bmatrix} 2+i & 2i \\ 2i & 2-i \end{bmatrix}.$

$$\overline{A} = \frac{1}{3}\begin{bmatrix} 2-i & -2i \\ -2i & 2+i \end{bmatrix}.$$

and $A^* = \overline{(A)}^T = \dfrac{1}{3}\begin{bmatrix} 2-i & -2i \\ -2i & 2+i \end{bmatrix}$.

$$\therefore AA^* = \dfrac{1}{9}\begin{bmatrix} 2+i & 2i \\ 2i & 2-i \end{bmatrix}\begin{bmatrix} 2-i & -2i \\ -2i & 2+i \end{bmatrix}$$

$$= \dfrac{1}{9}\begin{bmatrix} 4+1+4 & -4i+2+4i-2 \\ 4i+2-4i-2 & 4+4+1 \end{bmatrix}$$

$$= \dfrac{1}{9}\begin{bmatrix} 9 & 0 \\ 0 & 9 \end{bmatrix}$$

$$= \begin{bmatrix} 1 & 0 \\ 0 & 1 \end{bmatrix} = I$$

This shows that $A^* = A^{-1}$. Therefore, $A$ is a unitary matrix.

**Example:** Show that the matrix $\begin{bmatrix} a+ic & -b+id \\ b+id & a-ic \end{bmatrix}$ is unitary if $a^2+b^2+c^2+d^2 = 1$.

**Solution:** Let $A = \begin{bmatrix} a+ic & -b+id \\ b+id & a-ic \end{bmatrix}$

$\Rightarrow \overline{A} = \begin{bmatrix} a-ic & -b-id \\ b-id & a+ic \end{bmatrix}$

$\Rightarrow A^* = \overline{(A)}^T = \begin{bmatrix} a-ic & b-id \\ -b-id & a+ic \end{bmatrix}$

Now

$$AA^* = \begin{bmatrix} a+ic & -b+id \\ b+id & a-ic \end{bmatrix}\begin{bmatrix} a-ic & b-id \\ -b-id & a+ic \end{bmatrix}$$

$$= \begin{bmatrix} (a+ic)(a-ic)+(-b+id)(-b-id) & (a+ic)(b-id)+(-b-id)(a+ic) \\ (b+id)(a-ic)+(a-ic)(-b-id) & (b+id)9b-id)+(a-ic)(a+ic) \end{bmatrix}$$

$$= \begin{bmatrix} a^2+b^2+c^2+d^2 & 0 \\ 0 & a^2+b^2+c^2+d^2 \end{bmatrix}$$

Therefore $AA^* = I$ if and only if $a^2+b^2+c^2+d^2 = 1$

i.e., $A$ is unitary if and only if $a^2 + b^2 + c^2 + d^2 = 1$.
Hence, proved.

**Example:** Prove that $\begin{bmatrix} i & 0 & 0 \\ 0 & 0 & i \\ 0 & i & 0 \end{bmatrix}$ is a skew Hermitian matrix and also unitary.

**Solution:** Let $A = \begin{bmatrix} i & 0 & 0 \\ 0 & 0 & i \\ 0 & i & 0 \end{bmatrix}$.

$$\overline{A} = \begin{bmatrix} -i & 0 & 0 \\ 0 & 0 & -i \\ 0 & -i & 0 \end{bmatrix} = -\begin{bmatrix} i & 0 & 0 \\ 0 & 0 & i \\ 0 & i & 0 \end{bmatrix} = -A^T.$$

This shows that $A$ is a skew-Hermitian matrix.
Also we find that $A^* = \overline{A}^T = -A$ so that

$$AA^* = -A^2 = -\begin{bmatrix} i & 0 & 0 \\ 0 & 0 & i \\ 0 & i & 0 \end{bmatrix}\begin{bmatrix} i & 0 & 0 \\ 0 & 0 & i \\ 0 & i & 0 \end{bmatrix}$$

$$= -\begin{bmatrix} -1 & 0 & 0 \\ 0 & -1 & 0 \\ 0 & 0 & -1 \end{bmatrix} = \begin{bmatrix} 1 & 0 & 0 \\ 0 & 1 & 0 \\ 0 & 0 & 1 \end{bmatrix}$$

$\therefore A^* = A^{-1}$. This means that $A$ is unitary.

**Example:** If $A = \begin{bmatrix} 0 & 1+2i \\ -1+2i & 0 \end{bmatrix}$ then show that $(I-A)(I+A)^{-1}$ is a unitary matrix.

**Solution:** Given $A = \begin{bmatrix} 0 & 1+2i \\ -1+2i & 0 \end{bmatrix}$.

$$I - A = \begin{bmatrix} 1 & 0 \\ 0 & 1 \end{bmatrix} - \begin{bmatrix} 0 & 1+2i \\ -1+2i & 0 \end{bmatrix} = \begin{bmatrix} 1 & -1-2i \\ 1-2i & 1 \end{bmatrix}$$

$$I + A = \begin{bmatrix} 1 & 0 \\ 0 & 1 \end{bmatrix} + \begin{bmatrix} 0 & 1+2i \\ -1+2i & 0 \end{bmatrix} = \begin{bmatrix} 1 & 1+2i \\ -1+2i & 1 \end{bmatrix}$$

Also $|I + A| = \begin{vmatrix} 1 & 1+2i \\ -1+2i & 1 \end{vmatrix} = 1 - (1+2i)(-1+2i) = 1 + 1 + 4 = 6.$

$$\therefore (I+A)^{-1} = \frac{1}{6}\begin{bmatrix} 1 & -1-2i \\ 1-2i & 1 \end{bmatrix}$$

such that

$$\therefore (I-A)(I+A)^{-1} = \frac{1}{6}\begin{bmatrix} 1 & -1-2i \\ 1-2i & 1 \end{bmatrix}\begin{bmatrix} 1 & -1-2i \\ 1-2i & 1 \end{bmatrix}$$

$$= \frac{1}{6}\begin{bmatrix} 1-(1+4) & -1-2i-1-2i \\ 1-2i+1+2i & -(1+4)+1 \end{bmatrix}$$

$$= \frac{1}{6}\begin{bmatrix} -4 & -2-4i \\ 2-4i & -4 \end{bmatrix} = B \text{ (say)}$$

Now $\overline{B} = \frac{1}{6}\begin{bmatrix} -4 & -2+4i \\ 2+4i & -4 \end{bmatrix}$

and $B^* = (\overline{B})^T = \frac{1}{6}\begin{bmatrix} -4 & 2+4i \\ -2+4i & -4 \end{bmatrix}.$

To prove $B = (I-A)(I+A)^{-1}$ is unitary, verify $BB^* = I$.

Now

$$BB^* = \frac{1}{36}\begin{bmatrix} -4 & -2+4i \\ 2+4i & -4 \end{bmatrix}\begin{bmatrix} -4 & 2+4i \\ -2+4i & -4 \end{bmatrix}$$

$$= \frac{1}{36}\begin{bmatrix} 36 & 0 \\ 0 & 36 \end{bmatrix}$$

$$= \begin{bmatrix} 1 & 0 \\ 0 & 1 \end{bmatrix} = I.$$

$\therefore B$ is unitary.

Hence, $(I-A)(I+A)^{-1}$ is unitary.

**Example:** If $S = \begin{bmatrix} 1 & 1 & 1 \\ 1 & \alpha^2 & \alpha \\ 1 & \alpha & \alpha^2 \end{bmatrix}$, where $\alpha = e^{2\pi i/3}$, prove that $S$ is non-singular and $S^{-1} = \dfrac{1}{3}\overline{S}$.

**Solution:** From the given $\alpha$, we find that
$$\alpha^3 = e^{2\pi i} = \cos 2\pi + i \sin 2\pi = 1$$
and $\overline{\alpha} = e^{-2\pi i/3} = \dfrac{1}{\alpha} = \dfrac{\alpha^2}{\alpha^3} = \alpha^2$

so that $\overline{(\alpha^2)} = (\overline{\alpha}) = \alpha$.

$$\therefore \overline{S} = \begin{bmatrix} 1 & 1 & 1 \\ 1 & \overline{\alpha^2} & \overline{\alpha} \\ 1 & \overline{\alpha} & \overline{\alpha^2} \end{bmatrix} = \begin{bmatrix} 1 & 1 & 1 \\ 1 & \alpha & \alpha^2 \\ 1 & \alpha^2 & \alpha \end{bmatrix}$$

and

$$S\overline{S} = \begin{bmatrix} 1 & 1 & 1 \\ 1 & \alpha^2 & \alpha \\ 1 & \alpha & \alpha^2 \end{bmatrix} \begin{bmatrix} 1 & 1 & 1 \\ 1 & \alpha & \alpha^2 \\ 1 & \alpha^2 & \alpha \end{bmatrix}$$

$$= \begin{bmatrix} 1+1+1 & 1+\alpha+\alpha^2 & 1+\alpha^2+\alpha \\ 1+\alpha^2+\alpha & 1+\alpha^3+\alpha^3 & 1+\alpha^4+\alpha^2 \\ 1+\alpha+\alpha^2 & 1+\alpha^2+\alpha^4 & 1+\alpha^3+\alpha^3 \end{bmatrix} \quad \cdots\cdots \quad (1)$$

Use $\alpha^3 = 1$ so that
$$1 + \alpha^2 + \alpha^4 = 1 + \alpha^2 + \alpha = \dfrac{1-\alpha^3}{1-\alpha} = 0.$$

$\therefore$ (1) becomes

$$S\overline{S} = \begin{bmatrix} 3 & 0 & 0 \\ 0 & 3 & 0 \\ 0 & 0 & 3 \end{bmatrix} = 3 \begin{bmatrix} 1 & 0 & 0 \\ 0 & 1 & 0 \\ 0 & 0 & 1 \end{bmatrix} = 3I$$

$$\Rightarrow S\left(\dfrac{1}{3}\overline{S}\right) = I.$$

This shows that $S$ is non-singular and
$$S^{-1} = \dfrac{1}{3}\overline{S}.$$

# Chapter 3

# Iterative Methods

## 3.1 Introduction

**Definition:** An equation $f(x) = 0$, where $f(x) = a_0 x^n + a_1 x^{n-1} + a_2 x^{n-2} + \cdots + a_n$ where $a_0, a_1, a_2, \cdots, a_n$ are real numbers, $a_0 \neq 0$, is called an $n^{th}$ degree *algebraic equation*.

Examples: $x^2 - 5x + 6 = 0$, $x^3 - 5 = 0$, $x^4 - 5x^3 - 3x^2 + 6x - 9 = 0$.

If $f(x)$ contains some other functions namely trigonometric, logarithmic, exponential etc, then the equation $f(x) = 0$ is called a *transcendental equation*.

Examples: $x - e^x + 5 \log x = 0$, $xe^x + \log x = 0$, $x - e^x + 5 \sin x = 0$.

### 3.1.1 Root of an Equation

A real number $\alpha$ is called the root of the equation $f(x) = 0$ if and only if $f(\alpha) = 0$.

Geometrically, the real root of the equation $f(x) = 0$ is the value of $x$ at which the graph of $f(x)$ meets the $x$-axis.

To locate the root of an equation $f(x) = 0$, we use the following well known theorem in calculus.

### 3.1.2 Intermediate Value Theorem

If $f(x)$ is continuous in the interval $[a, b]$ and if $f(a)$ and $f(b)$ are of opposite signs, (i.e., $f(a) < 0$ and $f(b) > 0$ or $f(a) > 0$ and $f(b) < 0$) then the equation $f(x) = 0$

has at least one root lying between $a$ and $b$.

### 3.1.3 Absolute Error and Relative Error

If $\alpha$ is an approximate value of a quantity whose exact value is $a$, then the difference $\epsilon = \alpha - a$ is called the **absolute error** of $\alpha$ or simply the **error** of $\alpha$.

The **relative error** $\epsilon_r$ is defined by $\epsilon_r = \dfrac{\alpha - a}{a}$ provided $a \neq 0$. The percentage error $\epsilon_p$ is defined by $\epsilon_p = 100\epsilon_r$.

## 3.2 Bisection Method

The bisection method is one of the bracketing methods for finding roots of an equation. For a given a function $f(x)$, guess an interval which might contain a root and perform a number of iterations, where, in each iteration the interval containing the root is get halved.

Suppose we want to find the solution to the equation $f(x) = 0$, where $f$ is continuous.

Find $a$ and $b$ for which $f(a) < 0$ and $f(b) > 0$.
Let $a < b$.

1. Find $x_1 = \dfrac{a+b}{2}$. Calculate $f(x_1)$. If $f(x_1) = 0$, then $x_1$ becomes the root of the equation $f(x) = 0$, otherwise

    (a) If $f(x_1) < 0$, then the root lies between $x_1$ and $b$.
    Find $x_2 = \dfrac{x_1 + b}{2}$. Calculate $f(x_2)$ and so on.

    (b) If $f(x_1) > 0$, then the root lies between $a$ and $x_1$.
    Find $x_3 = \dfrac{a + x_1}{2}$. Calculate $f(x_3)$ and so on.

    We proceed in this way until the two successive approximations are approximately equal.

## Merits of bisection method:

✠ The iteration using bisection method always produces a root, since the method brackets the root between two values.

✠ As iterations are conducted, the length of the interval gets halved. So one can guarantee the convergence in case of the solution of the equation.

✠ The Bisection Method is simple to program in a computer.

## Demerits of bisection method:

✠ The convergence of the bisection method is slow as it is simply based on having the interval.

✠ Bisection method cannot be applied over an interval where there is a discontinuity.

✠ Bisection method cannot be applied over an interval where the function takes always values of the same sign.

✠ The method fails to determine complex roots.

**Example:** Find a root of the equation $x^3 - 4x - 9 = 0$, using bisection method correct to three decimal places.

**Solution:** Let $f(x) = x^3 - 4x - 9$

$$f(0) = -9 < 0$$
$$f(1) = -12 < 0$$
$$f(2) = -9 < 0$$
$$f(3) = 6 > 0$$

∴ One of the root lies between 2 and 3.

By bisection method, the first approximation root is
$$x_1 = \frac{2+3}{2} = 2.5$$
$$f(2.5) = -3.375 < 0$$

Since $f(2.5) < 0$ and $f(3) > 0$, the root lies between 2.5 and 3
$$x_2 = \frac{2.5 + 3}{2} = 2.75$$
$$f(2.75) = 0.7969 > 0$$

Since $f(2.5) < 0$ and $f(2.75) > 0$, the root lies between 2.5 and 2.75
$$x_3 = \frac{2.5 + 2.75}{2} = 2.625$$
$$f(2.625) = -1.4121 < 0$$
Since $f(2.625) < 0$ and $f(2.75) > 0$, the root lies between 2.625 and 2.75
$$x_4 = \frac{2.625 + 2.75}{2} = 2.6875$$
Repeating this process, the successive approximation are
$$x_5 = 2.71875, \quad x_6 = 2.70313, \quad x_7 = 2.71094$$
$$x_8 = 2.70703, \quad x_9 = 2.70508, \quad x_{10} = 2.70605$$
$$x_{11} = 2.70654, \quad x_{12} = 2.70642$$
Since 11th and 12th approximations are coinciding upto three decimal places. Therefore, the value of the root correct to three decimal places is 2.7064.

**Example:** Find a real root of the equation $x^3 - x - 11 = 0$, using bisection method correct to three decimal places.

**Solution:** Let $f(x) = x^3 - x - 11$
$$f(0) = -11 < 0$$
$$f(1) = -11 < 0$$
$$f(2) = -5 < 0$$
$$f(3) = 13 > 0$$
$\therefore$ One of the root lies between 2 and 3.

By bisection method, the first approximation root is
$$x_1 = \frac{2+3}{2} = 2.5$$
$$f(2.5) = 2.125 > 0$$
Since $f(2) < 0$ and $f(2.5) > 0$, the root lies between 2 and 2.5
$$x_2 = \frac{2 + 2.5}{2} = 2.25$$
$$f(2.25) = -1.8594 < 0$$
Since $f(2.25) < 0$ and $f(2.5) > 0$, the root lies between 2.25 and 2.5
$$x_3 = \frac{2.25 + 2.5}{2} = 2.375$$

$$f(2.375) = 0.0215 > 0$$

Since $f(2.25) < 0$ and $f(2.375) > 0$, the root lies between 2.25 and 2.375
$$x_4 = \frac{2.25 + 2.375}{2} = 2.3125$$
Repeating this process, the successive approximation are
$$x_5 = 2.3438, \quad x_6 = 2.3594, \quad x_7 = 2.3672$$
$$x_8 = 2.3711, \quad x_9 = 2.3731, \quad x_{10} = 2.3741$$
$$x_{11} = 2.3736, \quad x_{12} = 2.3739$$
Since 11th and 12th approximations are coinciding upto three decimal places. Therefore, the value of the root correct to three decimal places is 2.3739.

**Example:** Using bisection method find an approximate root of the equation $\sin x = \dfrac{1}{x}$ that lies between 1 and 1.5.

**Solution:** Let $\sin x = \dfrac{1}{x} \Rightarrow f(x) = x \sin x - 1$
$$f(1) = -0.1585 < 0$$
$$f(1.5) = 0.49624 > 0$$
Since $f(1) < 0$ and $f(1.5) > 0$. The root lies between 1 and 1.5.
By bisection method, the first approximation root is
$$x_1 = \frac{1 + 1.5}{2} = 1.25$$
$$f(1.25) = 0.1862 > 0$$
Since $f(1) < 0$ and $f(1.25) > 0$, the root lies between 1 and 1.25
$$x_2 = \frac{1 + 1.25}{2} = 1.125$$
$$f(1.125) = 0.01505 > 0$$
Since $f(1) < 0$ and $f(1.125) > 0$, the root lies between 1 and 1.25
$$x_3 = \frac{1 + 1.125}{2} = 1.0625$$
$$f(1.0625) = -0.07118 < 0$$
Since $f(1.0625) < 0$ and $f(1.125) > 0$, the root lies between 1.0625 and 1.125
$$x_4 = \frac{1.0625 + 1.125}{2} = 1.09375$$
$$f(1.09375) = -0.02836 < 0$$

Since $f(1.09375) < 0$ and $f(1.125) > 0$, the root lies between 1.09375 and 1.125
$$x_5 = \frac{1.09375 + 1.125}{2} = 1.109375$$
$$f(1.109375) = -0.0066428 < 0$$
Since $f(1.109375) < 0$ and $f(1.125 > 0$, the root lies between 1.109375 and 1.125
$$x_6 = \frac{1.109375 + 1.125}{2} = 1.1171875$$
$$f(1.1171875) = 0.0042080 > 0$$
Since $f(1.109375) < 0$ and $f(1.1171875) > 0$, the root lies between 1.109375 and 1.1171875
$$x_7 = \frac{1.109375 + 1.1171875}{2} = 1.113281$$
$$f(1.113281) = -0.00121 < 0$$
Since $f(1.113281) < 0$ and $f(1.1171875) > 0$, the root lies between 1.113281 and 1.1171875
$$x_8 = \frac{1.113281 + 1.1171875}{2} = 1.115234$$
$$f(1.115234) = 0.001495 > 0$$
Since 7th and 8th approximations are coinciding upto two decimal places. Therefore, the value of the root correct to two decimal places is 1.115234.

**Example:** Using bisection method, obtain an approximate root of the equation $x^3 - x - 1 = 0$.

**Solution:** Let $f(x) = x^3 - x - 1$
$$f(0) = -1 < 0$$
$$f(1) = -1 < 0$$
$$f(2) = 5 > 0$$
$\therefore$ One of the root lies between 1 and 2.

By bisection method, the first approximation root is
$$x_1 = \frac{1+2}{2} = 1.5$$
$$f(1.5) = 0.875 > 0$$
Since $f(1) < 0$ and $f(1.5) > 0$, the root lies between 1 and 1.5

$$x_2 = \frac{1+1.5}{2} = 1.25$$
$$f(1.25) = -0.296875 < 0$$

Since $f(1.25) < 0$ and $f(1.5) > 0$, the root lies between 1.25 and 1.5
$$x_3 = \frac{1.25+1.5}{2} = 1.375$$
$$f(1.375) = 0.224609 > 0$$

Since $f(1.25) < 0$ and $f(1.375) > 0$, the root lies between 1.25 and 1.375
$$x_4 = \frac{1.25+1.375}{2} = 1.3125$$
$$f(1.3125) = -0.05151 < 0$$

Since $f(1.3125) < 0$ and $f(1.375) > 0$, the root lies between 1.3125 and 1.375
$$x_5 = \frac{1.3125+1.375}{2} = 1.34375$$
$$f(1.34375) = 0.082611 > 0$$

Since $f(1.3125) < 0$ and $f(1.34375) > 0$, the root lies between 1.3125 and 1.34375
$$x_6 = \frac{1.3125+1.34375}{2} = 1.328125$$
$$f(1.328125) = 0.80145576 > 0$$

Since $f(1.3125) < 0$ and $f(1.328125) > 0$, the root lies between 1.3125 and 1.328125
$$x_7 = \frac{1.3125+1.328125}{2} = 1.32032$$
$$f(1.32032) = -0.0186 < 0$$

Since 6th and 7th approximations are coinciding upto two decimal places. Therefore, the value of the root correct to 2 decimal places is 1.32.

**Example:** Applying bisection method to find a root of the equation $x - \cos x = 0$.

**Solution:** Let $f(x) = x - \cos x$
$$f(0) = -1 < 0$$
$$f(1) = 0.459698 > 0$$

$\therefore$ One of the root lies between 0 and 1.

By bisection method, the first approximation root is
$$x_1 = \frac{0+1}{2} = 0.5$$
$$f(0.5) = -0.377583 < 0$$

Since $f(0.5) < 0$ and $f(1) > 0$, the root lies between 0.5 and 1
$$x_2 = \frac{0.5 + 1}{2} = 0.75$$
$$f(0.75) = 0.0183111 > 0$$
Since $f(0.5) < 0$ and $f(0.75) > 0$, the root lies between 0.5 and 0.75
$$x_3 = \frac{0.5 + 0.75}{2} = 0.625$$
$$f(0.625) = -0.185963 < 0$$
Since $f(0.625) < 0$ and $f(0.75) > 0$, the root lies between 0.625 and 0.75
$$x_4 = \frac{0.625 + 0.75}{2} = 0.6875$$
$$f(0.6875) = -0.08533 < 0$$
Since $f(0.6875) < 0$ and $f(0.75) > 0$, the root lies between 0.6875 and 0.75
$$x_5 = \frac{0.6875 + 0.75}{2} = 0.71875$$
$$f(0.71875) = -0.03388 < 0$$
Since $f(0.71875) < 0$ and $f(0.75) > 0$, the root lies between 0.71875 and 0.75
$$x_6 = \frac{0.71875 + 0.75}{2} = 0.734375$$
$$f(0.734375) = -0.00877 < 0$$
Since $f(0.734375) < 0$ and $f(0.75) > 0$, the root lies between 0.734375 and 0.75
$$x_7 = \frac{0.734375 + 0.75}{2} = 0.742188$$
$$f(0.742188) = 0.00519 > 0$$
Since $f(0.734375) < 0$ and $f(0.742188) > 0$, the root lies between 0.734375 and 0.742188
$$x_8 = \frac{0.734375 + 0.742188}{2} = 0.738281$$
$$f(0.738281) = -0.001345 < 0$$
Since 7th and 8th approximations are coinciding upto one decimal places. Therefore, the value of the root correct to one decimal places is 0.738281.

**Example:** Assuming that root of the equation $x^3 - 9x + 1 = 0$ lies in the interval $(2, 4)$, find that root by bisection method correct upto two decimal places.

**Solution:** Let $f(x) = x^3 - 9x + 1$

$$f(2) = -9 < 0$$
$$f(4) = 29 > 0$$
∴ One of the root lies between 2 and 4.

By bisection method, the first approximation root is
$$x_1 = \frac{2+4}{2} = 3$$
$$f(3) = 1 > 0$$

Since $f(2) < 0$ and $f(3) > 0$, the root lies between 2 and 3
$$x_2 = \frac{2+3}{2} = 2.5$$
$$f(2.5) = -5.875 < 0$$

Since $f(2.5) < 0$ and $f(3) > 0$, the root lies between 2.5 and 3
$$x_3 = \frac{2.5+3}{2} = 2.75$$
$$f(2.75) = -2.95313 < 0$$

Since $f(2.75) < 0$ and $f(3) > 0$, the root lies between 2.75 and 3
$$x_4 = \frac{2.75+3}{2} = 2.875$$
$$f(2.875) = -1.11133 < 0$$

Since $f(2.875) < 0$ and $f(3) > 0$, the root lies between 2.875 and 3
$$x_5 = \frac{2.875+3}{2} = 2.9375$$
$$f(2.9375) = -0.09009 < 0$$

Since $f(2.9375) < 0$ and $f(3) > 0$, the root lies between 2.9375 and 3
$$x_6 = \frac{2.9375+3}{2} = 2.96875$$
$$f(2.96875) = 0.446259 > 0$$

Since $f(2.9375) < 0$ and $f(2.96875) > 0$, the root lies between 2.9375 and 2.96875
$$x_7 = \frac{2.9375+2.96875}{2} = 2.95313$$
$$f(2.95313) = 0.0176008 > 0$$

Since $f(2.9375) < 0$ and $f(2.95313) > 0$, the root lies between 2.9375 and 2.95313
$$x_8 = \frac{2.9375+2.95313}{2} = 2.94532$$
$$f(2.94532) = 0.042505$$

Since $f(2.9375) < 0$ and $f(2.94532) > 0$, the root lies between 2.9375 and 2.94532
$$x_9 = \frac{2.9375 + 2.94532}{2} = 2.94141$$
$$f(2.94141) = -0.023926 < 0$$
Since 8th and 9th approximations are coinciding upto two decimal places. Therefore, the value of the root correct to two decimal places is 2.94141.

**Example:** By using bisection method find a square root of 26.

**Solution:** Let $x = \sqrt{26}$.

$\therefore x^2 = 26$. Hence $x^2 - 26 = 0$.

Let $f(x) = x^2 - 26$

$f(5) = -1 < 0$

$f(6) = 10 > 0$

$\therefore$ One of the root lies between 5 and 6.

By bisection method, the first approximation root is
$$x_1 = \frac{5+6}{2} = 5.5$$
$$f(5.5) = 4.25 > 0$$

Since $f(5) < 0$ and $f(5.5) > 0$, the root lies between 5 and 5.5
$$x_2 = \frac{5+5.5}{2} = 5.25$$
$$f(5.25) = 1.5625 > 0$$

Since $f(5) < 0$ and $f(5.25) > 0$, the root lies between 5 and 5.25
$$x_3 = \frac{5+5.25}{2} = 5.125$$
$$f(5.125) = 0.2656 > 0$$

Since $f(5) < 0$ and $f(5.125) > 0$, the root lies between 5 and 5.125
$$x_4 = \frac{5+5.125}{2} = 5.0625$$
$$f(5.0625) = -0.137109 < 0$$

Since $f(5.0625) < 0$ and $f(5.125) > 0$, the root lies between 5.0625 and 5.125
$$x_5 = \frac{5.0625 + 5.125}{2} = 5.09375$$
$$f(5.09375) = -0.0537 < 0$$

Since $f(5.09375) < 0$ and $f(5.125) > 0$, the root lies between 5.09375 and 5.125
$$x_6 = \frac{5.09375 + 5.125}{2} = 5.109375$$
$$f(5.109375) = 0.1057 > 0$$
Since $f(5.093755) < 0$ and $f(5.109375) > 0$, the root lies between 5.09375 and 5.109375
$$x_7 = \frac{5.09375 + 5.109375}{2} = 5.10156$$
$$f(5.10156) = 0.0259 > 0$$
Since $f(5.093755) < 0$ and $f(5.10156) > 0$, the root lies between 5.093755 and 5.10156
$$x_8 = \frac{5.093755 + 5.10156}{2} = 5.097655$$
$$f(5.097655) = -0.0139 < 0$$
Since $f(5.097655) < 0$ and $f(5.10156) > 0$, the root lies between 5.097655 and 5.10156
$$x_9 = \frac{5.097655 + 5.10156}{2} = 5.0996$$
Since 8th and 9th approximations are coinciding upto two decimal places. Therefore, the value of the root correct to two decimal places is 5.0996.

**Example:** Using bisection method, find the root between 2&3 of the equation $x^4 - x^3 - 2x^2 - 6x - 4 = 0$ up to three decimals.

**Solution:** Let $f(x) = x^4 - x^3 - 2x^2 - 6x - 4$
$$f(2) = -16 < 0$$
$$f(3) = 14 > 0$$
$\therefore$ One of the root lies between 2 and 3.

By bisection method, the first approximation root is
$$x_1 = \frac{2+3}{2} = 2.5$$
$$f(2.5) = -8.0625 < 0$$
Since $f(2.5) < 0$ and $f(3) > 0$, the root lies between 2.5 and 3
$$x_2 = \frac{2.5 + 3}{2} = 2.75$$
$$f(2.75) = 0.76 > 0$$
Since $f(2.5) < 0$ and $f(2.75) > 0$, the root lies between 2.5 and 2.75

$$x_3 = \frac{2.5 + 2.75}{2} = 2.625$$

$$f(2.625) = -4.1384 < 0$$

Since $f(2.625) < 0$ and $f(2.75) > 0$, the root lies between 2.625 and 2.75

$$x_4 = \frac{2.625 + 2.75}{2} = 2.6875$$

$$f(2.6875) = -1.8144 < 0$$

Since $f(2.6875) < 0$ and $f(2.75) > 0$, the root lies between 2.6875 and 2.75

$$x_5 = \frac{2.6875 + 3}{2.75} = 2.71875$$

$$f(2.71875) = -0.5558 < 0$$

Since $f(2.71875) < 0$ and $f(2.75) > 0$, the root lies between 2.71875 and 2.75

$$x_6 = \frac{2.71875 + 2.75}{2} = 2.734375$$

$$f(2.734375) = 0.09838 > 0$$

Since $f(2.71875) < 0$ and $f(2.734375) > 0$, the root lies between 2.71875 and 2.74375

$$x_7 = \frac{2.71875 + 2.734375}{2} = 2.72665$$

$$f(2.72665) = -0.227169 < 0$$

Since $f(2.72665) < 0$ and $f(2.734375) > 0$, the root lies between 2.72665 and 2.734375

$$x_8 = \frac{2.72665 + 2.734375}{2} = 2.73194$$

$$f(2.73194) = 0.042505$$

Since $f(2.72665) < 0$ and $f(2.73194) > 0$, the root lies between 2.726655 and 2.73194

$$x_9 = \frac{2.726655 + 2.73194}{2} = 2.73315$$

$$f(2.73315) = 0.046481 > 0$$

Since $f(2.72665) < 0$ and $f(2.73315) > 0$, the root lies between 2.726655 and 2.73315

$$x_{10} = \frac{2.726655 + 2.73315}{2} = 2.732545$$

Since 9th and 10th approximations are coinciding upto two decimal places. Therefore, the value of the root correct to two decimal places is 2.7325.

**Example:** Find the real root of the $x = e^{-x}$ using bisection method. **(Dec-2021)**

**Solution:** Let $f(x) = xe^x - 1$

$$f(0) = -1 < 0$$

$$f(1) = 1.7182 > 0$$

Since $f(0) < 0$ and $f(1) > 0$. The root lies between 0 and 1.

By bisection method, the first approximation root is
$$x_1 = \frac{0+1}{2} = 0.5$$
$$f(0.5) = -0.1756 < 0$$

Since $f(0.5) < 0$ and $f(1) > 0$, the root lies between 0.5 and 1
$$x_2 = \frac{0.5+1}{2} = 0.75$$
$$f(0.75) = 0.5877 > 0$$

Since $f(0.5) < 0$ and $f(0.75) > 0$, the root lies between 0.5 and 0.75
$$x_3 = \frac{0.5 + 0.75}{2} = 0.625$$
$$f(0.625) = 0.1676 > 0$$

Since $f(0.5) < 0$ and $f(0.625) > 0$, the root lies between 0.5 and 0.625
$$x_4 = \frac{0.5 + 0.625}{2} = 0.5625$$
$$f(0.5625) = -0.012 < 0$$

Since $f(0.5625) < 0$ and $f(0.625) > 0$, the root lies between 0.5625 and 0.625
$$x_5 = \frac{0.5625 + 0.625}{2} = 0.59375$$
$$f(0.59375) = 0.0075 > 0$$

Since $f(0.5625) < 0$ and $f(0.58375) > 0$, the root lies between 0.5625 and 0.59375
$$x_6 = \frac{0.5625 + 0.59375}{2} = 0.57812$$
$$f(0.57812) = 0.030 > 0$$

Since $f(0.5625) < 0$ and $f(0.57812) > 0$, the root lies between 0.5625 and 0.57812
$$x_7 = \frac{0.5625 + 0.57812}{2} = 0.57031$$
$$f(2.7421875) = 0.0013445 > 0$$

Hence, the root of the given equation $x = e^{-x}$ is 0.57.

**Example:** Using bisection method find an approximate root of the equation $x \log_{10} x = 1.2$ that lies between 2 and 3.

**Solution:** Let $f(x) = x \log_{10} x - 1.2$

$$f(2) = -0.5979 < 0$$
$$f(3) = 0.23136 > 0$$

Since $f(2) < 0$ and $f(3) > 0$. The root lies between 2 and 3.
By bisection method, the first approximation root is
$$x_1 = \frac{2+3}{2} = 2.5$$
$$f(2.5) = -0.20514 < 0$$

Since $f(2.5) < 0$ and $f(3) > 0$, the root lies between 2.5 and 3
$$x_2 = \frac{2.5+3}{2} = 2.75$$
$$f(2.75) = 0.00816490 > 0$$

Since $f(2.5) < 0$ and $f(2.75) > 0$, the root lies between 2.5 and 2.75
$$x_3 = \frac{2.5+2.75}{2} = 2.625$$
$$f(2.625) = -0.09978 < 0$$

Since $f(2.625) < 0$ and $f(2.75) > 0$, the root lies between 2.625 and 2.75
$$x_4 = \frac{2.625+2.75}{2} = 2.6875$$
$$f(2.6875) = -0.0461 < 0$$

Since $f(2.6875) < 0$ and $f(2.75) > 0$, the root lies between 2.6875 and 2.75
$$x_5 = \frac{2.6875+2.75}{2} = 2.71875$$
$$f(2.71875) = -0.01905 < 0$$

Since $f(2.71875) < 0$ and $f(2.75) > 0$, the root lies between 2.71875 and 2.75
$$x_6 = \frac{2.71875+2.75}{2} = 2.734375$$
$$f(2.734375) = -0.0054662 < 0$$

Since $f(2.734375) < 0$ and $f(2.75) > 0$, the root lies between 2.734375 and 2.75
$$x_7 = \frac{2.734375+2.75}{2} = 2.7421875$$
$$f(2.7421875) = 0.0013445 > 0$$

Since $f(2.734375) < 0$ and $f(2.7421875) > 0$, the root lies between 2.734375 and 2.7421875
$$x_8 = \frac{2.734375+2.7421875}{2} = 2.7382$$

$$f(2.7382) = 0.001495 > 0.$$

Therefore, the value of the root correct to two decimal places is 2.7382.

**EXERCISE**

1. Find the real root of the equation $2x - \log_{10} x = 7$ using bisection method.

   (Nov-2017)

2. Find the root of the equation $x^4 - 10 = x$ using Bisection method.

   **Ans:** 1.859375

3. Find the Real root of $e^x \sin x = 1$ using Bisection method.

4. Find the Real root of $x^3 - x - 2 = 0$ using Bisection method.

5. Find the Real root of the equation $x^3 + 2x^2 + 10x - 20 = 0$ using Bisection method.

   **Ans:** 1.3688081

6. Find the Real root of the equation $xe^x = \cos x$ using Bisection method.

7. Find the Real root of the equation $3x = 1 + \cos x$ using Bisection method.

   (Mar-2022)

8. Find the Real root of the equation $x^3 - 8x - 4 = 0$ using Bisection method.

   (Mar-2022)

## 3.3 Regula-False Method

This is the oldest method for finding the real root of an equation and closely resembles the bisection method.

Suppose an equation $f(x) = 0$ is known to have only one real root $m$ between $a$ and $b$. This Happens when $f(a)$ and $f(b)$ are of opposite signs and the curve $y = f(x)$ crosses the $x$-axis only once at the point $x = m$ lying between the points $x = a$ and $x = b$.

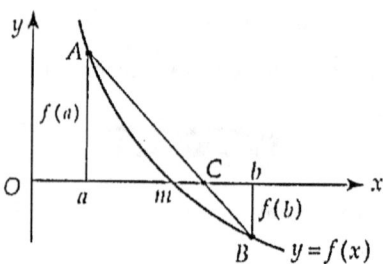

Consider the points $A(a, f(a))$ and $B(b, f(b))$ on the curve $y = f(x)$. Then the equation of the chord $AB$ is

$$\frac{y - f(a)}{x - a} = \frac{f(b) - f(a)}{b - a}. \qquad \cdots \quad (1)$$

At the point $C$ where the chord $AB$ crosses the $x$-axis, we have $y = 0$ and equation (1) yields

$$x = a - \frac{b - a}{f(b) - f(a)} f(a). \qquad \cdots \quad (2)$$

This gives the $x$-coordinate of the point $C$. If the interval $[a, b]$ is sufficiently small, the $x$-coordinate of the point $C$ is sufficiently close to the point $x = m$ (where the curve $y = f(x)$ crosses $x$-axis). In other words, $x$ given by (2) serves as an approximate value of $m$ when $b - a$ sufficiently small. The method of finding an approximate value of $m$ on the basis of this observation is known as the **Method of False position** or the **Regula-Falsi method**.

If $f(x_0)$ and $f(x_1)$ are of opposite signs, then root lies between $x_0$ and $x_1$.

Put $x_1 = x_2$, then we get next approximation. If $f(x_1)$ and $f(x_2)$ are of opposite signs, then root lies between $x_1$ and $x_2$.

Put $x_0 = x_2$, we get next approximation. We proceed in this way till the two successive approximations are equal.

The iteration formula for this method is given by

$$x_{n+1} = x_{n-1} - \frac{x_n - x_{n-1}}{f(x_n) - f(x_{n-1})} f(x_{n-1})$$

$$= \frac{x_{n-1}f(x_n) - x_n f(x_{n-1})}{f(x_n) - f(x_{n-1})}.$$

**Rate of Convergence:** This method has linear rate of convergence which is faster than that of the bisection method.

**Example:** Solve $x^3 - 2x - 5 = 0$ for a positive root by regula-falsi.

**Solution:** Let $f(x) = x^3 - 2x - 5$.  (Dec-2016), (Apr-2022)

$$f(0) = -5 < 0$$
$$f(1) = -6 < 0$$
$$f(2) = -1 < 0$$
$$f(3) = 16 > 0$$

Since $f(2) < 0$ and $f(3) > 0$. The root lies between 2 and 3.

Let $a = 2$ and $b = 3$.

By Regula-Falsi method,

$$x_1 = \frac{af(b) - bf(a)}{f(b) - f(a)}$$

$$x_1 = \frac{2f(3) - 3f(2)}{f(3) - f(2)} = 2.0588.$$

$$f(2.0588) = -0.3910 < 0$$

Since $f(2.0588) < 0$ and $f(3) > 0$. The root lies between 2.0588 and 3.

$$x_2 = \frac{(2.0588)f(3) - 3f(2.0588)}{f(3) - f(2.0588)} = 2.08125.$$

$$f(2.08125) = -0.1473 < 0$$

Since $f(2.08125) < 0$ and $f(3) > 0$. The root lies between 2.08125 and 3.

$$x_3 = \frac{(2.08125)f(3) - 3f(2.08125)}{f(3) - f(2.08125)} = 2.0896.$$

$$f(2.0896) = -0.05511 < 0$$

Since $f(2.0896) < 0$ and $f(3) > 0$. The root lies between 2.0896 and 3.

$$x_4 = \frac{(2.0896)f(3) - 3f(2.0896)}{f(3) - f(2.0896)} = 2.09272.$$

$$f(2.09272) = -0.0204 < 0$$

Since $f(2.09272) < 0$ and $f(3) > 0$. The root lies between 2.09272 and 3.
$$x_5 = \frac{(2.09272)f(3) - 3f(2.09272)}{f(3) - f(2.09272)} = 2.0938.$$
$f(2.0938) = -0.05838 < 0$

Since $f(2.0938) < 0$ and $f(3) > 0$. The root lies between 2.0938 and 3.
$$x_6 = \frac{(2.0938)f(3) - 3f(2.0938)}{f(3) - f(2.0938)} = 2.0942.$$
$f(2.0942) = -0.05838 < 0$

Since $f(2.0942) < 0$ and $f(3) > 0$. The root lies between 2.0942 and 3.
$$x_7 = \frac{(2.0942)f(3) - 3f(2.0942)}{f(3) - f(2.0942)} = 2.0944.$$
$f(2.0944) = -0.01696 < 0$

Since $f(2.0944) < 0$ and $f(3) > 0$. The root lies between 2.0944 and 3.
$$x_8 = \frac{(2.0944)f(3) - 3f(2.0944)}{f(3) - f(2.0944)} = 2.0944.$$
Hence the root of the equation 2.0944

**Example:** Find the root of the equation $\cos x = xe^x$ using the Regula-Falsi method correct to four decimal places.

**Solution:** Let $f(x) = \cos x - xe^x$.
$$f(0) = 1 > 0$$
$$f(1) = -2.1779 < 0$$
Since $f(0) > 0$ and $f(1) < 0$. The root lies between 0 and 1.
Let $a = 0$ and $b = 1$.
By Regula-Falsi method,
$$x_1 = \frac{af(b) - bf(a)}{f(b) - f(a)}$$
$$x_1 = \frac{0f(1) - 1f(0)}{f(1) - f(0)} = 0.31467.$$
$f(0.31467) = 0.51987 > 0$

Since $f(0.31467) > 0$ and $f(1) < 0$. The root lies between 0.31467 and 1.
$$x_2 = \frac{(0.31467)f(1) - 1f(0.31467)}{f(1) - f(0.31467)} = 0.44673.$$
$f(0.44673) = 0.20354 > 0$

Since $f(0.44673) > 0$ and $f(1) < 0$. The root lies between 0.44673 and 1.

$$x_3 = \frac{(0.44673)f(1) - 1f(0.44673)}{f(1) - f(0.44673)} = 0.49402.$$
$$f(0.49402) = 0.0706 > 0$$

Since $f(0.49402) > 0$ and $f(1) < 0$. The root lies between 0.49402 and 1.
$$x_4 = \frac{(0.49402)f(1) - 1f(0.49402)}{f(1) - f(0.49402)} = 0.50995.$$
$$f(0.50995) = 0.0235 > 0$$

Since $f(0.50995 > 0$ and $f(1) < 0$. The root lies between 0.50995 and 1.
$$x_5 = \frac{(0.50995)f(1) - 1f(0.50995)}{f(1) - f(0.50995)} = 0.51520.$$
$$f(0.51520) = 0.007763 > 0$$

Since $f(0.51520) > 0$ and $f(1) < 0$. The root lies between 0.51520 and 1.
$$x_6 = \frac{(0.51520)f(1) - 1f(0.51520)}{f(1) - f(0.51520)} = 0.51692.$$
$$f(0.51692) = 0.002545 > 0$$

Since $f(0.51692) > 0$ and $f(1) < 0$. The root lies between 0.51692 and 1.
$$x_7 = \frac{(0.51692)f(1) - 1f(0.51692)}{f(1) - f(0.51692)} = 0.51748.$$
$$f(0.51748) = 0.008435 > 0$$

Since $f(0.51748) > 0$ and $f(1) < 0$. The root lies between 0.51748 and 1.
$$x_8 = \frac{(0.51748)f(1) - 1f(0.51748)}{f(1) - f(0.51748)} = 0.51767.$$
$$f(0.51767) = 0.002657 > 0$$

Since $f(0.51767) > 0$ and $f(1) < 0$. The root lies between 0.51767 and 1.
$$x_9 = \frac{(0.51767)f(1) - 1f(0.51767)}{f(1) - f(0.51767)} = 0.51773.$$
$$f(0.51773) = 0.00008324 > 0$$

Since $f(0.51773) > 0$ and $f(1) < 0$. The root lies between 0.51773 and 1.
$$x_{10} = \frac{(0.51773)f(1) - 1f(0.51773)}{f(1) - f(0.51773)} = 0.51774.$$

Hence the root is 0.5177 correct to 4 decimal places.

**Example:** By using Regula-Falsi method, find an approximate root of the equation $x^4 - x - 10 = 0$ that lies between 1.8 and 2. Carry out three approximations.

**Solution:** Let $f(x) = x^4 - x - 10$, $a = 1.8$, $b = 2$.
$$f(1.8) = -1.3024 < 0$$

$$f(2) = 4 > 0$$

Since $f(1.8) < 0$ and $f(2) > 0$. The root lies between 1.8 and 2.

By Regula-Falsi method,

$$x_1 = \frac{af(b) - bf(a)}{f(b) - f(a)}$$

$$x_1 = \frac{(1.8)f(2) - 2f(1.8)}{f(2) - f(1.8)} = 1.8491.$$

$$f(1.8491) = -0.1583 < 0$$

Since $f(1.8491) < 0$ and $f(2) > 0$. The root lies between 1.8491 and 2.

$$x_2 = \frac{(1.8491)f(2) - 2f(1.8491)}{f(2) - f(1.8491)} = 1.8548.$$

$$f(1.8548) = -0.01925 < 0$$

Since $f(1.8548) < 0$ and $f(2) > 0$. The root lies between 1.8548 and 2.

$$x_3 = \frac{(1.8548)f(2) - 2f(1.8548)}{f(2) - f(1.8548)} = 1.8554.$$

Therefore, the root of the equation 1.8554.

**Example:** Solve the equation $x \tan x = -1$ by Regula-Falsi method staring with $a = 2.5$ and $b = 3$ correct to 3 decimal places.

**Solution:** Let $f(x) = x \tan x + 1 = 0$, $a = 2.5$, $b = 3$.

$$f(2.5) = -0.8675 < 0$$

$$f(3) = 0.5723 > 0$$

Since $f(2.5) < 0$ and $f(3) > 0$. The root lies between 2.5 and 3.

By Regula-Falsi method,

$$x_1 = \frac{af(b) - bf(a)}{f(b) - f(a)}$$

$$x_1 = \frac{(2.5)f(3) - 3f(2.5)}{f(3) - f(2.5)} = 2.80125$$

$$f(2.80125) = 0.0080144 > 0$$

Since $f(2.5) < 0$ and $f(2.80125) > 0$. The root lies between 2.5 and 2.80125.

$$x_2 = \frac{(2.5)f(2.80125) - (2.80125)f(2.5)}{f(2.80125) - f(2.5)} = 2.79849.$$

$$f(2.79849) = 0.000291 > 0$$

Since $f(2.5) < 0$ and $f(2.79849) > 0$. The root lies between 2.5 and 2.79849.

$$x_3 = \frac{(2.5)f(2.79849) - (2.79849)f(2.5)}{f(2.79849) - f(2.5)} = 2.79839.$$

$$f(2.79839) = 0.000010654 > 0$$

Since $f(2.5) < 0$ and $f(2.79839) > 0$. The root lies between 2.5 and 2.79839.

$$x_4 = \frac{(2.5)f(2.79839) - (2.79839)f(2.5)}{f(2.79839) - f(2.5)} = 2.79839.$$

Thus, the values of 3rd and 4th approximations are coinciding exactly. Therefore, the root of the equation 2.79839.

**Example:** Using Regula-Falsi method, find a real root of $f(x) = 2x^7 + x^5 + 1 = 0$ correct upto two decimal places using $a = -1$ and $b = 1$.

**Solution:** Let $f(x) = 2x^7 + x^5 + 1$, $a = -1$, $b = 1$.

$$f(-1) = -2 < 0$$
$$f(1) = 4 > 0$$

Since $f(-1) < 0$ and $f(1) > 0$. The root lies between $-1$ and $1$.

By Regula-Falsi method,

$$x_1 = \frac{af(b) - bf(a)}{f(b) - f(a)}$$

$$x_1 = \frac{(-1)f(1) - 1f(-1)}{f(1) - f(-1)} = -0.33333$$

$$f(-0.33333) = 0.99497 > 0$$

Since $f(-1) < 0$ and $f(-0.33333) > 0$. The root lies between $-1$ and $-0.33333$.

$$x_2 = \frac{(-1)f(-0.33333) - (-0.33333)f(-1)}{f(-0.33333) - f(-1)} = -0.554807.$$

$$f(-0.554807) = 0.9150723 > 0$$

Since $f(-1) < 0$ and $f(-0.554807) > 0$. The root lies between $-1$ and $-0.554807$.

$$x_3 = \frac{(-1)f(-0.554807) - (-0.554807)f(-1)}{f(-0.554807) - f(-1)} = -0.694558.$$

$$f(-0.694558) = 0.682411 > 0$$

Since $f(-1) < 0$ and $f(-0.694558) > 0$. The root lies between $-1$ and $-0.694558$.

$$x_4 = \frac{(-1)f(-0.694558) - (-0.694558)f(-1)}{f(-0.694558) - f(-1)} = -0.772263.$$

$$f(-0.772263) = -0.397688 > 0$$

Since $f(-1) < 0$ and $f(-0.772263) > 0$. The root lies between $-1$ and $-0.772263$.

$$x_5 = \frac{(-1)f(-0.772263) - (-0.772263)f(-1)}{f(-0.772263) - f(-1)} = -0.810036.$$

$$f(-0.810036) = 0.682411 > 0$$

Since $f(-1) < 0$ and $f(-0.810036) > 0$. The root lies between $-1$ and $-0.810036$.

$$x_6 = \frac{(-1)f(-0.810036) - (-0.810036)f(-1)}{f(-0.810036) - f(-1)} = -0.826799.$$

$$f(-0.826799) = 0.0853954 > 0$$

Since $f(-1) < 0$ and $f(-0.826799) > 0$. The root lies between $-1$ and $-0.826799$.

$$x_7 = \frac{(-1)f(-0.826799) - (-0.826799)f(-1)}{f(-0.826799) - f(-1)} = -0.833891.$$

$$f(-0.833891) = 0.03599 > 0$$

Since $f(-1) < 0$ and $f(-0.833891) > 0$. The root lies between $-1$ and $-0.833891$.

$$x_8 = \frac{(-1)f(-0.833891) - (-0.833891)f(-1)}{f(-0.833891) - f(-1)} = -0.836828.$$

$$f(-0.836828) = 0.014869 > 0$$

Since $f(-1) < 0$ and $f(-0.836828) > 0$. The root lies between $-1$ and $-0.836828$.

$$x_9 = \frac{(-1)f(-0.836828) - (-0.836828)f(-1)}{f(--0.836828) - f(-1)} = -0.838032.$$

Thus, the values of 8th and 9th approximations are coinciding upto 2 decimal places.

Therefore, the root of the equation is $-0.838032$.

**Example:** Find the real root of the equation $x \log_{10} x = 1.2$ using False position method.

(Nov-2017), (Aug-2022)

**Solution:** Let $f(x) = x \log_{10} x - 1.2$

$$f(1) = -1.2 < 0$$
$$f(2) = -0.597940 < 0$$
$$f(3) = 0.23136 > 0$$

Since $f(2) < 0$ and $f(3) > 0$. The root lies between 2 and 3.

Let $a = 2$ and $b = 3$.

By Regula-Falsi method,

$$x_1 = \frac{af(b) - bf(a)}{f(b) - f(a)}$$

$$x_1 = \frac{2f(3) - 3f(2)}{f(3) - f(2)} = 2.7210.$$

$$f(2.7210) = -0.01710 < 0$$

Since $f(2.7210) < 0$ and $f(3) > 0$. The root lies between 2.7210 and 3.

$$x_2 = \frac{(2.7210)f(3) - 3f(2.7210)}{f(3) - f(2.7210)} = 2.74020$$

$$f(2.74020) = -0.0003890 < 0$$

Since $f(2.74020) < 0$ and $f(3) > 0$. The root lies between 2.74020 and 3.

$$x_3 = \frac{(2.74020)f(3) - 3f(2.74020)}{f(3) - f(2.74020)} = 2.74064.$$

Since 2nd and 3rd approximation coincide upto two decimal places. Therefore, the root of the equation 2.74020.

**Example:** By using Regula-Falsi method, find the real root of $x^3 - 8x - 40 = 0$ correct upto three decimal places using $a = 4$ and $b = 5$.

**Solution:** Let $f(x) = x^3 - 8x - 40$, $a = 4$ and $b = 5$.

$$f(4) = -8 < 0$$
$$f(5) = 45 > 0$$

Since $f(4) < 0$ and $f(5) > 0$. The root lies between 4 and 5.

By Regula-Falsi method,

$$x_1 = \frac{af(b) - bf(a)}{f(b) - f(a)}$$

$$x_1 = \frac{4f(5) - 5f(4)}{f(5) - f(4)} = 4.15094.$$

$$f(4.15094) = -1.685567 < 0$$

Since $f(4.15094) < 0$ and $f(5) > 0$. The root lies between 4.15094 and 5.

$$x_2 = \frac{(4.15094)f(5) - 5f(4.15094)}{f(5) - f(4.15094)} = 4.1816$$

$$f(4.1816) = -0.334268 < 0$$

Since $f(4.1816) < 0$ and $f(5) > 0$. The root lies between 4.1816 and 5.

$$x_3 = \frac{(4.1816)f(5) - 5f(4.1816)}{f(5) - f(4.1816)} = 4.18763.$$

$$f(4.18763) = -0.0657342 < 0$$

Since $f(4.18763) < 0$ and $f(5) > 0$. The root lies between 4.18763 and 5.

$$x_4 = \frac{(4.18763)f(5) - 5f(4.18763)}{f(5) - f(4.18763)} = 4.18882.$$

$$f(4.18882) = -0.01263 < 0$$

Since $f(4.18882) < 0$ and $f(5) > 0$. The root lies between 4.18882 and 5.
$$x_5 = \frac{(4.18882)f(5) - 5f(4.18882)}{f(5) - f(4.18882)} = 4.18905.$$
$$f(4.18905) = -0.0023645 < 0$$

Since $f(4.18905) < 0$ and $f(5) > 0$. The root lies between 4.18905 and 5.
$$x_6 = \frac{(4.18905)f(5) - 5f(4.18905)}{f(5) - f(4.18905)} = 4.18909$$

Since 5th and 6th approximation coincide upto four decimal places. Therefore, the root of the equation 4.18909.

**Example:** Use the method of false position, to find the fourth root of 32 correct upto three decimal places.

**Solution:** Let $x = (32)^{1/4}$ so that $x^4 - 32 = 0$.

Take $f(x) = x^4 - 32$

$$f(2) = -16 < 0$$
$$f(3) = 49 > 0$$

Since $f(2) < 0$ and $f(3) > 0$. The root lies between 2 and 3.

Let $a = 2$ and $b = 3$.

By Regula-Falsi method,
$$x_1 = \frac{af(b) - bf(a)}{f(b) - f(a)}$$
$$x_1 = \frac{2f(3) - 3f(2)}{f(3) - f(2)} = 2.2462.$$
$$f(2.2462) = -6.5437 < 0$$

Since $f(2.2462) < 0$ and $f(3) > 0$. The root lies between 2.2462 and 3.
$$x_2 = \frac{(2.2462)f(3) - 3f(2.2462)}{f(3) - f(2.2462)} = 2.335$$
$$f(2.335) = -2.2732 < 0$$

Since $f(2.335) < 0$ and $f(3) > 0$. The root lies between 2.335 and 3.
$$x_3 = \frac{(2.335)f(3) - 3(4.1816)}{f(5) - f(4.1816)} = 2.3645.$$
$$f(2.3645) = -0.7422 < 0$$

Since $f(2.3645) < 0$ and $f(3) > 0$. The root lies between 2.3645 and 3.

$$x_4 = \frac{(2.3645)f(3) - 3f(2.3645)}{f(3) - f(2.3645)} = 2.3740.$$
$$f(2.3740) = -0.2369 < 0$$

Since $f(2.3740) < 0$ and $f(3) > 0$. The root lies between 2.3740 and 3.
$$x_5 = \frac{(2.3740)f(3) - 3f(2.3740)}{f(3) - f(2.3740)} = 2.3770.$$
$$f(2.3770) = -0.076 < 0$$

Since $f(2.3770) < 0$ and $f(3) > 0$. The root lies between 2.3770 and 3.
$$x_6 = \frac{(2.3770)f(3) - 3f(2.3770)}{f(3) - f(2.3770)} = 2.3779$$

Since 5th and 6th approximation coincide upto three decimal places, we take $(32)^{1/4} = 2.3779$.

### EXERCISE

1. Find a real root of $xe^x = 3$ using Regula-Falsi method. **(Nov-2017), (Apr-2022)**

   **Ans:** 1.0498

2. Find a real root of $xe^x = 2$ using Regula-Falsi method. **(Mar-2022)**

   **Ans:** 0.8526

3. Find a real root of the equation $x^3 - 5x - 7 = 0$ using False position method.

   **(Nov-2017), (Mar-2022)**

4. Using Regula-falsi method, find the real root of $2x - \log x = 6$ correct to three decimal places. **(Dec-2016), (Apr-2022)**

   **Ans:** 3.257

5. Find a real root of the equation $x^3 - 4x - 9 = 0$ using False position method correct to three decimal places. **(Dec-2016), (May-2017)**

   **Ans:** 2.70650

6. Find the root of the equation $2x - \log_{10} x = 7$ using False position method.

   **Ans:** 3.7892 **(May-2018)**

7. Find the root of the equation $3x = 1 + \cos x$ using False position method.

   **Ans:** 0.6071 **(May-2018)**

8. Find the root of the equation $4\sin x = e^x$ using False position method.
   Ans: 0.36718 (May-2018)
9. Find the Real root of $e^x \sin x = 2$ using False position method. (May-2018)
10. Find the Real root of $e^x - 3x = 0$ using False position method.
    Ans: 6.089
11. Find the Real root of $x^4 - x - 9 = 0$ using False position method.
12. Find the root of the equation $x^3 - x - 11 = 0$ using False position method.
    Ans: 2.375
13. Find a real root for $e^x \sin x = 1$ using Regula Falsi method. (Apr-2022)
    Ans: 0.5885
14. Find a real root for $x^3 - 6x - 4 = 0$ by false position method. (Aug-2022)
15. Find the real root of the equation $2x = 3\sin x + 5$ using False position method. (Aug-2022)
16. Find the real root of the equation $2x - \log_{10} x = 7$ using False position method. (Aug-2022)

## 3.4 Iteration Method

Let $f(x) = 0$ be the given equation, algebraic or transcendental. Suppose the equation can be expressed in the form

$$x = \phi(x)$$

where $\phi(x)$ is a continuous function.

Let $x = x_0$ be an initial approximation of the desired root $\alpha$. Then, the first approximation $x_1$ is given by $x_1 = \phi(x_0)$.

Now treating $x_1$ as the initial value, the second approximation is $x_2 = \phi(x_1)$.

Proceeding in this way the $n$th approximation is given by $x_n = \phi(x_{n-1})$. We proceed in this way until the two successive approximations are approximately equal. Here $x_{n+1} = \phi(x_n)$ is called the iterative formula.

The sequence $\{x_n\}$ converges to $\alpha$, then we say that the iteration process is convergent.

This method is convergent if $|\phi'(x)| < 1$, $\forall x \in I$, where $I$ is the interval which contains a root of the equation.

**Example:** Find a positive root of the equation by iteration method $2x = 3 + \cos x$.

**Solution:** Let $f(x) = 2x - 3 - \cos x$

$$f(0) = -4 < 0$$
$$f(1) = -1.5403 < 0$$
$$f(2) = 1.41615 > 0.$$

Since $f(1) < 0$ and $f(2) > 0$. The root lies between 1 and 2.

The given equation can be written as $x = \phi(x)$ in many ways such that

$$2x = 3 + \cos x$$
$$\Rightarrow x = \frac{3 + \cos x}{2} = \phi(x)$$
$$\phi'(x) = \frac{-\sin x}{2}$$
$$|\phi'(1)| = \left|\frac{-\sin 1}{2}\right| = 0.42073 < 1$$
$$|\phi'(2)| = \left|\frac{-\sin 2}{2}\right| = 0.4546 < 1$$

$$\therefore |\phi'(x)| < 1, \forall x \in (1, 2).$$

Therefore, the iteration method can be applicable and hence iteration formula is given by

$$x_{n+1} = \phi(x_n)$$

Let us take the initial approximation as $x_0 = 1.0$.

The iteration scheme is

$$x_{n+1} = \frac{3 + \cos x_n}{2}$$

$$x_1 = \frac{3 + \cos x_0}{2} = 1.77015$$

$$x_2 = \frac{3 + \cos x_1}{2} = 1.40098$$

$$x_3 = \frac{3 + \cos x_2}{2} = 1.5845$$

$$x_4 = \frac{3 + \cos x_3}{2} = 1.49315$$

$$x_5 = \frac{3 + \cos x_4}{2} = 1.53878$$

$$x_6 = \frac{3 + \cos x_5}{2} = 1.51601$$

$$x_7 = \frac{3 + \cos x_6}{2} = 1.52738$$

$$x_8 = \frac{3 + \cos x_7}{2} = 1.5217$$

The root of the equation correct to two decimal places is 1.5217.

**Example:** Find the real root of the equation $3x = 1 + \cos x$ using iteration method.

**Solution:** Let $f(x) = 3x - 1 - \cos x$                       (Nov-2017)

$$f(0) = -2 < 0$$

$$f(1) = 1.4596 > 0$$

Since $f(0) < 0$ and $f(1) > 0$. The root lies between 0 and 1.

The given equation can be written as $x = \phi(x)$ in many ways such that

$$3x = 1 + \cos x$$

$$\Rightarrow x = \frac{1 + \cos x}{3} = \phi(x)$$

$$\phi'(x) = \frac{-\sin x}{3}$$

$$|\phi'(0)| = \left|\frac{-\sin 0}{3}\right| = 0 < 1$$

$$|\phi'(1)| = \left|\frac{-\sin 1}{3}\right| = 0.2804 < 1$$

$$\therefore |\phi'(x)| < 1, \forall x \in (0, 1).$$

Therefore, the iteration method can be applicable and hence iteration formula is given by

$$x_{n+1} = \phi(x_n) = \frac{1 + \cos x_n}{3}.$$

Let us take the initial approximation as $x_0 = 0.5$.
The iteration scheme is

$$x_{n+1} = \frac{1 + \cos x_n}{3}$$

$$x_1 = \frac{1 + \cos x_0}{3} = 0.6259$$

$$x_2 = \frac{1 + \cos x_1}{3} = 0.6034$$

$$x_3 = \frac{1 + \cos x_2}{3} = 0.6078$$

$$x_4 = \frac{1 + \cos x_3}{3} = 0.6070$$

$$x_5 = \frac{1 + \cos x_4}{3} = 0.6071$$

$$x_6 = \frac{1 + \cos x_5}{3} = 0.6071.$$

The root of the equation is 0.6071.

**Example:** Solve $x^3 = 2x + 5$ for a positive root by iteration method.

**Solution:** Let $f(x) = x^3 - 2x - 5$

$f(0) = -5 < 0$

$f(1) = -6 < 0$

$f(2) = -1 < 0.$

$f(3) = 16 > 0.$

Since $f(2) < 0$ and $f(3) > 0$. The root lies between 2 and 3.
The given equation can be written as $x = \phi(x)$ in many ways such that

$$x^3 = 2x + 5$$

$$\Rightarrow x = (2x + 5)^{\frac{1}{3}} = \phi(x)$$

$$\phi'(x) = \tfrac{2}{3}(2x + 5)^{\frac{-2}{3}}$$

$$|\phi'(2)| = \left|\tfrac{2}{3}(4+5)^{\frac{-2}{3}}\right| = 0.1540 < 1$$

$$|\phi'(3)| = \left|\tfrac{2}{3}(6+5)^{\frac{-2}{3}}\right| = 0.1347 < 1$$
$$\Rightarrow |\phi'(x)| < 1, \forall x \in (2,3).$$

Therefore, the iteration method is applicable and hence iteration formula is given by

$$x_{n+1} = \phi(x_n)$$
$$\Rightarrow x_{n+1} = (2x_n + 5)^{\frac{1}{3}}$$

Let us take the initial approximation for the root as $x_0 = 2.0$.
The iteration scheme is

$$x_1 = (2x_0 + 5)^{\frac{1}{3}} = 2.08008$$
$$x_2 = (2x_1 + 5)^{\frac{1}{3}} = 2.09235$$
$$x_3 = (2x_2 + 5)^{\frac{1}{3}} = 2.09422$$
$$x_4 = (2x_3 + 5)^{\frac{1}{3}} = 2.09450$$
$$x_5 = (2x_4 + 5)^{\frac{1}{3}} = 2.09454$$
$$x_6 = (2x_5 + 5)^{\frac{1}{3}} = 2.09455$$
$$x_7 = (2x_6 + 5)^{\frac{1}{3}} = 2.09455$$

Since $x_6$ and $x_7$ are identical, we take 2.09455 as the required root, correct to 5 decimal places.

**Example:** Solve the equation $x^3 + x^2 - 1 = 0$ for a positive root by iteration method.

**Solution:** Let $f(x) = x^3 + x^2 - 1$
$$f(0) = -1 < 0$$
$$f(1) = 1 > 0$$

Since $f(0) < 0$ and $f(1) > 0$. The root lies between 0 and 1.

The given equation can be written as $x = \phi(x)$ in many ways such that
$$x^3 + x^2 - 1 = 0$$
$$\Rightarrow x^2(x+1) = 1$$

$$\Rightarrow x = \frac{1}{(1+x)^{\frac{1}{2}}} = (1+x)^{\frac{-1}{2}} = \phi(x)$$

$$\phi'(x) = \frac{-1}{2}(1+x)^{\frac{-3}{2}}$$

$$|\phi'(0)| = |\frac{-1}{2}(1)^{\frac{-3}{2}}| = \frac{1}{2} < 1$$

$$|\phi'(1)| = |\frac{-1}{2}(2)^{\frac{-3}{2}}| = 0.17678 < 1$$

$$\Rightarrow |\phi'(x)| < 1, \ \forall x \in (0,1).$$

Therefore, the iteration method is applicable and hence iteration formula is given by

$$x_{n+1} = \phi(x_n)$$
$$\Rightarrow x_{n+1} = (1+x_n)^{\frac{-1}{2}}$$

Let us take the initial approximation of the root as $x_0 = 0$.

The iteration scheme is

$$x_1 = (1+x_0)^{\frac{-1}{2}} = 1$$
$$x_2 = (1+x_1)^{\frac{-1}{2}} = 0.70711$$
$$x_3 = (1+x_2)^{\frac{-1}{2}} = 0.76537$$
$$x_4 = (1+x_3)^{\frac{-1}{2}} = 0.75263$$
$$x_5 = (1+x_4)^{\frac{-1}{2}} = 0.75536$$
$$x_6 = (1+x_5)^{\frac{-1}{2}} = 0.75477$$

The root of the equation correct to two decimal places is 0.75.

**Example:** Use the method of iteration to find the real root lying between 1 and 2 of the equation $x^3 - 3x + 1 = 0$.

**Solution:** Let $f(x) = x^3 - 3x + 1$

$$f(1) = -1 < 0$$
$$f(2) = 3 > 0$$

Since $f(1) < 0$ and $f(2) > 0$. One of the root lies between 1 and 2.

The given equation can be written as $x = \phi(x)$ in many ways such that

$$x^3 - 3x + 1 = 0$$
$$\Rightarrow x^3 = 3x - 1$$

$$\Rightarrow x = (3x-1)^{1/3} = \phi(x)$$
$$\phi'(x) = \frac{1}{3}(3x-1)^{-2/3}(3) = \frac{1}{(3x-1)^{2/3}}$$
$$|\phi'(1)| = \left|\frac{1}{2^{2/3}}\right| = 0.62996 < 1$$
$$|\phi'(2)| = \left|\frac{1}{5^{2/3}}\right| = 0.34199 < 1$$
$$\Rightarrow |\phi'(x)| < 1, \; \forall x \in (1,2).$$

Therefore, the iteration method is applicable and hence iteration formula is given by

$$x_{n+1} = \phi(x_n)$$
$$\Rightarrow x_{n+1} = (3x_n - 1)^{1/3}$$

Let us take the initial approximation of the root as $x_0 = 1.5$.
The iteration scheme is

$$x_1 = (3x_0 - 1)^{1/3} = 1.5183$$
$$x_2 = (3x_1 - 1)^{1/3} = 1.5262$$
$$x_3 = (3x_2 - 1)^{1/3} = 1.5296$$
$$x_4 = (3x_3 - 1)^{1/3} = 1.5310$$
$$x_5 = (3x_4 - 1)^{1/3} = 1.5316$$
$$x_6 = (3x_5 - 1)^{1/3} = 1.5319$$
$$x_7 = (3x_6 - 1)^{1/3} = 1.5320$$
$$x_8 = (3x_7 - 1)^{1/3} = 1.5321$$
$$x_9 = (3x_8 - 1)^{1/3} = 1.5321$$

Since $x_8 = x_9$, upto 4 decimal places. The root of the equation is 1.5321.

**Example:** Find a real root of the equation $2x - \log_{10} x = 7$ correct to four decimal places using iteration method.

**Solution:** Let $f(x) = 2x - \log_{10} x - 7$
$$f(1) = -5 < 0$$
$$f(2) = -3.3010 < 0$$

$$f(3) = -1.4771 < 0$$
$$f(4) = 0.3979 > 0$$

Since $f(3) < 0$ and $f(4) > 0$. The root lies between 3 and 4.

The given equation can be written as $x = \phi(x)$ in many ways such that
$$2x - \log_{10} x = 7$$
$$\Rightarrow x = \frac{7 + \log_{10} x}{2} = \phi(x) \text{ (say)}$$
$$\phi'(x) = \frac{1}{2}\left(\frac{\log_{10} e}{x}\right) = \frac{1}{2} \cdot \frac{0.43429}{x} = \frac{0.2171}{x}.$$
$$|\phi'(3)| = \left|\frac{0.2171}{3}\right| = 0.0736 < 1$$
$$|\phi'(4)| = \left|\frac{0.2171}{4}\right| = 0.05428 < 1$$
$$\Rightarrow |\phi'(x)| < 1, \forall x \in (3,4).$$

Therefore, the iteration method is applicable and hence iteration formula is given by
$$x_{n+1} = \phi(x_n)$$
$$\Rightarrow x_{n+1} = \frac{7 + \log_{10} x_n}{2}.$$

Let us take the initial approximation of the required root as $x_0 = 3.5$.
The iteration scheme is
$$x_1 = \phi(x_0) = 3.77203$$
$$x_2 = \phi(x_1) = 3.78829$$
$$x_3 = \phi(x_2) = 3.78922$$
$$x_4 = \phi(x_3) = 3.78927$$

The root of the equation correct to two decimal places is 3.7892

**Example:** Find a real root of $f(x) = x + \log x - 2$ staring from $x_0 = 1.0$, using iteration method.

**Solution:** Given $f(x) = x + \log x - 2$.
The given equation can be written as $x = \phi(x)$ in many ways such that
$$\phi(x) = 2 - \log x$$

Given $x_0 = 1.0$.

The iteration formula is given by

$$x_{n+1} = \phi(x_n)$$
$$\Rightarrow x_{n+1} = 2 - \log x_n$$
$$x_1 = \phi(x_0) = 2$$
$$x_2 = \phi(x_1) = 1.30685$$
$$x_3 = \phi(x_2) = 1.73238$$
$$x_4 = \phi(x_3) = 1.45050$$
$$x_5 = \phi(x_4) = 1.6289$$
$$x_6 = \phi(x_5) = 1.51259$$
$$x_7 = \phi(x_6) = 1.58617$$
$$x_8 = \phi(x_7) = 1.53867$$
$$x_9 = \phi(x_8) = 1.56908$$
$$x_{10} = \phi(x_9) = 1.54951$$
$$x_{11} = \phi(x_{10}) = 1.56206$$
$$x_{12} = \phi(x_{11}) = 1.55399$$
$$x_{13} = \phi(x_{12}) = 1.55917$$
$$x_{14} = \phi(x_{13}) = 1.55585$$
$$x_{15} = \phi(x_{14}) = 1.557977$$

The root of the equation is $1.557977$.

**Example:** Solve $x = 1 + \tan^{-1} x$ by iteration method. (Dec-2020)

**Solution:** Let $f(x) = x - 1 - \tan^{-1} x$

$$f(0) = -1 < 0$$
$$f(1) = -0.7853 < 0$$
$$f(2) = -0.1071 < 0$$

$$f(3) = 0.7509 > 0$$

Since $f(2) < 0$ and $f(3) > 0$. The root lies between 2 and 3.

The given equation can be written as $x = \phi(x)$ in many ways such that

$$x = 1 + \tan^{-1} x = \phi(x) \text{ (say)}$$
$$\phi'(x) = \frac{1}{1+x^2}.$$
$$|\phi'(2)| = 0.2 < 1$$
$$|\phi'(3)| = 0.1 < 1$$
$$\Rightarrow |\phi'(x)| < 1, \; \forall x \in (2,3).$$

Therefore, the iteration method is applicable and hence iteration formula is given by

$$x_{n+1} = \phi(x_n)$$
$$\Rightarrow x_{n+1} = 1 + \tan^{-1} x_n.$$

Let us take the initial approximation of the required root as $x_0 = 2.5$.

The iteration scheme is

$$x_1 = \phi(x_0) = 2.1903$$
$$x_2 = \phi(x_1) = 2.1425$$
$$x_3 = \phi(x_2) = 2.1341$$
$$x_4 = \phi(x_3) = 2.1326$$
$$x_5 = \phi(x_4) = 2.1323$$
$$x_6 = \phi(x_5) = 2.1323$$

Since $x_5 = x_6$, the desired root is 2.1323

**Example:** Find a real root of $x \tan x + 1 = 0$ using iteration method. **(Dec-2020)**

**Solution:** Let $f(x) = x \tan x + 1$

$$f(0) = 1 > 0$$
$$f(1) = 2.5574 > 0$$
$$f(2.5) = -3.3700 < 0$$

$$f(3) = 0.57236 > 0$$

Since $f(2) < 0$ and $f(3) > 0$. The root lies between 2 and 3.

The given equation can be written as $x = \phi(x)$ in many ways such that
$$x = 1 + \tan^{-1} x = \phi(x) \text{ (say)}$$
$$\phi'(x) = \frac{1}{1+x^2}.$$
$$|\phi'(2)| = 0.2 < 1$$
$$|\phi'(3)| = 0.1 < 1$$
$$\Rightarrow |\phi'(x)| < 1, \forall x \in (2, 3).$$

Therefore, the iteration method is applicable and hence iteration formula is given by
$$x_{n+1} = \phi(x_n)$$
$$\Rightarrow x_{n+1} = 1 + \tan^{-1} x_n.$$

Let us take the initial approximation of the required root as $x_0 = 2.5$.
The iteration scheme is
$$x_1 = \phi(x_0) = 2.1903$$
$$x_2 = \phi(x_1) = 2.1425$$
$$x_3 = \phi(x_2) = 2.1341$$
$$x_4 = \phi(x_3) = 2.1326$$
$$x_5 = \phi(x_4) = 2.1323$$
$$x_6 = \phi(x_5) = 2.1323$$

Since $x_5 = x_6$, the desired root is 2.1323.

## EXERCISE

1. Find the root of the equation $x^3 - 6x - 4 = 0$ using iteration method.
2. Find the Real root of $x^3 - x - 1 = 0$ using Iteration method.

**Ans:** 1.3247

3. Find the Real root of $x = 2\sin x$ using Iteration method.

4. Find the root of the equation $x^4 - x - 10 = 0$ using Iteration method.

**Ans:** 1.8556

## 3.5 Newton-Raphson Method

The Newton-Raphson method, or Newton method, is a powerful technique for solving equations numerically. Like so much of the differential calculus, it is based on the simple idea of linear approximation.

The Newton-Raphson method is a more advanced method in finding the root of the equation $f(x) = 0$.

Let $f(x) = 0$ be the given equation.

Let $x_0$ be an approximate root of the equation $f(x) = 0$. Let $x_1 = x_0 + h$ be the exact root, where $h$ is very small, positive or negative.

$$\therefore f(x_1) = 0. \quad \cdots \quad (1)$$

By Taylor's series expansion, we have

$$f(x_1) = f(x_0 + h) = f(x_0) + hf'(x_0) + \frac{h^2}{2!}f''(x_0) + \cdots.$$

Since $f(x_1) = 0$ and $h$ is very small, $h^2$ and higher powers of $h$ can be neglected. Hence

$$f(x_0) + hf'(x_0) = 0$$

$$\Rightarrow h = -\frac{f(x_0)}{f'(x_0)} \quad \text{if } f'(x_0) \neq 0.$$

Hence $x_1 = x_0 - \dfrac{f(x_0)}{f'(x_0)}$ is a first approximation to the root.

Similarly starting with $x_1$, we get the next approximation to the root given by

$$x_2 = x_1 - \frac{f(x_1)}{f'(x_1)}.$$

Proceeding like this, we obtain successive approximations $x_3, x_4, \cdots$.

The following is the formula for the $(n+1)^{\text{th}}$-order approximation in terms of the $n^{\text{th}}$-order approximation:

$$x_{n+1} = x_n - \frac{f(x_n)}{f'(x_n)}, \quad n = 0, 1, 2, \cdots$$

and this formula is known as the **Newton-Raphson formula** or the **Newton's iteration formula**. The process of computing an approximate root of an equation $f(x) = 0$ by using this formula is known as the **Newton-Raphson method**, or just **Newton method**.

### 3.5.1 Convergence of Newton-Raphson Method

By Newton's formula

$$x_{n+1} = x_n - \frac{f(x_n)}{f'(x_n)}$$

Comparing this with iteration formula

$$x_{n+1} = \phi(x_n)$$

$$\Rightarrow \phi(x_n) = x_n - \frac{f(x_n)}{f'(x_n)}$$

$$\Rightarrow \phi(x) = x - \frac{f(x)}{f'(x)}$$

Differentiating

$$\phi'(x) = 1 - \frac{[(f'(x))^2 - f(x)f''(x)]}{[f'(x)]^2}$$

$$= \frac{f(x)f''(x)}{[f'(x)]^2}$$

We know that iteration method converges if $|\phi'(x)| < 1$ i.e.,

$$\left| \frac{f(x)f''(x)}{[f'(x)]^2} \right| < 1$$

$$\Rightarrow \left| f(x)f''(x) \right| < \left| [f'(x)]^2 \right|.$$

**Example:** Find the positive root of $x^4 - x = 10$ using Newton-Raphson method.

**Solution:** Let $f(x) = x^4 - x - 10$

$$f(0) = -10 < 0$$
$$f(1) = -10 < 0$$
$$f(2) = 4 > 0$$

Since $f(1) < 0$ and $f(2) > 0$. Therefore, the root lies between 1 and 2.
Let us take $x_0 = 1.5$. Also $f'(x) = 4x^3 - 1$.
The Newton-Raphson formula is given by

$$x_{n+1} = x_n - \frac{f(x_n)}{f'(x_n)} = x_n - \frac{x_n^4 - x_n - 10}{4x_n^3 - 1} = \frac{3x_n^4 + 10}{4x_n^3 - 1}$$

$$x_1 = \frac{3x_0^4 + 10}{4x_0^3 - 1} = 2.015$$

$$x_2 = \frac{3x_1^4 + 10}{4x_1^3 - 1} = 1.8741$$

$$x_3 = \frac{3x_2^4 + 10}{4x_2^3 - 1} = 1.8559$$

$$x_4 = \frac{3x_3^4 + 10}{4x_3^3 - 1} = 1.8556$$

$$x_5 = \frac{3x_4^4 + 10}{4x_4^3 - 1} = 1.8556$$

Here $x_4 = x_5$. Hence the desired root of the equation is 1.8556 correct to four decimal places.

**Example:** Using Newton-Raphson method find a positive real root of the equation $x^3 - x - 10 = 0$ with $x_0 = 1.0$.

**Solution::** Let $f(x) = x^3 - x - 10$

$$f'(x) = 3x^2 - 1$$

Taking $x_0 = 1.0$

The Newton-Raphson formula is given by

$$x_{n+1} = x_n - \frac{f(x_n)}{f'(x_n)} = x_n - \frac{x_n^3 - x_n - 10}{3x_n^2 - 1} = \frac{2x_n^3 + 10}{3x_n^2 - 1}$$

$$x_1 = \frac{2x_0^3 + 10}{3x_0^2 - 1} = 6$$

$$x_2 = \frac{2x_1^3 + 10}{3x_1^2 - 1} = 4.13084$$

$$x_3 = \frac{2x_2^3 + 10}{3x_2^2 - 1} = 3.008$$

$$x_4 = \frac{2x_3^3 + 10}{3x_3^2 - 1} = 2.4645$$

$$x_5 = \frac{2x_4^3 + 10}{3x_4^2 - 1} = 2.3190$$

$$x_6 = \frac{2x_5^3 + 10}{3x_5^2 - 1} = 2.30895$$

$$x_7 = \frac{2x_6^3 + 10}{3x_6^2 - 1} = 2.30891$$

$$x_8 = \frac{2x_7^3 + 10}{3x_7^2 - 1} = 2.30891$$

The root of the equation is 2.30891.

**Example:** Find a real root for $x \tan x + 1 = 0$ using Newton-Raphson method.

**Solution:** Let $f(x) = x \tan x + 1 = x \sin x + \cos x$. (Mar-2022)

$$f(0) = 1 > 0$$
$$f(1) = 1.3817 > 0$$
$$f(2) = 1.4024 > 0$$
$$f(3) = -0.5666 < 0$$

Since $f(2) > 0$ and $f(3) < 0$. Therefore, the root lies between 2 and 3.
Let us take $x_0 = 2.5$. Also $f'(x) = x \cos x$.

The Newton-Raphson formula is given by

$$x_{n+1} = x_n - \frac{f(x_n)}{f'(x_n)} = x_n - \frac{x_n \sin x_n + \cos x_n}{x_n \cos x_n} = \frac{x_n^2 \cos x_n - x_n \sin x_n - \cos x_n}{x_n \cos x_n}$$

$$x_1 = \frac{x_0^2 \cos x_0 - x_0 \sin x_0 - \cos x_0}{x_0 \cos x_0} = 2.8470$$

$$x_2 = \frac{x_1^2 \cos x_1 - x_1 \sin x_1 - \cos x_1}{x_1 \cos x_1} = 2.7991$$

$$x_3 = \frac{x_2^2 \cos x_2 - x_2 \sin x_2 - \cos x_2}{x_2 \cos x_2} = 2.7983$$

$$x_4 = \frac{x_3^2 \cos x_3 - x_3 \sin x_3 - \cos x_3}{x_3 \cos x_3} = 2.7983$$

Therefore, the root of the equation is 2.7983.

**Example:** By Newton-Raphson method, find a real root of the equation $f(x) = x + \log x - 2$ up to 4 decimal places.

**Solution:** Let $f(x) = x + \log x - 2$

$$f(1) = -1 < 0$$
$$f(2) = 0.6931 > 0$$

Since $f(1) < 0$ and $f(2) > 0$. Therefore, the root lies between 1 and 2.
Let us take $x_0 = 1$. Also $f'(x) = 1 + \frac{1}{x}$.
The Newton-Raphson formula is given by

$$x_{n+1} = x_n - \frac{f(x_n)}{f'(x_n)} = x_n - \frac{x_n + \log x_n - 2}{1 + \frac{1}{x_n}} = \frac{3x_n - x_n \log x_n}{x_n + 1}$$

$$x_1 = \frac{3x_0 - x_0 \log x_0}{x_0 + 1} = 1.5$$

$$x_2 = \frac{3x_1 - x_1 \log x_1}{x_1 + 1} = 1.5567$$

$$x_3 = \frac{3x_2 - x_2 \log x_2}{x_2 + 1} = 1.5571$$

$$x_4 = \frac{3x_3 - x_3 \log x_3}{x_3 + 1} = 1.5571$$

Therefore, the root of the equation is 1.5571.

**Example:** Find the real root of the equation $x = e^{-x}$ using Newton-Raphson method. (Nov-2017)

**Solution:** Let $f(x) = xe^x = 1 \Rightarrow xe^x - 1$

$$f(0) = -1 < 0$$
$$f(1) = 1.718 > 0$$

Since $f(0) < 0$ and $f(1) > 0$. Therefore, the root lies between 0 and 1.
Let us take $x_0 = 0.5$. Also $f'(x) = xe^x + e^x$.
The Newton-Raphson formula is given by

$$x_{n+1} = x_n - \frac{f(x_n)}{f'(x_n)} = x_n - \frac{x_n e^{x_n} - 1}{x_n e^{x_n} + e^{x_n}} = \frac{x_n^2 e^{x_n} + 1}{x_n e^{x_n} + e^{x_n}}$$

$$x_1 = \frac{x_0^2 e^{x_0} + 1}{x_0 e^{x_0} + e^{x_0}} = 0.5710$$

$$x_2 = \frac{x_1^2 e^{x_1} + 1}{x_1 e^{x_1} + e^{x_1}} = 0.5672$$

$$x_3 = \frac{x_2^2 e^{x_2} + 1}{x_2 e^{x_2} + e^{x_2}} = 0.5671$$

$$x_4 = \frac{x_3^2 e^{x_3} + 1}{x_3 e^{x_3} + e^{x_3}} = 0.5671$$

Therefore, the root of the equation is 0.5671.

**Example:** Find the real root of the equation $e^x \sin x = 1$ using Newton-Raphson method. (Nov-2017)

**Solution:** Let $f(x) = e^x \sin x - 1$

$$f(0) = -1 < 0$$
$$f(1) = 1.2873 > 0$$

Since $f(0) < 0$ and $f(1) > 0$. Therefore, the root lies between 0 and 1.
Let us take $x_0 = 0.5$. Also $f'(x) = e^x \cos x + e^x \sin x$.

The Newton-Raphson formula is given by

$$x_{n+1} = x_n - \frac{f(x_n)}{f'(x_n)}$$

$$= x_n - \frac{e^{x_n} \sin x_n - 1}{e^{x_n} \cos x_n + e^{x_n} \sin x_n}$$

$$= \frac{x_n e^{x_n} \sin x_n + x_n e^{x_n} \cos x_n - e^{x_n} \sin x_n + 1}{e^{x_n} \cos x_n + e^{x_n} \sin x_n}$$

$$x_1 = \frac{x_0 e^{x_0} \sin x_0 + x_0 e^{x_0} \cos x_0 - e^{x_0} \sin x_0 + 1}{e^{x_0} \cos x_0 + e^{x_0} \sin x_0} = 0.593665$$

$$x_2 = \frac{x_1 e^{x_1} \sin x_1 + x_1 e^{x_1} \cos x_1 - e^{x_1} \sin x_1 + 1}{e^{x_1} \cos x_1 + e^{x_1} \sin x_1} = 0.58854$$

$$x_3 = \frac{x_2 e^{x_2} \sin x_2 + x_2 e^{x_2} \cos x_2 - e^{x_2} \sin x_2 + 1}{e^{x_2} \cos x_2 + e^{x_2} \sin x_2} = 0.58853$$

$$x_4 = \frac{x_3 e^{x_3} \sin x_3 + x_3 e^{x_3} \cos x_3 - e^{x_3} \sin x_3 + 1}{e^{x_3} \cos x_3 + e^{x_3} \sin x_3} = 0.58853$$

Therefore, the root of the equation is 0.58853.

**Example:** Find a real root of the equation $\log x - \cos x = 0$ near $x = 1$ correct to 3 decimal places using Newton-Raphson method.

**Solution:** Let $f(x) = \log x - \cos x$

$$f'(x) = \frac{1}{x} + \sin x = \frac{1 + x \sin x}{x}.$$

Let $x_0 = 1$.

The Newton-Raphson formula is given by

$$x_{n+1} = x_n - \frac{f(x_n)}{f'(x_n)}$$

$$= x_n - \frac{\log x_n - \cos x_n}{\frac{1 + x_n \sin x_n}{x_n}}$$

$$= x_n - \frac{x_n \log x_n - x_n \cos x_n}{1 + x_n \sin x_n}$$

$$= \frac{x_n + x_n^2 \sin x_n - x_n \log x_n + x_n \cos x_n}{1 + x_n \sin x_n}$$

$$x_1 = x_0 - \frac{f(x_0)}{f'(x_0)} = 1.2934$$

$$x_2 = x_1 - \frac{f(x_1)}{f'(x_1)} = 1.3029$$

$$x_3 = x_2 - \frac{f(x_2)}{f'(x_2)} = 1.3029$$

Therefore, the root of the equation is 1.3029.

**Example:** Find a real root of the equation $3x = \cos x + 1$ by using Newton-Raphson method.

**Solution:** Let $f(x) = 3x - \cos x - 1$

$$f(0) = -2 < 0$$

$$f(1) = 1.4597 > 0$$

Since $f(0) < 0$ and $f(1) > 0$. Therefore, the root lies between 0 and 1.

Let $x_0 = 0.5$. Also $f'(x) = 3 + \sin x$.

The Newton-Raphson formula is given by

$$x_{n+1} = x_n - \frac{f(x_n)}{f'(x_n)} = x_n - \frac{3x_n - \cos x_n - 1}{3 + \sin x_n} = \frac{x_n \sin x_n + \cos x_n + 1}{3 + \sin x_n}$$

$$x_1 = \frac{x_0 \sin x_0 + \cos x_0 + 1}{3 + \sin x_0} = 0.6085$$

$$x_2 = \frac{x_1 \sin x_1 + \cos x_1 + 1}{3 + \sin x_1} = 0.6071$$

$$x_3 = \frac{x_2 \sin x_2 + \cos x_2 + 1}{3 + \sin x_2} = 0.6071$$

Here $x_2 = x_3$. Hence the desired root of the equation is 0.6071 correct to four decimal places.

**Example:** Using Newton-Raphson method, find the real root of the equation $x \log_{10} x = 1.2$ correct to five decimal places.

**Solution:** Let $f(x) = x \log_{10} x - 1.2$

$$f(1) = -1.2 < 0$$
$$f(2) = -0.5979 < 0$$
$$f(3) = 0.2313 > 0$$

Since $f(2) < 0$ and $f(3) > 0$. Therefore, the root lies between 2 and 3.

Let us take $x_0 = 2$. Also $f'(x) = \log_{10} x + x \cdot \dfrac{1}{x} \log_{10} e = \log_{10} x + 0.43429$.

The Newton-Raphson formula is given by

$$x_{n+1} = x_n - \dfrac{f(x_n)}{f'(x_n)} = x_n - \dfrac{x_n \log_{10} x_n - 1.2}{\log_{10} x_n + 0.43429} = \dfrac{0.43429 x_n + 1.2}{\log_{10} x_n + 0.43429}$$

$$x_1 = \dfrac{0.43429 x_0 + 1.2}{\log_{10} x_0 + 0.43429} = 2.8132$$

$$x_2 = \dfrac{0.43429 x_1 + 1.2}{\log_{10} x_1 + 0.43429} = 2.7411$$

$$x_3 = \dfrac{0.43429 x_2 + 1.2}{\log_{10} x_2 + 0.43429} = 2.7406$$

$$x_4 = \dfrac{0.43429 x_3 + 1.2}{\log_{10} x_3 + 0.43429} = 2.7406$$

Here $x_4 = x_5$. Hence the required root of the equation is 2.7406 correct to four decimal places.

**Example:** Using Newton-Raphson method, find a real root of $xe^x - \cos x = 0$ correct to four decimal places. (Apr-2022)

**Solution:** Let $f(x) = xe^x - \cos x$

$$f(0) = -1 < 0$$
$$f(1) = 2.1779 > 0$$

Since $f(0) < 0$ and $f(1) > 0$. Therefore, the root lies between 0 and 1.

Let us take $x_0 = 0.5$. Also $f'(x) = xe^x + e^x + \sin x$.

The Newton-Raphson formula is given by

$$x_{n+1} = x_n - \frac{f(x_n)}{f'(x_n)} = x_n - \frac{x_n e^{x_n} - \cos x_n}{x_n e^{x_n} + e^{x_n} + \sin x_n} = \frac{x_n^2 e^{x_n} + x_n \sin x_n + \cos x_n}{x_n e^{x_n} + e^{x_n} + \sin x_n}$$

$$x_1 = \frac{x_0^2 e^{x_0} + x_0 \sin x_0 + \cos x_0}{x_0 e^{x_0} + e^{x_0} + \sin x_0} = 0.5180$$

$$x_2 = \frac{x_1^2 e^{x_1} + x_1 \sin x_1 + \cos x_1}{x_1 e^{x_1} + e^{x_1} + \sin x_1} = 0.5177$$

$$x_3 = \frac{x_2^2 e^{x_2} + x_2 \sin x_2 + \cos x_2}{x_2 e^{x_2} + e^{x_2} + \sin x_2} = 0.5177$$

Therefore, the root of the equation is 0.5177.

**EXERCISE**

1. Using Newton-Raphson method, find the root of the equation $x \log_{10} x = 3.375$ correct to four significant figures. **(Apr-2022)**

   **Ans:** 2.911

2. Find a real root of $x^4 - x - 9 = 0$ using Newton-Raphson method. **(Apr-2022)**

   **Ans:** 1.8134

3. Find the real root of of the equation $x = \cos x$ by Newton-Raphson method. **(Aug-2022)**

4. Find a real root of the equation $2x = 3 + \cos x$ using Newton-Raphson method. **(Aug-2022)**

## 3.5.2 Some Deductions from the Newton-Raphson formula

Below we deduce some useful formulas from the Newton-Raphson formula.

**Example:** Develop an Iterative formula to find (a) square root of a number (b) reciprocal of a number using Newton-Raphson method.

**Solution:** (a) Let $N$ be the number and $x = \sqrt{N} \Rightarrow x^2 = N$. Hence $x^2 - N = 0$
Let $f(x) = x^2 - N$ and hence $f'(x) = 2x$.

The Newton-Raphson formula is

$$x_{n+1} = x_n - \frac{f(x_n)}{f'(x_n)} = x_n - \frac{x_n^2 - N}{2x_n} = \frac{1}{2}\left(x_n + \frac{N}{x_n}\right).$$

(b) Let $x = \frac{1}{N} \Rightarrow \frac{1}{x} - N = 0$.

Let $f(x) = \frac{1}{x} - N$ we have $f'(x) = \frac{-1}{x^2} = -x^{-2}$.

The Newton-Raphson formula is

$$x_{n+1} = x_n - \frac{f(x_n)}{f'(x_n)}$$

$$= x_n - \frac{(1/x_n - N)}{-x_n^{-2}}$$

$$= x_n + \left(\frac{1}{x_n} - N\right)x_n^2$$

$$= x_n + x_n - Nx_n^2$$

$$= x_n(2 - Nx_n).$$

**Example:** Using Newton-Raphson method compute $\sqrt{41}$ correct to 4 decimal places.

**Solution:** Let $f(x) = x^2 - 41$.

$$f(6) = -5 < 0$$

$$f(7) = 8 > 0$$

Since $f(6) < 0$ and $f(7) > 0$. Therefore, the root lies between 6 and 7.
Let us take $x_0 = 6$. Also $f'(x) = 2x$.

The Newton-Raphson formula is given by

$$x_{n+1} = x_n - \frac{f(x_n)}{f'(x_n)} = x_n - \frac{x_n^2 - 41}{2x_n} = \frac{x_n^2 + 41}{2x_n}$$

$$x_1 = \frac{x_0^2 + 41}{2x_0} = 6.4166$$

$$x_2 = \frac{x_1^2 + 41}{2x_1} = 6.4031$$

$$x_3 = \frac{x_2^2 + 41}{2x_2} = 6.4031$$

Since $x_2 = x_3$ upto 4 decimal places, we have $\sqrt{41} = 6.4031$.

**Example:** Develop an Iterative formula to find (a) reciprocal of the square root (b) $p^{\text{th}}$ root of a number using Newton-Raphson method.

**Solution:** (a) Suppose we wish to find the reciprocal of the square root of a positive number $N$. That is, we wish to find $x$ such that $x = \dfrac{1}{\sqrt{N}} \Rightarrow x^2 = \dfrac{1}{N}$. Hence

$$N - \dfrac{1}{x^2} = 0$$

Let $f(x) = N - \dfrac{1}{x^2}$ and hence $f'(x) = \dfrac{2}{x^3}$.

The Newton-Raphson formula is

$$x_{n+1} = x_n - \dfrac{f(x_n)}{f'(x_n)} = x_n - \dfrac{N - (1/x_n^2)}{(2/x_n^3)} = x_n - \dfrac{Nx_n^3 - x_n}{2} = \dfrac{x_n}{2}\left(3 - Nx_n^2\right).$$

(b) Suppose we wish to find the $p^{\text{th}}$ root of a positive number $N$. That is, we wish to find $x$ such that $x = N^{1/p} \Rightarrow x^p - N = 0$.

Let $f(x) = x^p - N$ we have $f'(x) = px^{p-1}$.

The Newton-Raphson formula is

$$x_{n+1} = x_n - \dfrac{f(x_n)}{f'(x_n)}$$

$$= x_n - \dfrac{x_n^p - N}{px_n^{p-1}}$$

$$= \dfrac{(p-1)x_n^p + N}{px_n^{p-1}}$$

$$= \dfrac{1}{p}\left\{(p-1)x_n + \dfrac{N}{x_n^{p-1}}\right\}$$

**Example:** Evaluate $\dfrac{1}{\sqrt{12}}$ using Newton-Raphson method. (Dec-2021)

**Solution:** We wish to find $x$ such that $x = \dfrac{1}{\sqrt{12}} \Rightarrow x^2 = \dfrac{1}{12}$. Hence $12 - \dfrac{1}{x^2} = 0$

Let $f(x) = 12 - \dfrac{1}{x^2}$ and hence $f'(x) = \dfrac{2}{x^3}$.

We note that the positive number nearest to 12 which is a perfect square is 9 and $\dfrac{1}{\sqrt{9}} = \dfrac{1}{3}$.

The Newton-Raphson formula is

$$x_{n+1} = x_n - \frac{f(x_n)}{f'(x_n)} = x_n - \frac{12 - (1/x_n^2)}{(2/x_n^3)} = x_n - \frac{12x_n^3 - x_n}{2} = \frac{x_n}{2}\left(3 - 12x_n^2\right)$$

$$x_1 = \frac{x_0}{2}\left(3 - 12x_0^2\right) = 0.27778$$

$$x_2 = \frac{x_1}{2}\left(3 - 12x_1^2\right) = 0.28806$$

$$x_3 = \frac{x_2}{2}\left(3 - 12x_2^2\right) = 0.28867$$

$$x_4 = \frac{x_3}{2}\left(3 - 12x_3^2\right) = 0.28867$$

Since $x_3 = x_4$ upto 4 decimal places, we have $\frac{1}{\sqrt{12}} = 0.28867$.

**Example:** Write Newton-Raphson formula to obtain the cube root of $N$.

**Example:** Evaluate $\sqrt[3]{24}$ using Newton-Raphson method. **(Aug-2022)**

## 3.6 Solution of Non-Linear Simultaneous Equations

The real solutions of simultaneous algebraic and transcendental equations in several unknowns can be founded by Newton-Raphson method. We restrict ourselves to the case of two unknowns.

**Method-I:** Whenever it is possible, one of the variables may be eliminated from the given equations giving a single polynomial equation in the other variable. Then find this variable to desired degree of accuracy by Newton-Raphson method. Sometimes the above polynomial equation is seen to have a root by trail. If so, reduce this equation to the next lower degree equation and find its other root by Newton-Raphson method.

**Example:** Solve the system of nonlinear equations $x^2 + y = 11$, $y^2 + x = 7$ by Newton-Raphson method. **(Dec-2020)**

**Solution:** Given $x^2 + y = 11$ ...... (i)

and $y^2 + x = 7$ ...... (ii)

Eliminating $y$ from the given equations, we have

(i) $\Rightarrow y = 11 - x^2$.

$\therefore$ (ii) $\Rightarrow (11 - x^2)^2 + x = 7 \Rightarrow x^4 - 22x^2 + x - 7 = 0$.

By trail, $x = 3$ is its root.

$\therefore$ The required equation is $x^3 + 3x^2 - 13x - 38 = 0$.

$$\text{Let } f(x) = x^3 + 3x^2 - 13x - 38$$

$$f(0) = -38 < 0$$

$$f(1) = -47 < 0$$

$$f(2) = -44 < 0$$

$$f(3) = -23 < 0$$

$$f(4) = 22 > 0$$

Since $f(3) < 0$ and $f(4) > 0$. One of the root lies between 3 and 4.

Also $f'(x) = 3x^2 + 6x - 13$. Let us take $x_0 = 3.5$.

Using Newton-Raphson's formula, we have

$$x_{n+1} = x_n - \frac{f(x_n)}{f'(x_n)} = x_n - \frac{x_n^3 + 3x_n^2 - 13x_n - 38}{3x_n^2 + 6x_n - 13} = \frac{2x_n^3 + 3x_n^2 + 38}{3x_n^2 + 6x_n - 13}$$

$$x_1 = \frac{2x_0^3 + 3x_0^2 + 38}{3x_0^2 + 6x_0 - 13} = 3.5866$$

$$x_1 = \frac{2x_1^3 + 3x_1^2 + 38}{3x_1^2 + 6x_1 - 13} = 3.5844$$

$$x_3 = \frac{2x_2^3 + 3x_2^2 + 38}{3x_2^2 + 6x_2 - 13} = 3.5844$$

Since $x_2 = x_3$ upto four decimal places.

$$\therefore x = 3.5844$$

Thus $y = 11 - x^2 = -1.848$

Also $y = 2$ for $x = 3$.

**Example:** Newton-Raphson method to solve the system of nonlinear equations $x^2 - y^2 = 4$ and $x^2 + y^2 = 16$.

**Solution:** Given $x^2 - y^2 = 4$ ...... (i)

and $x^2 + y^2 = 16$ ...... (ii)

Eliminating $y$ from the given equations, we have

(ii) $\Rightarrow y^2 = 16 - x^2$.

$\therefore$ (i) $\Rightarrow x^2 - 16 + x^2 = 4 \Rightarrow 2x^2 = 20 \Rightarrow x^2 = 10 \Rightarrow x^2 - 10 = 0$.

$$\text{Let } f(x) = x^2 - 10$$
$$f(3) = -1 < 0$$
$$f(4) = 6 > 0.$$

Since $f(3) < 0$ and $f(4) > 0$. One of the root lies between 3 and 4. Also $f'(x) = 2x$. Let us take $x_0 = 3.5$.

Using Newton-Raphson's formula, we have

$$x_{n+1} = x_n - \frac{f(x_n)}{f'(x_n)} = x_n - \frac{x_n^2 - 10}{2x_n} = \frac{x_n^2 + 10}{2x_n}$$

$$x_1 = \frac{x_0^2 + 10}{2x_0} = 3.1786$$

$$x_2 = \frac{x_1^2 + 10}{2x_1} = 3.1623$$

$$x_3 = \frac{x_2^2 + 10}{2x_2} = 3.1623$$

Since $x_2 = x_3$ upto four decimal places.

$$\therefore x = 3.1623$$

Thus $y^2 = x^2 - 4 \Rightarrow y = 2.4495$.

Therefore, $x = 3.1623$ and $y = 2.4495$ is an approximate solution to the given system.

**Example:** Solve the system of nonlinear equations by Newton-Raphson method $x^2 + y^2 - 1 = 0$ and $y - x^2 = 0$. (Apr-2022)

**Solution:** Given $x^2 + y^2 - 1 = 0$ ...... (i)

and $y - x^2 = 0$ ...... (ii)

Eliminating $x$ from the given equations, we have

(ii) $\Rightarrow x^2 = y$.

$\therefore$ (i) $\Rightarrow y + y^2 - 1 = 0 \Rightarrow y^2 + y - 1 = 0$.

$$\text{Let } f(y) = y^2 + y - 1$$
$$f(0) = -1 < 0$$
$$f(1) = 1 > 0.$$

Since $f(0) < 0$ and $f(1) > 0$. One of the root lies between 0 and 1. Also $f'(y) = 2y + 1$. Let us take $y_0 = 0.5$.

Using Newton-Raphson's formula, we have

$$y_{n+1} = y_n - \frac{f(y_n)}{f'(y_n)} = y_n - \frac{y_n^2 + y_n - 1}{2y_n + 1} = \frac{y_n^2 + 1}{2y_n + 1}$$

$$y_1 = \frac{y_0^2 + 1}{2y_0 + 1} = 0.625$$

$$y_2 = \frac{y_1^2 + 1}{2y_1 + 1} = 0.6181$$

$$y_3 = \frac{y_2^2 + 1}{2y_2 + 1} = 0.6080$$

$$y_4 = \frac{y_3^2 + 1}{2y_3 + 1} = 0.6080$$

Since $y_3 = y_4$ upto four decimal places.

$$\therefore y = 0.6080$$

Thus $x^2 = y = 0.6080 \Rightarrow x = 0.7861$.

Therefore, $x = 0.7861$ and $y = 0.6080$ is an approximate solution to the given system.

**Method-II:** Consider the system of equations

$$f(x, y) = 0, \quad g(x, y) = 0. \quad \ldots \ldots \quad (1)$$

Let $(x_0, y_0)$ be an initial approximation solution of (1).

Let $x_1 = x_0 + h$ and $y_1 = y_0 + k$ be the next approximation.

Expanding $f$ and $g$ by Taylor's theorem for a function of two variables around the point $(x_1, y_1)$, we have

$$f(x_1, y_1) = f(x_0 + h, y_0 + k) = f(x_0, y_0) + h\left(\frac{\partial f}{\partial x}\right)_{(x_0, y_0)} + k\left(\frac{\partial f}{\partial y}\right)_{(x_0, y_0)}$$

$$g(x_1, y_1) = g(x_0 + h, y_0 + k) = g(x_0, y_0) + h\left(\frac{\partial g}{\partial x}\right)_{(x_0, y_0)} + k\left(\frac{\partial g}{\partial y}\right)_{(x_0, y_0)}$$

(omitting higher powers of $h$ and $k$)

If $(x_1, y_1)$ is a solution of (1), then $f(x_1, y_1) = 0$ and $g(x_1, y_1) = 0$.

Hence $f(x_0, y_0) + h f_x(x_0, y_0) + k f_y(x_0, y_0) = 0$

and $g(x_0, y_0) + h g_x(x_0, y_0) + k g_y(x_0, y_0) = 0$

$$\therefore \quad h f_x(x_0, y_0) + k f_y(x_0, y_0) = -f(x_0, y_0) \quad \cdots\cdots \quad (2)$$

$$h g_x(x_0, y_0) + k g_y(x_0, y_0) = -g(x_0, y_0) \quad \cdots\cdots \quad (3)$$

If the Jacobian $J = \begin{vmatrix} f_x & f_y \\ g_x & g_y \end{vmatrix} \neq 0$, then the equations (2) and (3) provide a unique solution for $h$ and $k$. Now $x_1 = x_0 + h$ and $y_1 = y_0 + k$ give a new approximation to the solution.

By repeating this process we obtain the required solution to the desired accuracy.

**Note:** 1. We find the initial approximation by trail and error or by graphical method so that we can improve the accuracy and the convergence is guaranteed.

2. The above method can be extended to simultaneous equations in any number of unknowns.

**Example:** Solve the equations $x = x^2 + y^2$, $y = x^2 - y^2$ using Newton-Raphson method with $x_0 = 0.8$ and $y_0 = 0.4$.

**Solution:** Let $f(x, y) = x - x^2 - y^2$; $g(x, y) = y - x^2 + y^2$.

$$\therefore \quad f_x = \frac{\partial f}{\partial x} = 1 - 2x; \quad f_y = \frac{\partial f}{\partial y} = -2y$$

$$g_x = \frac{\partial g}{\partial x} = -2x; \quad g_y = \frac{\partial g}{\partial y} = 1 + 2y.$$

Here $x_0 = 0.8$ and $y_0 = 0.4$.

$$\therefore \quad f(x_0, y_0) = x_0 - x_0^2 - y_0^2 = 0$$

$$g(x_0, y_0) = y_0 - x_0^2 + y_0^2 = -0.08$$

$$f_x(x_0, y_0) = 1 - 2x_0 = -0.6$$

$$f_y(x_0, y_0) = -2y_0 = -0.8$$

$$g_x(x_0, y_0) = -2x_0 = -1.6$$

$$g_y(x_0, y_0) = 1 + 2y_0 = 1.8$$

The Newton-Raphson equations are

$$f(x_0, y_0) + h f_x(x_0, y_0) + k f_y(x_0, y_0) = 0 \quad \cdots\cdots \quad (1)$$

$$g(x_0, y_0) + h g_x(x_0, y_0) + k g_y(x_0, y_0) = 0 \quad \cdots\cdots \quad (2)$$

i.e., $\quad 0 - 0.6h - 0.8k = 0 \Rightarrow 0.6h + 0.8k = 0$

$$-0.08 - 1.6h + 1.8k = 0 \Rightarrow -1.6h + 1.8k = 0.08$$

Solving these equations, we get

$$h = -0.027 \quad \text{and} \quad k = 0.02.$$

The next approximation to the solution is

$$x_1 = x_0 + h = 0.773 \quad \text{and} \quad y_1 = y_0 + k = 0.42$$

Now $\quad f(x_1, y_1) = x_1 - x_1^2 - y_1^2 = -0.0009$

$$g(x_1, y_1) = y_1 - x_1^2 + y_1^2 = -0.0011$$

$$f_x(x_1, y_1) = 1 - 2x_1 = -0.546$$

$$f_y(x_1, y_1) = -2y_1 = -0.84$$

$$g_x(x_1, y_1) = -2x_1 = -1.546$$

$$g_y(x_1, y_1) = 1 + 2y_1 = 1.84$$

Replacing $(x_0, y_0)$ by $(x_1, y_1)$ in (1) and (2) we get the Newton-Raphson equations for the second approximation as

$$-0.0009 - 0.546h - 0.84k = 0 \Rightarrow -0.546h - 0.84k = 0.0009$$

$$-0.0011 - 1.546h + 1.84k = 0 \Rightarrow -1.546h + 1.84k = 0.0011$$

Solving for $h$ and $k$, we get $h = -0.0011$ and $k = -0.0003$.

$$\therefore \quad x_2 = x_1 + h = 0.7719$$

$$y_2 = y_1 + k = 0.4197$$

Thus $x = 0.7719$ and $y = 0.4197$ is an approximate solution to the given system.

**Example:** Use Newton-Raphson method to solve the equations $x^2 - y^2 = 4$, $x^2 + y^2 = 16$ with $x_0 = y_0 = 2.828$.

**Solution:** Let $f(x, y) = x^2 - y^2 - 4$ and $g(x, y) = x^2 + y^2 - 16$.

$$\therefore \quad f_x = \frac{\partial f}{\partial x} = 2x; \quad f_y = \frac{\partial f}{\partial y} = -2y$$

$$g_x = \frac{\partial g}{\partial x} = 2x; \quad g_y = \frac{\partial g}{\partial y} = 2y.$$

Here $x_0 = 2.828$ and $y_0 = 2.828$.

$$\therefore \quad f(x_0, y_0) = x_0^2 - y_0^2 - 4 = -4$$

$$g(x_0, y_0) = x_0^2 + y_0^2 - 16 = -0.0048$$

$$f_x(x_0, y_0) = 2x_0 = 5.656$$

$$f_y(x_0, y_0) = -2y_0 = -5.656$$

$$g_x(x_0, y_0) = 2x_0 = 5.656$$

$$g_y(x_0, y_0) = 2y_0 = 5.656$$

The Newton-Raphson equations are

$$f(x_0, y_0) + h f_x(x_0, y_0) + k f_y(x_0, y_0) = 0 \quad \cdots \cdots \quad (1)$$

$$g(x_0, y_0) + h g_x(x_0, y_0) + k g_y(x_0, y_0) = 0 \quad \cdots \cdots \quad (2)$$

i.e., $\quad -4 + 5.656h - 5.656k = 0 \Rightarrow h - k = 0.707$

$\quad -0.0048 + 5.656h - 5.656k = 0 \Rightarrow h + k = 0.0048$

Solving these equations, we get

$$h = 0.356 \text{ and } k = -0.351.$$

The next approximation to the solution is

$$x_1 = x_0 + h = 3.184 \text{ and } y_1 = y_0 + k = 2.477$$

Now $f(x_1, y_1) = x_1^2 - y_1^2 - 4 = -0.0023$

$g(x_1, y_1) = x_1^2 + y_1^2 - 16 = 0.2734$

$f_x(x_1, y_1) = 2x_1 = 6.368$

$f_y(x_1, y_1) = -2y_1 = -4.954$

$g_x(x_1, y_1) = 2x_1 = 6.368$

$g_y(x_1, y_1) = 2y_1 = 4.954$

Replacing $(x_0, y_0)$ by $(x_1, y_1)$ in (1) and (2) we get the Newton-Raphson equations for the second approximation as

$$-0.0023 + 6.368h - 4.954k = 0 \Rightarrow 6.368h - 4.954k = 0.0023$$

$$0.2734 + 6.368h + 4.954k = 0 \Rightarrow 6.368h + 4.954k = -0.2734$$

Solving for $h$ and $k$, we get $h = -0.021$ and $k = -0.028$.

$$\therefore \quad x_2 = x_1 + h = 3.163$$

$$y_2 = y_1 + k = 2.449$$

Thus $x = 3.163$ and $y = 2.449$ is an approximate solution to the given system.

**Example:** Solve the system of equations $\sin x - y + 1.32 = 0$, $x - \cos y - 0.85 = 0$ with $x_0 = 0.6$ and $y_0 = 1.9$ by Newton-Raphson method. **(Apr-2022)**

**Solution:** Let $f(x, y) = \sin x - y + 1.32$ and $g(x, y) = x - \cos y - 0.85$.

$$\therefore \quad f_x = \frac{\partial f}{\partial x} = \cos x; \quad f_y = \frac{\partial f}{\partial y} = -1$$

$$g_x = \frac{\partial g}{\partial x} = 1; \quad g_y = \frac{\partial g}{\partial y} = \sin y.$$

Here $x_0 = 0.6$ and $y_0 = 1.9$.

$$\therefore \quad f(x_0, y_0) = \sin x_0 - y_0 + 1.32 = -0.0154$$

$$g(x_0, y_0) = x_0 - \cos y_0 - 0.85 = 0.0733$$

$$f_x(x_0, y_0) = \cos x_0 = 0.8253$$

$$f_y(x_0, y_0) = -1$$

$$g_x(x_0, y_0) = 1$$

$$g_y(x_0, y_0) = \sin y_0 = 0.9463$$

The Newton-Raphson equations are

$$f(x_0, y_0) + h f_x(x_0, y_0) + k f_y(x_0, y_0) = 0 \quad \cdots \cdots \quad (1)$$

$$g(x_0, y_0) + h g_x(x_0, y_0) + k g_y(x_0, y_0) = 0 \quad \cdots \cdots \quad (2)$$

i.e., $\quad -0.0154 + 0.8253h - k = 0 \Rightarrow 0.8253h - k = 0.0154$

$$0.0733 + h + 0.9463k = 0 \Rightarrow h + 0.9463k = -0.0733$$

Solving these equations, we get

$$h = -0.033 \quad \text{and} \quad k = -0.043.$$

The next approximation to the solution is

$$x_1 = x_0 + h = 0.567 \quad \text{and} \quad y_1 = y_0 + k = 1.857$$

Now $\quad f(x_1, y_1) = \sin x_1 - y_1 + 1.32 = -0.000104$

$$g(x_1, y_1) = x_1 - \cos y_1 - 0.85 = -0.00069$$

$$f_x(x_1, y_1) = \cos x_1 = 0.844$$

$$f_y(x_1, y_1) = -1$$

$$g_x(x_1, y_1) = 1$$

$$g_y(x_1, y_1) = \sin y_1 = 0.959$$

Replacing $(x_0, y_0)$ by $(x_1, y_1)$ in (1) and (2) we get the Newton-Raphson equations for the second approximation as

$$0.000104 + 0.844h - k = 0 \Rightarrow 0.844h - k = -0.000104$$

$$-0.00069 + h + 0.959k = 0 \Rightarrow h + 0.959k = 0.00069$$

Solving for $h$ and $k$, we get $h = 0.00033$ and $k = -0.00038$.

$$\therefore \quad x_2 = x_1 + h = 0.567$$

$$y_2 = y_1 + k = 1.857$$

Thus $x = 0.567$ and $y = 1.857$ is an approximate solution to the given system.

### EXERCISE

1. Perform two iterations of the Newton-Raphson method to solve the system of equations $x^2 + y^2 + xy = 7$ and $x^3 + y^3 = 9$. **(Nov-2017), (Dec-2020), (Apr-2022)**
2. Perform two iterations of the Newton-Raphson method to solve the system of equations $x^2 + 3y^2 = 4$ and $x^2 + 3x + y = 5$. **(Nov-2017), (Dec-2020), (Apr-2022)**
3. Solve the system of nonlinear equations $xy = x + 9$ and $y^2 = x^2 + 7$ by Newton-Raphson method. **(Dec-2020)**

## 3.7 Gauss-Jacobi Iteration Method

Consider the system of equations

$$\left.\begin{array}{l} a_1 x + b_1 y + c_1 z = d_1 \\ a_2 x + b_2 y + c_2 z = d_2 \\ a_3 x + b_3 y + c_3 z = d_3 \end{array}\right\} \quad \cdots \quad (1)$$

where the diagonal coefficients are not zero and are large in absolute value compared to other coefficients. Such system is called a diagonally dominant system. If these are not so, then on interchanging the equation we can make the leading diagonal dominant diagonal.

Let $a_1, b_2, c_3$ are large as compared to other coefficients, then the system can be written as

$$\left. \begin{array}{l} x = \dfrac{1}{a_1}\left[d_1 - b_1 y - c_1 z\right] \\ y = \dfrac{1}{b_2}\left[d_2 - a_2 x - c_2 z\right] \\ z = \dfrac{1}{c_3}\left[d_3 - a_3 x - b_3 y\right] \end{array} \right\} \quad \cdots \quad (2)$$

If we start with the initial approximations $x^{(0)}, y^{(0)}, z^{(0)}$ for the values of $x, y, z$ respectively then substituting these on the right sides of (2), the first approximations are given by

$$x^{(1)} = \dfrac{1}{a_1}\left[d_1 - b_1 y^{(0)} - c_1 z^{(0)}\right]$$
$$y^{(1)} = \dfrac{1}{b_2}\left[d_2 - a_2 x^{(0)} - c_2 z^{(0)}\right]$$
$$z^{(1)} = \dfrac{1}{c_3}\left[d_3 - a_3 x^{(0)} - b_3 y^{(0)}\right]$$

Again substituting the values $x^{(1)}, y^{(1)}, z^{(1)}$ on the right sides of (2), the second approximations are given by

$$x^{(2)} = \dfrac{1}{a_1}\left[d_1 - b_1 y^{(1)} - c_1 z^{(1)}\right]$$
$$y^{(2)} = \dfrac{1}{b_2}\left[d_2 - a_2 x^{(1)} - c_2 z^{(1)}\right]$$
$$z^{(2)} = \dfrac{1}{c_3}\left[d_3 - a_3 x^{(1)} - b_3 y^{(1)}\right]$$

Continue this process till the difference between two consecutive approximations is negligible.

**Note:** In the absence of any better estimates for $x^{(0)}, y^{(0)}, z^{(0)}$, these may each be taken as zero.

**Example:** Solve by Jacobi's iteration method, the equations $10x + y - z = 11.19, x + 10y + z = 28.08, -x + y + 10z = 35.61$, correct to two decimal places.

**Solution:** The given system is diagonally dominant, and we rewrite as

$$\left. \begin{array}{l} x = \dfrac{1}{10}[11.19 - y + z] \\ y = \dfrac{1}{10}[28.08 - x - z] \\ z = \dfrac{1}{10}[35.61 + x - y] \end{array} \right\} \quad \cdots \quad (1)$$

We start the iteration by taking $x^{(0)} = y^{(0)} = z^{(0)} = 0$ in (1).

**First Iteration:**

$$x^{(1)} = \frac{1}{10}[11.19 - y^{(0)} + z^{(0)}] = \frac{11.19}{10} = 1.119$$

$$y^{(1)} = \frac{1}{10}[28.08 - x^{(0)} - z^{(0)}] = \frac{28.08}{10} = 2.808$$

$$z^{(1)} = \frac{1}{10}[35.61 + x^{(0)} - y^{(0)}] = \frac{35.61}{10} = 3.561$$

**Second Iteration:**

$$x^{(2)} = \frac{1}{10}[11.19 - y^{(1)} + z^{(1)}] = 1.19$$

$$y^{(2)} = \frac{1}{10}[28.08 - x^{(1)} - z^{(1)}] = 2.34$$

$$z^{(2)} = \frac{1}{10}[35.61 + x^{(1)} - y^{(1)}] = 3.39$$

**Third Iteration:**

$$x^{(3)} = \frac{1}{10}[11.19 - y^{(2)} + z^{(2)}] = 1.22$$

$$y^{(3)} = \frac{1}{10}[28.08 - x^{(2)} - z^{(2)}] = 2.35$$

$$z^{(3)} = \frac{1}{10}[35.61 + x^{(2)} - y^{(2)}] = 3.45$$

**Fourth Iteration:**

$$x^{(4)} = \frac{1}{10}[11.19 - y^{(3)} + z^{(3)}] = 1.23$$

$$y^{(4)} = \frac{1}{10}[28.08 - x^{(3)} - z^{(3)}] = 2.34$$

$$z^{(4)} = \frac{1}{10}[35.61 + x^{(3)} - y^{(3)}] = 3.45$$

**Fifth Iteration:**
$$x^{(5)} = \frac{1}{10}\left[11.19 - y^{(4)} + z^{(4)}\right] = 1.23$$
$$y^{(5)} = \frac{1}{10}\left[28.08 - x^{(4)} - z^{(4)}\right] = 2.34$$
$$z^{(5)} = \frac{1}{10}\left[35.61 + x^{(4)} - y^{(4)}\right] = 3.45$$

Since the values of $4^{th}$ and $5^{th}$ iterations are similar, we get the solutions as $x = 1.23$, $y = 2.34$ and $z = 3.45$.

**Example:** Solve the system of equations $20x + y - 2z = 17, 3x + 20y - z = -18, 2x - 3y + 20z = 25$ by Jacobi method. **(Apr-2017), (Mar-2022)**

**Solution:** The given system is diagonally dominant, and we rewrite as
$$x = \frac{1}{20}[17 - y + 2z]$$
$$y = \frac{1}{20}[-18 - 3x + z]$$
$$z = \frac{1}{20}[25 - 2x + 3y]$$

We start the iteration by taking $x^{(0)} = y^{(0)} = z^{(0)} = 0$.

**First Iteration:**
$$x^{(1)} = \frac{1}{20}\left[17 - y^{(0)} + 2z^{(0)}\right] = \frac{17}{20} = 0.85$$
$$y^{(1)} = \frac{1}{20}\left[-18 - 3x^{(0)} + z^{(0)}\right] = \frac{-18}{20} = -0.9$$
$$z^{(1)} = \frac{1}{20}\left[25 - 2x^{(0)} + 3y^{(0)}\right] = \frac{25}{20} = 1.25$$

**Second Iteration:**
$$x^{(2)} = \frac{1}{20}\left[17 - y^{(1)} + 2z^{(1)}\right] = 1.02$$
$$y^{(2)} = \frac{1}{20}\left[-18 - 3x^{(1)} + z^{(1)}\right] = -0.965$$
$$z^{(2)} = \frac{1}{20}\left[25 - 2x^{(1)} + 3y^{(1)}\right] = 1.03$$

**Third Iteration:**

$$x^{(3)} = \frac{1}{20}\left[17 - y^{(2)} + 2z^{(2)}\right] = 1.00125$$

$$y^{(3)} = \frac{1}{20}\left[-18 - 3x^{(2)} + z^{(2)}\right] = -1.0015$$

$$z^{(3)} = \frac{1}{20}\left[25 - 2x^{(2)} + 3y^{(2)}\right] = 1.00325$$

**Fourth Iteration:**

$$x^{(4)} = \frac{1}{20}\left[17 - y^{(3)} + 2z^{(3)}\right] = 1.0004$$

$$y^{(4)} = \frac{1}{20}\left[-18 - 3x^{(3)} + z^{(3)}\right] = -1.000025$$

$$z^{(4)} = \frac{1}{20}\left[25 - 2x^{(3)} + 3y^{(3)}\right] = 0.99965$$

**Fifth Iteration:**

$$x^{(5)} = \frac{1}{20}\left[17 - y^{(4)} + 2z^{(4)}\right] = 0.9999$$

$$y^{(5)} = \frac{1}{20}\left[-18 - 3x^{(4)} + z^{(4)}\right] = -1.0000$$

$$z^{(5)} = \frac{1}{20}\left[25 - 2x^{(4)} + 3y^{(4)}\right] = 1.0000$$

**Sixth Iteration:**

$$x^{(6)} = \frac{1}{20}\left[17 - y^{(5)} + 2z^{(5)}\right] = 1.0000$$

$$y^{(6)} = \frac{1}{20}\left[-18 - 3x^{(5)} + z^{(5)}\right] = -0.99999$$

$$z^{(6)} = \frac{1}{20}\left[25 - 2x^{(5)} + 3y^{(5)}\right] = 1.0000$$

Since the values of $5^{th}$ and $6^{th}$ iterations are similar, we get the solutions as $x = 1$, $y = -1$ and $z = 1$.

**Example:** Solve the following system of equations using Gauss-Jacobi method $8x - 3y + 2z = 20, 4x + 11y - z = 33, 6x + 3y + 12z = 35$. **(Sep-2021), (Aug-2022)**

**Solution:** The given system is diagonally dominant, and we rewrite as

$$x = \frac{1}{8}[20 + 3y - 2z]$$
$$y = \frac{1}{11}[33 - 4x + z]$$
$$z = \frac{1}{12}[35 - 6x - 3y]$$

We start the iteration by taking $x^{(0)} = y^{(0)} = z^{(0)} = 0$.

**First Iteration:**

$$x^{(1)} = \frac{1}{8}[20 + 3y^{(0)} - 2z^{(0)}] = \frac{20}{8} = 2.5$$
$$y^{(1)} = \frac{1}{11}[33 - 4x^{(0)} + z^{(0)}] = \frac{33}{11} = 3$$
$$z^{(1)} = \frac{1}{12}[35 - 6x^{(0)} - 3y^{(0)}] = \frac{35}{12} = 2.917$$

**Second Iteration:**

$$x^{(2)} = \frac{1}{8}[20 + 3y^{(1)} - 2z^{(1)}] = 2.895$$
$$y^{(2)} = \frac{1}{11}[33 - 4x^{(1)} + z^{(1)}] = 2.356$$
$$z^{(2)} = \frac{1}{12}[35 - 6x^{(1)} - 3y^{(1)}] = 0.917$$

**Third Iteration:**

$$x^{(3)} = \frac{1}{8}[20 + 3y^{(2)} - 2z^{(2)}] = 3.154$$
$$y^{(3)} = \frac{1}{11}[33 - 4x^{(2)} + z^{(2)}] = 2.030$$
$$z^{(3)} = \frac{1}{12}[35 - 6x^{(2)} - 3y^{(2)}] = 0.880$$

**Fourth Iteration:**

$$x^{(4)} = \frac{1}{8}[20 + 3y^{(3)} - 2z^{(3)}] = 3.041$$
$$y^{(4)} = \frac{1}{11}[33 - 4x^{(3)} + z^{(3)}] = 1.933$$
$$z^{(4)} = \frac{1}{12}[35 - 6x^{(3)} - 3y^{(3)}] = 0.832$$

**Fifth Iteration:**
$$x^{(5)} = \frac{1}{8}\left[20 + 3y^{(4)} - 2z^{(4)}\right] = 3.017$$
$$y^{(5)} = \frac{1}{11}\left[33 - 4x^{(4)} + z^{(4)}\right] = 1.970$$
$$z^{(5)} = \frac{1}{12}\left[35 - 6x^{(4)} - 3y^{(4)}\right] = 0.913$$

**Sixth Iteration:**
$$x^{(6)} = \frac{1}{8}\left[20 + 3y^{(5)} - 2z^{(5)}\right] = 3.011$$
$$y^{(6)} = \frac{1}{11}\left[33 - 4x^{(5)} + z^{(5)}\right] = 1.986$$
$$z^{(6)} = \frac{1}{12}\left[35 - 6x^{(5)} - 3y^{(5)}\right] = 0.916$$

**Seventh Iteration:**
$$x^{(7)} = \frac{1}{8}\left[20 + 3y^{(6)} - 2z^{(6)}\right] = 3.016$$
$$y^{(7)} = \frac{1}{11}\left[33 - 4x^{(6)} + z^{(6)}\right] = 1.988$$
$$z^{(7)} = \frac{1}{12}\left[35 - 6x^{(6)} - 3y^{(6)}\right] = 0.915$$

**Eighth Iteration:**
$$x^{(8)} = \frac{1}{8}\left[20 + 3y^{(7)} - 2z^{(7)}\right] = 3.016$$
$$y^{(8)} = \frac{1}{11}\left[33 - 4x^{(7)} + z^{(7)}\right] = 1.986$$
$$z^{(8)} = \frac{1}{12}\left[35 - 6x^{(7)} - 3y^{(7)}\right] = 0.912$$

**Ninth Iteration:**
$$x^{(9)} = \frac{1}{8}\left[20 + 3y^{(8)} - 2z^{(8)}\right] = 3.017$$
$$y^{(9)} = \frac{1}{11}\left[33 - 4x^{(8)} + z^{(8)}\right] = 1.986$$
$$z^{(9)} = \frac{1}{12}\left[35 - 6x^{(8)} - 3y^{(8)}\right] = 0.912$$

**Tenth Iteration:**
$$x^{(10)} = \frac{1}{8}\left[20 + 3y^{(9)} - 2z^{(9)}\right] = 3.017$$
$$y^{(10)} = \frac{1}{11}\left[33 - 4x^{(9)} + z^{(9)}\right] = 1.986$$
$$z^{(10)} = \frac{1}{12}\left[35 - 6x^{(9)} - 3y^{(9)}\right] = 0.912$$

Since the values of $9^{th}$ and $10^{th}$ iterations are similar, we stop here.

∴ The solution is $x = 3.017$, $y = 1.986$ and $z = 0.912$.

**Example:** Solve the system of following equations using Gauss-Jacobi method
$27x + 6y - z = 85, x + y + 54z = 110, 6x + 15y + 2z = 72$.  (Nov-2021)

**Solution:** The given system is not diagonally dominant, and we rewrite as $27x + 6y - z = 85, 6x + 15y + 2z = 72, x + y + 54z = 110$.

The given system is diagonally dominant, and we rewrite as

$$x = \frac{1}{27}[85 - 6y + z]$$

$$y = \frac{1}{15}[72 - 6x - 2z]$$

$$z = \frac{1}{54}[110 - x - y]$$

We start the iteration by taking $x^{(0)} = y^{(0)} = z^{(0)} = 0$.

**First Iteration:**

$$x^{(1)} = \frac{1}{27}[85 - 6y^{(0)} + z^{(0)}] = \frac{85}{27} = 3.148$$

$$y^{(1)} = \frac{1}{15}[72 - 6x^{(0)} - 2z^{(0)}] = \frac{72}{15} = 4.8$$

$$z^{(1)} = \frac{1}{54}[110 - x^{(0)} - y^{(0)}] = \frac{110}{54} = 2.037$$

**Second Iteration:**

$$x^{(2)} = \frac{1}{27}[85 - 6y^{(1)} + z^{(1)}] = 2.1569$$

$$y^{(2)} = \frac{1}{15}[72 - 6x^{(1)} - 2z^{(1)}] = 3.2694$$

$$z^{(2)} = \frac{1}{54}[110 - x^{(1)} - y^{(1)}] = 1.8898$$

**Third Iteration:**

$$x^{(3)} = \frac{1}{27}[85 - 6y^{(2)} + z^{(2)}] = 2.4916$$

$$y^{(3)} = \frac{1}{15}[72 - 6x^{(2)} - 2z^{(2)}] = 3.6852$$

$$z^{(3)} = \frac{1}{54}[110 - x^{(2)} - y^{(2)}] = 1.9365$$

**Fourth Iteration:**

$$x^{(4)} = \frac{1}{27}\left[85 - 6y^{(3)} + z^{(3)}\right] = 2.4009$$

$$y^{(4)} = \frac{1}{15}\left[72 - 6x^{(3)} - 2z^{(3)}\right] = 3.5451$$

$$z^{(4)} = \frac{1}{54}\left[110 - x^{(3)} - y^{(3)}\right] = 1.9225$$

**Fifth Iteration:**

$$x^{(5)} = \frac{1}{27}\left[85 - 6y^{(4)} + z^{(4)}\right] = 2.4315$$

$$y^{(5)} = \frac{1}{15}\left[72 - 6x^{(4)} - 2z^{(4)}\right] = 3.5832$$

$$z^{(5)} = \frac{1}{54}\left[110 - x^{(4)} - y^{(4)}\right] = 1.9269$$

**Sixth Iteration:**

$$x^{(6)} = \frac{1}{27}\left[85 - 6y^{(5)} + z^{(5)}\right] = 2.4232$$

$$y^{(6)} = \frac{1}{15}\left[72 - 6x^{(5)} - 2z^{(5)}\right] = 3.5704$$

$$z^{(6)} = \frac{1}{54}\left[110 - x^{(5)} - y^{(5)}\right] = 1.9266$$

**Seventh Iteration:**

$$x^{(7)} = \frac{1}{27}\left[85 - 6y^{(6)} + z^{(6)}\right] = 2.4260$$

$$y^{(7)} = \frac{1}{15}\left[72 - 6x^{(6)} - 2z^{(6)}\right] = 3.5739$$

$$z^{(7)} = \frac{1}{54}\left[110 - x^{(6)} - y^{(6)}\right] = 1.9260$$

**Eighth Iteration:**

$$x^{(8)} = \frac{1}{27}\left[85 - 6y^{(7)} + z^{(7)}\right] = 2.4252$$

$$y^{(8)} = \frac{1}{15}\left[72 - 6x^{(7)} - 2z^{(7)}\right] = 3.5727$$

$$z^{(8)} = \frac{1}{54}\left[110 - x^{(7)} - y^{(7)}\right] = 1.9259$$

Since the values of $7^{th}$ and $8^{th}$ iterations are similar, we get the solutions as $x = 2.425$, $y = 3.573$ and $z = 1.926$ (correct to 2 decimal places).

## EXERCISE

1. Solve the system of equations $10x + 2y + z = 9, x + 20y - 2z = -44$ and $-2x + 3y + 20z = 32$ using Jacobi method.  (Dec-2020)
   Ans: $x = 1.108, y = -2.053, z = 2.018$.

2. Solve the system of equations using Gauss-Jacobi method:
   $4x + 2y + z = 11, -x + 2y + z = 3$ and $2x + y + 4z = 16$.  (Dec-2021)

3. Solve the system of equations using Gauss-Jacobi method:
   $9x + 4y + z = -17, x - 2y - 6z = 14, x + 6y = 4$.  (Mar-2022)

4. Solve the system of following equations using Gauss-Jacobi iteration method:
   $28x + 4y - z = 32, x + 3y + 10z = 24, 2x + 17y + 4z = 35$.  (Aug-2022)

5. Solve the following system of equations using Gauss-Jacobi method:
   $10x + y + z = 12, 2x + 10y + z = 13, 2x + 2y + 10z = 14$.  (Aug-2022)

## 3.8 Gauss-Seidel Iteration Method

Gauss-Seidel method is a modification of Jacobi's method. In place of substituting the same set of values in all three equations, we use in each step the values obtained in the earlier step.

Consider the system of equations

$$\left.\begin{array}{l} a_{11}x + a_{12}y + a_{13}z = b_1 \\ a_{21}x + a_{22}y + a_{23}z = b_2 \\ a_{31}x + a_{32}y + a_{33}z = b_3 \end{array}\right\} \quad \cdots\cdots \quad (1)$$

where the diagonal coefficients are not zero and are large in absolute value compared to other coefficients. Such system is called a diagonally dominant system.

After dividing by suitable constants and transposition, the equations may be written as

$$x = c_1 - k_{12}y - k_{13}z \quad \cdots \quad (i)$$
$$y = c_2 - k_{21}x - k_{23}z \quad \cdots \quad (ii)$$
$$z = c_3 - k_{31}x - k_{32}y \quad \cdots \quad (iii)$$

**Step:1** As a first approximation, let us put $y = 0$ and $z = 0$ in (i) and then we get $x = c_1$.

Then in equation (ii), we put this value of $x = c_1$ and $z = 0$ and then find the value of $y$.

Therefore, $y = c_2 - k_{21}c_1$.

Then in equation (iii), we put these values of $x$ and $y$ to obtain the value of $z$ as

$$z = c_3 - k_{31}c_1 - k_{32}(c_2 - k_{21}c_1).$$

**Step 2:** We repeat the above procedure. In the first equation, we put the values of $y$ and $z$ obtained in step 1 and redetermine $x$. By using the new value of $x$ and the value of $z$ obtained in step 1 we redetermine $y$ and so on.

This method uses the latest values of the unknowns in each steps. The process is repeated till two successive solutions are equal or nearly equal.

**Example:** Use Gauss-Seidel iteration method to solve the equations $27x + 6y - z = 85$, $6x + 15y + 2z = 72$, $x + y + 54z = 110$. **(Apr-2022)**

**Solution:** The given system is diagonally dominant, and we rewrite as

$$x = \frac{1}{27}[85 - 6y + z] \quad \cdots \quad (1)$$
$$y = \frac{1}{15}[72 - 6x - 2z] \quad \cdots \quad (2)$$
$$z = \frac{1}{54}[110 - x - y] \quad \cdots \quad (3)$$

**First Iteration:**

Assuming first approximation $x^{(0)} = 0, y^{(0)} = 0$ and $z^{(0)} = 0$.

Substituting these values into (1), we get

$$x^{(1)} = \frac{1}{27}[85 - 6y^{(0)} + z^{(0)}] = \frac{1}{27}(85) = 3.15$$

Now substitute $x^{(1)} = 3.15$ and $z^{(0)} = 0$ in (2), we get

$$y^{(1)} = \frac{1}{15}[72 - 6x^{(1)} - 2z^{(0)}] = \frac{1}{15}[72 - 6(3.15) - 0] = 3.54$$

Substitute $x^{(1)} = 3.15$ and $y^{(1)} = 3.54$ in (3), we get

$$z^{(1)} = \frac{1}{54}[110 - x^{(1)} - y^{(1)}] = \frac{1}{54}[110 - 3.15 - 3.54] = 1.91$$

**Second Iteration:**

Now substitute $y^{(1)} = 3.54$ and $z^{(1)} = 1.91$ in (1), we get

$$x^{(2)} = \frac{1}{27}[85 - 6y^{(1)} + z^{(1)}] = \frac{1}{27}[85 - 6(3.54) + 1.91] = 2.43$$

Substitute $x^{(2)} = 2.43$ and $z^{(1)} = 1.91$ in (2), we get

$$y^{(2)} = \frac{1}{15}[72 - 6x^{(2)} - 2z^{(1)}] = \frac{1}{15}[72 - 6(2.43) - 2(1.91)] = 3.57$$

Substitute $x^{(2)} = 2.43$ and $y^{(2)} = 3.57$ in (3), we get

$$z^{(2)} = \frac{1}{54}[110 - x^{(2)} - y^{(2)}] = \frac{1}{54}[110 - 2.43 - 3.57] = 1.926$$

**Third Iteration:**

Now substitute $y^{(2)} = 3.57$ and $z^{(2)} = 1.926$ in (1), we get

$$x^{(3)} = \frac{1}{27}[85 - 6y^{(2)} + z^{(2)}] = \frac{1}{27}[85 - 6(3.57) + 1.926] = 2.426$$

Substitute $x^{(3)} = 2.426$ and $z^{(2)} = 1.926$ in (2), we get

$$y^{(3)} = \frac{1}{15}[72 - 6x^{(3)} - 2z^{(2)}] = \frac{1}{15}[72 - 6(2.426) - 2(1.926)] = 3.572$$

Substitute $x^{(3)} = 2.426$ and $y^{(3)} = 3.572$ in (3), we get

$$z^{(3)} = \frac{1}{54}[110 - x^{(3)} - y^{(3)}] = \frac{1}{54}[110 - 2.426 - 3.572] = 1.926$$

**Fourth Iteration:**

Now substitute $y^{(3)} = 3.572$ and $z^{(3)} = 1.926$ in (1), we get

$$x^{(4)} = \frac{1}{27}\left[85 - 6y^{(3)} + z^{(3)}\right] = \frac{1}{27}\left[85 - 6(3.572) + 1.926\right] = 2.425$$

Substitute $x^{(4)} = 2.425$ and $z^{(3)} = 1.926$ in (2), we get

$$y^{(4)} = \frac{1}{15}\left[72 - 6x^{(4)} - 2z^{(3)}\right] = \frac{1}{15}\left[72 - 6(2.425) - 2(1.926)\right] = 3.573$$

Substitute $x^{(4)} = 2.425$ and $y^{(4)} = 3.573$ in (3), we get

$$z^{(4)} = \frac{1}{54}\left[110 - x^{(4)} - y^{(4)}\right] = \frac{1}{54}\left[110 - 2.425 - 3.573\right] = 1.926$$

**Fifth Iteration:**

Now substitute $y^{(4)} = 3.573$ and $z^{(4)} = 1.926$ in (1), we get

$$x^{(5)} = \frac{1}{27}\left[85 - 6y^{(4)} + z^{(4)}\right] = \frac{1}{27}\left[85 - 6(3.573) + 1.926\right] = 2.425$$

Substitute $x^{(5)} = 2.425$ and $z^{(4)} = 1.926$ in (2), we get

$$y^{(5)} = \frac{1}{15}\left[72 - 6x^{(5)} - 2z^{(4)}\right] = \frac{1}{15}\left[72 - 6(2.425) - 2(1.926)\right] = 3.573$$

Substitute $x^{(5)} = 2.425$ and $y^{(5)} = 3.573$ in (3), we get

$$z^{(5)} = \frac{1}{54}\left[110 - x^{(5)} - y^{(5)}\right] = \frac{1}{54}\left[110 - 2.425 - 3.573\right] = 1.926$$

Since the values of $4^{th}$ and $5^{th}$ iterations are similar, we get the solutions as $x = 2.425$, $y = 3.573$ and $z = 1.926$.

**Example:** Solve the system of equations $10x + y + z = 12$, $2x + 10y + z = 13$, $2x + 2y + 10z = 14$ by Gauss-Seidel method.  (Apr-2017), (Aug-2022)

**Solution:** The given system is diagonally dominant, and we rewrite as

$$x = \frac{1}{10}[12 - y - z] \quad \cdots\cdots \quad (1)$$

$$y = \frac{1}{10}[13 - 2x - z] \quad \cdots\cdots \quad (2)$$

$$z = \frac{1}{10}[14 - 2x - 2y] \quad \cdots\cdots \quad (3)$$

## First Iteration:

Let us start with initial approximation $x^{(0)} = 0, y^{(0)}$ and $z^{(0)} = 0$ in (1), we get

$$x^{(1)} = \frac{1}{10}\left[12 - y^{(0)} - z^{(0)}\right] = \frac{1}{10}(12) = 1.2$$

Now substitute $x^{(1)} = 1.2$ and $z^{(0)} = 0$ in (2), we get

$$y^{(1)} = \frac{1}{10}\left[13 - 2x^{(1)} - z^{(0)}\right] = \frac{1}{10}\left[13 - 2(1.2) - 0\right] = 1.06$$

Substitute $x^{(1)} = 1.2$ and $y^{(1)} = 1.06$ in (3), we get

$$z^{(1)} = \frac{1}{10}\left[14 - 2x^{(1)} - 2y^{(1)}\right] = \frac{1}{10}\left[14 - 2(1.2) - 2(1.06)\right] = 0.95$$

## Second Iteration:

Now substitute $y^{(1)} = 1.06$ and $z^{(1)} = 0.95$ in (1), we get

$$x^{(2)} = \frac{1}{10}\left[12 - y^{(1)} - z^{(1)}\right] = \frac{1}{10}\left[12 - 1.06 - 0.95\right] = 0.999$$

Substitute $x^{(2)} = 0.999$ and $z^{(1)} = 0.95$ in (2), we get

$$y^{(2)} = \frac{1}{10}\left[13 - 2x^{(2)} - z^{(1)}\right] = \frac{1}{10}\left[13 - 2(0.999) - 0.95\right] = 1.005$$

Substitute $x^{(2)} = 0.999$ and $y^{(2)} = 1.005$ in (3), we get

$$z^{(2)} = \frac{1}{10}\left[14 - 2x^{(2)} - 2y^{(2)}\right] = \frac{1}{10}\left[14 - 2(0.999) - 2(1.005)\right] = 0.9999$$

## Third Iteration:

Now substitute $y^{(2)} = 1.005$ and $z^{(2)} = 0.999$ in (1), we get

$$x^{(3)} = \frac{1}{10}\left[12 - y^{(2)} - z^{(2)}\right] = \frac{1}{10}\left[12 - 1.005 - 0.999\right] = 0.9996 = 1.00$$

Substitute $x^{(3)} = 1.00$ and $z^{(2)} = 0.999$ in (2), we get

$$y^{(3)} = \frac{1}{10}\left[13 - 2x^{(3)} - z^{(2)}\right] = \frac{1}{10}\left[13 - 2(1.00) - 0.999\right] = 1.00$$

Substitute $x^{(3)} = 1.00$ and $y^{(3)} = 1.00$ in (3), we get

$$z^{(3)} = \frac{1}{10}[14 - 2x^{(3)} - 2y^{(3)}] = \frac{1}{10}[14 - 2(1.00) - 2(1.00)] = 1.00$$

**Fourth Iteration:**

Now substitute $y^{(3)} = 1.00$ and $z^{(3)} = 1.00$ in (1), we get

$$x^{(4)} = \frac{1}{10}[12 - y^{(3)} - z^{(3)}] = \frac{1}{10}[12 - 1.00 - 1.00] = 1.00$$

Substitute $x^{(4)} = 1.00$ and $z^{(3)} = 1.00$ in (2), we get

$$y^{(4)} = \frac{1}{10}[13 - 2x^{(4)} - z^{(3)}] = \frac{1}{10}[13 - 2(1.00) - 1.00] = 1.00$$

Substitute $x^{(4)} = 1.00$ and $y^{(4)} = 1.00$ in (3), we get

$$z^{(4)} = \frac{1}{10}[14 - 2x^{(4)} - 2y^{(4)}] = \frac{1}{10}[14 - 2(1.00) - 2(1.00)] = 1.00$$

Since the values of $3^{th}$ and $4^{th}$ iterations are similar, we get the solutions as $x = 1.00$, $1.00$ and $z = 1.00$.

**Example:** Solve the system of equations by the Gauss-Seidel method.

**(Apr-2022)**

$$10x_1 - 2x_2 - x_3 - x_4 = 3$$
$$-2x_1 + 10x_2 - x_3 - x_4 = 15$$
$$-x_1 - x_2 + 10x_3 - 2x_4 = 27$$
$$-x_1 - x_2 - 2x_3 + 10x_4 = -9$$

**Solution:** The given system is diagonally dominant, and we rewrite as

$$x_1 = \frac{1}{10}[3 + 2x_2 + x_3 + x_4] \quad \cdots \cdots \quad (1)$$
$$x_2 = \frac{1}{10}[15 + 2x_1 + x_3 + x_4] \quad \cdots \cdots \quad (2)$$
$$x_3 = \frac{1}{10}[27 + x_1 + x_2 + 2x_4] \quad \cdots \cdots \quad (3)$$
$$x_4 = \frac{1}{10}[-9 + x_1 + x_2 + 2x_3] \quad \cdots \cdots \quad (4)$$

## First Iteration:

Now substitute $x_2^{(0)} = 0, x_3^{(0)} = 0, x_4^{(0)} = 0$ in (1), we get

$$x_1^{(1)} = \frac{1}{10}\left[3 + 2x_2^{(0)} + x_3^{(0)} + x_4^{(0)}\right] = \frac{1}{10}(3) = 0.3$$

Substitute $x_1^{(1)} = 0.3, x_3^{(0)} = 0, x_4^{(0)} = 0$ in (2), we get

$$x_2^{(1)} = \frac{1}{10}\left[15 + 2x_1^{(1)} + x_3^{(0)} + x_4^{(0)}\right] = \frac{1}{10}\left[15 + 2(0.3) + 0 + 0\right] = 1.56$$

Substitute $x_1^{(1)} = 0.3, x_2^{(1)} = 1.56, x_4^{(0)} = 0$ in (3), we get

$$x_3^{(1)} = \frac{1}{10}\left[27 + x_1^{(1)} + x_2^{(1)} + 2x_4^{(0)}\right] = \frac{1}{10}\left[27 + 0.3 + 1.56 + 0\right] = 2.886$$

Substitute $x_1^{(1)} = 0.3, x_2^{(1)} = 1.56, x_3^{(1)} = 2.886$ in (4), we get

$$x_4^{(1)} = \frac{1}{10}\left[-9 + x_1^{(1)} + x_2^{(1)} + 2x_3^{(1)}\right] = \frac{1}{10}\left[-9 + 0.3 + 1.56 + 2(2.886)\right] = -0.1368$$

## Second Iteration:

Now substitute $x_2^{(1)} = 1.56, x_3^{(1)} = 2.886, x_4^{(1)} = -0.1368$ in (1), we get

$$x_1^{(2)} = \frac{1}{10}\left[3 + 2x_2^{(1)} + x_3^{(1)} + x_4^{(1)}\right] = \frac{1}{10}\left[3 + 2(1.56) + 2.886 - 0.1368\right] = 0.8869$$

Substitute $x_1^{(2)} = 0.8869, x_3^{(1)} = 2.886, x_4^{(1)} = -0.1368$ in (2), we get

$$x_2^{(2)} = \frac{1}{10}\left[15 + 2x_1^{(2)} + x_3^{(1)} + x_4^{(1)}\right] = \frac{1}{10}\left[15 + 2(0.8869) + 2.886 - 0.1368\right] = 1.9523$$

Substitute $x_1^{(2)} = 0.8869, x_2^{(2)} = 1.9523, x_4^{(1)} = -0.1368$ in (3), we get

$$x_3^{(2)} = \frac{1}{10}\left[27 + x_1^{(2)} + x_2^{(2)} + 2x_4^{(1)}\right] = \frac{1}{10}\left[27 + 0.8869 + 1.9523 + 2(-0.1368)\right] = 2.9566$$

Substitute $x_1^{(2)} = 0.8869, x_2^{(2)} = 1.9523, x_3^{(2)} = 2.9566$ in (4), we get

$$x_4^{(2)} = \frac{1}{10}\left[-9 + x_1^{(2)} + x_2^{(2)} + 2x_3^{(2)}\right] = \frac{1}{10}\left[-9 + 0.8869 + 1.9523 + 2(2.9566)\right] = -0.0248$$

## Third Iteration:

Now substitute $x_2^{(2)} = 1.9523, x_3^{(2)} = 2.9566, x_4^{(2)} = -0.0248$ in (1), we get

$$x_1^{(3)} = \frac{1}{10}\left[3 + 2x_2^{(2)} + x_3^{(2)} + x_4^{(2)}\right] = \frac{1}{10}\left[3 + 2(1.9523) + 2.9566 - 0.0248\right] = 0.9836$$

Substitute $x_1^{(3)} = 0.9836, x_3^{(2)} = 2.9566, x_4^{(2)} = -0.0248$ in (2), we get

$$x_2^{(3)} = \frac{1}{10}\left[15 + 2x_1^{(3)} + x_3^{(2)} + x_4^{(2)}\right] = \frac{1}{10}\left[15 + 2(0.9836) + 2.9566 - 0.0248\right] = 1.9899$$

Substitute $x_1^{(3)} = 0.9836, x_2^{(3)} = 1.9899, x_4^{(2)} = -0.0248$ in (3), we get

$$x_3^{(3)} = \frac{1}{10}\left[27 + x_1^{(3)} + x_2^{(3)} + 2x_4^{(2)}\right] = \frac{1}{10}\left[27 + 0.9836 + 1.9899 + 2(-0.0248)\right] = 2.9924$$

Substitute $x_1^{(3)} = 0.9836, x_2^{(3)} = 1.9800, x_3^{(3)} = 2.9924$ in (4), we get

$$x_4^{(3)} = \frac{1}{10}\left[-9 + x_1^{(3)} + x_2^{(3)} + 2x_3^{(3)}\right] = \frac{1}{10}\left[-9 + 0.9836 + 1.9899 + 2(2.9924)\right] = -0.0042$$

## Fourth Iteration:

Now substitute $x_2^{(3)} = 1.9899, x_3^{(3)} = 2.9924, x_4^{(3)} = -0.0042$ in (1), we get

$$x_1^{(4)} = \frac{1}{10}\left[3 + 2x_2^{(3)} + x_3^{(2)} + x_4^{(3)}\right] = \frac{1}{10}\left[3 + 2(1.9899) + 2.9924 - 0.0024\right] = 0.9968$$

Substitute $x_1^{(4)} = 0.9986, x_3^{(3)} = 2.9924, x_4^{(3)} = -0.0024$ in (2), we get

$$x_2^{(4)} = \frac{1}{10}\left[15 + 2x_1^{(4)} + x_3^{(3)} + x_4^{(3)}\right] = \frac{1}{10}\left[15 + 2(0.9968) + 2.9924 - 0.0024\right] = 1.9982$$

Substitute $x_1^{(4)} = 0.9986, x_2^{(4)} = 1.9982, x_4^{(3)} = -0.0024$ in (3), we get

$$x_3^{(4)} = \frac{1}{10}\left[27 + x_1^{(4)} + x_2^{(4)} + 2x_4^{(3)}\right] = \frac{1}{10}\left[27 + 0.9968 + 1.9982 + 2(-0.0024)\right] = 2.9987$$

Substitute $x_1^{(4)} = 0.9986, x_2^{(4)} = 1.9982, x_3^{(4)} = 2.9987$ in (4), we get

$$x_4^{(4)} = \frac{1}{10}\left[-9 + x_1^{(4)} + x_2^{(4)} + 2x_3^{(4)}\right] = \frac{1}{10}\left[-9 + 0.9968 + 1.9982 + 2(2.9987)\right] = -0.0008$$

## Fifth Iteration:

Now substitute $x_2^{(4)} = 1.9982, x_3^{(4)} = 2.9987, x_4^{(4)} = -0.0008$ in (1), we get

$$x_1^{(5)} = \frac{1}{10}[3 + 2x_2^{(4)} + 2x_3^{(4)} + x_4^{(4)}] = \frac{1}{10}[3 + 2(1.9982) + 2(2.9987) - 0.0008] = 0.9994$$

Substitute $x_1^{(5)} = 0.9994, x_3^{(4)} = 2.9987, x_4^{(4)} = -0.0008$ in (2), we get

$$x_2^{(5)} = \frac{1}{10}[15 + 2x_1^{(5)} + x_3^{(4)} + x_4^{(4)}] = \frac{1}{10}[15 + 2(0.9994) + 2.9987 - 0.0008] = 1.9997$$

Substitute $x_1^{(5)} = 0.9994, x_2^{(5)} = 1.9997, x_4^{(4)} = -0.0008$ in (3), we get

$$x_3^{(5)} = \frac{1}{10}[27 + x_1^{(5)} + x_2^{(5)} + 2x_4^{(4)}] = \frac{1}{10}[27 + 0.9997 + 1.9997 + 2(-0.0008)] = 2.9998$$

Substitute $x_1^{(5)} = 0.9994, x_2^{(5)} = 1.9997, x_3^{(5)} = 2.9998$ in (4), we get

$$x_4^{(5)} = \frac{1}{10}[-9 + x_1^{(5)} + x_2^{(5)} + 2x_3^{(5)}] = \frac{1}{10}[-9 + 0.9994 + 1.9997 + 2(2.9998)] = -0.0001$$

## Sixth Iteration:

Now substitute $x_2^{(5)} = 1.9997, x_3^{(5)} = 2.9998, x_4^{(5)} = -0.0001$ in (1), we get

$$x_1^{(6)} = \frac{1}{10}[3 + 2x_2^{(5)} + x_3^{(5)} + x_4^{(5)}] = \frac{1}{10}[3 + 2(1.9997) + 2.9998 - 0.0001] = 0.9999$$

Substitute $x_1^{(6)} = 0.9999, x_3^{(4)} = 2.9998, x_4^{(4)} = -0.0001$ in (2), we get

$$x_2^{(6)} = \frac{1}{10}[15 + 2x_1^{(6)} + x_3^{(5)} + x_4^{(5)}] = \frac{1}{10}[15 + 2(0.9999) + 2.9998 - 0.0001] = 1.9999$$

Substitute $x_1^{(6)} = 0.9999, x_2^{(6)} = 1.9999, x_4^{(5)} = -0.0001$ in (3), we get

$$x_3^{(6)} = \frac{1}{10}[27 + x_1^{(6)} + x_2^{(6)} + 2x_4^{(5)}] = \frac{1}{10}[27 + 0.9999 + 1.9999 + 2(-0.0001)] = 2.9999$$

Substitute $x_1^{(6)} = 0.9999, x_2^{(6)} = 1.9999, x_3^{(6)} = 2.9999$ in (4), we get

$$x_4^{(6)} = \frac{1}{10}[-9 + x_1^{(6)} + x_2^{(6)} + 2x_3^{(6)}] = \frac{1}{10}[-9 + 0.9999 + 1.9999 + 2(2.9999)] = -0.0001$$

We find that the $5^{th}$ and $6^{th}$ iterates are close to each other.

Hence, the solution is $x_1 = 1, x_2 = 2, x_3 = 3, x_4 = 0$.

**Example:** Solve the system $10x - 2y - z - w = 3$; $-2x + 10y - z - w = 15$; $-x - y + 10z - 2w = 15$; $-x - y - 2z + 10w = -9$ using Gauss Seidel method.

**Solution:** The given system is diagonally dominant, and we rewrite as  (Apr-2022)

$$x = \frac{1}{10}[3 + 2y + z + w] \quad \cdots\cdots \quad (1)$$

$$y = \frac{1}{10}[15 + 2x + z + w] \quad \cdots\cdots \quad (2)$$

$$z = \frac{1}{10}[15 + x + y + 2w] \quad \cdots\cdots \quad (3)$$

$$w = \frac{1}{10}[-9 + x + y + 2z] \quad \cdots\cdots \quad (4)$$

**First Iteration:**

Now substitute $y^{(0)} = 0, z^{(0)} = 0, w^{(0)} = 0$ in (1), we get

$$x^{(1)} = \frac{1}{10}[3 + 2y^{(0)} + z^{(0)} + w^{(0)}] = \frac{1}{10}[3 + 2(0) + (0) + (0)] = \frac{1}{10}(3) = 0.3$$

Substitute $x^{(1)} = 0.3, z^{(0)} = 0, w^{(0)} = 0$ in (2), we get

$$y^{(1)} = \frac{1}{10}[15 + 2x^{(1)} + z^{(0)} + w^{(0)}] = \frac{1}{10}[15 + 2(0.3) + 0 + 0] = 1.56$$

Substitute $x^{(1)} = 0.3, y^{(1)} = 1.56, w^{(0)} = 0$ in (3), we get

$$z^{(1)} = \frac{1}{10}[15 + x^{(1)} + y^{(1)} + 2w^{(0)}] = \frac{1}{10}[15 + 0.3 + 1.56 + 2(0)] = 1.686$$

Substitute $x^{(1)} = 0.3, y^{(1)} = 1.56, z^{(1)} = 1.686$ in (4), we get

$$w^{(1)} = \frac{1}{10}[-9 + x^{(1)} + y^{(1)} + 2z^{(1)}] = \frac{1}{10}[-9 + (0.3) + (1.56) - 2(1.686)] = -1.0512$$

**Second Iteration:**

Now substitute $y^{(1)} = 1.56, z^{(1)} = 1.686, w^{(1)} = -1.0512$ in (1), we get

$$x^{(2)} = \frac{1}{10}[3 + 2y^{(1)} + z^{(1)} + w^{(1)}] = \frac{1}{10}[3 + 2(1.56) + (1.686) + (-1.0512)] = 0.6755$$

Substitute $x^{(2)} = 0.6755, z^{(1)} = 1.686, w^{(1)} = -1.0512$ in (2), we get

$$y^{(2)} = \frac{1}{10}[15 + 2x^{(2)} + z^{(1)} + w^{(1)}] = \frac{1}{10}[15 + 2(0.6755) + (1.686) + (-1.0512)] = 1.6986$$

Substitute $x^{(2)} = 0.6755, y^{(2)} = 1.6986, w^{(1)} = -1.0512$ in (3), we get

$$z^{(2)} = \frac{1}{10}\left[27 + x^{(2)} + y^{(2)} + 2w^{(1)}\right] = \frac{1}{10}\left[15 + (0.6755) + (1.6986) + 2(-1.0512)\right] = 1.5272$$

Substitute $x^{(2)} = 0.6755, y^{(2)} = 1.6986, z^{(2)} = 1.5272$ in (4), we get

$$w^{(2)} = \frac{1}{10}\left[-9 + x^{(2)} + y^{(2)} + 2w^{(2)}\right] = \frac{1}{10}\left[-9 + (0.6755) + (1.6986) - 2(1.5272)\right] = -0.968$$

**Third Iteration:**

Now substitute $y^{(2)} = 1.6986, z^{(2)} = 1.5272, w^{(2)} = -0.968$ in (1), we get

$$x^{(3)} = \frac{1}{10}\left[3 + 2y^{(2)} + z^{(2)} + w^{(2)}\right] = \frac{1}{10}\left[3 + 2(1.6986) + (1.5272) + (-0.968)\right] = 0.6956$$

Substitute $x^{(3)} = 0.6956, z^{(2)} = 1.5272, w^{(2)} = -0.968$ in (2), we get

$$y^{(3)} = \frac{1}{10}\left[15 + 2x^{(3)} + z^{(2)} + w^{(2)}\right] = \frac{1}{10}\left[15 + 2(0.6956) + (1.5272) + (-0.968)\right] = 1.695$$

Substitute $x^{(3)} = 0.6956, y^{(3)} = 1.695, w^{(2)} = -0.968$ in (3), we get

$$z^{(3)} = \frac{1}{10}\left[15 + x^{(3)} + y^{(3)} + 2w^{(2)}\right] = \frac{1}{10}\left[15 + (0.6956) + (1.695) + 2(-0.968)\right] = 1.5455$$

Substitute $x^{(3)} = 0.6956, y^{(3)} = 1.695, z^{(3)} = 1.5455$ in (4), we get

$$w^{(3)} = \frac{1}{10}\left[-9 + x^{(3)} + y^{(3)} + 2z^{(3)}\right] = \frac{1}{10}\left[-9 + (0.6956) + (1.695) - 2(1.5455)\right] = -0.97$$

**Fourth Iteration:**

Now substitute $y^{(3)} = 1.695, z^{(3)} = 1.5455, w^{(3)} = -0.97$ in (1), we get

$$x^{(4)} = \frac{1}{10}\left[3 + 2y^{(3)} + z^{(2)} + w^{(3)}\right] = \frac{1}{10}\left[3 + 2(1.695) + (1.5455) + (-0.97)\right] = 0.6966$$

Substitute $x^{(4)} = 0.6966, z^{(3)} = 1.5455, w^{(3)} = -0.97$ in (2), we get

$$y^{(4)} = \frac{1}{10}\left[15 + 2x^{(4)} + z^{3)} + w^{(3)}\right] = \frac{1}{10}\left[15 + 2(0.6966) + (1.5455) + (-0.97)\right] = 1.6969$$

Substitute $x^{(4)} = 0.6966, y^{(4)} = 1.6969, w^{(3)} = -0.97$ in (3), we get

$$z^{(4)} = \frac{1}{10}\left[15 + x^{(4)} + y^{(4)} + 2w^{(3)}\right] = \frac{1}{10}\left[15 + (0.6966) + (1.6969) + 2(-0.97)\right] = 1.5453$$

Substitute $x^{(4)} = 0.6966, y^{(4)} = 1.6969, z^{(4)} = 1.5453$ in (4), we get

$$w^{(4)} = \frac{1}{10}\left[-9+x^{(4)}+y^{(4)}+2z^{(4)}\right] = \frac{1}{10}\left[-9+(0.6966)+(1.6969)-2(1.5453)\right] = -0.9697$$

**Fifth Iteration:**

Now substitute $y^{(4)} = 1.6969, z^{(4)} = 1.5453, w^{(4)} = -0.9697$ in (1), we get

$$x^{(5)} = \frac{1}{10}\left[3+2y^{(4)}+2z^{(4)}+w^{(4)}\right] = \frac{1}{10}\left[3+2(1.6969)+(1.5453)+(-0.9697)\right] = 0.6969$$

Substitute $x^{(5)} = 0.6969, z^{(4)} = 1.5453, w^{(4)} = -0.9697$ in (2), we get

$$y^{(5)} = \frac{1}{10}\left[15+2x^{(5)}+z^{(4)}+w^{(4)}\right] = \frac{1}{10}\left[15+2(0.6969)+(1.5453)+(-0.9697)\right] = 1.6969$$

Substitute $x^{(5)} = 0.6969, y^{(5)} = 1.6969, w^{(4)} = -0.9697$ in (3), we get

$$z^{(5)} = \frac{1}{10}\left[15+x^{(5)}+y^{(5)}+2w^{(4)}\right] = \frac{1}{10}\left[15+(0.6969)+(1.6969)+2(-0.9697)\right] = 1.5454$$

Substitute $x^{(5)} = 0.6969, y^{(5)} = 1.6969, z^{(5)} = 1.5454$ in (4), we get

$$w^{(5)} = \frac{1}{10}\left[-9+x^{(5)}+y^{(5)}+2w^{(5)}\right] = \frac{1}{10}\left[-9+(0.6969)+(1.6969)-2(1.5454)\right] = -0.9697$$

We find that the $4^{th}$ and $5^{th}$ iterates are close to each other.

Hence, the solution is $x = 0.7, y = 1.7, z = 1.55, w = -0.97$.

## EXERCISE

1. Solve the system of equations $x + 10y + z = 6, 10x + y + z = 6, x + y + 10z = 6$ by Gauss-Seidel method. **(Apr-2017)**

**Ans:**

| Iteration | $x$ | $y$ | $z$ |
| --- | --- | --- | --- |
| 1 | 0.6 | 0.54 | 0.486 |
| 2 | 0.4974 | 0.5017 | 0.4992 |
| 3 | 0.4998 | 0.5 | 0.5 |
| 4 | 0.5 | 0.5 | 0.5 |

Thus the solution is $x = 0.5, y = 0.5, z = 0.5$.

2. Solve the system of equations $8x-3y+2z = 20, 4x+11y-z = 33$ and $6x+3y+12z = 35$ using Gauss-Seidel method. **(July-2015), (Apr-2018), (Aug-2022)**

**Ans:**

| Iterations | $x$ | $y$ | $z$ |
|---|---|---|---|
| 1 | 2.5 | 2.0909 | 1.1439 |
| 2 | 2.9981 | 2.0138 | 0.9142 |
| 3 | 3.0266 | 1.9825 | 0.9077 |
| 4 | 3.0165 | 1.9856 | 0.912 |
| 5 | 3.017 | 1.986 | 0.912 |
| 6 | 3.0166 | 1.986 | 0.9119 |

Thus the solution is $x = 3.02, y = 1.99, z = 0.91$.

3. Solve the system of equations using Gauss-Seidel method: $6x + y + z = 105, 4x + 8y + 3z = 155, 5x + 4y - 10z = 65$.

**Ans:**

| Iteration | $x$ | $y$ | $z$ |
|---|---|---|---|
| 1 | 17.5 | 10.625 | 6.5 |
| 2 | 14.6458 | 9.6146 | 4.6688 |
| 3 | 15.1194 | 10.0645 | 5.0855 |
| 4 | 14.975 | 9.9804 | 4.9797 |
| 5 | 15.0066 | 10.0043 | 5.005 |
| 6 | 14.9984 | 9.9989 | 4.9988 |
| 7 | 15.0004 | 10.0003 | 5.0003 |
| 8 | 14.9999 | 9.9999 | 4.999 |

Thus the solution is $x = 15, y = 10, z = 5$.

4. Solve the system of equations using Gauss-Seidel method: $2x + y + 6z = 9, 8x + 3y + 2z = 13, x + 5y + z = 7$.

**Ans:**

| Iteration | x | y | z |
|---|---|---|---|
| 1 | 1.625 | 1.075 | 0.7792 |
| 2 | 1.0271 | 1.0387 | 0.9845 |
| 3 | 0.9893 | 1.0052 | 1.0027 |
| 4 | 0.9974 | 1 | 1.0009 |
| 5 | 0.9998 | 0.9999 | 1.0001 |
| 6 | 1 | 1 | 1 |

Thus $x = 1, y = 1, z = 1$

5. Using Gauss Seidel method to solve $25x + 2y + 2z = 69$, $2x + 10y + 9z = -23$ and $2x - 7y - 20z = -57$. **(Dec-2020)**

**Ans:**

| Iteration | x | y | z |
|---|---|---|---|
| 1 | 2.76 | -2.852 | 4.1242 |
| 2 | 2.6582 | -6.5434 | 5.406 |
| 3 | 2.851 | -7.7356 | 5.8426 |
| 4 | 2.9114 | -8.1406 | 5.9904 |
| 5 | 2.932 | -8.2777 | 6.0404 |
| 6 | 2.939 | -8.3242 | 6.0574 |
| 7 | 2.9413 | -8.3399 | 6.0631 |
| 8 | 2.9421 | -8.3452 | 6.065 |
| 9 | 2.9424 | -8.347 | 6.0657 |
| 10 | 2.9425 | -8.3476 | 6.0659 |

6. Solve the system of equations $20x + 2y + 6z = 28$, $x + 20y + 9z = -23$, $2x - 7y - 20z = -57$ using Gauss Seidel method. **(Dec-2020)**

**Ans:**

| Iteration | $x$ | $y$ | $z$ |
|---|---|---|---|
| 1 | 1.4 | -1.22 | 3.417 |
| 2 | 0.4969 | -2.7125 | 3.8491 |
| 3 | 0.5165 | -2.9079 | 3.9194 |
| 4 | 0.515 | -2.9395 | 3.9303 |
| 5 | 0.5149 | -2.9444 | 3.932 |
| 6 | 0.5148 | -2.9452 | 3.9323 |

Thus $x = 0.5149, y = -2.9451, z = 3.9323$.

7. Solve the system of equations $10x + 2y + z = 9, 2x + 20y - 2z = -44$ and $-2x + 3y + 10z = 22$ using Gauss Seidel method. **(Dec-2020)**

**Ans:**

| Iteration | $x$ | $y$ | $z$ |
|---|---|---|---|
| 1 | 0.9 | -2.29 | 3.067 |
| 2 | 1.0513 | -1.9984 | 3.0098 |
| 3 | 0.9987 | -1.9989 | 2.9994 |
| 4 | 0.9998 | -2 | 3 |
| 5 | 1 | -2 | 3 |

Thus $x = 1, y = -2, z = 3$

8. Solve the system of following equations using Gauss Seidel method:

$2x - 7y - 10z = 17, 5x + y + 3z = 14$ and $x + 10y + 9z = 7$. **(Nov-2021)**

**Ans:**

| Iteration | $x$ | $y$ | $z$ |
|---|---|---|---|
| 1 | 2.8 | 0.42 | -1.434 |
| 2 | 3.5764 | 1.633 | -2.1278 |
| 3 | 3.7501 | 2.24 | -2.518 |
| 4 | 3.8628 | 2.5799 | -2.7334 |
| 5 | 3.924 | 2.7676 | -2.8525 |
| 6 | 3.958 | 2.8715 | -2.9184 |
| 7 | 3.9768 | 2.9289 | -2.9549 |
| 8 | 3.9872 | 2.9607 | -2.975 |
| 9 | 3.9929 | 2.9783 | -2.9862 |
| 10 | 3.9961 | 2.988 | -2.9924 |
| 11 | 3.9978 | 2.9933 | -2.9958 |
| 12 | 3.9988 | 2.9963 | -2.9977 |
| 13 | 3.9993 | 2.998 | -2.9987 |
| 14 | 3.9996 | 2.9989 | -2.9993 |

Thus $x = 3.9996, y = 2.9989, z = -2.9993$

9. Solve the system of following equations using Gauss Seidel method:

$28x + 4y - z = 32, x + 3y + 10z = 24$ and $2x + 17y + 4z = 35$. (July-2021)

Ans:

| Iteration | $x$ | $y$ | $z$ |
|---|---|---|---|
| 1 | 1.1429 | 1.9244 | 1.7084 |
| 2 | 0.929 | 1.5476 | 1.8428 |
| 3 | 0.9876 | 1.509 | 1.8485 |
| 4 | 0.9933 | 1.507 | 1.8486 |
| 5 | 0.9936 | 1.507 | 1.8485 |

Thus $x = 0.99, y = 1.51, z = 1.85$.

10. Solve the system of following equations using Gauss Seidel method:

$2x + 10y + z = 51, 10x + y + 2z = 44, x + 2y + 10z = 61$. (Mar-2022)

Ans:

| Iteration | $x$ | $y$ | $z$ |
|---|---|---|---|
| 1 | 4.4 | 4.22 | 4.816 |
| 2 | 3.0148 | 4.0154 | 4.9954 |
| 3 | 2.9994 | 4.0006 | 4.9999 |
| 4 | 3 | 4 | 5 |

Thus $x = 3, y = 4, z = 5$.

11. Use Gauss Seidel method to solve $25x+2y+2z = 69, 2x+10y+z = 63, x+y+z = 43$.

**Ans:**

| Iteration | $x$ | $y$ | $z$ |
|---|---|---|---|
| 1 | 2.76 | 5.748 | 34.492 |
| 2 | -0.4592 | 2.9426 | 40.5166 |
| 3 | -0.7167 | 2.3917 | 41.325 |
| 4 | -0.7373 | 2.315 | 41.4224 |
| 5 | -0.739 | 2.3056 | 41.4334 |
| 6 | -0.7391 | 2.3045 | 41.4346 |
| 7 | -0.7391 | 2.3044 | 41.4348 |

Thus $x = -0.74, y = 2.3, z = 41.43$.

12. Solve by Gauss Seidal method, the equations $9x-2y+z-t = 50, x-7y+3z+t = 20, -2x+2y+7z+2t = 22, x+y-2z+6t = 10$.

13. Solve the equation $5x + y + z + w = 4, x + 7y + z + w = 12, x + y + 6z + w = -5, x + y + z + 4w = -6$ by Gauss-Seidal method.

14. Solve the system of following equations using Gauss-Seidel iteration method:
$10x - 5y - 2z = 3, 4x - 10y + 3z = -3, x + 6y + 10z = -3$. **(Aug-2022)**

# Chapter 4

# Interpolation

## 4.1 Finite Differences

Let $y = f(x)$. The values, which the independent variable $x$ takes, are called *arguments* and the corresponding values of $f(x)$ are called *entries*. The difference between consecutive values of $x$ is called the *interval of differencing*.

Suppose we are given the following values of $y = f(x)$ for a set of values of $x$:

$x: \quad x_0 \quad x_1 \quad x_2 \cdots x_n$

$y: \quad y_0 \quad y_1 \quad y_2 \cdots y_n$

Then the process of finding the values of $y$ corresponding to any value of $x = x_i$ between $x_0$ and $x_n$ is called *interpolation*.

### 4.1.1 Forward Differences

Consider the function $y = f(x)$ of an independent variable $x$.
Let $y_0, y_1, y_2, \cdots, y_r, \cdots$ be the values of $y$ corresponding to the values of $x_0, x_1, \cdots, x_r, \cdots$ of $x$ respectively which are equally spaced with step length $h$. We define

$$\boxed{\Delta[f(x)] = f(x+h) - f(x)}$$

Thus $\Delta y_0 = f(x_0 + h) - f(x_0) = f(x_1) - f(x_0) = y_1 - y_0$.
Similarly
$$\Delta y_1 = y_2 - y_1$$
$$\Delta y_2 = y_3 - y_2$$
$$\vdots$$
$$\Delta y_r = y_{r+1} - y_r$$

Here the symbol $\Delta$ is called the **forward difference operator** and $\Delta y_0, \Delta y_1, \cdots, \Delta y_r$ are called the **first forward differences** of the function $y = f(x)$.

The **second order forward differences** of the function are defined by
$$\Delta^2 y_0 = \Delta y_1 - \Delta y_0$$
$$\Delta^2 y_1 = \Delta y_2 - \Delta y_1$$
$$\vdots$$
$$\Delta^2 y_r = \Delta y_{r+1} - \Delta y_r$$

In a similar manner higher order differences can be defined. In general, the $n^{\text{th}}$ order forward differences are defined by the formula
$$\Delta^n y_r = \Delta^{n-1} y_{r+1} - \Delta^{n-1} y_r, \quad r = 0, 1, 2 \cdots.$$

**Note:** If $f(x)$ is a constant function, i.e., $f(x) = c$, then $y_0 = y_1 = y_2 = \cdots = c$ and we have $\Delta^n y_r = 0$ for $r = 1, 2, \cdots$.

These differences of the function $y = f(x)$ can be systematically represented in the form a table called **forward difference table**.

**Forward Difference Table:**

| Argument $x$ | Entry $y = f(x)$ | $\Delta y$ | $\Delta^2 y$ | $\Delta^3 y$ | $\Delta^4 y$ |
|---|---|---|---|---|---|
| $x_0$ | $y_0$ | | | | |
| | | $\Delta y_0$ | | | |
| $x_1 = x_0 + h$ | $y_1$ | | $\Delta^2 y_0$ | | |
| | | $\Delta y_1$ | | $\Delta^3 y_0$ | |
| $x_2 = x_0 + 2h$ | $y_2$ | | $\Delta^2 y_1$ | | $\Delta^4 y_0$ |
| | | $\Delta y_2$ | | $\Delta^3 y_1$ | |
| $x_3 = x_0 + 3h$ | $y_3$ | | $\Delta^2 y_2$ | | |
| | | $\Delta y_3$ | | | |
| $x_4 = x_0 + 4h$ | $y_4$ | | | | |

In a difference table, the values of $x$ are called arguments and the corresponding values of $y$ are called entries. The first entry $y_0$ is called the **leading term** and the differences $\Delta y_0, \Delta^2 y_0, \Delta^3 y_0, \cdots$ are called the **leading differences**.

**Example:** The following table gives a set of values of $x$ and the corresponding values of $y = f(x)$.

| $x$ | 10 | 15 | 20 | 25 | 30 | 35 |
|---|---|---|---|---|---|---|
| $y$ | 19.97 | 21.51 | 22.47 | 23.52 | 24.65 | 25.89 |

Form the forward difference table and write down the values of $\Delta f(10)$, $\Delta^2 f(10)$, $\Delta^3 f(15)$, $\Delta^4 f(15)$.

**Solution:** The forward difference table for the given values of $x$ and $y$ as shown below:

| $x$ | $y$ | $\Delta y$ | $\Delta^2 y$ | $\Delta^3 y$ | $\Delta^4 y$ | $\Delta^5 y$ |
|---|---|---|---|---|---|---|
| 10 | 19.97 | | | | | |
| | | 1.54 | | | | |
| 15 | 21.51 | | −0.58 | | | |
| | | 0.96 | | 0.67 | | |
| 20 | 22.47 | | 0.09 | | −0.68 | |
| | | 1.05 | | −0.01 | | 0.72 |
| 25 | 23.52 | | 0.08 | | 0.04 | |
| | | 1.13 | | 0.03 | | |
| 30 | 24.65 | | 0.11 | | | |
| | | 1.24 | | | | |
| 35 | 25.89 | | | | | |

From the table we find that

$$\Delta f(10) = \Delta y_0 = 1.54$$

$$\Delta^2 f(10) = \Delta^2 y_0 = -0.58$$

$$\Delta^3 f(15) = \Delta^3 y_1 = -0.0$$

$$\Delta^4 f(15) = \Delta^4 y_1 = 0.04$$

**Example:** By constructing a difference table and taking the second order differences as constant find the sixth term of the series $8, 12, 19, 29, 42, \cdots$.

**Solution:** Let $K$ be the sixth term of the series. The difference table is

| $x$ | $y$ | $\Delta y$ | $\Delta^2 y$ |
|---|---|---|---|
| 1 | 8 | | |
| | | 4 | |
| 2 | 12 | | 3 |
| | | 7 | |
| 3 | 19 | | 3 |
| | | 10 | |
| 4 | 29 | | 3 |
| | | 13 | |
| 5 | 42 | | $K - 55$ |
| | | $K - 42$ | |
| 6 | $K$ | | |

The second differences are constant.

$\therefore K - 55 = 3 \Rightarrow K = 58$.

The sixth term of the series is 58.

**EXERCISE**

1. Construct a difference table for $y = f(x) = x^3 + 2x + 1$ for $x = 1, 2, 3, 4, 5$.

2. Construct the Difference Table for the values of $x$ and $y$ given in the following Table:

| $x$ | 35 | 36 | 37 | 38 | 39 | 40 | 41 |
|---|---|---|---|---|---|---|---|
| $y$ | 14.298 | 14.144 | 13.986 | 13.825 | 13.661 | 13.495 | 13.328 |

Show that the third differences are constants. Also find $\Delta f(39)$ and $\Delta^2 f(38)$.

**Ans:** $-0.166, -0.002$.

3. If $y_0 = 3, y_1 = 12, y_2 = 81, y_3 = 200, y_4 = 100$, find $\Delta^4 y_0$ by constructing the difference table.

**Ans:** $-259$.

## Properties of the operator $\Delta$

**1.** If $c$ is a constant then $\Delta c = 0$.
**Proof:** Let $f(x) = c$. Hence $f(x+h) = c$, where $h$ is the interval of differencing. Hence $\Delta f(x) = f(x+h) - f(x) = c - c = 0$ or $\Delta c = 0$.

**2.** $\Delta$ is linear, i.e., $\Delta[af(x) + bg(x)] = a\Delta f(x) + b\Delta g(x)$ where $a, b$ are constants.
**Proof:**
$$\Delta[af(x) + bg(x)] = [af(x+h) + bg(x+h)] - [af(x) + bg(x)]$$
$$= a[f(x+h) - f(x)] + b[g(x+h) - g(x)]$$
$$= a\Delta f(x) + b\Delta g(x).$$

**3.** $\Delta^m \Delta^n f(x) = \Delta^{m+n} f(x)$, $m$ and $n$ being positive integers.
**Proof:**
$$\Delta^m \Delta^n f(x) = (\Delta\Delta \cdots m \text{ times})(\Delta\Delta \cdots n \text{ times}) f(x)$$
$$= [\Delta\Delta \cdots (m+n) \text{ times}] f(x)$$
$$= \Delta^{m+n} f(x).$$

**4.** $\Delta[f(x)g(x)] = f(x+h)\Delta[g(x)] + g(x)\Delta[f(x)]$.  (April-2022)
**Proof:**
$$\Delta[f(x)g(x)] = f(x+h)g(x+h) - f(x)g(x)$$
$$= [f(x+h)g(x+h) - f(x+h)g(x)] + [f(x+h)g(x) - f(x)g(x)]$$
$$= f(x+h)[g(x+h) - g(x)] + g(x)[f(x+h) - f(x)]$$
$$= f(x+h)\Delta[g(x)] + g(x)\Delta[f(x)].$$

**5.** $\Delta\left[\dfrac{f(x)}{g(x)}\right] = \dfrac{g(x)\Delta[f(x)] - f(x)\Delta[g(x)]}{g(x+h)g(x)}.$

**Proof:**

$$\Delta\left[\frac{f(x)}{g(x)}\right] = \frac{f(x+h)}{g(x+h)} - \frac{f(x)}{g(x)}$$

$$= \frac{f(x+h)g(x) - f(x)g(x+h)}{g(x)g(x+h)}$$

$$= \frac{f(x+h)g(x) - f(x)g(x) + f(x)g(x) - f(x)g(x+h)}{g(x)g(x+h)}$$

$$= \frac{g(x)\big[f(x+h) - f(x)\big] - f(x)\big[g(x+h) - g(x)\big]}{g(x)g(x+h)}$$

$$= \frac{g(x)\Delta\big[f(x)\big] - f(x)\Delta\big[g(x)\big]}{g(x+h)g(x)}$$

or

$$\Delta\left(\frac{f_i}{g_i}\right) = \frac{f_{i+1}}{g_{i+1}} - \frac{f_i}{g_i}$$

$$= \frac{f_{i+1}g_i - f_i g_{i+1}}{g_i g_{i+1}}$$

$$= \frac{f_{i+1}g_i - f_i g_i + f_i g_i - f_i g_{i+1}}{g_i g_{i+1}}$$

$$= \frac{g_i(f_{i+1} - f_i) - f_i(g_{i+1} - g_i)}{g_i g_{i+1}}$$

$$= \frac{g_i \Delta f_i - f_i \Delta g_i}{g_i g_{i+1}}$$

## 4.1.2 Backward Differences

Let $y_0, y_1, \cdots, y_r, \cdots$ be the values of a function $y = f(x)$, corresponding to the values $x_0, x_1, \cdots, x_r, \cdots$ of $x$, respectively which are equally spaced with step length $h$. We define

$$\boxed{\nabla f(x) = f(x) - f(x-h)}$$

Thus,
$$\nabla y_1 = y_1 - y_0$$
$$\nabla y_2 = y_2 - y_1$$
$$\cdots \quad \cdots$$
$$\nabla y_n = y_n - y_{n-1}.$$

$\nabla$ is called the **backward difference operator** and $\nabla y_1, \nabla y_2, \cdots, \nabla y_n$ are called the **first order backward differences** of the function $y = f(x)$.

The **second order differences** of the function are defined by

$$\nabla^2 y_2 = \nabla y_2 - \nabla y_1$$
$$\nabla^2 y_3 = \nabla y_3 - \nabla y_2$$
$$\cdots \quad \cdots$$
$$\nabla^2 y_n = \nabla y_n - \nabla y_{n-1}$$

In a similar manner higher order differences can be defined. In general the $n^{th}$ order backward differences are defined by the formula

$$\boxed{\nabla^n y_r = \nabla^{n-1} y_r - \nabla^{n-1} y_{r-1}.}$$

**Note:** 1. If $y = f(x)$ is a constant function, then $\nabla^n y_r = 0, \ \forall n$.

2. $\nabla \big[ f(x+h) \big] = \Delta f(x)$.

These differences of the function $y = f(x)$ can be systematically represented in the form a table called **backward difference table**.

**Backward Difference Table:**

| Argument $x$ | Entry $y = f(x)$ | $\nabla y$ | $\nabla^2 y$ | $\nabla^3 y$ | $\nabla^4 y$ |
|---|---|---|---|---|---|
| $x_0$ | $y_0$ | | | | |
| | | $\nabla y_1$ | | | |
| $x_1 = x_0 + h$ | $y_1$ | | $\nabla^2 y_2$ | | |
| | | $\nabla y_2$ | | $\nabla^3 y_3$ | |
| $x_2 = x_0 + 2h$ | $y_2$ | | $\nabla^2 y_3$ | | $\nabla^4 y_4$ |
| | | $\nabla y_3$ | | $\nabla^3 y_4$ | |
| $x_3 = x_0 + 3h$ | $y_3$ | | $\nabla^2 y_4$ | | |
| | | $\nabla y_4$ | | | |
| $x_4 = x_0 + 4h$ | $y_4$ | | | | |

**Example:** Construct the backward difference table for the following data

| $x$ | 0 | 1 | 2 | 3 | 4 |
|---|---|---|---|---|---|
| $f(x)$ | 4 | 12 | 13 | 76 | 156 |

and find the values of $\nabla f(3)$, $\nabla^2 f(2)$, $\nabla^4 f(4)$.

**Solution:** The backward difference table is

| $x$ | $y$ | $\nabla y$ | $\nabla^2 y$ | $\nabla^3 y$ | $\nabla^4 y$ |
|---|---|---|---|---|---|
| 0 | 4 | | | | |
| | | 8 | | | |
| 1 | 12 | | -7 | | |
| | | 1 | | 69 | |
| 2 | 13 | | 62 | | -114 |
| | | 63 | | -45 | |
| 3 | 76 | | 17 | | |
| | | 80 | | | |
| 4 | 156 | | | | |

From the table we find that
$$\nabla f(3) = 63$$
$$\nabla^2 f(2) = -7$$
$$\nabla^4 f(4) = -114$$

**EXERCISE**

1. Form the difference table for the following data:

| $x$ | 40 | 50 | 60 | 70 | 80 | 90 |
|---|---|---|---|---|---|---|
| $f(x)$ | 184 | 204 | 226 | 250 | 276 | 304 |

Also, write down the value of $\nabla y(80)$, $\nabla^2 y(70)$ and $\nabla^5 y(90)$.

**Ans:** 26, 2, 0.

2. Given $f(x) = x^3 - x + 1$, form the difference table taking $x = 0, 1, 2$ and $3$. Hence find $\nabla f(2)$ and $\nabla^2 f(3)$.

**Ans:** 6, 12

### 4.1.3 Central Differences

Let $y_0, y_1, \cdots, y_r, \cdots$ be the values of a function $y = f(x)$, corresponding to the values $x_0, x_1, \cdots, x_r, \cdots$ of $x$, respectively which are equally spaced with step length $h$. We define **Central difference operator** $\delta$ as

$$\boxed{\delta f(x) = f\left(x + \frac{h}{2}\right) - f\left(x - \frac{h}{2}\right)}$$

Thus if $f(x_i) = y_i$, then we have

$$\delta y_{1/2} = y_1 - y_0$$
$$\delta y_{3/2} = y_2 - y_1$$
$$\delta y_{5/2} = y_3 - y_2$$
$$\vdots$$
$$\delta y_{r-\frac{1}{2}} = y_r - y_{r-1}$$

are known as first central differences.

The higher order differences can be defined similar to forward and backward differences.

$$\delta^2 y_1 = \delta y_{3/2} - \delta y_{1/2}$$

$$\delta^2 y_2 = \delta y_{5/2} - \delta y_{3/2}$$

$$\vdots$$

$$\delta^2 y_r = \delta y_{r+1/2} - \delta y_{r-1/2}$$

**Note:** From forward and backward difference formulas, we have

$$\delta y_{1/2} = \Delta y_0 = \nabla y_1,$$

$$\delta y_{3/2} = \Delta y_1 = \nabla y_2,$$

$$\delta y_{5/2} = \Delta y_2 = \nabla y_3,$$

$$\vdots$$

These differences of the function $y = f(x)$ can be systematically represented in the form a table called **central difference table**.

**Central Difference Table:**

| Argument $x$ | Entry $y = f(x)$ | $\delta y$ | $\delta^2 y$ | $\delta^3 y$ | $\delta^4 y$ |
|---|---|---|---|---|---|
| $x_0$ | $y_0$ | | | | |
| | | $\delta y_{1/2}$ | | | |
| $x_1 = x_0 + h$ | $y_1$ | | $\delta^2 y_1$ | | |
| | | $\delta y_{3/2}$ | | $\delta^3 y_{3/2}$ | |
| $x_2 = x_0 + 2h$ | $y_2$ | | $\delta^2 y_2$ | | $\delta^4 y_4$ |
| | | $\delta y_{5/2}$ | | $\delta^3 y_{5/2}$ | |
| $x_3 = x_0 + 3h$ | $y_3$ | | $\delta^2 y_3$ | | |
| | | $\delta y_{7/2}$ | | | |
| $x_4 = x_0 + 4h$ | $y_4$ | | | | |

## 4.1.4  Differences of a Polynomial

If $f(x)$ is a polynomial of degree $n$ and the values of $x$ are equally spaced then $\Delta^n f(x)$ is a constant and all higher order differences are zero.

i.e., if $f(x) = a_0 x^n + a_1 x^{n-1} + \cdots + a_{n-1} x + a_n$, where $a_0, a_1, \cdots, a_{n-1}, a_n$ are constants and $a_0 \neq 0$, then

$$\Delta^n f(x) = a_0 n! h^n.$$

**Note:** Since $\Delta^n f(x)$ is a constant, it follows that $\Delta^{n+1} f(x) = 0, \Delta^{n+2} f(x) = 0, \cdots$.

**Example:** Evaluate $\Delta^3 \big[(1-x)(1-2x)(1-3x)\big]$.

**Solution:** If $f(x) = a_0 x^n + a_1 x^{n-1} + \cdots + a_{n-1} x + a_n$ is polynomial of degree $n$, then

$$\Delta^n f(x) = a_0 n! h^n \text{ and } \Delta^{n+1} f(x) = 0.$$

Here $f(x) = (1-x)(1-2x)(1-3x)$ which is a polynomial of degree 3.

$a_0 =$ coefficient of higher degree $= (-1)(-2)(-3) = -6$, $n = 3$, $h = 1$.

$\therefore \Delta^3 f(x) = (-6) 3! (1)^3 = -36.$

**Example:** Evaluate $\Delta^{10} \big[(1-ax)(1-bx^2)(1-cx^3)(1-dx^4)\big]$.

**Solution:** Here $f(x) = (1-ax)(1-bx^2)(1-cx^3)(1-dx^4)$ which is a polynomial of degree 10.

$a_0 =$ coefficient of higher degree $= (-a)(-b)(-c)(-d) = abcd$, $n = 4$, $h = 1$.

$$\Delta^{10}\big[(1-ax)(1-bx^2)(1-cx^3)(1-dx^4)\big] = \Delta^{10}\big[abcd\, x^{10} + ()x^9 + ()x^8 + \cdots + 1\big]$$
$$= abcd\,\Delta^{10}(x^{10}) \quad \big[\because \Delta^{10}(x^n) = 0 \text{ for } n < 10\big]$$
$$= abcd(10!)$$

**Example:** Show that $\Delta^{10}\big[(1-x)(1-2x^2)(1-3x^3)(1-4x^4)\big] = 24 \times 2^{10} \times 10!$ if the interval of differencing is 2.

**Solution:** Let $f(x) = (1-x)(1-2x^2)(1-3x^3)(1-4x^4)$.

If $f(x) = a_0 x^n + a_1 x^{n-1} + \cdots + a_{n-1} x + a_n$ is polynomial of degree $n$, then

$\Delta^n f(x) = a_0 n! h^n$ and $\Delta^{n+1} f(x) = 0$.

$\Delta^{10}\left[(1-x)(1-2x^2)(1-3x^3)(1-4x^4)\right]$

$= \Delta^{10}\left[(-1)(-2)(-3)(-4)x^{10} + \text{terms containing powers of } x \text{ less than } 10\right]$

$= \Delta^{10}(24x^{10}) + 0$

$= 24(10!)2^{10}$

### EXERCISE

1. Find $\Delta^3\left[(1-ax)(1-bx^2)\right]$  (Feb-2022), (Mar-2022)
2. Prove the following for $h = 1$.
   (i) $\Delta^3\left(ax^3 + bx^2 + cx + d\right) = 6a$
   (ii) $\Delta^4\left[(1-x)(1-2x)(1-3x)(1-4x)\right] = 576$
   (iii) $\Delta^6\left[3x^6 + 4x^5 + 2x^4 - x^3 + 2x^2 - 3x + 11\right] = 2160$
3. If $\Delta^3\left[(1+\alpha x)(1-2x)(1+4x)\right] = -114$ for $h = 1$, show that $\alpha = -3$.

## 4.2 Other Difference Operators

We have already introduced the operators $\Delta, \nabla$ and $\delta$. Besides these, there are the operators $E$ and $\mu$, which we define below:

**Mean Or Average Operator ($\mu$):**

The average operator $\mu$ is defined as

$$\mu f(x) = \frac{1}{2}\left[f\left(x + \frac{h}{2}\right) + f\left(x - \frac{h}{2}\right)\right]$$

or

$$\mu y_r = \frac{1}{2}\left(y_{r+\frac{1}{2}} + y_{r-\frac{1}{2}}\right).$$

If $f(x) = c$, a constant, we have

$$f\left(x + \frac{h}{2}\right) = f\left(x - \frac{h}{2}\right) = c, \text{ therefore } \mu f(x) = c.$$

**Shift Operator (E):** Let $y = f(x)$ be function of $x$ and $x, x+h, x+2h, x+3h, \cdots$, etc., be the consecutive values of $x$. Then shift (increment) operator $E$ is defined as

$$Ef(x) = f(x+h).$$

or

$$Ey_r = y_{r+1}.$$

Hence, shift operator shifts the function value $y_r$ to the next higher value $y_{r+1}$.

In general,

$$E^n f(x) = f(x+nh) \quad \text{or} \quad E^n y_r = y_{r+n}.$$

The inverse shift operator is defined as

$$E^{-1} f(x) = f(x-h).$$

or

$$E^{-1} y_r = y_{r-1}.$$

In general,

$$E^{-n} f(x) = f(x-nh) \quad \text{or} \quad E^{-n} y_r = y_{r-n}.$$

**Note:** In the difference calculus $E$ is regarded as the fundamental operator and $\Delta, \nabla, \delta, \mu$ can be expressed in terms of $E$.

**Differential operator $D$:**

The differential operator is usually denoted by $D$, where

$$Df(x) = \frac{d}{dx} f(x) = f'(x)$$

$$D^2 f(x) = \frac{d^2}{dx^2} f(x) = f''(x)$$

## 4.3 Relations Between the Operators

We shall now establish the following identities:

1. $\Delta \equiv E - 1$
2. $\nabla = 1 - E^{-1}$
3. $\delta \equiv E^{1/2} - E^{-1/2}$
4. $\mu \equiv \frac{1}{2}(E^{1/2} + E^{-1/2})$
5. $\Delta = E\nabla = \nabla E = \delta E^{1/2}$
6. $E = e^{hD}$

**Proof:** 1. We have $\Delta y_r = y_{r+1} - y_r$
$$= Ey_r - y_r$$
$$= (E - 1)y_r$$
$$\Rightarrow \Delta \equiv E - 1 \text{ or } E = 1 + \Delta.$$

2. We have $\nabla y_r = y_r - y_{r-1}$
$$= y_r - E^{-1}y_r \quad [\because E^n y_r = y_{r+n}]$$
$$= (1 - E^{-1})y_r$$
$$\Rightarrow \nabla \equiv 1 - E^{-1}$$

3. We have $\delta y_r = y_{r+\frac{1}{2}} - y_{r-\frac{1}{2}}$
$$= E^{1/2}y_r - E^{-1/2}y_r$$
$$= (E^{1/2} - E^{-1/2})y_r$$
$$\Rightarrow \delta \equiv E^{1/2} - E^{-1/2}$$

4. We have $\mu y_r = \frac{1}{2}\left(y_{r+\frac{1}{2}} + y_{r-\frac{1}{2}}\right)$
$$= \frac{1}{2}\left(E^{1/2}y_r + E^{-1/2}y_r\right)$$
$$= \frac{1}{2}\left(E^{1/2} + E^{-1/2}\right)y_r$$
$$\Rightarrow \mu \equiv \frac{1}{2}(E^{1/2} + E^{-1/2})$$

5. $\nabla E = (1 - E^{-1})E = E - 1 = \Delta$ and $E\nabla = E(1 - E^{-1}) = E - 1 = \Delta$.
Also $\delta E^{1/2} = \left(E^{1/2} - E^{-1/2}\right)E^{1/2} = E - 1 = \Delta$.

Hence $\Delta = E\nabla = \nabla E = \delta E^{1/2}$.

6. Using Taylor's series, we have

$$Ef(x) = f(x+h) = f(x) + hf'(x) + \frac{h^2}{2!}f''(x) + \cdots + \frac{h^n}{n!}f^{(n)}(x) + \cdots$$
$$= f(x) + hDf(x) + \frac{h^2}{2!}D^2f(x) + \cdots + \frac{h^n}{n!}D^nf(x) + \cdots$$
$$= \left[1 + hD + \frac{h^2}{2!}D^2 + \cdots + \frac{h^n}{n!}D^n + \cdots\right]f(x)$$
$$= e^{hD}f(x)$$
$$\Rightarrow E = e^{hD}.$$

### 4.3.1 Examples

**Example:** Prove the following

(i) $\Delta\nabla = \Delta - \nabla = \delta^2$ (April-2022)

(ii) $\frac{\Delta}{\nabla} - \frac{\nabla}{\Delta} = \Delta + \nabla$ (April-2022)

(iii) $(1 + \Delta)(1 - \nabla) = 1$

(iv) $\mu^2 \equiv 1 + \frac{1}{4}\delta^2$

(v) $1 + \mu^2\delta^2 = \left(1 + \frac{\delta^2}{2}\right)^2$ (Feb-2022)

(vi) $\mu\delta = \frac{1}{2}(\Delta + \nabla) = \frac{1}{2}\Delta E^{-1} + \frac{1}{2}\Delta$

(vii) $\Delta = \frac{1}{2}\delta^2 + \delta\sqrt{1 + \frac{1}{4}\delta^2}$

(viii) $hD = \log(1 + \Delta) = -\log(1 - \nabla) = \sinh^{-1}(\mu\delta)$. (Mar-2022)

**Proof:** (i). We know that $\Delta = E - 1$, $\nabla = 1 - E^{-1}$ and $\delta = E^{1/2} - E^{-1/2}$.

$\therefore \Delta\nabla = (E-1)(1-E^{-1}) = E - 1 - 1 + E^{-1} = (E-1) - (1-E^{-1}) = \Delta - \nabla$

and $\Delta - \nabla = (E-1) - (1-E^{-1}) = E + E^{-1} - 2 = \left(E^{1/2} - E^{-1/2}\right)^2 = \delta^2$.

Hence $\Delta\nabla = \Delta - \nabla = \delta^2$.

(ii). $\frac{\Delta}{\nabla} - \frac{\nabla}{\Delta} = \frac{\Delta^2 - \nabla^2}{\Delta\nabla} = \frac{(\Delta + \nabla)(\Delta - \nabla)}{\Delta - \nabla} = \Delta - \nabla$

(iii). $(1 + \Delta)(1 - \nabla) = EE^{-1} = 1$.

(iv). We have

$$\mu = \frac{1}{2}(E^{1/2} + E^{-1/2})$$
$$\Rightarrow \mu^2 = \frac{1}{4}(E^{1/2} + E^{-1/2})^2$$
$$\Rightarrow \mu^2 = \frac{1}{4}\left[(E^{1/2} - E^{-1/2})^2 + 4E^{1/2}E^{-1/2}\right]$$
$$\Rightarrow \mu^2 = \frac{1}{4}\left[\delta^2 + 4\right]$$
$$\Rightarrow \mu^2 = 1 + \frac{1}{4}\delta^2$$

(v). We know that $\mu = \frac{1}{2}(E^{1/2} + E^{-1/2})$ and $\delta = E^{1/2} - E^{-1/2}$.

$$\therefore 1 + \mu^2\delta^2 = 1 + \frac{1}{4}(E^{1/2} + E^{-1/2})^2(E^{1/2} - E^{-1/2})^2$$
$$= 1 + \frac{1}{4}(E - E^{-1})^2$$
$$= \frac{4 + (E - E^{-1})^2}{4}$$
$$= \frac{(E + E^{-1})^2}{2} \quad \ldots\ldots \quad (1)$$

and

$$\left(1 + \frac{\delta^2}{2}\right)^2 = \left[1 + \frac{(E^{1/2} - E^{-1/2})^2}{2}\right]^2$$
$$= \left[1 + \frac{E + E^{-1} - 2}{2}\right]^2$$
$$= \left[\frac{2 + E + E^{-1} - 2}{2}\right]^2$$
$$= \frac{(E + E^{-1})^2}{2} \quad \ldots\ldots \quad (2)$$

From (1) and (2), we get $1 + \mu^2\delta^2 = \left(1 + \frac{\delta^2}{2}\right)^2$.

(vi). $\mu\delta = \frac{1}{2}(E^{1/2} + E^{-1/2})(E^{1/2} - E^{-1/2})$

$= \frac{1}{2}(E - E^{-1})$

$= \frac{1}{2}[(E - 1) + (1 - E^{-1})]$

$= \frac{1}{2}(\Delta + \nabla)$

$= \frac{1}{2}\nabla + \frac{1}{2}\Delta$

$= \frac{1}{2}(1 - E^{-1}) + \frac{1}{2}\Delta$

$= \frac{1}{2}(E - 1)E^{-1} + \frac{1}{2}\Delta$

$= \frac{1}{2}\Delta E^{-1} + \frac{1}{2}\Delta.$

(vii). $\frac{1}{2}\delta^2 + \delta\sqrt{1 + \frac{\delta^2}{4}} = \frac{1}{2}\delta\left[\delta + 2\sqrt{1 + \frac{\delta^2}{4}}\right]$

$= \frac{1}{2}\delta\left[\delta + \sqrt{4 + \delta^2}\right]$

$= \frac{1}{2}\delta\left[\left(E^{1/2} - E^{-1/2}\right) + \sqrt{4 + \left(E^{1/2} - E^{-1/2}\right)^2}\right]$

$= \frac{1}{2}\delta\left[\left(E^{1/2} - E^{-1/2}\right) + \sqrt{\left(E^{1/2} + E^{-1/2}\right)^2}\right]$

$= \frac{1}{2}\delta\left[\left(E^{1/2} - E^{-1/2}\right) + \left(E^{1/2} + E^{-1/2}\right)\right]$

$= \frac{1}{2}\delta\left[2E^{1/2}\right]$

$= \delta E^{1/2}$

$= \left(E^{1/2} - E^{-1/2}\right)E^{1/2}$

$= E - 1$

$= \Delta.$

(viii). We know that $e^{hD} = E = 1 + \Delta$.

Taking log on both sides, we have

$$hD = \log(1+\Delta) \quad \ldots\ldots \quad (1)$$

Also $\nabla = 1 - E^{-1} \Rightarrow E^{-1} = 1 - \nabla$

i.e, $e^{-hD} = E^{-1} = 1 - \nabla$

$$\Rightarrow -hD = \log(1-\nabla)$$

$$\Rightarrow hD = -\log(1-\nabla) \quad \ldots\ldots \quad (2)$$

We know that $\mu = \frac{1}{2}\left(E^{1/2} + E^{-1/2}\right)$ and $\delta = E^{1/2} - E^{-1/2}$.

$$\therefore \mu\delta = \frac{1}{2}\left(E^{1/2} + E^{-1/2}\right)\left(E^{1/2} - E^{-1/2}\right)$$

$$= \frac{1}{2}\left(E - E^{-1}\right) = \frac{1}{2}\left(e^{hD} - e^{-hD}\right) = \sinh(hD)$$

$$\Rightarrow hD = \sinh^{-1}(\mu\delta)$$

Hence $hD = \log(1+\Delta) = -\log(1-\nabla) = \sinh^{-1}(\mu\delta)$.

**Example:** Evaluate (a) $\Delta \cos x$ (b) $\Delta \tan^{-1} x$ (c) $\Delta \log f(x)$ (d) $\Delta \sin(ax+b)$ (e) $\Delta(e^{ax} \log bx)$ (f) $\Delta(^nC_{r+1})$ (g) $\Delta^2(3e^x)$ (h) $\Delta^2 ab^{cx}$ if $h = 1$.

**Solution:** We have $\Delta f(x) = f(x+h) - f(x)$.

(a) $\Delta \cos x = \cos(x+h) - \cos x$

$$= -2\sin\left(\frac{x+h+x}{2}\right) \sin\left(\frac{x+h-x}{2}\right)$$

$$= -2\sin\left(\frac{h}{2}\right) \sin\left(x + \frac{h}{2}\right)$$

(b) $\Delta \tan^{-1} x = \tan^{-1}(x+h) - \tan^{-1} x$

$$= \tan^{-1}\left[\frac{(x+h) - x}{1 + (x+h)x}\right]$$

$$= \tan^{-1}\left[\frac{h}{1 + x(x+h)}\right]$$

(c) $\Delta \log f(x) = \log f(x+h) - \log f(x)$

$$= \log \left[\frac{f(x+h)}{f(x)}\right]$$

$$= \log \left[\frac{f(x) + \Delta f(x)}{f(x)}\right] \quad \left[\because \Delta f(x) = f(x+h) - f(x)\right]$$

$$= \log \left[1 + \frac{\Delta f(x)}{f(x)}\right].$$

(d) $\Delta \sin(ax+b) = \sin(a(x+h)+b) - \sin(ax+b)$

$$= \sin(ax+b+ah) - \sin(ax+b)$$

$$= 2\cos\left(ax+b+\frac{ah}{2}\right)\sin\left(\frac{ah}{2}\right)$$

(e) Let $f(x) = e^{ax}$ and $g(x) = \log bx$.

Hence

$$\Delta f(x) = e^{a(x+h)} - e^{ax} = e^{ax}\left(e^{ah} - 1\right)$$

$$\Delta g(x) = \log b(x+h) - \log bx = \log b\left(1 + \frac{h}{x}\right)$$

Also

$$\Delta\bigl(f(x) \cdot g(x)\bigr) = f(x+h)\Delta g(x) + g(x) \cdot \Delta f(x)$$

$$= e^{a(x+h)} \log b(1+h/x) + \log bx \cdot e^{ax}\left(e^{ah}-1\right)$$

$$= e^{ax}\left[e^{ah} \log b(1+h/x) + \left(e^{ah}-1\right)\log bx\right].$$

(f) $\Delta(^nC_{r+1}) = {}^{n+1}C_{r+1} - {}^nC_{r+1}$

$$= \frac{(n+1)!}{(r+1)!(n-r)!} - \frac{n!}{(r+1)!(n-r-1)!}$$

$$= \frac{n!}{(r+1)!(n-r-1)!}\left[\frac{n+1}{n-r} - 1\right]$$

$$= \frac{n!}{(r+1)!(n-r-1)!}\left[\frac{r+1}{n-r}\right]$$

$$= \frac{n!}{r!(n-r)!} = {}^nC_r.$$

(g) $\Delta^2(3e^x) = 3\Delta^2 e^x = 3\Delta[\Delta e^x]$
$= 3\Delta\left[e^{x+h} - e^x\right] = 3\Delta\left[e^x(e^h - 1)\right]$
$= 3(e^h - 1)[\Delta e^x]$
$= 3(e^h - 1)[(e^h - 1)e^x]$
$= 3(e^h - 1)^2 e^x.$

(h) $\Delta(ab^{cx}) = a\Delta(b^{cx}) = a\left[b^{c(x+1)} - b^{cx}\right] = ab^{cx}(b^c - 1)$
$\Delta^2(ab^{cx}) = \Delta[\Delta(ab^{cx})] = a(b^c - 1)\Delta(b^{cx}) = a(b^c - 1)b^{cx}(b^c - 1) = a(b^c - 1)^2 b^{cx}$

**Example:** Prove that $\Delta^3 y_2 = \nabla^3 y_5$.

**Solution:**

$\Delta^3 y_2 = (E-1)^3 y_2 = (E^3 - 3E^2 + 3E - 1)y_2 = y_5 - 3y_4 + 3y_3 - y_2$ ... (1)

$\nabla^3 y_5 = (1 - E^{-1})^3 y_5 = (1 - 3E^{-1} + 3E^{-2} - E^{-3})y_5 = y_5 - 3y_4 + 3y_3 - y_2$ ... (2)

From (1) and (2), we have $\Delta^3 y_2 = \nabla^3 y_5$.

**Example:** Evaluate $\Delta^2(\cos 2x)$.

**Solution:**

$\Delta^2(\cos 2x) = (E-1)^2 \cos 2x$
$= (E^2 - 2E + 1)\cos 2x$
$= E^2 \cos 2x - 2E \cos 2x + \cos 2x$
$= \cos 2(x + 2h) - 2\cos 2(x + h) + \cos 2x$
$= \cos 2(x + 2h) - \cos 2(x + h) - \cos 2(x + h) + \cos 2x$
$= -2\sin(2x + 3h)\sin h + 2\sin(2x + h)\sin h$
$= -2\sin h\left[\sin(2x + 3h) - \sin(2x + h)\right]$
$= -2\sin h\left[2\cos 2(x + h)\sin h\right]$
$= -4\sin^2 h \cos 2(x + h)$

**Example:** Prove that $\Delta^n e^{ax+b} = \left(e^{ah} - 1\right)^n e^{ax+b}$. (Dec-2020)

**Solution:** We know that $\Delta f(x) = f(x+h) - f(x)$.

We prove the result by induction. For $n = 1$,

$$\Delta e^{ax+b} = e^{a(x+h)+b} - e^{ax+b}$$
$$= e^{ax+b}e^{ah} - e^{ax+b}$$
$$= \left(e^{ah} - 1\right)e^{ax+b}$$

It is true.

Assume that the result holds true for $n = k$ so that $\Delta^k e^{ax+b} = \left(e^{ah} - 1\right)^k e^{ax+b}$.

$$\Delta^{k+1} e^{ax+b} = \Delta\left[\Delta^k e^{ax+b}\right]$$
$$= \Delta\left[\left(e^{ah} - 1\right)^k e^{ax+b}\right]$$
$$= \left(e^{ah} - 1\right)^k \left[\Delta e^{ax+b}\right]$$
$$= \left(e^{ah} - 1\right)^k \left(e^{ah} - 1\right) e^{ax+b}$$
$$= \left(e^{ah} - 1\right)^{k+1} e^{ax+b}.$$

By the principle of mathematical induction, the result is true for all $n \in \mathbb{N}$.

$$\therefore \Delta^n e^{ax+b} = \left(e^{ah} - 1\right)^n e^{ax+b}.$$

**Example:** Prove the following for $h = 1$

$$\Delta^n \left(\frac{1}{x}\right) = \frac{(-1)^n n!}{x(x+1)(x+2)\cdots(x+n)}$$

**Solution:** We know that $\Delta f(x) = f(x+h) - f(x)$.

$$\Delta\left(\frac{1}{x}\right) = \frac{1}{x+1} - \frac{1}{x}$$
$$= \frac{-1}{x(x+1)} = \frac{(-1)^1 1!}{x(x+1)}$$

$$\Delta^2\left(\frac{1}{x}\right) = \Delta\left[\Delta\left(\frac{1}{x}\right)\right]$$
$$= \Delta\left[\frac{-1}{x(x+1)}\right]$$
$$= \frac{-1}{(x+1)(x+2)} + \frac{1}{x(x+1)}$$
$$= \frac{1}{x+1}\left(\frac{-1}{x+2} + \frac{1}{x}\right)$$
$$= \frac{2}{x(x+1)(x+2)} = \frac{(-1)^2 2!}{x(x+1)(x+2)}$$

Similarly,

$$\Delta^3\left(\frac{1}{x}\right) = \Delta\left[\Delta^2\left(\frac{1}{x}\right)\right]$$
$$= \Delta\left[\frac{2}{x(x+1)(x+2)}\right]$$
$$= \frac{2}{(x+1)(x+2)(x+3)} - \frac{2}{x(x+1)(x+2)}$$
$$= \frac{2}{(x+1)(x+2)}\left(\frac{1}{x+3} - \frac{1}{x}\right)$$
$$= \frac{(2)(-3)}{x(x+1)(x+2)(x+3)} = \frac{(-1)^3 3!}{x(x+1)(x+2)(x+3)}$$

Proceeding as above and applying the principle of mathematical induction, we get

$$\Delta^n\left(\frac{1}{x}\right) = \frac{(-1)^n n!}{x(x+1)(x+2)\cdots(x+n)}.$$

**Example:** If $h$ is the interval of differncing prove that **(Dec-2020), (April-2022)**

$$\Delta^n \sin(ax+b) = \left(2\sin\frac{ah}{2}\right)^n \sin\left(ax+b+\frac{nah+n\pi}{2}\right).$$

**Solution:**

$$\Delta \sin(ax+b) = \sin\left[(a(x+h)+b\right] - \sin(ax+b)$$

$$= 2\cos\left[\frac{ax+ah+b+ax+b}{2}\right]\sin\left[\frac{ax+ah+b-ax-b}{2}\right]$$

$$= 2\cos\left(ax+b+\frac{ah}{2}\right)\sin\left(\frac{ah}{2}\right)$$

$$= 2\sin\left(\frac{ah}{2}\right)\sin\left(\frac{\pi}{2}+\left(ax+b+\frac{ah}{2}\right)\right)$$

$$= 2\sin\left(\frac{ah}{2}\right)\sin\left(ax+b+\frac{ah+\pi}{2}\right)$$

$$\Delta^2 \sin(ax+b) = 2\sin\left(\frac{ah}{2}\right)\sin\left(a(x+h)+b+\frac{ah+\pi}{2}\right) - 2\sin\left(\frac{ah}{2}\right)\sin\left(ax+b+\frac{ah+\pi}{2}\right)$$

$$= \left(2\sin\frac{ah}{2}\right)\left[2\cos\left(ax+b+\frac{2ah+\pi}{2}\right)\sin\left(\frac{ah}{2}\right)\right]$$

$$= \left(2\sin\frac{ah}{2}\right)^2\sin\left(\frac{\pi}{2}+ax+b+\frac{2ah+\pi}{2}\right)$$

$$= \left(2\sin\frac{ah}{2}\right)^2\sin\left(ax+b+\frac{2ah+2\pi}{2}\right)$$

Proceeding as above and applying the principle of mathematical induction, we get

$$\Delta^n \sin(ax+b) = \left(2\sin\frac{ah}{2}\right)^n \sin\left(ax+b+\frac{nah+n\pi}{2}\right).$$

**Example:** Evaluate $\Delta\left(\dfrac{x}{\sin x}\right)$.

**Solution:** We know that $\Delta\left[\dfrac{f(x)}{g(x)}\right] = \dfrac{g(x)\Delta[f(x)] - f(x)\Delta[g(x)]}{g(x+h)g(x)}.$

$\Delta(x) = x+h-x = h$

$\Delta(\sin x) = \sin(x+h) - \sin x = 2\sin\left(\dfrac{2x+h}{2}\right)\cos\dfrac{h}{2}$

$$\therefore \Delta\left(\frac{x}{\sin x}\right) = \frac{\sin \Delta(x) - x\Delta \sin x}{\sin x \sin(x+h)}$$

$$= \frac{h\sin x - 2x\sin\left(\dfrac{2x+h}{2}\right)\cos\dfrac{h}{2}}{\sin x \sin(x+h)}$$

**Example:** Evaluate $\Delta\left(\dfrac{x^2}{\cos 2x}\right)$.

**Solution:**

$$\Delta\left(\dfrac{x^2}{\cos 2x}\right) = \dfrac{(x+h)^2}{\cos 2(x+h)} - \dfrac{x^2}{\cos 2x}$$

$$= \dfrac{(x+h)^2 \cos 2x - x^2 \cos 2(x+h)}{\cos 2(x+h) \cos 2x}$$

$$= \dfrac{\left[(x+h)^2 - x^2\right]\cos 2x + x^2\left[\cos 2x - \cos 2(x+h)\right]}{\cos 2(x+h)\cos 2x}$$

$$= \dfrac{(2hx + h^2)\cos 2x + 2x^2 \sin h \sin(2x+h)}{\cos 2(x+h)\cos 2x}$$

**Example:** Evaluate $\left(\dfrac{\Delta^2}{E}\right)e^x$.

**Solution:** We have

$$\left(\dfrac{\Delta^2}{E}\right)e^x = \left(\Delta^2 E^{-1}\right)e^x$$

$$= (E-1)^2 E^{-1} e^x$$

$$= (E^2 - 2E + 1)E^{-1}e^x$$

$$= (E - 2 + E^{-1})e^x$$

$$= Ee^x - 2e^x + E^{-1}e^x$$

$$= e^{x+h} - 2e^x + e^{x-h}$$

**Example:** Evaluate $\left(\dfrac{\Delta^2}{E}\right)x^3$.

**Solution:** We have
$$\left(\dfrac{\Delta^2}{E}\right)x^3 = (\Delta^2 E^{-1})x^3$$
$$= (E-1)^2 E^{-1} x^3$$
$$= (E^2 - 2E + 1)^2 E^{-1} x^3$$
$$= (E - 2 + E^{-1})x^3$$
$$= Ex^3 - 2x^3 + E^{-1}x^3$$
$$= (x+h)^3 - 2x^3 + (x-h)^3$$
$$= x^3 + 3x^2 h + 3xh^2 + h^3 - 2x^3 + x^3 - 3x^2 h + 3xh^2 - h^3$$
$$= 6xh^2.$$

**Note:** If $h = 1$, then $\left(\dfrac{\Delta^2}{E}\right)x^3 = 6x$.

**Example:** Find: (i) $\left(\dfrac{\Delta^2}{E}\right)f(x)$ (ii) $\dfrac{\Delta^2 f(x)}{Ef(x)}$. Hence deduce that
$$e^x = \dfrac{\Delta^2}{E}e^x \cdot \dfrac{Ee^x}{\Delta^2 e^x}.$$

**Solution:** We have

(i) $\left(\dfrac{\Delta^2}{E}\right)f(x) = \dfrac{(E-1)^2}{E}f(x)$
$$= (E-1)^2 E^{-1} f(x)$$
$$= (E^2 - 2E + 1)E^{-1} f(x)$$
$$= (E - 2 + E^{-1})f(x)$$
$$= f(x+h) - 2f(x) + f(x-h).$$

(ii) $\dfrac{\Delta^2 f(x)}{Ef(x)} = \dfrac{(E-1)^2 f(x)}{Ef(x)}$
$$= \dfrac{(E^2 - 2E + 1)f(x)}{Ef(x)}$$
$$= \dfrac{f(x+2h) - 2f(x+h) + f(x)}{f(x+h)}.$$

Next, taking $f(x) = e^x$ in the above results, we get

$$\frac{\Delta^2}{E}e^x = e^{x+h} - 2e^x + e^{x-h}$$

and $$\frac{\Delta^2 e^x}{Ee^x} = \frac{e^{x+2h} - 2e^{x+h} + e^x}{e^{x+h}}$$

These give

$$\frac{\Delta^2}{E}e^x \cdot \frac{Ee^x}{\Delta^2 e^x} = \frac{(e^{x+h} - 2e^x + e^{x-h})e^{x+h}}{e^{x+2h} - 2e^{x+h} + e^x}$$

$$= \frac{e^x(e^{x+2h} - 2e^{x+h} + e^x)}{e^{x+2h} - 2e^{x+h} + e^x}$$

$$= e^x.$$

**Example:** Prove the following:
(i) $f(2) = f(0) + 2\Delta f(0) + \Delta^2 f(0)$  (ii) $f(4) = f(3) + \Delta f(2) + \Delta^2 f(1) + \Delta^3 f(1)$.

**Solution:**

(i) $\Delta f(0) = f(1) - f(0)$

$\Delta^2 f(0) = (E-1)^2 f(0) = (E^2 - 2E + 1)f(0) = f(2) - 2f(1) + f(0)$

$\therefore f(0) + 2\Delta f(0) + \Delta^2 f(0) = f(0) + 2[f(1) - f(0)] + [f(2) - 2f(1) + f(0)]$

$= f(2).$

(ii) $\Delta f(2) = f(3) - f(2)$

$\Delta^2 f(1) = (E-1)^2 f(1) = (E^2 - 2E + 1)f(1)$

$= f(3) - 2f(2) + f(1)$

$\Delta^3 f(1) = (E-1)^3 f(1) = (E^3 - 3E^2 + 3E - 1)f(1)$

$= f(4) - 3f(3) + 3f(2) - f(1)$

$\therefore f(3) + \Delta f(2) + \Delta^2 f(1) + \Delta^3 f(1) = f(3) + f(3) - f(2) + f(3) - 2f(2) + f(1)$
$\qquad\qquad + f(4) - 3f(3) + 3f(2) - f(1)$

$= f(4).$

**Example:** Find the first term of the series whose second and subsequent terms are $8, 3, 0, -1, 0$.

**Solution:** Given $f(2) = 8, f(3) = 3, f(4) = 0, f(5) = -1, f(6) = 0$, we are to find $f(1)$.

We construct the difference table with the given values:

| $x$ | $f(x)$ | $\Delta f(x)$ | $\Delta^2 f(x)$ | $\Delta^3 f(x)$ | $\Delta^4 f(x)$ |
|---|---|---|---|---|---|
| 2 | 8 | | | | |
| | | $-5$ | | | |
| 3 | 3 | | 2 | | |
| | | $-3$ | | 0 | |
| 4 | 0 | | 2 | | 0 |
| | | $-1$ | | 0 | |
| 5 | $-1$ | | 2 | | |
| | | 1 | | | |
| 6 | 0 | | | | |

We have $\Delta^3 f(x) = \Delta^4 f(x) = \cdots = 0$.

Using the Shift operator

$$f(1) = E^{-1} f(2) = (1 + \Delta)^{-1} f(2)$$
$$= (1 - \Delta + \Delta^2 - \Delta^3 + \cdots) f(2)$$
$$= f(2) - \Delta f(2) + \Delta^2 f(2) - \Delta^3 f(2) + \cdots$$
$$= 8 - (-5) + 2 = 15$$

$\therefore f(1) = 15.$

**Example:** Given $u_0 = 1, u_1 = 11, u_2 = 21, u_3 = 28$ and $u_4 = 29$ find $\Delta^4 u_0$ without forming difference table.

**Solution:** We have

$$\Delta^4 u_0 = (E-1)^4 u_0 = (E^4 - {}^4C_1 E^3 + {}^4C_2 E^2 - {}^4C_3 E + 1) u_0$$

$$= (E^4 u_0 - 4E^3 u_0 + 6E^2 u_0 - 4E u_0 + u_0)$$

$$= u_4 - 4u_3 + 6u_2 - 4u_1 + u_0$$

$$= 29 - 4(28) + 6(21) - 4(11) + 1$$

$$= 29 - 112 + 126 - 44 + 1$$

$$= 0.$$

**Example:** Given $u_0 = 3, u_1 = 12, u_2 = 81, u_3 = 200, u_4 = 100$, and $u_5 = 8$, find $\Delta^5 u_0$.

**Solution:**

$$\Delta^5 u_0 = (E-1)^5 u_0$$

$$= (E^5 - 5E^4 + 10E^3 - 10E^2 + 5E - 1) u_0$$

$$= u_5 - 5u_4 + 10u_3 - 10u_2 + 5u_1 - u_0$$

$$= 8 - 500 + 2000 - 810 + 60 - 3$$

$$= 755.$$

**Example:** Use the method of separation of symbols to prove the following identities:

(a). $u_0 - u_1 + u_2 - \cdots = \dfrac{1}{2} u_0 - \dfrac{1}{4} \Delta u_0 + \dfrac{1}{8} \Delta^2 u_0 - \cdots$

(b). $e^x \left( u_0 + x \Delta u_0 + \dfrac{x^2}{2!} \Delta^2 u_0 + \cdots \right) = u_0 + u_1 x + u_2 \dfrac{x^2}{2!} + \cdots$

(c). $(u_1 - u_0) - x(u_2 - u_1) + x^2(u_3 - u_2) - \cdots = \dfrac{\Delta u_0}{1+x} - x \dfrac{\Delta^2 u_0}{(1+x)^2} + x^2 \dfrac{\Delta^3 u_0}{(1+x)^3}$

**Solution:**

(a) $u_0 - u_1 + u_2 - \cdots = u_0 - Eu_0 + E^2 u_0 - \cdots$

$$= (1 - E + E^2 - \cdots) u_0$$
$$= (1 + E)^{-1} u_0$$
$$= (1 + 1 + \Delta)^{-1} u_0$$
$$= (2 + \Delta)^{-1} u_0$$
$$= 2^{-1}\left(1 + \frac{\Delta}{2}\right)^{-1} u_0$$
$$= \frac{1}{2}\left(1 - \frac{\Delta}{2} + \frac{\Delta^2}{2^2} - \frac{\Delta^3}{2^3} + \cdots\right) u_0$$
$$= \frac{1}{2}\left(u_0 - \frac{\Delta u_0}{2} + \frac{\Delta^2 u_0}{2^2} - \frac{\Delta^3 u_0}{2^3} + \cdots\right)$$

$$\therefore u_0 - u_1 + u_2 - \cdots = \frac{1}{2} u_0 - \frac{1}{4} \Delta u_0 + \frac{1}{8} \Delta^2 u_0 - \cdots$$

(b) $e^x \left( u_0 + x \Delta u_0 + \frac{x^2}{2!} \Delta^2 u_0 + \cdots \right) = e^x \left( 1 + x\Delta + \frac{x^2}{2!} \Delta^2 + \cdots \right) u_0$

$$= e^x \left( e^{x\Delta} \right) u_0$$
$$= e^x e^{x(E-1)} u_0$$
$$= e^{xE} u_0$$
$$= \left( 1 + xE + \frac{x^2}{2!} E^2 + \cdots \right) u_0$$
$$= u_0 + xE u_0 + \frac{x^2}{2!} E^2 u_0 + \cdots$$
$$= u_0 + x u_1 + \frac{x^2}{2!} u_2 + \cdots$$

(c) L. H. S $= (u_1 - u_0) - x(u_2 - u_1) + x^2(u_3 - u_2) - \cdots$

$$= \Delta u_0 - x \Delta u_1 + x^2 \Delta u_2 - \cdots$$
$$= \Delta u_0 - x \Delta E u_0 + x^2 \Delta E^2 u_0 - \cdots$$
$$= (1 - xE + x^2 E^2 - \cdots) \Delta u_0$$
$$= (1 + xE)^{-1} \Delta u_0$$

$$\text{R. H. S} = \frac{\Delta u_0}{1+x} - x\frac{\Delta^2 u_0}{(1+x)^2} + x^2\frac{\Delta^3 u_0}{(1+x)^3} - \cdots$$

$$= \frac{1}{x}\left[\frac{x\Delta}{1+x} - \frac{x^2\Delta^2}{(1+x)^2} + \frac{x^3\Delta^3}{(1+x)^3} - \cdots\right]u_0$$

$$= \frac{1}{x}\cdot\frac{x\Delta}{1+x}\left[1 - \frac{x\Delta}{1+x} + \frac{x^2\Delta^2}{(1+x)^2} - \cdots\right]u_0$$

$$= \frac{\Delta}{1+x}\left[1 + \frac{x\Delta}{1+x}\right]^{-1}u_0$$

$$= \frac{\Delta}{1+x}\left[\frac{1+x+x\Delta}{1+x}\right]^{-1}u_0$$

$$= \left(\frac{\Delta}{1+x+x\Delta}\right)u_0$$

$$= \left(\frac{\Delta}{1+x(1+\Delta)}\right)u_0$$

$$= \left(\frac{\Delta}{1+xE}\right)u_0$$

$$= (1+xE)^{-1}\Delta u_0 = \text{L. H. S}$$

**Example:** Prove that

$$u_1 x + u_2 x^2 + u_3 x^3 + \cdots = \left(\frac{x}{1-x}\right)u_1 + \left(\frac{x}{1-x}\right)^2 \Delta u_1 + \left(\frac{x}{1-x}\right)^3 \Delta^2 u_1 + \cdots$$

**Solution:**

$$\text{L.H.S} = u_1 x + u_2 x^2 + u_3 x^3 + \cdots$$

$$= x u_1 + x^2 E u_1 + x^3 E^2 u_1 + \cdots$$

$$= x\big(1 + (xE) + (xE)^2 + \cdots\big) u_1$$

$$= x\big(1 - xE\big)^{-1} u_1$$

$$= \frac{x}{1 - xE} u_1$$

$$= \frac{x}{1 - x - x\Delta} u_1 \quad \big[\because E = 1 + \Delta\big]$$

$$= \left[\frac{\frac{x}{1-x}}{1 - \frac{x}{1-x}\Delta}\right] u_1$$

$$= \frac{x}{1-x}\left(1 - \frac{x}{1-x}\Delta\right)^{-1} u_1$$

$$= \frac{x}{1-x}\left(1 + \frac{x}{1-x}\Delta + \left(\frac{x}{1-x}\right)^2 \Delta^2 + \cdots\right) u_1$$

$$= \frac{x}{1-x} u_1 + \left(\frac{x}{1-x}\right)^2 \Delta u_1 + \left(\frac{x}{1-x}\right)^3 \Delta^2 u_1 + \cdots$$

**Example:** If the interval of differencing is unity, prove that $\Delta\left(\dfrac{2^x}{x!}\right) = \dfrac{2^x(1-x)}{(x+1)!}$.

**Solution:** If the interval of differencing is unity (that is $h = 1$), we have

$$\Delta f(x) = f(x+1) - f(x).$$

Let $f(x) = \dfrac{2^x}{x!}$.

$$\therefore \Delta f(x) = \frac{2^{x+1}}{(x+1)!} - \frac{2^x}{x!}$$

$$= \frac{2^x \cdot 2}{(x+1)x!} - \frac{2^x}{x!}$$

$$= \frac{2^x}{x!}\left(\frac{2}{x+1} - 1\right)$$

$$= \frac{2^x}{x!} \cdot \frac{1-x}{x+1}$$

$$= \frac{2^x(1-x)}{(x+1)!}.$$

**Example:** If the interval of differencing is unity, prove that
$$\Delta[x(x+1)(x+2)(x+3)] = 4(x+1)(x+2)(x+3). \quad \text{(Nov-2017)}$$
**Solution:** We have $\Delta f(x) = f(x+h) - f(x)$.
Here $h = 1$. Let $f(x) = x(x+1)(x+2)(x+3)$.

$$\Delta f(x) = (x+1)(x+2)(x+3)(x+4) - x(x+1)(x+2)(x+3)$$
$$= (x+1)(x+2)(x+3)[x+4-x]$$
$$= 4(x+1)(x+2)(x+3)$$

**Example:** If the interval of differencing is unity, prove that $\Delta\left(\dfrac{1}{f(x)}\right) = -\dfrac{\Delta f(x)}{f(x)f(x+1)}$.

**Solution:** We have $\Delta f(x) = f(x+h) - f(x)$.
Here $h = 1$.

$$\therefore \Delta\left[\dfrac{1}{f(x)}\right] = \dfrac{1}{f(x+1)} - \dfrac{1}{f(x)}$$
$$= \dfrac{f(x) - f(x+1)}{f(x)f(x+1)}$$
$$= -\dfrac{f(x+1) - f(x)}{f(x)f(x+1)}$$
$$= -\dfrac{\Delta f(x)}{f(x)f(x+1)}$$

**Example:** If the interval of differencing is unity, prove that

(Dec-2020), (Aug-2021), (April-2022)

$$\Delta \tan^{-1}\left(\dfrac{n-1}{n}\right) = \tan^{-1}\left(\dfrac{1}{2n^2}\right).$$

**Solution:** We have $\Delta f(x) = f(x+h) - f(x)$.
Let $f(n) = \tan^{-1}\left(\dfrac{n-1}{n}\right)$. Here $h = 1$.

Thus, we have

$$\Delta \tan^{-1}\left(\frac{n-1}{n}\right) = \Delta \tan^{-1}\left(1 - \frac{1}{n}\right)$$

$$= \tan^{-1}\left(1 - \frac{1}{n+1}\right) - \tan^{-1}\left(1 - \frac{1}{n}\right)$$

$$= \tan^{-1}\left[\frac{\left(1 - \frac{1}{n+1}\right) - \left(1 - \frac{1}{n}\right)}{1 + \left(1 - \frac{1}{n+1}\right)\left(1 - \frac{1}{n}\right)}\right]$$

$$= \tan^{-1}\left[\frac{\frac{n}{n+1} - \frac{n-1}{n}}{1 + \left(\frac{n}{n+1}\right)\left(\frac{n-1}{n}\right)}\right]$$

$$= \tan^{-1}\left[\frac{\frac{1}{n(n+1)}}{\frac{2n}{n+1}}\right]$$

$$= \tan^{-1}\left(\frac{1}{2n^2}\right).$$

**Example:** Find the second difference of the polynomial $x^4 - 12x^3 + 42x^2 - 30x + 9$ with interval of differencing $h = 2$. **(April-2022)**

**Solution:** We know that $\Delta f(x) = f(x+h) - f(x)$
Let $f(x) = x^4 - 12x^3 + 42x^2 - 30x + 9$. Here $h = 2$.

$$\Delta f(x) = f(x+2) - f(x)$$
$$= \left[(x+2)^4 - 12(x+2)^3 + 42(x+2)^2 - 30(x+2) + 9\right]$$
$$\quad - \left[x^4 - 12x^3 + 42x^2 - 30x + 9\right]$$
$$= \left[x^4 + 4x^3(2) + 6x^2(4) + 4x(8) + 16\right] - 12\left[x^3 + 3x^2(2) + 3x(4) + 8\right]$$
$$\quad + 42\left[x^2 + 4x + 4\right] - 30\left[x + 2\right] + 9 - x^4 + 12x^3 - 42x^2 + 30x - 9$$
$$= 0x^4 + 8x^3 - 48x^2 + 56x + 28$$
$$= 8x^3 - 48x^2 + 56x + 28.$$

$$\Delta^2 f(x) = \Delta(\Delta f(x))$$
$$= \left[8(x+2)^2 - 48(x+2)^2 + 56(x+2) + 28\right] - \left[8x^3 - 48x^2 + 56x + 28\right]$$
$$= 8\left[x^3 + 3x^2(2) + 3x(4) + 8\right] - 48\left[x^2 + 4x + 4\right] + 56\left[x+2\right] + 28$$
$$- 8x^3 + 48x^2 - 56x - 28$$
$$= 0x^3 + 48x^2 - 96x - 16$$
$$= 48x^2 - 96x - 16$$

**Example:** Evaluate the following taking the interval of differencing being unity.

(i) $\Delta^2 \left[\dfrac{1}{x^2 + 5x + 6}\right]$  (ii) $\Delta^2 \left[\dfrac{5x + 12}{x^2 + 5x + 6}\right]$ (Dec-2021)

**Solution:** We have $\Delta f(x) = f(x+h) - f(x)$.

Here $h = 1$.
$$\Delta f(x) = f(x+1) - f(x)$$

(i) $\Delta \left[\dfrac{1}{x^2 + 5x + 6}\right] = \Delta \left[\dfrac{1}{(x+2)(x+3)}\right]$
$$= \Delta \left[\dfrac{1}{x+2} - \dfrac{1}{x+3}\right]$$
$$= \Delta\left(\dfrac{1}{x+2}\right) - \Delta\left(\dfrac{1}{x+3}\right)$$
$$= \left(\dfrac{1}{x+3} - \dfrac{1}{x+2}\right) - \left(\dfrac{1}{x+4} - \dfrac{1}{x+3}\right)$$
$$= \dfrac{-1}{(x+2)(x+3)} + \dfrac{1}{(x+3)(x+4)}$$

$$\Delta^2\left[\frac{1}{x^2+5x+6}\right] = (-1)\Delta\left[\frac{1}{(x+2)(x+3)}\right] + \Delta\left[\frac{1}{(x+3)(x+4)}\right]$$

$$= (-1)\left[\frac{1}{(x+3)(x+4)} - \frac{1}{(x+2)(x+3)}\right]$$

$$+ \left[\frac{1}{(x+4)(x+5)} - \frac{1}{(x+3)(x+4)}\right]$$

$$= \frac{(-1)(-2)}{(x+2)(x+3)(x+4)} + \frac{-2}{(x+3)(x+4)(x+5)}$$

$$= 2\left[\frac{x+5-x-2}{(x+2)(x+3)(x+4)(x+5)}\right]$$

$$= \frac{6}{(x+2)(x+3)(x+4)(x+5)}$$

(ii) $\Delta^2\left[\frac{5x+12}{x^2+5x+6}\right] = \Delta^2\left[\frac{5x+12}{(x+2)(x+3)}\right]$

$$= \Delta^2\left[\frac{2}{x+2} + \frac{3}{x+3}\right]$$

$$= \Delta\left[\Delta\left(\frac{2}{x+2}\right) + \Delta\left(\frac{3}{x+3}\right)\right]$$

$$= \Delta\left[2\left(\frac{1}{x+3} - \frac{1}{x+2}\right) + 3\left(\frac{1}{x+4} - \frac{1}{x+3}\right)\right]$$

$$= -2\Delta\left[\frac{1}{(x+2)(x+3)}\right] - 3\Delta\left[\frac{1}{(x+3)(x+4)}\right]$$

$$= -2\left[\frac{1}{(x+3)(x+4)} - \frac{1}{(x+2)(x+3)}\right]$$

$$- 3\left[\frac{1}{(x+4)(x+5)} - \frac{1}{(x+3)(x+4)}\right]$$

$$= \frac{4}{(x+2)(x+3)(x+4)} + \frac{6}{(x+3)(x+4)(x+5)}$$

$$= \frac{10x+32}{(x+5)(x+4)(x+3)(x+2)}$$

$$= \frac{2(5x+16)}{(x+5)(x+4)(x+3)(x+2)}$$

**Example:** Prove the relations:

(i) $\Delta f_i^2 = (f_i + f_{i+1})\Delta f_i$

(ii) $\sum_{k=0}^{n-1} \Delta^2 f_k = \Delta f_n - \Delta f_0$.

**Solution:**

(i) $\Delta f_i^2 = f_{i+1}^2 - f_i^2$

$= (f_{i+1} - f_i)(f_{i+1} + f_i)$

$= (f_{i+1} + f_i)\Delta f_i.$

(ii) $\sum_{k=0}^{n-1} \Delta^2 f_k = \sum_{k=0}^{n-1} \Delta(\Delta f_k)$

$= \sum_{k=0}^{n-1} \Delta(f_{k+1} - f_k)$

$= \sum_{k=0}^{n-1} (\Delta f_{k+1} - \Delta f_k)$

$= (\Delta f_1 - \Delta f_0) + (\Delta f_2 - \Delta f_1) + (\Delta f_3 - \Delta f_2) + \cdots + (\Delta f_n - \Delta f_{n-1})$

$= \Delta f_n - \Delta f_0.$

**Example:** Given $u_0 + u_8 = 1.9243$, $u_1 + u_7 = 1.9590$
$u_2 + u_6 = 1.9823$, $u_3 + u_5 = 1.9956$.

Find $u_4$.

**Solution:** Since 8 entries are given, we have $\Delta^8 u_0 = 0$

i.e., $(E-1)^8 u_0 = 0$

$\Rightarrow \left( {}^8C_0 E^8 - {}^8C_1 E^7 + {}^8C_2 E^6 - {}^8C_3 E^5 + {}^8C_4 E^4 - {}^8C_5 E^3 \right.$
$\left. + {}^8C_6 E^2 - {}^8C_7 E + {}^8C_8 \right) u_0 = 0$

$\Rightarrow \left( E^8 - 8E^7 + 28E^6 - 56E^5 + 70E^4 - 56E^3 + 28E^2 - 8E + 1 \right) u_0 = 0$

$\Rightarrow u_8 - 8u_7 + 28u_6 - 56u_5 + 70u_4 - 56u_3 + 28u_2 - 8u_1 + u_0 = 0$

$\Rightarrow (u_8 + u_0) - 8(u_7 + u_1) + 28(u_6 + u_2) - 56(u_5 + u_3) + 70u_4 = 0$

$\Rightarrow 1.9243 - 8(1.9590) + 28(1.9823) - 56(1.9956) + 70u_4 = 0$

$\Rightarrow -69.9969 + 70u_4 = 0$

$\Rightarrow u_4 = \dfrac{69.9969}{70} = 0.999955.$

**Example:** If $y_x$ is the value of $y$ at $x$ for which the fifth differences are constant and $y_1 + y_7 = -784$, $y_2 + y_6 = 686$, $y_3 + y_5 = 1088$, then find $y_4$.

**Solution:** Since fifth differences are constants, we have $\Delta^6 y_x = 0$. For $x = 1$, this gives

$$\Delta^6 y_1 = 0$$
$$\Rightarrow (E-1)^6 y_1 = 0$$
$$\Rightarrow \left(E^6 - {}^6C_1 E^5 + {}^6C_2 E^4 - {}^6C_3 E^3 + {}^6C_4 E^2 - {}^6C_5 E + {}^6C_6\right) y_1 = 0$$
$$\Rightarrow \left(E^6 - 6E^5 + 15E^4 - 20E^3 + 15E^2 - 6E + 1\right) y_1 = 0$$
$$\Rightarrow y_7 - 6y_6 + 15y_5 - 20y_4 + 15y_3 - 6y_2 + y_1 = 0$$
$$\Rightarrow (y_7 + y_1) - 6(y_2 + y_6) + 15(y_3 + y_5) - 20 y_4 = 0$$
$$\Rightarrow y_4 = \frac{1}{20}\Big[(y_1 + y_7) - 6(y_2 + y_6) + 15(y_3 + y_5)\Big]$$
$$= \frac{1}{20}\Big[-784 - 6(686) + 15(1088)\Big] = 571$$

**Example:** Given that $u_0 + u_8 = 80$; $u_1 + u_7 = 10$; $u_2 + u_6 = 5$; $u_3 + u_5 = 10$, find $u_4$.

**Ans:** 6.

**Example:** Evaluate $\Delta^2 \left[\dfrac{4x^2 - 25x + 11}{(x-1)(x-2)(x-3)}\right]$.

**Ans:** $\dfrac{5}{x+1} - \dfrac{7}{x} - \dfrac{5}{x-1} + \dfrac{11}{x-2} - \dfrac{4}{x-3}$.

### 4.3.2 Missing Term Techniques

When one or more values of $y = f(x)$ corresponding to the equidistant values of $x$ are missing, we can find these using any of the following methods:

**Method I:** We assume the missing term or terms as $a, b, etc.$ and form the difference table. Assuming the last differences as zero, we solve these equations for $a, b$. These give the missing term/terms.

**Method II:** If $n$ entries of $y$ are given, $f(x)$ can be represented by a $(n-1)$th degree polynomial i.e., $\Delta^n y = 0$. Since $\Delta = E - 1$, therefore $(E-1)^n y = 0$. Now expanding $(E-1)^n$ and substituting the given values, we obtain the missing term/terms.

**Example:** Find the missing term in the following table

| x | 1 | 2 | 3 | 4 | 5 |
|---|---|---|---|---|---|
| y | 2 | 5 | 7 | – | 32 |

**Solution:** Since the values are known at four points, the maximum degree of the polynomial that can be fit is three. For such a polynomial the fourth order difference is zero. Let the missing value be $y_3$. Writing the table of difference.

| x | y | $\Delta y$ | $\Delta^2 y$ | $\Delta^3 y$ | $\Delta^4 y$ |
|---|---|---|---|---|---|
| 1 | $y_0 = 2$ | | | | |
| | | 3 | | | |
| 2 | $y_1 = 5$ | | $-1$ | | |
| | | 2 | | $y_3 - 8$ | |
| 3 | $y_2 = 7$ | | $y_3 - 9$ | | $56 - 4y_3$ |
| | | $y_3 - 7$ | | $48 - 3y_3$ | |
| 4 | $y_3$ | | $39 - 2y_3$ | | |
| | | $32 - y_3$ | | | |
| 5 | $y_4 = 32$ | | | | |

Equating the fourth order difference term to zero, we get
$$\Delta^4 y_0 = 0 \Rightarrow 56 - 4y_3 \Rightarrow y_3 = 14.$$
Therefore, the missing term $y_3 = 14$.

**Example:** Find the missing term in the following table

| x | 0 | 1 | 2 | 3 | 4 |
|---|---|---|---|---|---|
| y | 1 | 3 | 9 | – | 81 |

**Solution:** Since the values are known at four points, the maximum degree of the polynomial that can be fit is three. For such a polynomial the fourth order difference is zero. Let the missing value be $y_3$. Writing the table of difference.

| $x$ | $y$ | $\Delta y$ | $\Delta^2 y$ | $\Delta^3 y$ | $\Delta^4 y$ |
|---|---|---|---|---|---|
| 0 | $y_0 = 1$ | | | | |
| | | 2 | | | |
| 1 | $y_1 = 3$ | | 4 | | |
| | | 6 | | $y_3 - 19$ | |
| 2 | $y_2 = 9$ | | $y_3 - 15$ | | $124 - 4y_3$ |
| | | $y_3 - 9$ | | $105 - 3y_3$ | |
| 3 | $y_3$ | | $90 - 2y_3$ | | |
| | | $81 - y_3$ | | | |
| 4 | $y_4 = 81$ | | | | |

Equating the fourth order difference term to zero, we get
$$\Delta^4 y_0 = 0 \Rightarrow 124 - 4y_3 \Rightarrow y_3 = 31.$$
Therefore, the missing term $y_3 = 31$.

**Example:** Estimate the missing term in the following table.

| $x$ | 0 | 1 | 2 | 3 | 4 |
|---|---|---|---|---|---|
| $y$ | 4 | 3 | 4 | – | 12 |

**Solution:** We are given four values, so the third differences are constant and the fourth differences are zero. Hence $\Delta^4 f(x) = 0$ for all values of $x$, i.e.,

$$(E - 1)^4 = 0$$
$$\Rightarrow (E^4 - 4E^3 + 6E^2 - 4E + 1)f(x) = 0$$
$$\Rightarrow E^4 f(x) - 4E^3 f(x) + 6E^2 f(x) - 4E f(x) + f(x) = 0$$
$$\Rightarrow f(x+4) - 4f(x+3) + 6f(x+2) - 4f(x+1) + f(x) = 0$$

where the interval of differencing is 1.
Now substituting $x = 0$, we obtain

$$f(4) + 4f(3) + 6f(2) - 4f(1) + f(0) = 0$$
$$12 + 4f(3) + 6(4) - 4(3) + 4 = 0$$
$$f(3) = 7.$$

**Example:** Compute the missing values in the table

| $x$ | 45 | 50 | 55 | 60 | 65 |
|---|---|---|---|---|---|
| $y$ | 3.0 | – | 2 | – | –2.4 |

**Solution: Method I:** Since the values are known at three points, the maximum degree of the polynomial that can be fit is two. For such a polynomial the third order difference is zero. Let the missing values be $y_1$ and $y_3$. Writing the table of difference.

| $x$ | $y$ | $\Delta y$ | $\Delta^2 y$ | $\Delta^3 y$ |
|---|---|---|---|---|
| 45 | $y_0 = 3.0$ | | | |
| | | $y_1 - 3$ | | |
| 50 | $y_1$ | | $5 - 2y_1$ | |
| | | $2 - y_1$ | | $3y_1 + y_3 - 9$ |
| 55 | $y_2 = 2$ | | $y_1 + y_3 - 4$ | |
| | | $y_3 - 2$ | | $3.6 - y_1 - 3y_3$ |
| 60 | $y_3$ | | $-0.4 - 2y_3$ | |
| | | $-2.4 - y_3$ | | |
| 65 | $y_4 = -2.4$ | | | |

Equating the third order difference term to zero, we get

$$\Delta^3 y_0 = 0 \text{ and } \Delta^3 y_1 = 0.$$

$$3y_1 + y_3 - 9 = 0, 3.6 - y_1 - 3y_3 = 0.$$

$$\Rightarrow 3y_1 + y_3 = 9, y_1 + 3y_3 = 3.6.$$

Solving these, we get $y_1 = 2.925, y_3 = 0.225$.

Therefore, the missing terms are $y_1 = 2.925$ and $y_3 = 0.225$.

**Method II:** Since only three entries $y_0 = 3, y_2 = 2, y_4 = -2.4$ are given, the function $y$ can be represented by a second degree polynomial having third differences as zero.

$$\Delta^3 y_0 = 0 \text{ and } \Delta^3 y_1 = 0$$

$$\Rightarrow (E^3 - 3E^2 + 3E - 1)y_0 = 0 \text{ and } (E^3 - 3E^2 + 3E - 1)y_1 = 0$$

$$\Rightarrow y_3 - 3y_2 + 3y_1 - y_0 = 0; \quad y_4 - 3y_3 + 3y_2 - y_1 = 0$$

$$\Rightarrow y_3 + 3y_1 = 9; \; 3y_3 + y_1 = 3.6$$

Solving these, we get $y_1 = 2.925, y_2 = 0.225$.

**Example:** Find the missing entries in the following table.

| $x$ | 0 | 1 | 2 | 3 | 4 | 5 |
|---|---|---|---|---|---|---|
| $y$ | 1 | – | 11 | 28 | – | 116 |

**Solution:** Here, we are given $y_0 = 1, y_2 = 11, y_3 = 28$, and $y_5 = 116$. Since three values are known, we assume $y = f(x)$ as a polynomial of degree three. Hence

$$\Delta^4 y_0 = 0$$
$$\Rightarrow (E-1)^4 y_0 = 0$$
$$\Rightarrow (E^4 - 4E^3 + 6E^2 - 4E + 1) y_0 = 0$$
$$\Rightarrow y_4 - 4y_3 + 6y_2 - 4y_1 + y_0 = 0$$
$$\Rightarrow y_4 - 4(28) + 6(11) - 4y_1 + 1 = 0$$
$$\Rightarrow y_4 - 4y_1 = 45 \quad \cdots\cdots \quad (1)$$

$$\Delta^5 y_0 = 0$$
$$\Rightarrow (E-1)^5 y_0 = 0$$
$$\Rightarrow (E^5 - 5E^4 + 10E^3 - 10E^2 + 5E - 1) y_0 = 0$$
$$\Rightarrow y_5 - 5y_4 + 10y_3 - 10y_2 + 5y_1 - y_0 = 0$$
$$\Rightarrow 116 - 5y_4 + 10(28) - 10(11) + 5y_1 - 1 = 0$$
$$\Rightarrow -5y_4 + 5y_1 = -285 \quad \cdots\cdots \quad (2)$$

Solving Eqs.(1) and (2), we obtain $y_1 = 4$ and $y_4 = 61$.

**Example:** Find the missing entries in the following data **(Aug-2022)**

| $x$ | 0 | 1 | 2 | 3 | 4 | 5 | 6 |
|---|---|---|---|---|---|---|---|
| $y$ | 200 | 220 | 260 | – | 350 | – | 430 |

## 4.4 Newton's Forward Difference Interpolation Formula

Let the function $y = f(x)$ take the values $y_0, y_1, \cdots, y_n$ at the points $x_0, x_1, \cdots, x_n$, where $x_i = x_0 + ih$. Then Newton's forward interpolation formula is given by

$$y = f(x) = y_0 + p\Delta y_0 + \frac{p(p-1)}{2!}\Delta^2 y_0 + \frac{p(p-1)(p-2)}{3!}\Delta^3 y_0 + \cdots$$

where $p = \dfrac{x - x_0}{h}$.

**Note:** This formula is used for interpolating the values of $y$ near the beginning of the set of tabulated values or for extrapolating values of $y$ to the left of the beginning.

**Example:** A second degree polynomial passes through $(0,1), (1,3), (2,7)$ and $(3,13)$. Find the polynomial.

**Solution:** Since the given observations are at equal intervals of with unity. Construct the following difference table:

| $x$ | $f(x)$ | $\Delta f(x)$ | $\Delta^2 f(x)$ | $\Delta^3 f(x)$ |
|---|---|---|---|---|
| 0 | 1 | | | |
| | | 2 | | |
| 1 | 3 | | 2 | |
| | | 4 | | 0 |
| 2 | 7 | | 2 | |
| | | 6 | | |
| 3 | 13 | | | |

Given $x_0 = 0$, $h = 1$, $p = \dfrac{x - x_0}{h} = \dfrac{x - 0}{1} = x$.

By Newton's forward interpolation formula, we have

$$f(x) = f(0) + p\Delta f(0) + \frac{p(p-1)}{2!}\Delta^2 f(0) + \frac{p(p-1)(p-2)}{3!}\Delta^3 f(0)$$

$$= 1 + x(2) + \frac{x(x-1)}{2}(2) + 0$$

$$= 1 + 2x + x^2 - x$$

$$= x^2 + x + 1.$$

**Example:** Using Newton's forward formula, find the value of $f(1.6)$, if

| $x$ | 1 | 1.4 | 1.8 | 2.2 | 2.6 | 3.0 |
|---|---|---|---|---|---|---|
| $y$ | 3.49 | 4.82 | 5.96 | 6.5 | 7.2 | 8.4 |

**Solution:** The forward difference table is

| $x$ | $y$ | $\Delta y$ | $\Delta^2 y$ | $\Delta^3 y$ | $\Delta^4 y$ | $\Delta^5 y$ |
|---|---|---|---|---|---|---|
| 1 | 3.49($y_0$) | | | | | |
| | | 1.33($\Delta y_0$) | | | | |
| 1.4 | 4.82 | | -0.19($\Delta^2 y_0$) | | | |
| | | 1.14 | | -0.41($\Delta^3 y_0$) | | |
| 1.8 | 5.96 | | −0.6 | | 1.17($\Delta^4 y_0$) | |
| | | 0.54 | | 0.76 | | -1.59($\Delta^5 y_0$) |
| 2.2 | 6.5 | | 0.16 | | −0.42 | |
| | | 0.7 | | 0.34 | | |
| 2.6 | 7.2 | | 0.5 | | | |
| | | 1.2 | | | | |
| 3.0 | 8.4 | | | | | |

Given $x = 1.6$, $x_0 = 1$, $h = 0.4$, $p = \dfrac{x - x_0}{h} = \dfrac{1.6 - 1}{0.4} = \dfrac{0.6}{0.4} = 1.5$.

The Newton's forward interpolation formula is given by

$$y = f(x) = y_0 + p\Delta y_0 + \dfrac{p(p-1)}{2!}\Delta^2 y_0 + \dfrac{p(p-1)(p-2)}{3!}\Delta^3 y_0 + \dfrac{p(p-1)(p-2)(p-3)}{4!}\Delta^4 y_0$$
$$+ \dfrac{p(p-1)(p-2)(p-3)(p-4)}{5!}\Delta^5 y_0$$

$$f(1.6) = 3.49 + (1.5)(1.33) + \dfrac{(1.5)(1.5-1)}{2!}(-0.19) + \dfrac{(1.5)(1.5-1)(1.5-2)}{3!}(-0.41)$$
$$+ \dfrac{(1.5)(1.5-1)(1.5-2)(1.5-3)}{4!}(1.17)$$
$$+ \dfrac{(1.5)(1.5-1)(1.5-2)(1.5-3)(1.5-4)}{5!}(-1.59)$$

$$= 3.49 + 1.995 - 0.07125 + 0.025625 + 0.0274131 + 0.01861$$

$$= 5.4854$$

**Example:** From the following data, estimate the number of students who obtained the marks between 40 and 45:

| Marks: | 30 – 40 | 40 – 50 | 50 – 60 | 60 – 70 | 70 – 80 |
|---|---|---|---|---|---|
| No. of students: | 31 | 42 | 51 | 35 | 31 |

**Solution:** Let $y(x)$ be the number of students getting less than $x$ marks. Then, $y = 31$ for $x = 40$, $y = 31 + 42 = 73$ for $x = 50$, $y = 73 + 51 = 124$ for $x = 60$, $y = 124 + 35 = 159$ for $x = 70$ and $y = 159 + 31 = 190$ for $x = 80$. The less than cumulative frequency table for the given data is as shown below:

| Marks less than ($x$): | 40 | 50 | 60 | 70 | 80 |
|---|---|---|---|---|---|
| No. of students ($y$): | 31 | 73 | 124 | 159 | 190 |

We now construct the difference table for the data

| $x$ | $y$ | $\Delta y$ | $\Delta^2 y$ | $\Delta^3 y$ | $\Delta^4 y$ |
|---|---|---|---|---|---|
| 40 | **31**($y_0$) | | | | |
| | | 42($\Delta y_0$) | | | |
| 50 | 73 | | 9($\Delta^2 y_0$) | | |
| | | 51 | | -25($\Delta^3 y_0$) | |
| 60 | 124 | | −16 | | **37** ($\Delta^4 y_0$) |
| | | 35 | | 12 | |
| 70 | 159 | | −4 | | |
| | | 31 | | | |
| 80 | 190 | | | | |

Number of students who obtained marks between 40 and 45 is got from $y(45) - y(40)$.

We have $y(40) = 31$.

Now we shall find $y(45)$ i.e., number of students with marks less than 45.

Here $x_0 = 40, x = 45, h = 10$. Hence $p = \dfrac{x - x_0}{h} = \dfrac{45 - 40}{10} = \dfrac{5}{10} = 0.5$.

Using Newton's forward interpolation formula, we get

$$y = f(x) = y_0 + p\Delta y_0 + \frac{p(p-1)}{2!}\Delta^2 y_0 + \frac{p(p-1)(p-2)}{3!}\Delta^3 y_0 + \frac{p(p-1)(p-2)(p-3)}{4!}\Delta^4 y_0$$

$$f(45) = 31 + (0.5)(42) + \frac{(0.5)(0.5-1)}{2}(9) + \frac{(0.5)(0.5-1)(0.5-2)}{6}(-25)$$

$$+ \frac{(0.5)(0.5-1)(0.5-2)(0.5-3)}{24}(37)$$

$$= 31 + 21 - 1.125 - 1.5625 - 1.445$$

$$= 47.8675$$

$$= 48 \text{ (approximately)}$$

This shows that the number of students who obtained marks less than 45 marks is 48.

But the number of students who obtained less than 40 marks is 31.

Hence the number of students who obtained between 40 and 45 marks is $48 - 31 = 17$.

**Example:** Find the number of men getting wages between 10 and 15 from the following data.

| Wages in Rs: | 0-10 | 10-20 | 20-30 | 30-40 |
|---|---|---|---|---|
| Frequency: | 9 | 30 | 35 | 42 |

**Solution:** The difference table is as under:

| Wages in Rs $x$ | Frequency $y$ | $\Delta y$ | $\Delta^2 y$ | $\Delta^3 y$ |
|---|---|---|---|---|
| Under 10 | $9(y_0)$ | | | |
| | | $30(\Delta y_0)$ | | |
| Under 20 | 39 | | $5(\Delta^2 y_0)$ | |
| | | 35 | | $2(\Delta^3 y_0)$ |
| Under 30 | 74 | | 7 | |
| | | 42 | | |
| Under 40 | 116 | | | |

Number of persons getting wages between Rs 10 and 15 is got from $y(15) - y(10)$.

We have $y(10) = 9$.

Now we shall find $y(15)$ i.e., number of persons getting wages under Rs. 15.

Here $x_0 = 10, x = 15, h = 10$. Hence $p = \dfrac{x - x_0}{h} = \dfrac{15 - 10}{10} = \dfrac{5}{10} = 0.5$.

Using Newton's forward interpolation formula, we get

$$y = f(x) = y_0 + p\Delta y_0 + \frac{p(p-1)}{2!}\Delta^2 y_0 + \frac{p(p-1)(p-2)}{3!}\Delta^3 y_0$$

$$f(15) = 9 + (0.5)(30) + \frac{(0.5)(0.5-1)}{2}(5) + \frac{(0.5)(0.5-1)(0.5-2)}{6}(2)$$

$$= 9 + 15 - 0.625 + 0.125$$

$$= 23.5$$

$$= 24 \text{ (approximately)}$$

Therefore, number of persons getting wages between Rs 10 and Rs 15 is $24 - 9 = 15$.

**Example:** Apply Newton's forward interpolation, compute the value of $\sqrt{5.5}$, given that $\sqrt{5} = 2.236, \sqrt{6} = 2.449, \sqrt{7} = 2.646, \sqrt{8} = 2.828$.

**Solution:** Let the function $f(x) = \sqrt{x}$. **(Jan-2020), (Mar-2022)**

The forward difference table is

| $x$ | $y$ | $\Delta y$ | $\Delta^2 y$ | $\Delta^3 y$ |
|---|---|---|---|---|
| 5 | **2.236**$(y_0)$ | | | |
| | | **0.213**$(\Delta y_0)$ | | |
| 6 | 2.449 | | **-0.016**$(\Delta^2 y_0)$ | |
| | | 0.197 | | **0.001**$(\Delta^3 y_0)$ |
| 7 | 2.646 | | $-0.015$ | |
| | | 0.182 | | |
| 8 | 2.828 | | | |

Here $x_0 = 5, x = 5.5, h = 1$ and $p = \dfrac{x - x_0}{h} = \dfrac{5.5 - 5}{1} = \dfrac{0.5}{1} = 0.5.$

The Newton's forward interpolation formula is given by

$$y = f(x) = y_0 + p\Delta y_0 + \frac{p(p-1)}{2!}\Delta^2 y_0 + \frac{p(p-1)(p-2)}{3!}\Delta^3 y_0 + \frac{p(p-1)(p-2)(p-3)}{4!}\Delta^4 y_0$$

$$f(5.5) = 2.236 + (0.5)(0.213) + \frac{(0.5)(0.5 - 1)}{2}(-0.016) + \frac{(0.5)(0.5 - 1)(0.5 - 2)}{6}(0.001)$$

$$= 2.236 + 0.1065 + 0.00200 + 0.0000625$$

$$= 2.3445625$$

$$\simeq 2.345$$

**Example:** Using Newton's forward difference formula, estimate $y(0.12)$ from the following table

| $x$ | 0.10 | 0.15 | 0.20 | 0.25 | 0.30 |
|---|---|---|---|---|---|
| $y$ | 0.656 | 0.522 | 0.410 | 0.316 | 0.240 |

**Solution:** The forward difference table is

| $x$ | $y$ | $\Delta y$ | $\Delta^2 y$ | $\Delta^3 y$ | $\Delta^4 y$ |
|---|---|---|---|---|---|
| 0.10 | **0.656**$(y_0)$ | | | | |
| | | **-0.134**$(\Delta y_0)$ | | | |
| 0.15 | 0.522 | | **0.022**$(\Delta^2 y_0)$ | | |
| | | $-0.112$ | | **-0.004**$(\Delta^3 y_0)$ | |
| 0.20 | 0.410 | | 0.018 | | **0.004**$(\Delta^3 y_0)$ |
| | | $-0.094$ | | 0 | |
| 0.25 | 0.316 | | 0.018 | | |
| | | $-0.076$ | | | |
| 0.30 | 0.240 | | | | |

Given $x = 0.12$, $x_0 = 0.10$, $h = 0.05$, $p = \dfrac{x - x_0}{h} = \dfrac{0.12 - 0.10}{0.05} = 0.4$.

The Newton's forward interpolation formula is given by

$$y = f(x) = y_0 + p\Delta y_0 + \frac{p(p-1)}{2!}\Delta^2 y_0 + \frac{p(p-1)(p-2)}{3!}\Delta^3 y_0 + \frac{p(p-1)(p-2)(p-3)}{4!}\Delta^4 y_0$$

$$f(0.12) = 0.656 + (0.4)(-0.134) + \frac{(0.4)(0.4-1)}{2}(0.022) + \frac{(0.4)(0.4-1)(0.4-2)}{6}(-0.004)$$
$$+ \frac{(0.4)(0.4-1)(0.4-2)(0.4-3)}{24}(0.004)$$

$$= 0.656 - 0.0536 - 0.00264 - 0.000256 - 0.0001664$$

$$= 0.5994$$

**Example:** Given $\sin 45^0 = 0.7071, \sin 50^0 = 0.7660, \sin 55^0 = 0.8192, \sin 60^0 = 0.8660$, find $\sin 52^0$, by using Newton's forward formula.

**Solution:** Here

| $x$ | $45^0$ | $50^0$ | $55^0$ | $60^0$ |
|---|---|---|---|---|
| $y = \sin x$ | 0.7071 | 0.7660 | 0.8192 | 0.8660 |

**Solution:** The forward difference table is

| $x$ | $y$ | $\Delta y$ | $\Delta^2 y$ | $\Delta^3 y$ | $\Delta^4 y$ |
|---|---|---|---|---|---|
| 45 | **0.7071**($y_0$) | | | | |
| | | **0.0589**($\Delta y_0$) | | | |
| 50 | 0.7660 | | **-0.0057**($\Delta^2 y_0$) | | |
| | | 0.0532 | | **-0.0007**($\Delta^3 y_0$) | |
| 55 | 0.8192 | | $-0.0064$ | | |
| | | 0.0468 | | | |
| 60 | 0.8660 | | | | |

Here $x = 52^0$, $x_0 = 45^0$, $h = 5$, $p = \dfrac{x - x_0}{h} = \dfrac{52 - 45}{5} = \dfrac{7}{5} = 1.4$.

The Newton's forward interpolation formula is given by

$$y = f(x) = y_0 + p\Delta y_0 + \frac{p(p-1)}{2!}\Delta^2 y_0 + \frac{p(p-1)(p-2)}{3!}\Delta^3 y_0$$

$$\sin 52^0 = f(52) = 0.7071 + (1.4)(0.0589) + \frac{(1.4)(1.4-1)}{2}(-0.0057)$$

$$+ \frac{(1.4)(1.4-1)(1.4-2)}{6}(-0.0007)$$

$$= 0.7071 + 0.08246 - 0.001596 + 0.000392$$

$$= 0.7880032$$

**Example:** The values of $\sin x$ are given below for the different values of $x$. Find $\sin 32^0$.

| $x$ | $30^0$ | $35^0$ | $40^0$ | $45^0$ | $50^0$ |
|---|---|---|---|---|---|
| $y = \sin x$ | 0.5 | 0.5736 | 0.6428 | 0.7071 | 0.7660 |

**Solution:** The value of $32^0$ is near the beginning of the table. So, we use Newton's forward difference interpolation formula.

The forward difference table is

| $x$ | $y$ | $\Delta y$ | $\Delta^2 y$ | $\Delta^3 y$ | $\Delta^4 y$ |
|---|---|---|---|---|---|
| 30 | **0.5**($y_0$) | | | | |
| | | **0.0736**($\Delta y_0$) | | | |
| 35 | 0.5736 | | **-0.0044**($\Delta^2 y_0$) | | |
| | | 0.0692 | | **-0.0005**($\Delta^3 y_0$) | |
| 40 | 0.6428 | | $-0.0049$ | | **0**($\Delta^4 y_0$) |
| | | 0.0643 | | $-0.0005$ | |
| 45 | 0.7071 | | $-0.0054$ | | |
| | | 0.0589 | | | |
| 50 | 0.7660 | | | | |

Given $x = 32^0$, $x_0 = 30^0$, $h = 5$, $p = \dfrac{x - x_0}{h} = \dfrac{32 - 30}{5} = \dfrac{2}{5} = 0.4$.

The Newton's forward interpolation formula is given by

$$y = f(x) = y_0 + p\Delta y_0 + \frac{p(p-1)}{2!}\Delta^2 y_0 + \frac{p(p-1)(p-2)}{3!}\Delta^3 y_0$$

$$\sin 32^0 = f(32) = 0.5 + (0.4)(0.0736) + \frac{(0.4)(0.4-1)}{2}(-0.0044)$$

$$+ \frac{(0.4)(0.4-1)(0.4-2)}{6}(-0.0005)$$

$$= 0.5 + 0.02944 + 0.000528 - 0.000032$$

$$= 0.529936$$

**Example:** Find the cubic polynomial which takes the following values:

| $x$: | 0 | 1 | 2 | 3 |
|---|---|---|---|---|
| $f(x)$: | 1 | 2 | 1 | 10 |

Hence or otherwise evaluate $f(4)$.

**Solution:** Let us form the forward difference table first.

| $x$ | $f(x)$ | $\Delta f(x)$ | $\Delta^2 f(x)$ | $\Delta^3 f(x)$ |
|---|---|---|---|---|
| 0 | 1 | | | |
| | | 1 | | |
| 1 | 2 | | $-2$ | |
| | | $-1$ | | 12 |
| 2 | 1 | | 10 | |
| | | 9 | | |
| 3 | 10 | | | |

Here $x_0 = 0, h = 1, p = \dfrac{x - x_0}{h} = \dfrac{x - 0}{1} = x.$

Using Newton's forward interpolation formula, we get

$$f(x) = f(0) + p\Delta f(0) + \frac{p(p-1)}{2!}\Delta^2 f(0) + \frac{p(p-1)(p-2)}{3!}\Delta^3 f(0)$$

$$= 1 + x(1) + \frac{x(x-1)}{2}(-2) + \frac{x(x-1)(x-2)}{6}(12)$$

$$= 1 + x - x^2 + x - 2x(x^2 - 3x + 2)$$

$$= 2x^3 - 7x^2 + 6x + 1, \text{ which is the required polynomial.}$$

To compute $f(4)$, we take $x = 4$, so that $f(4) = 41$.

**Example:** The population of a town in the decimal census was given below. Estimate the population for the year 1895.

| Year | 1891 | 1901 | 1911 | 1921 | 1931 |
|---|---|---|---|---|---|
| Population $y$ in thousands | 46 | 66 | 81 | 93 | 101 |

**Solution:** Since the value of 1895 is nearer to beginning of the table. So, we use Newton's forward interpolation formula.

The forward difference table is

| $x$ | $y$ | $\Delta y$ | $\Delta^2 y$ | $\Delta^3 y$ | $\Delta^4 y$ |
|------|-----|------|------|------|------|
| 1891 | 46 | | | | |
| | | 20 | | | |
| 1901 | 66 | | -5 | | |
| | | 15 | | 2 | |
| 1911 | 81 | | -3 | | -3 |
| | | 12 | | -1 | |
| 1921 | 93 | | -4 | | |
| | | 8 | | | |
| 1931 | 101 | | | | |

Here $x = 1895$, $x_0 = 1891$, $h = 10$.

$$\therefore p = \frac{x - x_0}{h} = \frac{1895 - 1891}{10} = 0.4.$$

The Newton's forward interpolation formula is given by

$$y = f(x) = y_0 + p\Delta y_0 + \frac{p(p-1)}{2!}\Delta^2 y_0 + \frac{p(p-1)(p-2)}{3!}\Delta^3 y_0 + \frac{p(p-1)(p-2)(p-3)}{4!}\Delta^4 y_0$$

$$f(1895) = 46 + (0.4)(20) + \frac{(0.4)(0.4-1)}{2}(-5) + \frac{(0.4)(0.4-1)(0.4-2)}{6}(2)$$

$$+ \frac{(0.4)(0.4-1)(0.4-2)(0.4-3)}{24}(-3)$$

$$= 46 + 8 + 0.6 + 0.128 + 0.1248$$

$$= 54.8528$$

Hence the population for the year 1895 is 54853 approximately.

**Example:** The following table gives the population of a town during the last six censuses. Estimate using any suitable interpolation formula, the increase in the population during the period from 1976 to 1978.

| year | 1941 | 1951 | 1961 | 1971 | 1981 | 1991 |
|------|------|------|------|------|------|------|
| Population (in thousands) | 12 | 15 | 20 | 27 | 39 | 52 |

**Solution:** The forward difference table is as under

| $x$ | $y$ | $\Delta y$ | $\Delta^2 y$ | $\Delta^3 y$ | $\Delta^4 y$ | $\Delta^5 y$ |
|---|---|---|---|---|---|---|
| 1941 | **12**($y_0$) | | | | | |
| | | **3**($\Delta y_0$) | | | | |
| 1951 | 15 | | **2**($\Delta^2 y_0$) | | | |
| | | 5 | | **0**($\Delta^3 y_0$) | | |
| 1961 | 20 | | 2 | | **3**($\Delta^4 y_0$) | |
| | | 7 | | 3 | | **-10**($\Delta^5 y_0$) |
| 1971 | 27 | | 5 | | $-7$ | |
| | | 12 | | $-4$ | | |
| 1981 | 39 | | 1 | | | |
| | | 13 | | | | |
| 1991 | 52 | | | | | |

**Compute** $y(1976)$:

Here $x = 1976$, $x_0 = 1941$, $h = 10$, $p = \dfrac{x - x_0}{h} = \dfrac{1976 - 1941}{10} = 3.5$.

Applying Newton's formula for forward differences, we have

$$y = f(x) = y_0 + p\Delta y_0 + \frac{p(p-1)}{2!}\Delta^2 y_0 + \frac{p(p-1)(p-2)}{3!}\Delta^3 y_0 + \frac{p(p-1)(p-2)(p-3)}{4!}\Delta^4 y_0$$
$$+ \frac{p(p-1)(p-2)(p-3)(p-4)}{5!}\Delta^5 y_0$$

$$f(1976) = 12 + (3.5)(3) + \frac{(3.5)(3.5-1)}{2}(2) + \frac{(3.5)(3.5-1)(3.5-2)}{6}(0)$$
$$+ \frac{(3.5)(3.5-1)(3.5-2)(3.5-3)}{24}(3)$$
$$+ \frac{(3.5)(3.5-1)(3.5-2)(3.5-3)(3.5-4)}{120}(-10)$$

$$= 12 + 10.5 + 8.75 + 0 + 0.8203 + 0.734$$

$$= 32.3437$$

**To find** $y(1978)$

Here $x_0 = 1941, x = 1978, h = 10, p = \frac{x-x_0}{h} = \frac{1978-1941}{10} = 3.7$.

$$f(1978) = 12 + (3.7)(3) + \frac{(3.7)(3.7-1)}{2}(2) + \frac{(3.7)(3.7-1)(3.7-2)}{6}(0)$$
$$+ \frac{(3.7)(3.7-1)(3.7-2)(3.7-3)}{24}(3)$$
$$+ \frac{(3.7)(3.7-1)(3.7-2)(3.7-3)(3.7-4)}{120}(-10)$$

$= 12 + 11.1 + 9.99 + 0 + 1.4860 + 0.2972$

$= 34.8732$

Therefore increase in the population during the period from 1976 to 1978

$= 34.8732 - 32.3437$

$= 2.5295$ thousands

$= 2.53$ thousands approximately

$= 2530$.

**Example:** The area $A$ of a circle of diameter $d$ is given for the following values:

| d: | 80 | 85 | 90 | 95 | 100 |
|---|---|---|---|---|---|
| A: | 5026 | 5674 | 6362 | 7088 | 7854 |

Find approximate values for the areas of circles of diameters 82 and 91 respectively.

**Solution:** We shall find the values of $A$ corresponding to $d = 82$. Let $A = f(d)$. Now, the difference table is as shown below:

| $d$ | $f(d)$ | $\Delta f(d)$ | $\Delta^2 f(d)$ | $\Delta^3 f(d)$ | $\Delta^4 f(d)$ |
|---|---|---|---|---|---|
| 80 | **5026**($y_0$) | | | | |
| | | 648($\Delta y_0$) | | | |
| 85 | 5674 | | 40($\Delta^2 y_0$) | | |
| | | 688 | | -2($\Delta^3 y_0$) | |
| 90 | 6362 | | 38 | | 4($\Delta^4 y_0$) |
| | | 726 | | 2 | |
| 95 | 7088 | | 40 | | |
| | | 766 | | | |
| 100 | 7854 | | | | |

Here $x = 82$, $x_0 = 80$, $h = 5$, $p = \dfrac{x - x_0}{h} = \dfrac{82 - 80}{5} = 0.4$.

By Newton's forward interpolation formula, we have

$$y = f(x) = y_0 + p\Delta y_0 + \frac{p(p-1)}{2!}\Delta^2 y_0 + \frac{p(p-1)(p-2)}{3!}\Delta^3 y_0 + \frac{p(p-1)(p-2)(p-3)}{4!}\Delta^4 y_0$$

$$\Rightarrow f(82) = 5026 + (0.4)(648) + \frac{(0.4)(0.4-1)}{2}(40) + \frac{(0.4)(0.4-1)(0.4-2)}{6}(-2)$$

$$+ \frac{(0.4)(0.4-1)(0.4-2)(0.4-3)}{24}(4)$$

$$= 5026 = 259.2 - 4.8 - 0.128 - 0.1664$$

$$= 5280 \text{ approximately}$$

Similarly, we can find $f(91)$, i.e., the area of circle of diameter 91, which we get as $f(91) = 6504$.

**Example:** From the following table of half-yearly premium for policies maturing at different ages, estimate the premium for policies maturing at age of 46:

| Age: | 45 | 50 | 55 | 60 | 65 |
|---|---|---|---|---|---|
| Premium(in rupees): | 114.84 | 96.16 | 83.32 | 74.98 | 68.48 |

**Solution:** The difference table is as under:

| Age($x$) | Premium($y$) | $\Delta y$ | $\Delta^2 y$ | $\Delta^3 y$ | $\Delta^4 y$ |
|---|---|---|---|---|---|
| 45 | 114.8($y_0$) | | | | |
| | | -8.68($\Delta y_0$) | | | |
| 50 | 96.16 | | 5.84($\Delta^2 y_0$) | | |
| | | $-12.84$ | | -1.84($\Delta^3 y_0$) | |
| 55 | 83.32 | | 4 | | 0.68($\Delta^4 y_0$) |
| | | $-8.84$ | | $-1.16$ | |
| 60 | 74.48 | | 2.84 | | |
| | | $-6$ | | | |
| 65 | 68.48 | | | | |

Here $x = 46$, $x_0 = 45$, $h = 5$, $p = \dfrac{x - x_0}{h} = \dfrac{46 - 45}{5} = 0.2$.

By Newton's forward interpolation formula, we have

$$y = f(x) = y_0 + p\Delta y_0 + \frac{p(p-1)}{2!}\Delta^2 y_0 + \frac{p(p-1)(p-2)}{3!}\Delta^3 y_0 + \frac{p(p-1)(p-2)(p-3)}{4!}\Delta^4 y_0$$

$$\Rightarrow f(46) = 114.84 + (0.2)(-18.68) + \frac{(0.2)(0.2-1)}{2}(5.84) + \frac{(0.2)(0.2-1)(0.2-2)}{6}(-1.84)$$

$$+ \frac{(0.2)(0.2-1)(0.2-2)(0.2-3)}{24}(0.68)$$

$$= 114.84 - 3.736 - 0.4672 - 0.08832 - 0.022848$$

$$= 110.52$$

$\therefore f(46) =$ Rs. 110.52.

## EXERCISE

1. A second degree polynomial passes through the points $(1, -1), (2, -1), (3, 1)$ and $(4, 5)$. Find the polynomial.

   **Ans:** $y = x^2 - 3x + 1$.

2. Fit a polynomial for the following data $\hspace{2cm}$ **(Jan-2020), (Sep-2021)**

   $$y_0 = -5, y_1 = -1, y_2 = 9, y_3 = 25, y_4 = 55, y_5 = 105.$$

   **Ans:** $\dfrac{-1}{12}(x^5 - 22x^4 + 107x^3 - 218x^2 + 84x + 60)$

3. Fit a interpolating polynomial in $x$ for the following data $\hspace{2cm}$ **(July-2021)**

| x | 0 | 1 | 2 | 3 | 4 |
|---|---|---|---|---|---|
| y | -3 | 3 | 4 | 27 | 57 |

**Ans:** $-7x^4 + 60x^3 - 141x^2 + 4$.

4. Fit a interpolating polynomial in $x$ for the following data

| x | 0 | 1 | 2 | 3 | 4 |
|---|---|---|---|---|---|
| y | 3 | 5 | 6 | 9 | 17 |

**Ans:** $-\dfrac{x^4}{12} + x^3 - \dfrac{35x^2}{12} + 4x + 3$

5. Find $f(2.5)$ using Newton's forward formula from the following table:

| x | 0 | 1 | 2 | 3 | 4 | 5 | 6 |
|---|---|---|---|---|---|---|---|
| f(x) | 0 | 1 | 16 | 81 | 256 | 625 | 1296 |

**Ans:** $y(2.5) = 39.0625$

6. Find $f(1.2)$ from the following table: **(Jan-2020)**

| x | 1 | 1.4 | 1.8 | 2.2 |
|---|---|---|---|---|
| y | 3.49 | 4.82 | 5.96 | 6.5 |

**Ans:** 4.15313

7. Find $y(0.5)$ using from the following table:

| x | 0 | 1 | 2 | 3 | 4 | 5 | 6 |
|---|---|---|---|---|---|---|---|
| f(x) | 0 | 1 | 16 | 81 | 256 | 625 | 1296 |

**Ans:** $y(0.5) = 0.5625$

8. Find $f(2.5)$ using Newton's forward formula from the following table:

**Ans:** 39.0625 **(April-2022)**

| x | 0 | 1 | 2 | 3 | 4 | 5 | 6 |
|---|---|---|---|---|---|---|---|
| f(x) | 0 | 1 | 16 | 81 | 256 | 625 | 1296 |

9. Find $f(1.75)$ if $f(1.7) = 5.474$, $f(1.8) = 6.050$, $f(1.9) = 6.686$, $f(2) = 7.389$

**Ans:** 5.7868  (Jan-2020)

10. Find $f(17.5)$ if $f(17) = 574$, $f(18) = 605$, $f(19) = 668$, $f(20) = 738$.

(Aug-2022)

11. From the data given below, find the number of students whose weight is between 60 and 70  (Aug-2022)

| Weight: | 0 – 40 | 40 – 60 | 60 – 80 | 80 – 100 | 100 – 120 |
|---|---|---|---|---|---|
| No. of students: | 250 | 120 | 100 | 70 | 50 |

**Ans:** 54

12. Construct Newton's forward interpolation polynomial for the following data:

| x: | 4 | 6 | 8 | 10 |
|---|---|---|---|---|
| y: | 1 | 3 | 8 | 16 |

Hence evaluate $y$ for $x = 5$.

**Ans:** $y = 1 + (x - 4) + \frac{3}{8}(x^2 - 10x + 24)$, $y(5) = 1.625$.

13. Using Newtons Forward difference formula find $y(2)$ from the following table

| x: | 0 | 5 | 10 | 15 | 20 | 25 |
|---|---|---|---|---|---|---|
| y: | 7 | 11 | 14 | 18 | 24 | 32 |

**Ans:** 8.8896

14. Using Newtons forward formula compute $f(142)$ from the following table:

| x: | 140 | 150 | 160 | 170 | 180 |
|---|---|---|---|---|---|
| y: | 3.685 | 4.854 | 6.302 | 8.076 | 10.225 |

**Ans:** 3.898

15. The population of a nation in the decimal census was given below. Estimate the population in the year 1975 using appropriate interpolation formula  (Jan-2020)

| Year $x$: | 1961 | 1971 | 1981 | 1991 | 2001 |
|---|---|---|---|---|---|
| Population $y$ (thousands): | 66 | 76 | 81 | 93 | 105 |

**Ans:** 77.50

16. Consider following data.  (**Dec-2020**)

| $x$:    | 0.2    | 0.5    | 0.8    | 1.1    | 1.4    |
|---------|--------|--------|--------|--------|--------|
| $g(x)$: | 9.9833 | 4.9696 | 3.2836 | 2.4339 | 1.9177 |

Calculate approximately $g(0.25)$ using Newtons Forward Interpolation.

**Ans:** 3.92258

17. Construct difference table for the following data  (**April-2022**)

| $x$:    | 0.1   | 0.3   | 0.5   | 0.7   | 0.9   | 1.1   | 1.3   |
|---------|-------|-------|-------|-------|-------|-------|-------|
| $g(x)$: | 0.003 | 0.067 | 0.148 | 0.248 | 0.370 | 0.518 | 0.697 |

and evaluate $f(0.6)$.

**Ans:** 0.1955

18. Find $y(21)$ using Newton's forward formula for the following table (**Aug-2022**)

| $x$    | 20  | 25  | 30  | 35  | 40  | 45  |
|--------|-----|-----|-----|-----|-----|-----|
| $f(x)$ | 354 | 332 | 291 | 260 | 231 | 204 |

## 4.5 Newton's Backward Difference Interpolation Formula

Let the function $y = f(x)$ take the values $y_0, y_1, \cdots, y_n$ at the points $x_0, x_1, \cdots, x_n$ where $x_i = x_0 + ih$. Then Newton's backward interpolation formula is given by

$$y = f(x) = y_n + p\nabla y_n + \frac{p(p+1)}{2!}\nabla^2 y_n + \frac{p(p+1)(p+2)}{3!}\nabla^3 y_n + \cdots$$

where $p = \dfrac{x - x_n}{h}$.

**Note:** This formula is used to interpolate the values of $y$ near the end of the set of tabulated values or for extrapolating values of $y$ to the right of the last tabulated value $y_n$.

**Example:** Using Newton's backward formula, find $y(18)$ from the following table:

| $x$ | 0 | 5 | 10 | 15 | 20 | 25 |
|---|---|---|---|---|---|---|
| $f(x)$ | 8 | 11 | 14 | 18 | 24 | 32 |

**Solution:** The backward difference table is

| $x$ | $y$ | $\nabla y$ | $\nabla^2 y$ | $\nabla^3 y$ | $\nabla^4 y$ | $\nabla^5 y$ |
|---|---|---|---|---|---|---|
| 0 | 8 | | | | | |
| | | 3 | | | | |
| 5 | 11 | | 0 | | | |
| | | 3 | | 1 | | |
| 10 | 14 | | 1 | | 0 | |
| | | 4 | | 1 | | -1 |
| 15 | 18 | | 2 | | -1 | |
| | | 6 | | 0 | | |
| 20 | 24 | | 2 | | | |
| | | 8 | | | | |
| 25 | 32 | | | | | |

Here $x = 18$, $x_n = 25$, $h = 5$.
$$\therefore p = \frac{x - x_n}{h} = \frac{18 - 25}{5} = \frac{-7}{5} = -1.4.$$
The Newton's backward interpolation formula is given by

$$y = f(x) = y_n + p\nabla y_n + \frac{p(p+1)}{2!}\nabla^2 y_n + \frac{p(p+1)(p+2)}{3!}\nabla^3 y_n$$
$$+ \frac{p(p+1)(p+2)(p+3)}{4!}\nabla^4 y_n + \frac{p(p+1)(p+2)(p+3)(p+4)}{5!}\nabla^5 y_n$$
$$y(18) = 32 + (-1.4)(8) + \frac{(-1.4)(-1.4+1)}{2}(2) + \frac{(-1.4)(-1.4+1)(-1.4+2)}{6}(0)$$
$$+ \frac{(-1.4)(-1.4+1)(-1.4+2)(-1.4+3)}{24}(-1)$$
$$+ \frac{(-1.4)(-1.4+1)(-1.4+2)(-1.4+3)(-1.4+4)}{120}(-1)$$
$$= 32 - 11.2 + 0.56 + 0 - 0.0224 - 0.011648$$
$$= 21.3259.$$

**Example:** Find $f(1.28)$ if $f(1.15) = 1.0723$, $f(1.20) = 1.0954$, $f(1.25) = 1.1180$ and $f(1.30) = 1.1401$.

**Solution:** Since the value of $x(= 1.28)$ is near the end of the table. So, we use Newton's backward interpolation formula.

The backward difference table is

| $x$ | $y$ | $\nabla y$ | $\nabla^2 y$ | $\nabla^3 y$ |
|---|---|---|---|---|
| 1.15 | 1.0723 | | | |
| | | 0.0231 | | |
| 1.20 | 1.0954 | | −0.0005 | |
| | | 0.0226 | | 0 |
| 1.25 | 1.1180 | | **-0.0005** | |
| | | 0.0221 | | |
| 1.30 | **1.1401** | | | |

Here $x = 1.28$, $x_n = 1.30$, $h = 0.05$.

$$\therefore p = \frac{x - x_n}{h} = \frac{1.28 - 1.30}{0.05} = -0.4.$$

The Newton's backward interpolation formula is given by

$$y = f(x) = y_n + p\nabla y_n + \frac{p(p+1)}{2!}\nabla^2 y_n + \frac{p(p+1)(p+2)}{3!}\nabla^3 y_n$$

$$f(1.28) = 1.1401 + (-0.4)(0.0221) + \frac{(-0.4)(-0.4+1)}{2}(-0.0005)$$

$$= 1.1401 - 0.00884 + 0.00006$$

$$= 1.13132.$$

**Example:** Find $f(2.36)$ from the following table

| $x$ | 1.6 | 1.8 | 2.0 | 2.2 | 2.4 | 2.6 |
|---|---|---|---|---|---|---|
| $f(x)$ | 4.95 | 6.05 | 7.39 | 9.03 | 11.02 | 13.46 |

**Solution:** Since the value of $x(= 2.36)$ is near the end of the table. So, we use Newton's backward interpolation formula.

The backward difference table is

| $x$ | $y$ | $\nabla y$ | $\nabla^2 y$ | $\nabla^3 y$ | $\nabla^4 y$ | $\nabla^5 y$ |
|---|---|---|---|---|---|---|
| 1.6 | 4.95 | | | | | |
| | | 1.1 | | | | |
| 1.8 | 6.05 | | 0.24 | | | |
| | | 1.34 | | 0.06 | | |
| 2.0 | 7.39 | | 0.3 | | −0.01 | |
| | | 1.64 | | 0.05 | | **0.06** |
| 2.2 | 9.03 | | 0.35 | | **0.05** | |
| | | 1.99 | | **0.1** | | |
| 2.4 | 11.02 | | **0.45** | | | |
| | | **2.44** | | | | |
| 2.6 | **13.46** | | | | | |

Here $x = 2.36$, $x_n = 2.6$, $h = 0.2$.

$$\therefore p = \frac{x - x_n}{h} = \frac{2.36 - 2.6}{0.2} = -1.2.$$

The Newton's backward interpolation formula is given by

$$y = f(x) = y_n + p\nabla y_n + \frac{p(p+1)}{2!}\nabla^2 y_n + \frac{p(p+1)(p+2)}{3!}\nabla^3 y_n$$

$$+ \frac{p(p+1)(p+2)(p+3)}{4!}\nabla^4 y_n + \frac{p(p+1)(p+2)(p+3)(p+4)}{5!}\nabla^5 y_n$$

$$y(2.36) = 13.46 + (-1.2)(2.44) + \frac{(-1.2)(-1.2+1)}{2}(0.45) + \frac{(-1.2)(-1.2+1)(-1.2+2)}{6}(0.1)$$

$$+ \frac{(-1.2)(-1.2+1)(-1.2+2)(-1.2+3)}{24}(0.05)$$

$$+ \frac{(-1.2)(-1.2+1)(-1.2+2)(-1.2+3)(-1.2+4)}{120}(0.06)$$

$$= 13.46 - 2.928 + 0.054 + 0.0032 + 0.00072 + 0.0004838$$

$$= 10.5904.$$

**Example:** The population of a town in the decimal census was given below. Estimate the population for the year 1955.

| Year | 1921 | 1931 | 1941 | 1951 | 1961 |
|---|---|---|---|---|---|
| Population $y$ in thousands | 46 | 66 | 81 | 93 | 101 |

**Solution:** Since the value of $x = 1955$ is near the end of the table. So, we use Newton's backward interpolation formula.

The backward difference table is

| $x$ | $y$ | $\nabla y$ | $\nabla^2 y$ | $\nabla^3 y$ | $\nabla^4 y$ |
|---|---|---|---|---|---|
| 1921 | 46 | | | | |
| | | 20 | | | |
| 1931 | 66 | | −5 | | |
| | | 15 | | 2 | |
| 1941 | 81 | | −3 | | -3 |
| | | 12 | | -1 | |
| 1951 | 93 | | -4 | | |
| | | 8 | | | |
| 1961 | **101** | | | | |

Here $x = 1955$, $x_n = 1961$, $h = 10$.

$$\therefore p = \frac{x - x_n}{h} = \frac{1955 - 1961}{10} = -0.6.$$

The Newton's backward interpolation formula is given by

$$y = f(x) = y_n + p\nabla y_n + \frac{p(p+1)}{2!}\nabla^2 y_n + \frac{p(p+1)(p+2)}{3!}\nabla^3 y_n$$
$$+ \frac{p(p+1)(p+2)(p+3)}{4!}\nabla^4 y_n$$

$$y(1955) = 101 + (-0.6)(8) + \frac{(-0.6)(-0.6+1)}{2}(-4) + \frac{(-0.6)(-0.6+1)(-0.6+2)}{6}(-1)$$
$$+ \frac{(-0.6)(-0.6+1)(-0.6+2)(-0.6+3)}{24}(-3)$$

$$= 101 - 4.8 + 0.48 + 0.056 + 0.1008$$

$$= 96.8368.$$

Hence, the population for 1955 is estimated at 96.84 thousands.

**Example:** The population of a nation in the decimal census was given below. Estimate the population in the year 1925 using appropriate interpolation formula.

**(Dec-2020), (April-2022)**

| Year $x$ | 1891 | 1901 | 1911 | 1921 | 1931 |
|---|---|---|---|---|---|
| Population $y$ (thousands) | 46 | 66 | 81 | 93 | 101 |

**Solution:** Since the value of $x = 1925$ is near the end of the table. So, we use Newton's backward interpolation formula.

The backward difference table is

| $x$ | $y$ | $\nabla y$ | $\nabla^2 y$ | $\nabla^3 y$ | $\nabla^4 y$ |
|---|---|---|---|---|---|
| 1891 | 46 | | | | |
| | | 20 | | | |
| 1901 | 66 | | $-5$ | | |
| | | 15 | | 2 | |
| 1911 | 81 | | $-3$ | | $-3$ |
| | | 12 | | $-1$ | |
| 1921 | 93 | | $-4$ | | |
| | | 8 | | | |
| 1931 | **101** | | | | |

Here $x = 1925$, $x_n = 1931$, $h = 10$.

$$\therefore p = \frac{x - x_n}{h} = \frac{1925 - 1931}{10} = -0.6.$$

The Newton's backward interpolation formula is given by

$$y = f(x) = y_n + p\nabla y_n + \frac{p(p+1)}{2!}\nabla^2 y_n + \frac{p(p+1)(p+2)}{3!}\nabla^3 y_n$$
$$+ \frac{p(p+1)(p+2)(p+3)}{4!}\nabla^4 y_n$$

$$y(1925) = 101 + (-0.6)(8) + \frac{(-0.6)(-0.6+1)}{2}(-4) + \frac{(-0.6)(-0.6+1)(-0.6+2)}{6}(-1)$$
$$+ \frac{(-0.6)(-0.6+1)(-0.6+2)(-0.6+3)}{24}(-3)$$

$$= 101 - 4.8 + 0.48 + 0.056 + 0.1008$$

$$= 96.8368.$$

Hence, the population for 1925 is estimated at 96.84 thousands.

## EXERCISE

1. Calculate the value $f(7.5)$ for the table:

   | $x$ | 1 | 2 | 3 | 4 | 5 | 6 | 7 | 8 |
   |---|---|---|---|---|---|---|---|---|
   | $f(x)$ | 1 | 8 | 27 | 64 | 125 | 216 | 343 | 512 |

   **Ans: 421.875**

2. Estimate the value of $f(22)$ and $f(42)$ from the following table:

   | $x$ | 20 | 25 | 30 | 35 | 40 | 45 |
   |---|---|---|---|---|---|---|
   | $f(x)$ | 354 | 332 | 291 | 260 | 231 | 204 |

   **Ans: 352, 219.**

3. Find $y(46)$ using Newton's back ward formula for the following table **(Aug-2022)**

   | $x$ | 20 | 25 | 30 | 35 | 40 | 45 |
   |---|---|---|---|---|---|---|
   | $f(x)$ | 354 | 332 | 291 | 260 | 231 | 204 |

4. From the following data find $\theta$ at $x = 43$ and $x = 84$.

   | $x$ | 40 | 50 | 60 | 70 | 80 | 90 |
   |---|---|---|---|---|---|---|
   | $\theta$ | 184 | 204 | 226 | 250 | 276 | 304 |

   **Ans: 189.79, 286.96**

5. Evaluate $\sqrt{8.5}$ given that $\sqrt{5} = 2.236$, $\sqrt{6} = 2.449$, $\sqrt{7} = 2.646$, & $\sqrt{8} = 2.828$ by Newton Back word interpolation formula.

6. Given that $\sin 45^0 = 0.7077$, $\sin 50^0 = 0.766$, $\sin 55^0 = 0.8192$, $\sin 60^0 = 0.866$ find $\sin 65^0$ using Newtons Back ward difference formula. **(Dec-2021)**

7. Find $f(2.4)$ from the following data using appropriate interpolation method **(Dec-2020)**

   | $x$ | 1.0 | 1.5 | 2.0 | 2.5 |
   |---|---|---|---|---|
   | $f(x)$ | 3 | 3.375 | 5.0 | 12.072 |

8. Using Newtons backward formula compute $f(84)$ from the following table:

| $x$ | 40 | 50 | 60 | 70 | 80 | 90 |
|---|---|---|---|---|---|---|
| $f(x)$ | 184 | 204 | 226 | 250 | 276 | 304 |

**Ans:** $f(84) = 287$.

9. Find $f(5.5)$ using Newton's backward formula for the following table: **(Mar-2022)**

| $x$ | 0 | 1 | 2 | 3 | 4 | 5 | 6 |
|---|---|---|---|---|---|---|---|
| $f(x)$ | 0 | 1 | 16 | 81 | 256 | 625 | 1296 |

10. From the following table of half yearly premium for policies, estimate the premium for policies at the age of 63. **(April-2022)**

| Age $x$ | 45 | 50 | 55 | 60 | 65 |
|---|---|---|---|---|---|
| Premium $y$ | 114.84 | 96.16 | 83.32 | 74.48 | 68.48 |

11. Find $y(5)$ using Newtons Backward difference formula from the table **(April-2022)**

| $x$ | 1 | 2 | 3 | 4 |
|---|---|---|---|---|
| $y$ | 34 | 48 | 59 | 65 |

## 4.6 Gauss's Forward Interpolation formula

The Gauss forward interpolation formula is given by

$$y = f(x) = y_0 + p\Delta y_0 + \frac{p(p-1)}{2!}\Delta^2 y_{-1} + \frac{(p+1)p(p-1)}{3!}\Delta^3 y_{-1}$$
$$+ \frac{(p+1)p(p-1)(p-2)}{4!}\Delta^4 y_{-2} + \cdots$$

**Example:** Use Gauss's forward interpolation formula to find $f(30)$ given that $f(21) = 18.4708$, $f(25) = 17.8144$, $f(29) = 17.1070$, $f(33) = 16.3432$, $f(37) = 15.5154$.

**Solution:** Let us take $x_0 = 29$ and prepare the following difference table:

| $x$ | $y$ | $\Delta y$ | $\Delta^2 y$ | $\Delta^3 y$ | $\Delta^4 y$ |
|---|---|---|---|---|---|
| 21 | 18.4708($y_{-2}$) | | | | |
| | | −0.6564 | | | |
| 25 | 17.8144($y_{-1}$) | | −0.0510 | | |
| | | −0.7074 | | −0.0054 | |
| 29 | 17.1070($y_0$) ↘ | | -0.0564 ↘ | | -0.0022 |
| | | -0.7638 ↗ | | -0.0076 ↗ | |
| 33 | 16.3432 | | −0.0640 | | |
| | | −0.8278 | | | |
| 37 | 15.5154 | | | | |

Here $x = 30, x_0 = 29, h = 4$.
$$\therefore p = \frac{x - x_0}{h} = \frac{30 - 29}{4} = \frac{1}{4} = 0.25.$$

The Gauss forward interpolation formula is given by

$$y = f(x) = y_0 + p\Delta y_0 + \frac{p(p-1)}{2!}\Delta^2 y_{-1} + \frac{(p+1)p(p-1)}{3!}\Delta^3 y_{-1}$$
$$+ \frac{(p+1)p(p-1)(p-2)}{4!}\Delta^4 y_{-2}$$

$$f(30) = 17.1070 + (0.25)(-0.7638) + \frac{(0.25)(0.25-1)}{2}(-0.0564)$$
$$+ \frac{(0.25+1)(0.25)(0.25-1)}{6}(-0.0076)$$
$$+ \frac{(0.25+1)(0.25)(0.25-1)(0.25-2)}{24}(-0.0022)$$

$$= 17.1070 - 0.19095 + 0.0052875 + 0.0002968 - 0.00003759$$

$$= 16.9216$$

**Example:** Using Gauss' forward formula, find the value of $f(25)$ from the following data $f(20) = 24, f(24) = 32, f(28) = 35, f(32) = 40$. **(Nov-2017)**

**Solution:** Let us take $x_0 = 24$ and prepare the following difference table:

| $x$ | $y$ | $\Delta y$ | $\Delta^2 y$ | $\Delta^3 y$ |
|---|---|---|---|---|
| 20($x_{-1}$) | 24($y_{-1}$) | | | |
| | | 8 | | |
| 24($x_0$) | 32($y_0$) ↘ | | -5($\Delta^2 y_{-1}$) ↘ | |
| | | 3($\Delta y_0$) ↗ | | 7($\Delta^3 y_{-1}$) |
| 28 | 35 | | 2 | |
| | | 5 | | |
| 32 | 40 | | | |

Here $x = 25, x_0 = 24, h = 4$.
$$\therefore p = \frac{x - x_0}{h} = \frac{25 - 24}{4} = \frac{1}{4} = 0.25.$$
The Gauss forward interpolation formula is given by

$$y = f(x) = y_0 + p\Delta y_0 + \frac{p(p-1)}{2!}\Delta^2 y_{-1} + \frac{(p+1)p(p-1)}{3!}\Delta^3 y_{-1}$$

$$f(25) = 32 + (0.25)(3) + \frac{(0.25)(0.25-1)}{2}(-5) + \frac{(0.25+1)(0.25)(0.25-1)}{6}(7)$$

$$= 32 + 0.75 + 0.46875 - 0.2734$$

$$= 32.94535$$

**Example:** Find $f(22)$ from the Gauss forward formula

| $x$ | 20 | 25 | 30 | 35 | 40 | 45 |
|---|---|---|---|---|---|---|
| $f(x)$ | 354 | 332 | 291 | 260 | 231 | 204 |

**Solution:** Let us take $x_0 = 25$ and prepare the following difference table:

| $x$ | $y$ | $\Delta y$ | $\Delta^2 y$ | $\Delta^3 y$ | $\Delta^4 y$ | $\Delta^5 y$ |
|---|---|---|---|---|---|---|
| 20 | 354($y_{-1}$) | | | | | |
| | | −22($\Delta y_{-1}$) | | | | |
| 25 | **332**($y_0$) ↘ | | -19($\Delta^2 y_{-1}$) ↘ | | | |
| | | -41($\Delta y_0$) ↗ | | 29($\Delta^3 y_{-1}$) ↘ | | |
| 30 | 291 | | 10 | | -37 ↘ | |
| | | −31 | | −8 | | 45 |
| 35 | 260 | | 2 | | 8 | |
| | | −29 | | 0 | | |
| 40 | 231 | | 2 | | | |
| | | −27 | | | | |
| 45 | 204 | | | | | |

Here $x = 22, x_0 = 25, h = 5$.

$$\therefore p = \frac{x - x_0}{h} = \frac{22 - 25}{5} = \frac{-3}{5} = -0.6.$$

The Gauss forward interpolation formula is given by

$$y = f(x) = y_0 + p\Delta y_0 + \frac{p(p-1)}{2!}\Delta^2 y_{-1} + \frac{(p+1)p(p-1)}{3!}\Delta^3 y_{-1}$$
$$+ \frac{(p+1)p(p-1)(p-2)}{4!}\Delta^4 y_{-1} + \frac{(p+2)(p+1)p(p-1)(p-2)}{4!}\Delta^5 y_{-1}$$

$$f(22) = 332 + (-0.6)(-41) + \frac{(-0.6)(-0.6-1)}{2}(-19)$$
$$+ \frac{(-0.6+1)(-0.6)(-0.6-1)}{6}(29)$$
$$+ \frac{(-0.6+1)(-0.6)(-0.6-1)(-0.6-2)}{24}(-37)$$
$$+ \frac{(-0.6+2)(-0.6+1)(-0.6)(-0.6-1)(-0.6-2)}{120}(45)$$

$$= 332 + 24.6 - 9.12 + 1.856 + 1.5392 - 0.52416$$

$$= 350.351$$

**Example:** Estimate the value of $f(2.5)$ using Gauss's forward interpolation formula given that

Dr. A. Kameswara Rao, M.Sc, Ph.D                                Interpolation    401

| x:    | 1 | 2 | 3  | 4  |
|-------|---|---|----|----|
| f(x): | 1 | 8 | 27 | 64 |

**Solution:** Since $f(2.5)$ is to be computed, let us take $x_0 = 2$ and prepare the following difference table:

| $x$ | $y$ | $\Delta y$ | $\Delta^2 y$ | $\Delta^3 y$ |
|---|---|---|---|---|
| $1(x_{-1})$ | $1(y_{-1})$ | | | |
| | | 7 | | |
| $2(x_0)$ | $8(y_0)$ ↘ | | $12(\Delta^2 y_{-1})$ ↘ | |
| | | $19(\Delta y_0)$ ↗ | | $6(\Delta^3 y_{-1})$ |
| 3 | 27 | | 18 | |
| | | 37 | | |
| 4 | 64 | | | |

Here $x = 2.5, x_0 = 2, h = 1$.

$$\therefore p = \frac{x - x_0}{h} = \frac{2.5 - 2}{1} = 0.5.$$

The Gauss forward interpolation formula is given by

$$y = f(x) = y_0 + p\Delta y_0 + \frac{p(p-1)}{2!}\Delta^2 y_{-1} + \frac{(p+1)p(p-1)}{3!}\Delta^3 y_{-1}$$

$$f(2.5) = 8 + (0.5)(19) + \frac{(0.5)(0.5-1)}{2}(12) + \frac{(0.5+1)(0.5)(0.5-1)}{6}(6)$$

$$= 8 + 9.5 - 1.5 - 0.375$$

$$= 15.625$$

Hence the value of $f(x)$ at $x = 2.5$ is 15.625.

**Example:** Use Gauss's forward formula to find the value of $y$ when $x = 3.75$ from the following table:

| x | 2.5    | 3.0    | 3.5    | 4.0    | 4.5    | 5.0   |
|---|--------|--------|--------|--------|--------|-------|
| y | 24.145 | 22.043 | 20.225 | 18.644 | 17.262 | 16.47 |

**Solution:** Let us take $x_0 = 3.5$ and prepare the forward difference table:

| $x$ | $y$ | $\Delta y$ | $\Delta^2 y$ | $\Delta^3 y$ | $\Delta^4 y$ | $\Delta^5 y$ |
|---|---|---|---|---|---|---|
| 2.5 | 24.145 | | | | | |
| | | −2.102 | | | | |
| 3.0 | 22.043 | | 0.284 | | | |
| | | −1.818 | | −0.047 | | |
| 3.5 | **20.225**($y_0$) | | 0.237 | | 0.009 | |
| | | **-1.581** | | **-0.038** | | **-0.003** |
| 4.0 | 18.644 | | 0.199 | | 0.006 | |
| | | −1.382 | | −0.032 | | |
| 4.5 | 17.262 | | 0.167 | | | |
| | | −1.215 | | | | |
| 5.0 | 16.047 | | | | | |

Here $x = 3.5, x_0 = 3.75, h = 0.5$.

$$\therefore p = \frac{x - x_0}{h} = \frac{3.75 - 3.5}{0.5} = 0.6.$$

Using Gauss forward interpolation formula is given by

$$y = f(x) = y_0 + p\Delta y_0 + \frac{p(p-1)}{2!}\Delta^2 y_{-1} + \frac{(p+1)p(p-1)}{3!}\Delta^3 y_{-1}$$
$$+ \frac{(p+1)p(p-1)(p-2)}{4!}\Delta^4 y_{-2} + \frac{(p+2)(p+1)p(p-1)(p-2)}{4!}\Delta^5 y_{-2}$$

$$f(3.75) = 20.225 + (0.5)(-15811) + \frac{(0.5)(0.5-1)}{2}(0.37)$$
$$+ \frac{(0.5+1)(0.5)(0.5-1)}{6}(-0.038)$$
$$+ \frac{(0.5+1)(0.6)(0.5-1)(0.5-2)}{24}(0.009)$$
$$+ \frac{(0.5+2)(0.5+1)(0.5)(0.5-1)(0.5-2)}{120}(-0.003)$$

$$= 20.225 - 0.7905 - 0.029625 + 0.00238 + 0.0002109 - 0.00003516$$

$$= 19.4074$$

The value of $y$ for $x = 3.75$ is 19.40 approximately.

**Example:** Estimate $f\left(\dfrac{1}{2}\right)$ using Gauss's forward formula given that $f(2) =$

$10, f(1) = 8, f(0) = 5, f(-1) = 10.$

**Solution:** Let us take $x_0 = 0$ and prepare the forward difference table:

| $x$ | $y$ | $\Delta y$ | $\Delta^2 y$ | $\Delta^3 y$ |
|---|---|---|---|---|
| $-1$ | 10 | | | |
| | | $-5$ | | |
| 0 | $5(y_0)$ ↘ | | $8(\Delta^2 y_{-1})$ ↘ | |
| | | $3(\Delta y_0)$ ↗ | | $-9(\Delta^3 y_{-1})$ |
| 1 | 8 | | $-1$ | |
| | | 2 | | |
| 2 | 10 | | | |

Here $x = 0.5, x_0 = 0, h = 1$.

$$\therefore p = \frac{x - x_0}{h} = \frac{0.5 - 0}{1} = 0.5.$$

Using Gauss forward interpolation formula is given by

$$y = f(x) = y_0 + p\Delta y_0 + \frac{p(p-1)}{2!}\Delta^2 y_{-1} + \frac{(p+1)p(p-1)}{3!}\Delta^3 y_{-1}$$

$$f\left(\frac{1}{2}\right) = 5 + (0.5)(3) + \frac{(0.5)(0.5-1)}{2}(8) + \frac{(0.5+1)(0.5)(0.5-1)}{6}(-9)$$

$$= 5 + 1.5 - 1 + 0.5625$$

$$= 6.0625$$

### EXERCISE

1. Using Gauss's forward formula obtain the value of $e^{1.17}$ from the following data:

| $x$ | 1.00 | 1.05 | 1.10 | 1.15 | 1.20 | 1.25 | 1.30 |
|---|---|---|---|---|---|---|---|
| $e^x$ | 2.7183 | 2.8577 | 3.0042 | 3.1582 | 3.3201 | 3.4903 | 3.6613 |

**Ans:** 3.2221

2. Apply Gauss's forward interpolation formula to find the value of $u_9$ if $u_0 = 14, u_4 = 24, u_8 = 32, u_{12} = 35, u_{16} = 40$.

**Ans:** 33.1162

3. Apply Gauss's forward formula obtain the value of $f(32)$ from the following table

| $x$ | 25 | 30 | 35 | 40 |
|---|---|---|---|---|
| $y = f(x)$ | 0.2707 | 0.3027 | 0.3386 | 0.3794 |

**Ans:** 0.31653

4. Apply Gauss's forward formula to find $f(x)$ at $x = 3.5$ from the table below

| $x$ | 2 | 3 | 4 | 5 |
|---|---|---|---|---|
| $y = f(x)$ | 2.626 | 3.454 | 4.784 | 6.986 |

**Ans:** 4.033125

5. Find $y(23)$ for the following data using Gauss Forward interpolation formula.

| $x$ | 10 | 20 | 30 | 40 | 50 |
|---|---|---|---|---|---|
| $y$ | 9.21 | 17.54 | 31.82 | 55.32 | 92.51 |

**Ans:** 21.07375                                                                 (July-2021)

6. The population of a nation in the decimal census was given below. Estimate the population in the year 1975 using appropriate interpolation formula.

| Year $x$ | 1961 | 1971 | 1981 | 1991 | 2001 |
|---|---|---|---|---|---|
| Population $y$(thousands) | 66 | 76 | 81 | 93 | 105 |

(Dec-2020)

7. Find $f(3.5)$ using Gauss's forward formula for the following table  (Mar-2022)

| $x$ | 0 | 1 | 2 | 3 | 4 | 5 | 6 |
|---|---|---|---|---|---|---|---|
| $y$ | 1 | 3 | 14 | 22 | 35 | 48 | 56 |

## 4.7 Gauss's Backward Interpolation formula

The Gauss backward interpolation formula is given by

$$y = f(x) = y_0 + p\Delta y_{-1} + \frac{(p+1)p}{2!}\Delta^2 y_{-1} + \frac{(p+1)p(p-1)}{3!}\Delta^3 y_{-2}$$
$$+ \frac{(p+2)(p+1)p(p-1)}{4!}\Delta^4 y_{-2} + \cdots$$

**Example:** Given that $f(25) = 0.2707$, $f(30) = 0.3027$, $f(35) = 0.3386$ and $f(40) = 0.3794$. Use Gauss backward interpolation formula find $f(32)$.

**Solution:** Let us take $x_0 = 35$ and prepare the following difference table:

| $x$ | $y$ | $\Delta y$ | $\Delta^2 y$ | $\Delta^3 y$ |
|---|---|---|---|---|
| 25 | $0.2707(y_{-2})$ | | | |
| | | 0.0320 | | |
| 30 | $0.3027(y_{-1})$ | | 0.0039 | |
| | | $0.0359(\Delta y_{-1})$ ↘ | | $\mathbf{0.0010(\Delta^3 y_{-2})}$ |
| 35 | $\mathbf{0.3386}$ $(y_0)$ ↗ | | $\mathbf{0.0049(\Delta^2 y_{-1})}$ ↗ | |
| | | 0.0408 | | |
| 40 | 0.3794 | | | |

Here $x = 32, x_0 = 10, h = 5.$

$$\therefore p = \frac{x - x_0}{h} = \frac{32 - 35}{5} = \frac{-3}{5} = -0.6.$$

The Gauss backward interpolation formula is given by

$$y = f(x) = y_0 + p\Delta y_{-1} + \frac{(p+1)p}{2!}\Delta^2 y_{-1} + \frac{(p+1)p(p-1)}{3!}\Delta^3 y_{-2}$$

$$f(32) = 0.3386 + (-0.6)(0.0359) + \frac{(-0.6+1)(-0.6)}{2}(0.0049)$$

$$+ \frac{(-0.6+1)(-0.6)(-0.6-1)}{6}(0.0010)$$

$$= 0.3164$$

i.e., $f(32) = 0.3164$.

**Example:** Use Gauss backward interpolation formula to find $f(8)$ from the following table

| $x$ | 0 | 5 | 10 | 15 | 20 | 25 |
|---|---|---|---|---|---|---|
| $f(x)$ | 7 | 11 | 14 | 18 | 24 | 32 |

**Solution:** Let us take $x_0 = 10$ and prepare the following difference table:

| $x$ | $y$ | $\Delta y$ | $\Delta^2 y$ | $\Delta^3 y$ | $\Delta^4 y$ | $\Delta^5 y$ |
|---|---|---|---|---|---|---|
| 0 | 7 $(y_{-2})$ | | | | | |
| | | 4 | | | | |
| 5 | 11 $(y_{-1})$ | | −1 | | | |
| | | 3 ↘ | | 2 ↘ | | |
| 10 | 14 $(y_0)$ ↗ | | 1 ↗ | | -1 ↘ | |
| | | 4 | | 1 | | 0 |
| 15 | 18 | | 2 | | −1 | |
| | | 6 | | 0 | | |
| 20 | 24 | | 2 | | | |
| | | 8 | | | | |
| 25 | 32 | | | | | |

Here $x = 8, x_0 = 10, h = 5$.

$$\therefore p = \frac{x - x_0}{h} = \frac{8 - 10}{5} = \frac{-2}{5} = -0.4.$$

The Gauss backward interpolation formula is given by

$$y = f(x) = y_0 + p\Delta y_{-1} + \frac{(p+1)p}{2!}\Delta^2 y_{-1} + \frac{(p+1)p(p-1)}{3!}\Delta^3 y_{-2}$$
$$+ \frac{(p+2)(p+1)p(p-1)}{4!}\Delta^4 y_{-2}$$

$$f(8) = 14 + (-0.4)(3) + \frac{(-0.4+1)(-0.4)}{2}(1) + \frac{(-0.4+1)(-0.4)(-0.4-1)}{6}(2)$$
$$+ \frac{(-0.4+2)(-0.4+1)(-0.4)(-0.4-1)}{24}(-1)$$

$$= 14 - 1.2 - 0.12 + 0.112 - 0.0224$$

$$= 12.7696.$$

**Example:** Interpolate by Gauss backward formula, the sales of a concern for the year 1976 given that

| Year | 1940 | 1950 | 1960 | 1970 | 1980 | 1990 |
|---|---|---|---|---|---|---|
| Sales (in lakh of Rs) | 17 | 20 | 27 | 32 | 36 | 38 |

**Solution:** Let us take $x_0 = 1980$ and prepare the following difference table:

| $x$ | $y$ | $\Delta y$ | $\Delta^2 y$ | $\Delta^3 y$ | $\Delta^4 y$ | $\Delta^5 y$ |
|---|---|---|---|---|---|---|
| 1940 | $17(y_{-3})$ | | | | | |
| | | 3 | | | | |
| 1950 | $20\ (y_{-2})$ | | 4 | | | |
| | | 7 | | $-6$ | | |
| 1960 | $27\ (y_{-1})$ | | $-2$ | | 7 | |
| | | 5 | | 1 | | $-9$ |
| 1970 | $32(y_{-1})$ | | $-1$ | | $-2$ ↗ | |
| | | 4 ↘ | | $-1$ ↗ | | |
| 1980 | $36(y_0)$ ↗ | | $-2$ ↗ | | | |
| | | 2 | | | | |
| 1990 | $38(y_1)$ | | | | | |

Here $x = 1976, x_0 = 1980, h = 10$.

$$\therefore p = \frac{x - x_0}{h} = \frac{1976 - 1980}{10} = \frac{-4}{10} = -0.4.$$

The Gauss backward interpolation formula is given by

$$y = f(x) = y_0 + p\Delta y_{-1} + \frac{(p+1)p}{2!}\Delta^2 y_{-1} + \frac{(p+1)p(p-1)}{3!}\Delta^3 y_{-2}$$
$$+ \frac{(p+2)(p+1)p(p-1)}{4!}\Delta^4 y_{-3} + \frac{(p+2)(p+1)p(p-1)(p-2)}{5!}\Delta^5 y_{-4}$$

$$f(1976) = 36 + (-0.4)(4) + \frac{(-0.4+1)(-0.4)}{2}(-2) + \frac{(-0.4+1)(-0.4)(-0.4-1)}{6}(-1)$$
$$+ \frac{(-0.4+2)(-0.4+1)(-0.4)(-0.4-1)}{24}(-2)$$
$$+ \frac{(-0.4+2)(-0.4+1)(-0.4)(-0.4-1)(-0.4-2)}{120}(-9)$$

$$= 36 - 1.6 + 0.24 - 0.056 - 0.0448 + 0.096768$$

$$= 34.63596.$$

Sales in the year 1976 is 34.64 lakhs.

**Example:** The census of a town is as given in the table below. Using Gauss backward formula, find the population for the year 1974. **(Sep-2021)**

| Year $x$ | 1939 | 1949 | 1959 | 1969 | 1979 | 1989 |
|---|---|---|---|---|---|---|
| Population in thousands | 12 | 15 | 20 | 27 | 39 | 52 |

**Solution:** Let us take $x_0 = 1969$ and prepare the following difference table:

| $x$ | $y$ | $\Delta y$ | $\Delta^2 y$ | $\Delta^3 y$ | $\Delta^4 y$ | $\Delta^5 y$ |
|---|---|---|---|---|---|---|
| 1939 | 12($y_{-3}$) | | | | | |
| | | 3 | | | | |
| 1949 | 15 ($y_{-2}$) | | 2 | | | |
| | | 5 | | 0 | | |
| 1959 | 20 ($y_{-1}$) | | 2 | | 3 | |
| | | 7↘ | | 3↘ | | -10 |
| 1969 | **27**($y_0$) ↗ | | 5↗ | | -7↗ | |
| | | 12 | | -4 | | |
| 1979 | 39 | | 1 | | | |
| | | 13 | | | | |
| 1989 | 52 | | | | | |

Here $x = 1974, x_0 = 1969, h = 10$.
$$\therefore p = \frac{x - x_0}{h} = \frac{1974 - 1969}{10} = 0.5.$$
The Gauss backward interpolation formula is given by

$$y = f(x) = y_0 + p\Delta y_{-1} + \frac{(p+1)p}{2!}\Delta^2 y_{-1} + \frac{(p+1)p(p-1)}{3!}\Delta^3 y_{-2}$$
$$+ \frac{(p+2)(p+1)p(p-1)}{4!}\Delta^4 y_{-2} + \frac{(p+2)(p+1)p(p-1)(p-2)}{5!}\Delta^5 y_{-3}$$

$$f(1974) = 27 + (0.5)(7) + \frac{(0.5+1)(0.5)}{2}(5) + \frac{(0.5+1)(0.5)(0.5-1)}{6}(3)$$
$$+ \frac{(0.5+2)(0.5+1)(0.5)(0.5-1)}{24}(-7)$$
$$+ \frac{(0.5+2)(0.5+1)(0.5)(0.5-1)(0.5-2)}{120}(-10)$$

$$= 27 + 3.5 + 1.875 - 0.1875 + 0.2743 - 0.1172$$

$$= 32.345.$$

The population of a town for the year 1974 is 32. 345 thousands approx.

**Example:** From the following table find $y$ at $x = 1.35$. Apply Gauss' backward interpolation formula

| $x$ | 1   | 1.2    | 1.4    | 1.6   | 1.8   | 2 |
|-----|-----|--------|--------|-------|-------|---|
| $y$ | 0.0 | −0.112 | −0.016 | 0.336 | 0.992 | 2 |

**Solution:** Let us take $x_0 = 1.4$ and prepare the following difference table:

| $x$ | $y$ | $\Delta y$ | $\Delta^2 y$ | $\Delta^3 y$ | $\Delta^4 y$ | $\Delta^5 y$ |
|-----|-----|------------|--------------|--------------|--------------|--------------|
| 1   | $0(y_{-2})$ |  |  |  |  |  |
|     |     | −0.112 |  |  |  |  |
| 1.2 | $-0.112(y_{-1})$ |  | 0.208 |  |  |  |
|     |     | 0.096 ↘ |  | 0.048 ↘ |  |  |
| 1.4 | **-0.016**$(y_0)$ ↗ |  | 0.256 ↗ |  | 0 |  |
|     |     | 0.352 |  | 0.048 |  | 0 |
| 1.6 | 0.336 |  | 0.304 |  | 0 |  |
|     |     | 0.656 |  | 0.048 |  |  |
| 1.8 | 0.992 |  | 0.352 |  |  |  |
|     |     | 1.008 |  |  |  |  |
| 2   | 2   |  |  |  |  |  |

Here $x = 1.35$, $x_0 = 1.4$, $h = 0.2$.

$$\therefore p = \frac{x - x_0}{h} = \frac{1.35 - 1.4}{0.2} = \frac{-0.05}{0.2} = -0.25.$$

The Gauss backward interpolation formula is given by

$$y = f(x) = y_0 + p\Delta y_{-1} + \frac{(p+1)p}{2!}\Delta^2 y_{-1} + \frac{(p+1)p(p-1)}{3!}\Delta^3 y_{-2}$$

$$f(1.35) = -0.016 + (-0.25)(0.096) + \frac{(-0.25+1)(-0.25)}{2}(0.256)$$

$$+ \frac{(-0.25+1)(-0.25)(-0.25-1)}{6}(0.048)$$

$$= -0.016 - 0.024 - 0.024 - 0.001875$$

$$= -0.062125.$$

### EXERCISE

1. Given that $\sqrt{6500} = 80.6223, \sqrt{6510} = 80.6846, \sqrt{6520} = 80.7465, \sqrt{6530} = 80.8084$. Find $\sqrt{6526}$ by using Gauss's backward formula.

**Ans:** 80.7828

2. Apply Gauss' backward formula to obtain $\sin 54^0$ from the following data:

| $x^0$ | 20 | 30 | 40 | 50 | 60 | 70 |
|---|---|---|---|---|---|---|
| $\sin x^0$ | 0.34202 | 050200 | 0.64279 | 0.76604 | 0.86603 | 0.93969 |

**Ans:** 0.809166

## 4.8 Interpolation with Unequal Intervals

So far we have considered interpolation formulas for equally spaced values of $x$. We now develop two interpolation formulas for unequally spaced values of $x$.

### 4.8.1 Lagrange's Interpolation formula

In this section we consider an interpolation formula which can be used in the case where the specified values of $x$ are *not necessarily equally spaced*.

Let $x_0, x_1, x_2, \cdots, x_n$ be the $(n+1)$ values of $x$ which are not necessarily equally spaced. Let $y_0, y_1, y_2, \cdots, y_n$ be the corresponding values of $y = f(x)$. Then the Lagrange's interpolation formula is given by

$$y = f(x) = \frac{(x-x_1)(x-x_2)\cdots(x-x_n)}{(x_0-x_1)(x_0-x_2)\cdots(x_0-x_n)}f(x_0) + \frac{(x-x_0)(x-x_2)\cdots(x-x_n)}{(x_1-x_0)(x_1-x_2)\cdots(x_1-x_n)}f(x_1)$$
$$+ \cdots + \frac{(x-x_0)(x-x_1)\cdots(x-x_{n-1})}{(x_n-x_0)(x_n-x_1)\cdots(x_n-x_{n-1})}f(x_n)$$

This formula is known as the **Lagrange's interpolation formula**. This formula determines $y$ as a polynomial in $x$ and may be used to find an approximate value of $y_k$ for $y$ for any specified value $x_k$ of $x$.

**Example:** Find the cubic polynomial which takes the values

| $x$ | 0 | 1 | 2 | 5 |
|---|---|---|---|---|
| $f(x)$ | 2 | 3 | 12 | 147 |

**Solution:** Here the values of $x$ are not equally spaced.

We have, $x_0 = 0, x_1 = 1, x_2 = 2, x_3 = 5$ and $y_0 = 2, y_1 = 3, y_2 = 12, y_3 = 147$.

Using Lagrange's interpolation formula to this

$$y = f(x) = \frac{(x-x_1)(x-x_2)(x-x_3)}{(x_0-x_1)(x_0-x_2)(x_0-x_3)} f(x_0) + \frac{(x-x_0)(x-x_2)(x-x_3)}{(x_1-x_0)(x_1-x_2)(x_1-x_3)} f(x_1)$$
$$+ \frac{(x-x_0)(x-x_1)(x-x_3)}{(x_2-x_0)(x_2-x_1)(x_2-x_3)} f(x_2) + \frac{(x-x_0)(x-x_1)(x-x_2)}{(x_3-x_0)(x_3-x_1)(x_3-x_2)} f(x_3)$$
$$= \frac{(x-1)(x-2)(x-5)}{(0-1)(0-2)(0-5)}(2) + \frac{(x-0)(x-2)(x-5)}{(1-0)(1-2)(1-5)}(3)$$
$$+ \frac{(x-0)(x-1)(x-5)}{(2-0)(2-1)(2-5)}(12) + \frac{(x-0)(x-1)(x-2)}{(5-0)(5-1)(5-2)}(147)$$
$$= (x-1)(x-2)(x-5)\left(\frac{2}{-10}\right) + x(x-2)(x-5)\left(\frac{3}{4}\right)$$
$$+ x(x-1)(x-5)\left(\frac{12}{-6}\right) + x(x-1)(x-2)\left(\frac{147}{60}\right)$$
$$= (x-2)(x-5)\left[\frac{-1}{5}(x-1) + \frac{3}{4}x\right] + x(x-1)\left[-2(x-5) + (x-2)\frac{147}{60}\right]$$
$$= (x^2 - 7x + 10)\left(\frac{11x+4}{20}\right) + (x^2 - x)\left(\frac{27x+306}{60}\right)$$
$$= \frac{60x^3 + 60x^2 - 60x + 120}{60}$$
$$= x^3 + x^2 - x + 2$$

**Example:** Find the interpolating polynomial $f(x)$ from the table:

| $x$ | 0 | 1 | 4 | 5 |
|---|---|---|---|---|
| $f(x)$ | 4 | 3 | 24 | 39 |

**Solution:** Here $x_0 = 0, x_1 = 1, x_2 = 4, x_3 = 5$ and $y_0 = 4, y_1 = 3, y_2 = 24, y_3 = 39$.

Using Lagrange's interpolation formula to this

$$y = f(x) = \frac{(x-x_1)(x-x_2)(x-x_3)}{(x_0-x_1)(x_0-x_2)(x_0-x_3)}f(x_0) + \frac{(x-x_0)(x-x_2)(x-x_3)}{(x_1-x_0)(x_1-x_2)(x_1-x_3)}f(x_1)$$
$$+ \frac{(x-x_0)(x-x_1)(x-x_3)}{(x_2-x_0)(x_2-x_1)(x_2-x_3)}f(x_2) + \frac{(x-x_0)(x-x_1)(x-x_2)}{(x_3-x_0)(x_3-x_1)(x_3-x_2)}f(x_3)$$
$$= \frac{(x-1)(x-4)(x-5)}{(0-1)(0-4)(0-5)}(4) + \frac{(x-0)(x-4)(x-5)}{(1-0)(1-4)(1-5)}(3)$$
$$+ \frac{(x-0)(x-1)(x-5)}{(4-0)(4-1)(4-5)}(24) + \frac{(x-0)(x-1)(x-4)}{(5-0)(5-1)(5-4)}(39)$$
$$= (x-1)(x-4)(x-5)\left(\frac{4}{-20}\right) + x(x-4)(x-5)\left(\frac{3}{12}\right)$$
$$+ x(x-1)(x-5)\left(\frac{24}{-12}\right) + x(x-1)(x-4)\left(\frac{39}{20}\right)$$
$$= (x-4)(x-5)\left[\frac{-1}{5}(x-1) + \frac{1}{4}x\right] + x(x-1)\left[-2(x-5) + \frac{39}{20}(x-4)\right]$$
$$= (x^2 - 9x + 20)\left(\frac{x+4}{20}\right) + (x^2 - x)\left(\frac{44-x}{20}\right)$$
$$= \frac{40x^2 - 60x + 80}{20}$$
$$= 2x^2 - 3x + 4.$$

**Example:** Find the parabola passing through the points $(0, 1)$, $(1, 3)$ and $(3, 55)$, using the Lagrange's interpolation formula.

**Solution:** Given points are $(0, 1)$, $(1, 3)$ and $(3, 55)$.

Take $x_0 = 0, x_1 = 1, x_2 = 3$ and $y_0 = 1, y_1 = 3, y_2 = 55$.

Using Lagrange's interpolation formula to this

$$y = f(x) = \frac{(x-x_1)(x-x_2)}{(x_0-x_1)(x_0-x_2)}y_0 + \frac{(x-x_0)(x-x_2)}{(x_1-x_0)(x_1-x_2)}y_1 + \frac{(x-x_0)(x-x_1)}{(x_2-x_0)(x_2-x_1)}y_2$$

$$= \frac{(x-1)(x-3)}{(0-1)(0-3)}(1) + \frac{(x-0)(x-3)}{(1-0)(1-3)}(3) + \frac{(x-0)(x-1)}{(3-0)(3-1)}(55)$$

$$= (x-1)(x-3)\left(\frac{1}{3}\right) + x(x-3)\left(\frac{3}{-2}\right) + x(x-1)\left(\frac{55}{6}\right)$$

$$= (x^2-4x+3)\left(\frac{1}{3}\right) + (x^2-3x)\left(\frac{3}{-2}\right) + (x^2-x)\left(\frac{55}{6}\right)$$

$$= \frac{48x^2 - 36x + 6}{6}$$

$$= 8x^2 - 6x + 1.$$

This show that $y = 8x^2 - 6x + 1$ is the curve that passes through the given points. This curve is a parabola.

**Example:** A curve passes through the points $(0, 18)$, $(1, 10)$, $(3, -18)$ and $(6, 90)$. Find the slope of the curve at $x = 2$.

**Solution:** Given points are $(0, 18)$, $(1, 10)$, $(3, -18)$ and $(6, 90)$.

Take $x_0 = 0, x_1 = 1, x_2 = 3, x_3 = 6$ and $y_0 = 18, y_1 = 10, y_2 = -18, y_3 = 90$.

For this data, the Lagrange's interpolation formula gives

$$y = f(x) = \frac{(x-x_1)(x-x_2)(x-x_3)}{(x_0-x_1)(x_0-x_2)(x_0-x_3)}y_0 + \frac{(x-x_0)(x-x_2)(x-x_3)}{(x_1-x_0)(x_1-x_2)(x_1-x_3)}y_1$$
$$+ \frac{(x-x_0)(x-x_1)(x-x_3)}{(x_2-x_0)(x_2-x_1)(x_2-x_3)}y_2 + \frac{(x-x_0)(x-x_1)(x-x_2)}{(x_3-x_0)(x_3-x_1)(x_3-x_2)}y_3$$
$$= \frac{(x-1)(x-3)(x-6)}{(0-1)(0-3)(0-6)}(18) + \frac{(x-0)(x-3)(x-6)}{(1-0)(1-3)(1-6)}(10)$$
$$+ \frac{(x-0)(x-1)(x-6)}{(3-0)(3-1)(3-6)}(-18) + \frac{(x-0)(x-1)(x-3)}{(6-0)(6-1)(6-3)}(90)$$
$$= -(x-1)(x-3)(x-6) + x(x-3)(x-6)$$
$$+ x(x-1)(x-6) + x(x-1)(x-3)$$
$$= -(x^3 - 10x^2 + 27x - 18) + (x^3 - 9x^2 + 18x)$$
$$+ (x^3 - 7x^2 + 6x) + (x^3 - 4x^2 + 3x)$$
$$= 2x^3 - 10x^2 + 18.$$

This shows that the curve which passes through the given points is $y = 2x^3 - 10x^2 + 18$.
This gives $y' = 6x^2 - 20x$.
At $x = 2$, we get $y'(2) = -16$. Thus, the required slope is $-16$.

**Example:** Find the unique polynomial $P(x)$ of degree 2 or less such that $P(1) = 1, P(3) = 27, P(4) = 64$, using Lagrange's interpolation formula.

**Solution:** Here $x_0 = 0, x_1 = 3, x_2 = 4$ and $y_0 = P(1) = 1, y_1 = P(3) = 27, y_2 = P(4) = 64$. Using Lagrange's interpolation formula, we have

$$y = P(x) = \frac{(x-x_1)(x-x_2)}{(x_0-x_1)(x_0-x_2)}y_0 + \frac{(x-x_0)(x-x_2)}{(x_1-x_0)(x_1-x_2)}y_1 + \frac{(x-x_0)(x-x_1)}{(x_2-x_0)(x_2-x_1)}y_2$$
$$= \frac{(x-3)(x-4)}{(1-3)(1-4)}(1) + \frac{(x-1)(x-4)}{(3-1)(3-4)}(27) + \frac{(x-1)(x-3)}{(4-1)(4-3)}(64)$$
$$= \frac{1}{6}(x^2 - 7x + 12) - \frac{27}{2}(x^2 - 5x + 4) + \frac{64}{3}(x^2 - 4x + 3)$$
$$= \frac{48}{6}x^2 - \frac{114}{6}x + \frac{72}{6}$$
$$= 8x^2 - 19x + 12.$$

**Example:** Given $f(0) = -18, f(1) = 0, f(3) = 0, f(5) = -248, f(6) = 0, f(9) = 13104$, find $f(x)$.

**Solution:**

$$f(x) = \frac{(x-1)(x-3)(x-5)(x-6)(x-9)}{(0-1)(0-3)(0-5)(0-6)(0-9)}(-18) + \frac{(x-0)(x-3)(x-5)(x-6)(x-9)}{(1-0)(1-3)(1-5)(1-6)(1-9)}(0)$$
$$+ \frac{(x-0)(x-1)(x-5)(x-6)(x-9)}{(3-0)(3-1)(3-5)(3-6)(3-9)}(0) + \frac{(x-0)(x-1)(x-3)(x-6)(x-9)}{(5-0)(5-1)(5-3)(5-6)(5-9)}(-248)$$
$$+ \frac{(x-0)(x-1)(x-3)(x-5)(x-9)}{(6-0)(6-1)(6-3)(6-5)(6-9)}(0) + \frac{(x-0)(x-1)(x-3)(x-5)(x-6)}{(9-0)(9-1)(9-3)(9-5)(9-6)}(13104)$$

$$= (x-1)(x-3)(x-6)\left[\frac{(x-5)(x-9)}{45} - x(x-9)\frac{31}{20} + x(x-5)\frac{91}{36}\right]$$

$$= (x-1)(x-3)(x-6)\left[\frac{x-9}{180}\{4(x-5) - 279x\} + \frac{91}{36}x(x-5)\right]$$

$$= (x-1)(x-3)(x-6)\left[-\frac{(x-9)(55x+4)}{36} + \frac{91}{36}x(x-5)\right]$$

$$= (x-1)(x-3)(x-6)\left[\frac{36x^2 + 36x + 36}{36}\right]$$

$$= (x-1)(x-3)(x-6)(x^2 + x + 1).$$

**Example:** Using Lagrange's formula, fit a polynomial to the data

| $x$ | 0 | 1 | 3 | 4 |
|---|---|---|---|---|
| $f(x)$ | -12 | 0 | 6 | 12 |

Also find $y$ at $x = 2$.  (Dec-2020), (April-2022)

**Solution:** Here $x_0 = 0, x_1 = 1, x_2 = 3, x_3 = 4$ and $y_0 = -12, y_1 = 0, y_2 = 6, y_3 = 12$.

Using Lagrange's interpolation formula to this

$$y = f(x) = \frac{(x-x_1)(x-x_2)(x-x_3)}{(x_0-x_1)(x_0-x_2)(x_0-x_3)}f(x_0) + \frac{(x-x_0)(x-x_2)(x-x_3)}{(x_1-x_0)(x_1-x_2)(x_1-x_3)}f(x_1)$$
$$+ \frac{(x-x_0)(x-x_1)(x-x_3)}{(x_2-x_0)(x_2-x_1)(x_2-x_3)}f(x_2) + \frac{(x-x_0)(x-x_1)(x-x_2)}{(x_3-x_0)(x_3-x_1)(x_3-x_2)}f(x_3)$$
$$= \frac{(x-1)(x-3)(x-4)}{(0-1)(0-3)(0-4)}(-12) + \frac{(x-0)(x-3)(x-4)}{(1-0)(1-3)(1-4)}(0)$$
$$+ \frac{(x-0)(x-1)(x-4)}{(3-0)(3-1)(3-4)}(6) + \frac{(x-0)(x-1)(x-3)}{(4-0)(4-1)(4-3)}(12)$$
$$= (x-1)(x-3)(x-4)\left(\frac{-12}{12}\right) + x(x-3)(x-4)\left(\frac{0}{6}\right)$$
$$+ x(x-1)(x-4)\left(\frac{6}{-6}\right) + x(x-1)(x-3)\left(\frac{12}{12}\right)$$
$$= (x-1)(x-3)(x-4) - x(x-1)(x-4) + x(x-1)(x-3)$$
$$= (x-1)\big[(x-3)(x-4) - x(x-4) + x(x-3)\big]$$
$$= (x-1)\big[x^2 - 6x + 12\big]$$
$$= x^3 - 7x^2 + 18x - 12.$$

$\therefore y(x) = x^3 - 7x^2 + 18x - 12$ is the required polynomial.

Put $x = 2$, we get $y(2) = 8 - 28 + 36 - 12 = 4$.

**Example:** Using Lagrange's formula, fit a polynomial to the data and hence find $y(1)$.

| $x$ | $-1$ | $0$ | $2$ | $3$ |
|---|---|---|---|---|
| $y$ | $-8$ | $3$ | $1$ | $12$ |

(Aug-2022)

**Solution:** Here $x_0 = -1, x_1 = 0, x_2 = 2, x_3 = 13$ and $y_0 = -8, y_1 = 3, y_2 = 1, y_3 = 12$. Using Lagrange's interpolation formula to this

$$y = f(x) = \frac{(x-x_1)(x-x_2)(x-x_3)}{(x_0-x_1)(x_0-x_2)(x_0-x_3)}f(x_0) + \frac{(x-x_0)(x-x_2)(x-x_3)}{(x_1-x_0)(x_1-x_2)(x_1-x_3)}f(x_1)$$
$$+ \frac{(x-x_0)(x-x_1)(x-x_3)}{(x_2-x_0)(x_2-x_1)(x_2-x_3)}f(x_2) + \frac{(x-x_0)(x-x_1)(x-x_2)}{(x_3-x_0)(x_3-x_1)(x_3-x_2)}f(x_3)$$

$$= \frac{(x-0)(x-2)(x-3)}{(-1-0)(-1-2)(-1-3)}(-8) + \frac{(x+1)(x-2)(x-3)}{(0+1)(0-2)(0-3)}(3)$$
$$+ \frac{(x+1)(x-0)(x-3)}{(2+1)(2-0)(2-3)}(1) + \frac{(x+1)(x-0)(x-2)}{(3+1)(3-0)(3-2)}(12)$$
$$= x(x-2)(x-3)\left(\frac{-8}{12}\right) + (x+1)(x-2)(x-3)\left(\frac{3}{6}\right)$$
$$+ (x+1)x(x-3)\left(\frac{1}{-6}\right) + (x+1)x(x-2)\left(\frac{12}{12}\right)$$
$$= (x-2)(x-3)\left[\frac{2}{3}x + \frac{1}{2}(x+1)\right] + (x+1)x\left[\frac{-1}{6}(x-3) + (x-2)\right]$$
$$= (x^2 - 5x + 6)\left(\frac{7x+3}{6}\right) + (x^2 + x)\left(\frac{5x-9}{6}\right)$$
$$= \frac{12x^3 - 36x^2 + 18x + 18}{6}$$
$$= 2x^3 - 6x^2 + 3x + 3.$$

$\therefore y(x) = 2x^3 - 6x^2 + 3x + 3$ is the required polynomial.

Put $x = 1$, we get $y(1) = 2$.

**Example:** Given $\log_{10} 654 = 2.8156, \log_{10} 658 = 2.8182, \log_{10} 659 = 2.8189, \log_{10} 661 = 2.8202$, find by using Lagrange's formula, the value of $\log_{10} 656$.

**Solution:** Given

| $x$ | 654 | 658 | 659 | 661 |
|---|---|---|---|---|
| $\log_{10} x$ | 2.8156 | 2.8182 | 2.8189 | 2.8202 |

Here $f(x) = \log_{10} x$, $x_0 = 654, x_1 = 658, x_2 = 659, x_3 = 661$ and $y_0 = 2.8156, y_1 = 2.8182, y_2 = 2.8189, y_3 = 2.8202$.

Thus by Lagrange's interpolation formula, we have

$$f(x) = \frac{(x-x_1)(x-x_2)(x-x_3)}{(x_0-x_1)(x_0-x_2)(x_0-x_3)}f(x_0) + \frac{(x-x_0)(x-x_2)(x-x_3)}{(x_1-x_0)(x_1-x_2)(x_1-x_3)}f(x_1)$$
$$+ \frac{(x-x_0)(x-x_1)(x-x_3)}{(x_2-x_0)(x_2-x_1)(x_2-x_3)}f(x_2) + \frac{(x-x_0)(x-x_1)(x-x_2)}{(x_3-x_0)(x_3-x_1)(x_3-x_2)}f(x_3)$$

$$f(656) = \frac{(656-658)(656-659)(656-661)}{(654-658)(654-659)(654-661)} \times (2.8156)$$
$$+ \frac{(656-654)(656-659)(656-661)}{(658-654)(658-659)(658-661)} \times (2.8182)$$
$$+ \frac{(656-654)(656-658)(656-661)}{(659-654)(659-658)(659-661)} \times (2.8189)$$
$$+ \frac{(656-654)(656-658)(656-659)}{(661-654)(661-658)(661-669)} \times (2.8202)$$
$$= \frac{(-2)(-3)(-5)}{(-4)(-5)(-7)}(2.8156) + \frac{(2)(-3)(-5)}{(4)(-1)(-3)}(2.8182)$$
$$+ \frac{(2)(-2)(-5)}{(5)(1)(-2)}(2.8189) + \frac{(2)(-2)(-3)}{(7)(3)(2)}(2.8202)$$
$$= 0.6033 + 7.0458 - 5.6378 + 0.8058$$
$$= 2.8171$$

**Example:** Given the values

| $x$:    | 5   | 7   | 11   | 13   | 17   |
|---------|-----|-----|------|------|------|
| $f(x)$: | 150 | 392 | 1452 | 2366 | 5202 |

evaluate $f(9)$ using Lagrange's formula.

**Solution:** Here $x_0 = 5, x_1 = 7, x_2 = 11, x_3 = 13, x_4 = 17$ and $y_0 = 150, y_1 = 392, y_2 = 1452, y_3 = 2366, y_4 = 5202$.

Putting $x = 9$ and substituting the above values in Lagrange's interpolation formula, we obtain

$$f(x) = \frac{(x-x_1)(x-x_2)(x-x_3)(x-x_4)}{(x_0-x_1)(x_0-x_2)(x_0-x_3)(x_0-x_4)}f(x_0) + \frac{(x-x_0)(x-x_2)(x-x_3)(x-x_4)}{(x_1-x_0)(x_1-x_2)(x_1-x_3)(x_1-x_4)}f(x_1)$$
$$+ \frac{(x-x_0)(x-x_1)(x-x_3)(x-x_4)}{(x_2-x_0)(x_2-x_1)(x_2-x_3)(x_2-x_4)}f(x_2) + \frac{(x-x_0)(x-x_1)(x-x_2)(x-x_4)}{(x_3-x_0)(x_3-x_1)(x_3-x_2)(x_3-x_4)}f(x_3)$$
$$+ \frac{(x-x_0)(x-x_1)(x-x_2)(x-x_3)}{(x_4-x_0)(x_4-x_1)(x_4-x_2)(x_4-x_3)}f(x_4)$$

$$f(9) = \frac{(9-7)(9-11)(9-13)(9-17)}{(5-7)(5-11)(5-13)(5-17)}(150) + \frac{(9-5)(9-11)(9-13)(9-17)}{(7-5)(7-11)(7-13)(7-17)}(392)$$
$$+ \frac{(9-5)(9-7)(9-13)(9-17)}{(11-5)(11-7)(11-13)(11-17)}(1452) + \frac{(9-5)(9-7)(9-11)(9-17)}{(13-5)(13-7)(13-11)(13-17)}(2366)$$
$$+ \frac{(9-5)(9-7)(9-11)(9-13)}{(17-5)(17-7)(17-11)(17-13)}(5202)$$
$$= -\frac{50}{3} + \frac{3136}{15} + \frac{3872}{3} - \frac{2366}{3} + \frac{578}{5} = 810.$$

**Example:** The function $y = \sin x$ is tabulated below:

| $x$ | 0 | $\pi/4$ | $\pi/2$ |
|---|---|---|---|
| $y = \sin x$ | 0 | 0.70711 | 1.0 |

Using Lagrange's interpolation formula, find the value of $\sin(\pi/6)$.

**Solution:** Here $x_0 = 0, x_1 = \pi/4, x_2 = \pi/2$ and $y_0 = 0, y_1 = 0.70711, y_2 = 1.0$.

Using Lagrange's interpolation formula to this

$$f(x) = \frac{(x-x_1)(x-x_2)}{(x_0-x_1)(x_0-x_2)}f(x_0) + \frac{(x-x_0)(x-x_2)}{(x_1-x_0)(x_1-x_2)}f(x_1) + \frac{(x-x_0)(x-x_1)}{(x_2-x_0)(x_2-x_1)}f(x_2)$$

$$f(\pi/6) = \frac{(\pi/6-\pi/4)(\pi/6-\pi/2)}{(0-\pi/4)(0-\pi/2)}(0) + \frac{(\pi/6-0)(\pi/6-\pi/2)}{(\pi/4-0)(\pi/4-\pi/2)}(0.70711)$$
$$+ \frac{(\pi/6-0)(\pi/6-\pi/4)}{(\pi/2-0)(\pi/2-\pi/4)}(1.0)$$
$$= 0 + \frac{(\pi/6)(-\pi/3)}{(\pi/4)(-\pi/4)}(0.70711) + \frac{(\pi/6)(-\pi/12)}{(\pi/2)(\pi/4)}(1.0)$$
$$= 0.62854 - 0.11111$$
$$= 0.51743.$$

**Example:** Using Lagrange's interpolation formula find the value of $y(10)$ from the following table: (Jan-2020), (Dec-2020), (August-2021), (April-2022)

| $x$ | 5 | 6 | 9 | 11 |
|---|---|---|---|---|
| $f(x)$ | 12 | 13 | 14 | 16 |

**Solution:** Here $x_0 = 5, x_1 = 6, x_2 = 9, x_3 = 11$ and $y_0 = 12, y_1 = 13, y_2 = 14, y_3 = 16$.
Using Lagrange's interpolation formula to this

$$f(x) = \frac{(x-x_1)(x-x_2)(x-x_3)}{(x_0-x_1)(x_0-x_2)(x_0-x_3)}f(x_0) + \frac{(x-x_0)(x-x_2)(x-x_3)}{(x_1-x_0)(x_1-x_2)(x_1-x_3)}f(x_1)$$
$$+ \frac{(x-x_0)(x-x_1)(x-x_3)}{(x_2-x_0)(x_2-x_1)(x_2-x_3)}f(x_2) + \frac{(x-x_0)(x-x_1)(x-x_2)}{(x_3-x_0)(x_3-x_1)(x_3-x_2)}f(x_3)$$

$$f(x) = \frac{(x-6)(x-9)(x-11)}{(5-6)(5-9)(5-11)}(12) + \frac{(x-5)(x-9)(x-11)}{(6-5)(6-9)(6-11)}(13)$$
$$+ \frac{(x-5)(x-6)(x-11)}{(9-5)(9-6)(9-11)}(14) + \frac{(x-5)(x-6)(x-9)}{(11-5)(11-6)(11-9)}(16)$$

$$f(10) = \frac{(10-6)(10-9)(10-11)}{(-1)(-4)(-6)}(12) + \frac{(10-5)(10-9)(10-11)}{(1)(-3)(-5)}(13)$$
$$+ \frac{(10-5)(10-6)(10-11)}{(4)(3)(-2)}(14) + \frac{(10-5)(10-6)(10-9)}{(6)(5)(2)}(16)$$

$$= 2 - \frac{13}{3} + \frac{35}{3} + \frac{16}{3}$$

$$= 14.6666$$

**Example:** Given $u_1 = 22, u_2 = 30, u_4 = 82, u_7 = 106, u_8 = 206$ find $u_6$. Use Lagrange's interpolation formula.

**Solution:** Given data can be tabulated as follows:

| $x$ | 1 | 2 | 4 | 7 | 8 |
|---|---|---|---|---|---|
| $u(x)$ | 22 | 30 | 82 | 106 | 206 |

Here $x_0 = 1, x_1 = 2, x_2 = 4, x_3 = 7, x_4 = 8$ and $u_0 = 22, u_1 = 30, u_2 = 82, u_3 = 106, u_4 = 206$.

Using Lagrange's interpolation formula to this

$$u(x) = \frac{(x-x_1)(x-x_2)(x-x_3)(x-x_4)}{(x_0-x_1)(x_0-x_2)(x_0-x_3)(x_0-x_4)}u_0 + \frac{(x-x_0)(x-x_2)(x-x_3)(x-x_4)}{(x_1-x_0)(x_1-x_2)(x_1-x_3)(x_1-x_4)}u_1$$
$$+ \frac{(x-x_0)(x-x_1)(x-x_3)(x-x_4)}{(x_2-x_0)(x_2-x_1)(x_2-x_3)(x_2-x_4)}u_2 + \frac{(x-x_0)(x-x_1)(x-x_2)(x-x_4)}{(x_3-x_0)(x_3-x_1)(x_3-x_2)(x_3-x_4)}u_3$$
$$+ \frac{(x-x_0)(x-x_1)(x-x_2)(x-x_3)}{(x_4-x_0)(x_4-x_1)(x_4-x_3)(x_4-x_3)}u_4$$
$$= \frac{(x-2)(x-4)(x-7)(x-8)}{(1-2)(1-4)(1-7)(1-8)}(22) + \frac{(x-1)(x-4)(x-7)(x-8)}{(2-1)(2-4)(2-7)(2-8)}(30)$$
$$+ \frac{(x-1)(x-2)(x-7)(x-8)}{(4-1)(4-2)(4-7)(4-8)}(82) + \frac{(x-1)(x-2)(x-4)(x-8)}{(7-1)(7-2)(7-4)(7-8)}(106)$$
$$+ \frac{(x-1)(x-2)(x-4)(x-7)}{(8-1)(8-2)(8-4)(8-7)}(206)$$

$$f(6) = \frac{(6-2)(6-4)(6-7)(6-8)}{(-1)(-3)(-6)(-7)}(22) + \frac{(6-1)(6-4)(6-7)(6-8)}{(1)(-2)(-5)(-6)}(30)$$
$$+ \frac{(6-1)(6-2)(6-7)(6-8)}{(3)(2)(-3)(-4)}(82) + \frac{(6-1)(6-2)(6-4)(6-8)}{(6)(5)(3)(-1)}(106)$$
$$+ \frac{(6-1)(6-2)(6-4)(6-7)}{(7)(6)(4)(1)}(206)$$
$$= 2.7936 - 10 + 45.5 + 94.2 - 49.0476$$
$$= 83.446$$

**Example:** The following are the measurements of $t$ made on a curve recorded by an oscillograph representing a change of current $I$ due to a change in the conditions of an electric current.

| $t$ : | 1.2 | 2.0 | 2.5 | 3.0 |
|---|---|---|---|---|
| $I$ : | 1.36 | 0.58 | 0.34 | 0.20 |

Using Lagrange's formula, find $I$ at $t = 1.6$.

**Solution:** Here $t_0 = 1.2, t_1 = 2.0, t_2 = 2.5, t_3 = 3.0$ and $I_0 = 1.36, I_1 = 0.58, I_2 = 0.34, I_3 = 0.20$.

Using Lagrange's interpolation formula to this

$$I(x) = \frac{(t-t_1)(t-t_2)(t-t_3)}{(t_0-t_1)(t_0-t_2)(t_0-t_3)}I_0 + \frac{(t-t_0)(t-t_2)(t-t_3)}{(t_1-t_0)(t_1-t_2)(t_1-t_3)}I_1$$
$$+ \frac{(t-t_0)(t-t_1)(t-t_3)}{(t_2-t_0)(t_2-t_1)(t_2-t_3)}I_2 + \frac{(t-t_0)(t-t_1)(t-t_2)}{(t_3-t_0)(t_3-t_1)(t_3-t_2)}I_3$$

$$I(1.6) = \frac{(1.6-2.0)(1.6-2.5)(1.6-3.0)}{(1.2-2.0)(1.2-2.5)(1.2-3.0)} \times (1.36)$$
$$+ \frac{(1.6-1.2)(1.6-2.5)(1.6-3.0)}{(2.0-1.2)(2.0-2.5)(2.0-2.5)} \times (0.58)$$
$$+ \frac{(1.6-1.2)(1.6-2.0)(1.6-3.0)}{(2.5-1.2)(2.5-2.0)(2.5-3.0)} \times (0.34)$$
$$+ \frac{(1.6-1.2)(1.6-2.0)(1.6-2.5)}{(3.0-1.2)(3.0-2.0)(3.0-2.5)} \times (0.20)$$

$$= \frac{(-0.4)(-0.9)(-1.4)}{(-0.8)(-1.3)(-1.8)} \times (1.36) + \frac{(0.4)(0.9)(1.4)}{(0.8)(0.5)(-1.0)} \times (0.58)$$
$$+ \frac{(0.4)(-0.4)(-1.4)}{(1.3)(0.5)(-0.5)} \times (0.34) + \frac{(0.4)(-0.4)(-0.9)}{(1.8)(1.0)(0.5)} \times (0.20)$$

$$= \frac{0.68544}{1.872} + \frac{0.29232}{0.4} - \frac{0.07616}{0.325} + \frac{0.0288}{0.9}$$

$$= 0.36615 + 0.7308 - 0.23434 + 0.032$$

$$= 0.89461.$$

**EXERCISE**

1. Fit second degree polynomial for the data $(0,1), (1, 2.75), (2, 4)$ using Lagrange's interpolation formula. **(Dec-2021)**

**Ans:** $-0.25x^2 + 2x + 1$

2. Using Lagrange's interpolation formula calculate $f(3)$ from the following data:

| $x$ | 0 | 1 | 2 | 4 | 5 |
|---|---|---|---|---|---|
| $f(x)$ | 1 | 14 | 15 | 5 | 6 |

**(Jan-2020)**

**Ans:** 10.

3. Find the Lagrange's polynomial for the following data. **(Aug-2021)**

| $x$ | 0 | 2 | 3 | 6 |
|---|---|---|---|---|
| $f(x)$ | 6 | 7 | 9 | 12 |

**Ans:** $-\dfrac{x^3}{8} + \dfrac{9x^2}{8} - \dfrac{5x}{4} + 6$

4. Using Lagrange's interpolation formula calculate $f(6)$ from the following data:

| $x$    | 1  | 2  | 4  | 7   | 8   |
|--------|----|----|----|-----|-----|
| $f(x)$ | 22 | 30 | 82 | 106 | 206 |

**Ans:** 83.5237

5. Find the $y(4)$ for the following data  **(Sep-2021)**

| $x$ | 0   | 2   | 3   | 6   |
|-----|-----|-----|-----|-----|
| $y$ | 707 | 819 | 866 | 966 |

**Ans:** $y(4) = \dfrac{8158}{9}$

6. Certain corresponding values of $x$ and $\log_{10} x$ are: $(300, 2.4771), (304, 2.4829)$, $(305, 2.4843)$ and $(307, 2.4871)$. Find $\log_{10} 301$ by using the Lagrange's interpolation formula.

**Ans:** $\log_{10} 301 = 2.4786$.

7. Apply Lagranges formula to find $f(5)$ given that $f(1) = 2, f(2) = 4, f(3) = 8, f(4) = 16$ and $f(7) = 128$.

**Ans:** 32.9333

8. Compute $f(27)$ Using Lagranges formula from the following table:

| $x$    | 0   | 2   | 3   | 6   |
|--------|-----|-----|-----|-----|
| $f(x)$ | 648 | 704 | 729 | 792 |

**Ans:** $648 + 30x - x^2$

9. Using the Table:

| $x$    | 14   | 17   | 31   | 53   |
|--------|------|------|------|------|
| $f(x)$ | 68.7 | 64.0 | 44.0 | 39.1 |

find the polynomial approximation $f(x)$.

**Ans:** $0.000650575x^3 - 0.0322124x^2 - 1.03845x + 87.7667$.

10. Evaluate $f(10)$ given that the value of $f(x)$ at $x = 2, 6, 12$ are $16, 19, 33$ respectively.  **(Dec-2020)**

**Ans:** $f(10) = 26.4$

11. Find $u_3$ using Lagrange's formula given that  (Mar-2022)

$$u_0 = 580, u_1 = 556, u_2 = 520, \& u_4 = 385.$$

**Ans:** $u_3 = 465.25$

12. Using Lagrange's interpolation formula, find $y(8)$ from the following table

(April-2022)

| $x$ | 1 | 4 | 6 | 10 |
|---|---|---|---|---|
| $y$ | 3 | 5 | 9 | 11 |

**Ans:** $y(8) = 11.9185$

13. Using Lagrange's interpolation formula calculate $f(6)$ from the following data:

| $x$ | 3 | 5 | 7 | 9 | 11 |
|---|---|---|---|---|---|
| $f(x)$ | 6 | 24 | 58 | 108 | 74 |

**Ans:** 36.356

14. Using Lagrange's interpolation formula calculate $f(5)$ from the following data:

| $x$ | 0 | 1 | 3 | 8 |
|---|---|---|---|---|
| $f(x)$ | 1 | 3 | 13 | 128 |

**Ans:** 38.857  (Jan-2020)

15. Find the $y(4)$ from the following data:  (Apr-2022)

| $x$ | 1 | 2 | 5 | 7 |
|---|---|---|---|---|
| $y$ | 2 | 3 | 6 | 8 |

16. Using Lagrange's formula find $f(3)$ from the following data  (Aug-2022)

| $x$ | 0 | 1 | 4 | 5 |
|---|---|---|---|---|
| $f(x)$ | 4 | 3 | 24 | 39 |

17. Using Lagrange's formula find $f(4)$ from the following data  (Aug-2022)

| $x$ | 1 | 2 | 2.5 | 3 |
|---|---|---|---|---|
| $f(x)$ | −6 | −1 | 5 | 16 |

18. Evaluate $y(4)$ from the following table: (Aug-2022)

| x | 1 | 3 | 5 | 6 | 8 |
|---|---|---|---|---|---|
| y | 2 | 1.5 | 2.4 | 4 | 5.6 |

**Example:** Use lagrange's formula to express the following function $\dfrac{3x^2 + x + 1}{(x-1)(x-2)(x-3)}$ as the sum of partial fractions.

**Solution:** Let $f(x) = 3x^2 + x + 1$. The zeros of the determinant are at $x_0 = 1, x_1 = 2$ and $x_3 = 3$.

Evaluate $f(x)$ at these points

| x: | 1 | 2 | 3 |
|---|---|---|---|
| y: | 5 | 15 | 31 |

We fit a polynomial using Lagrange's formula

$$f(x) = \frac{(x-x_1)(x-x_2)}{(x_0-x_1)(x_0-x_2)}f(x_0) + \frac{(x-x_0)(x-x_2)}{(x_1-x_0)(x_1-x_2)}f(x_1) + \frac{(x-x_0)(x-x_1)}{(x_2-x_0)(x_2-x_1)}f(x_2)$$

$$= \frac{(x-2)(x-3)}{(1-2)(1-3)}(5) + \frac{(x-1)(x-3)}{(2-1)(2-3)}(15) + \frac{(x-1)(x-2)}{(3-1)(3-2))}(31)$$

$$= \frac{5}{2}(x-2)(x-3) - 15(x-1)(x-3) + \frac{31}{2}(x-1)(x-2)$$

Dividing both sides by $(x-1)(x-2)(x-3)$ we obtain

$$\frac{f(x)}{(x-1)(x-2)(x-3)} = \frac{3x^2 + x + 1}{(x-1)(x-2)(x-3)}$$

$$= \frac{5}{2} \cdot \frac{1}{x-1} - 15 \cdot \frac{1}{x-2} + \frac{31}{2} \cdot \frac{1}{x-3}.$$

**Example:** Express the following function $\dfrac{3x^2 - 12x + 11}{(x-1)(x-2)(x-3)}$ as the sum of partial fractions, using lagrange's formula.

**Solution:** Let $f(x) = 3x^2 - 12x + 11$. The zeros of the determinant are at $x_0 = 1, x_1 = 2$ and $x_3 = 3$.

Evaluate $f(x)$ at these points

| $x$: | 1 | 2 | 3 |
|---|---|---|---|
| $y$: | 2 | $-1$ | 2 |

We fit a polynomial using Lagrange's formula

$$f(x) = \frac{(x-x_1)(x-x_2)}{(x_0-x_1)(x_0-x_2)} f(x_0) + \frac{(x-x_0)(x-x_2)}{(x_1-x_0)(x_1-x_2)} f(x_1) + \frac{(x-x_0)(x-x_1)}{(x_2-x_0)(x_2-x_1)} f(x_2)$$

$$= \frac{(x-2)(x-3)}{(1-2)(1-3)}(2) + \frac{(x-1)(x-3)}{(2-1)(2-3)}(-1) + \frac{(x-1)(x-2)}{(3-1)(3-2)}(2)$$

$$= (x-2)(x-3) \cdot 1 + (x-1)(x-3) \cdot 1 + (x-1)(x-2) \cdot 1$$

Dividing both sides by $(x-1)(x-2)(x-3)$ we obtain

$$\frac{f(x)}{(x-1)(x-2)(x-3)} = \frac{3x^2 - 12x + 11}{(x-1)(x-2)(x-3)}$$

$$= \frac{1}{x-1} + \frac{1}{x-2} + \frac{1}{x-3}.$$

**Example:** Express the following function

$$\frac{x^2 + 6x - 1}{(x^2-1)(x-4)(x-6)}$$

as the sum of partial fractions, using lagrange's formula.

**Solution:** Let $f(x) = x^2 + 6x - 1$. The zeros of the determinant are at $x_0 = -1, x_1 = 1, x_2 = 4$ and $x_3 = 6$.

Evaluate $f(x)$ at these points

$f(x_0) = f(-1) = -6$
$f(x_1) = f(1) = 6$
$f(x_2) = f(4) = 39$
$f(x_3) = f(6) = 71$.

| $x$: | $-1$ | 1 | 4 | 6 |
|---|---|---|---|---|
| $y$: | $-6$ | 6 | 39 | 71 |

We fit a polynomial using Lagrange's formula

$$f(x) = \frac{(x-x_1)(x-x_2)(x-x_3)}{(x_0-x_1)(x_0-x_2)(x_0-x_3)}f(x_0) + \frac{(x-x_0)(x-x_2)(x-x_3)}{(x_1-x_0)(x_1-x_2)(x_1-x_3)}f(x_1)$$
$$+ \frac{(x-x_0)(x-x_1)(x-x_3)}{(x_2-x_0)(x_2-x_1)(x_2-x_3)}f(x_2) + \frac{(x-x_0)(x-x_1)(x-x_2)}{(x_3-x_0)(x_3-x_1)(x_3-x_2)}f(x_3)$$
$$= \frac{(x-1)(x-4)(x-6)}{(-1-1)(-1-4)(-1-6)}(-6) + \frac{(x+1)(x-4)(x-6)}{(1+1)(1-4)(1-6)}(6)$$
$$+ \frac{(x+1)(x-1)(x-6)}{(4+1)(4-1)(4-6)}(39) + \frac{(x+1)(x-1)(x-4)}{(6+1)(6-1)(6-4)}(71)$$
$$= \frac{3}{35}(x-1)(x-4)(x-6) + \frac{1}{5}(x+1)(x-1)(x-6) - \frac{13}{10}(x+1)(x-1)(x-6)$$
$$+ \frac{71}{70}(x+1)(x-1)(x-4)$$

Dividing both sides by $(x^2-1)(x-4)(x-6)$ we obtain

$$\frac{f(x)}{(x^2-1)(x-4)(x-6)} = \frac{x^2+6x-1}{(x^2-1)(x-4)(x-6)}$$
$$= \frac{3}{35}\frac{1}{x+1} + \frac{1}{5}\frac{1}{x-1} - \frac{13}{10}\frac{1}{x-4} + \frac{71}{70}\frac{1}{x-6}.$$

## EXERCISE

1. Using lagrange's formula to express the following function $\frac{x^2+x-3}{x^3-2x^2+x-2}$ as the sum of partial fractions.

   **Ans:** $\frac{1}{2} \cdot \frac{1}{x-1} - \frac{1}{2} \cdot \frac{1}{x+1} + \frac{1}{x-2}$.

2. Using lagrange's formula to express the following function $\frac{x^2+2x+4}{x^2-x+12}$ as the sum of partial fractions.

   **Ans:** $1 + \frac{3x-8}{(x-4)(x+3)} = 1 + \frac{4}{7} \cdot \frac{1}{x-4} + \frac{17}{7} \cdot \frac{1}{x+3}$.

3. Using lagrange's formula to express the following function $\frac{3x^2+x+1}{x^3-6x^2+11x-6}$ as the sum of partial fractions.

   **Ans:** $\frac{5}{2} \cdot \frac{1}{x-1} - \frac{15}{x-2} + \frac{31}{2} \cdot \frac{1}{x-3}$.

## 4.8.2 Inverse Interpolation

In the inverse interpolation for a given value of $y$, the corresponding value of $x$ is to be found. Interchange the roles of $x$ and $y = f(x)$ in Lagrange's interpolation formula we get Lagrange' inverse interpolation formula as

$$x = \frac{(y-y_1)(y-y_2)\cdots(y-y_n)}{(y_0-y_1)(y_0-y_2)\cdots(y_0-y_n)}x_0 + \frac{(y-y_0)(y-y_2)\cdots(y-y_n)}{(y_1-y_0)(y_1-y_2)\cdots(y_1-y_n)}x_1$$
$$+ \cdots + \frac{(y-y_0)(y-y_1)\cdots(y-y_{n-1})}{(y_n-y_0)(y_n-y_1)\cdots(y_n-y_{n-1})}x_n$$

**Example:** Determine $x(7)$ from the following data

| x: | 1 | 3 | 4 |
|---|---|---|---|
| y: | 4 | 12 | 19 |

**Solution:** Let $y = f(x)$ and $x_0 = 1, x_1 = 3, x_2 = 4$;

$$y_0 = 4, y_1 = 12, y_2 = 19.$$

By Lagrange's inverse interpolation formula, we have

$$x(y) = \frac{(y-y_1)(y-y_2)}{(y_0-y_1)(y_0-y_2)}x_0 + \frac{(y-y_0)(y-y_2)}{(y_1-y_0)(y_1-y_2)}x_1 + \frac{(y-y_0)(y-y_1)}{(y_2-y_0)(y_2-y_1)}x_2$$

$$x(7) = \frac{(7-12)(7-19)}{(4-12)(4-19)}(1) + \frac{(7-4)(7-19)}{(12-4)(12-19)}(3) + \frac{(7-4)(7-12)}{(19-4)(19-12)}(4)$$

$$= \frac{(-5)(-12)}{(-8)(-15)} + \frac{(3)(-12)}{(8)(-7)}(3) + \frac{(3)(-5)}{(15)(7)}(4)$$

$$= \frac{1}{2} + \frac{27}{14} - \frac{4}{7} = 1.8571.$$

**Example:** Compute the value of $x$ when $y = 8$ by Lagrange's inverse interpolation formula

| x: | −2 | −1 | 1 | 2 |
|---|---|---|---|---|
| y: | −7 | 2 | 0 | 11 |

**Solution:** From the given data, we have

$$x_0 = -2, x_1 = -1, x_2 = 1, x_3 = 2$$

$y_0 = -7, y_1 = 2, y_2 = 0, y_3 = 11$.

By Lagrange's inverse interpolation formula

$$x(y) = \frac{(y-y_1)(y-y_2)(y-y_3)}{(y_0-y_1)(y_0-y_2)(y_0-y_3)}x_0 + \frac{(y-y_0)(y-y_2)(y-y_3)}{(y_1-y_0)(y_1-y_2)(y_1-y_3)}x_1$$
$$+ \frac{(y-y_0)(y-y_1)(y-y_3)}{(y_2-y_0)(y_2-y_1)(y_2-y_3)}x_2 + \frac{(y-y_0)(y-y_1)(y-y_2)}{(y_3-y_0)(y_3-y_1)(y_3-y_2)}x_3$$
$$= \frac{(y-2)(y-0)(y-1)}{(-7-2)(-7-0)(-7-1)}(-2) + \frac{(y+7)(y-0)(y-11)}{(2+7)(2-0)(2-11)}(-1)$$
$$+ \frac{(y+7)(y-2)(y-11)}{(0+7)(0-2)(0-11)}(1) + \frac{(y+7)(y-2)(y-0)}{(11+7)(11-2)(11-0)}(11)$$

$$x(8) = \frac{(6)(8)(-3)}{(-9)(-7)(-18)}(-2) + \frac{(15)(8(-3))}{(9)(2)(-9)}(-1) + \frac{(15)(6)(-3)}{(7)(-2)(-11)}(1) + \frac{(15)(6)(8)}{(18)(9)(11)}(2)$$
$$= \frac{-8}{21} - \frac{20}{9} - \frac{135}{77} + \frac{80}{99} = -3.5483.$$

**Example:** Compute the value of $x$ when $y = 15$

| $x$: | 5 | 6 | 9 | 11 |
|---|---|---|---|---|
| $y$: | 12 | 13 | 14 | 16 |

**Ans: 11.5.**

## 4.8.3 Divided Differences

Let the function $y = f(x)$ be given at the points $x_0, x_1, \cdots, x_n$ (which need not be equally spaced).

Let $f(x_0), f(x_1), \cdots, f(x_n)$ denote the $(n+1)$ values of the function at the points $x_0, x_1, \cdots, x_n$.

Then the **first divided differences** of $f(x)$ for the arguments $x_0, x_1$ defined as

$$\frac{f(x_1) - f(x_0)}{x_1 - x_0}$$

It is denoted by $f(x_0, x_1)$ or $[x_0, x_1]$. We have

$$f(x_0, x_1) = \frac{f(x_1) - f(x_0)}{x_1 - x_0}.$$

Similarly we can define
$$f(x_1, x_2) = \frac{f(x_2) - f(x_1)}{x_2 - x_1}.$$
$$f(x_2, x_3) = \frac{f(x_3) - f(x_2)}{x_3 - x_2}.$$

The **second divided differences** for the arguments $x_0, x_1, x_2$ is defined as
$$f(x_0, x_1, x_2) = \frac{f(x_1, x_2) - f(x_0, x_1)}{x_2 - x_0}.$$

Similarly the **third divided differences** for the arguments $x_0, x_1, x_2, x_3$ is defined as
$$f(x_0, x_1, x_2, x_3) = \frac{f(x_1, x_2, x_3) - f(x_0, x_1, x_2)}{x_3 - x_0}.$$

In general, the $n$th order divided differences:
$$f(x_0, x_1, x_2, \cdots, x_n) = \frac{f(x_1, x_2, \cdots, x_n) - f(x_0, x_1, \cdots, x_{n-1})}{x_n - x_0}.$$

**Example:** If $f(x) = \dfrac{1}{x}$, then find $f(a, b)$ and $f(a, b, c)$.

**Solution:** We have $f(x) = \dfrac{1}{x}$.

$$\therefore f(a, b) = \frac{f(b) - f(a)}{b - a} = \frac{\frac{1}{b} - \frac{1}{a}}{b - a} = \frac{a - b}{ab(b - a)} = -\frac{1}{ab}$$

$$f(a, b, c) = \frac{f(b, c) - f(a, b)}{c - a}$$
$$= \frac{\frac{-1}{bc} - \left(\frac{-1}{ab}\right)}{c - a}$$
$$= \frac{1}{b}\left[\frac{-1}{c} + \frac{1}{a}\right]\frac{1}{c - a}$$
$$= \frac{1}{b}\left[\frac{-a + c}{ac}\right]\frac{1}{c - a}$$
$$= \frac{1}{abc}.$$

**Example:** If $f(x) = \dfrac{1}{x^2}$, then find the divided differences $f(a,b)$, $f(a,b,c)$ and $f(a,b,c,d)$.

**Solution:** Here $a, b, c$ are arguments for $f(x) = \dfrac{1}{x^2}$.

$$\therefore f(a) = \frac{1}{a^2}; \quad f(b) = \frac{1}{b^2} \text{ and } f(c) = \frac{1}{c^2}.$$

$$\therefore f(a,b) = \frac{f(b)-f(a)}{b-a} = \frac{\frac{1}{b^2}-\frac{1}{a^2}}{b-a} = -\frac{b^2-a^2}{(b-a)a^2b^2} = -\frac{a+b}{a^2b^2}$$

Similarly, we can show that $f(b,c) = -\dfrac{b+c}{b^2c^2}$.

Again

$$\begin{aligned}
f(a,b,c) &= \frac{f(b,c)-f(a,b)}{c-a} \\
&= \frac{-1}{c-a}\left[\frac{b+c}{b^2c^2} - \frac{a+b}{a^2b^2}\right] \\
&= \frac{-1}{c-a}\left[\frac{a^2(b+c)-c^2(a+b)}{a^2b^2c^2}\right] \\
&= \frac{-1}{c-a}\left[\frac{a^2b+a^2c-c^2a-c^2b}{a^2b^2c^2}\right] \\
&= -\left[\frac{b(a^2-c^2)+ac(a-c)}{(c-a)a^2b^2c^2}\right] \\
&= -\left[\frac{(a-c)\{b(a+c)+ac\}}{(c-a)a^2b^2c^2}\right] \\
&= \frac{ab+bc+ca}{a^2b^2c^2}
\end{aligned}$$

By symmetry, we have $f(b,c,d) = \dfrac{bc+cd+db}{b^2c^2d^2}$.

Again,

$$\begin{aligned}
f(a,b,c,d) &= \frac{f(b,c,d) - f(a,b,c)}{d-a} \\
&= \frac{1}{d-a}\left[\frac{bc+cd+db}{b^2c^2d^2} - \frac{ab+bc+ca}{a^2b^2c^2}\right] \\
&= \frac{1}{d-a}\left[\frac{a^2(bc+cd+db) - d^2(ab+bc+ca)}{a^2b^2c^2d^2}\right] \\
&= \frac{1}{d-a}\left[\frac{a^2bc + a^2cd + a^2db - d^2ab - d^2bc - d^2ca}{a^2b^2c^2d^2}\right] \\
&= \frac{1}{d-a}\left[\frac{bc(a^2 - d^2) + acd(a-d) + abd(a-d)}{a^2b^2c^2d^2}\right] \\
&= \frac{1}{d-a}\left[\frac{(a-d)(abc + bcd + acd + abd)}{a^2b^2c^2d^2}\right] \\
&= -\frac{abc + bcd + acd + abd}{a^2b^2c^2d^2}.
\end{aligned}$$

### Newton Divided Difference Table.

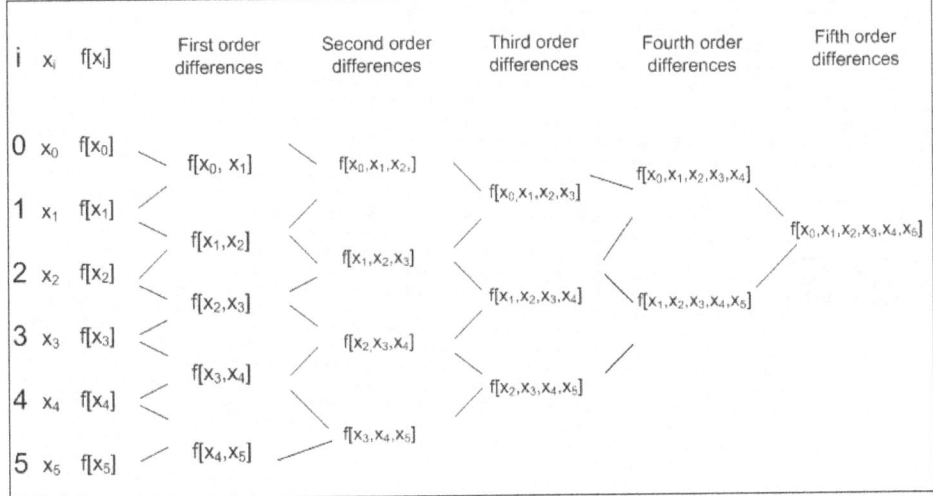

**Example:** Construct the Newton's divided difference table for the following data:

| $i$ | 0 | 1 | 2 | 3 | 4 |
|---|---|---|---|---|---|
| $x_i$ | 0 | 1 | 2 | 3 | 4 |
| $y_i = f(x_i)$ | 0 | 1 | 8 | 27 | 64 |

| i | $x_i$ | f[$x_i$] | 1ˢᵗ order differences | 2ⁿᵈ order differences | 3ʳᵈ order differences | 4ᵗʰ order differences |
|---|---|---|---|---|---|---|
| 0 | 0 | 0 | | | | |
| | | | $\frac{1-0}{1-0} = 1$ | | | |
| 1 | 1 | 1 | | $\frac{7-1}{2-0} = 3$ | | |
| | | | $\frac{8-1}{2-1} = 7$ | | $\frac{6-3}{3-0} = 1$ | |
| 2 | 2 | 8 | | $\frac{19-7}{3-1} = 6$ | | $\frac{1-1}{4-0} = 0$ |
| | | | $\frac{27-8}{3-2} = 19$ | | $\frac{9-6}{4-1} = 1$ | |
| 3 | 3 | 27 | | $\frac{37-19}{4-2} = 9$ | | |
| | | | $\frac{64-27}{4-3} = 37$ | | | |
| 4 | 4 | 64 | | | | |

**Example:** Prepare a divided difference table for the following data

| $x$: | 1 | 3 | 4 | 6 | 10 |
|---|---|---|---|---|---|
| $f(x)$: | 0 | 18 | 58 | 190 | 920 |

**Solution:** The divided difference table is

| $x$ | $y$ | first | second | third | fourth |
|---|---|---|---|---|---|
| 1 | 0 | | | | |
| 3 | 18 | $\dfrac{18-0}{3-1}=9$ | $\dfrac{40-9}{4-1}=10.33$ | | |
| 4 | 58 | $\dfrac{58-18}{4-3}=40$ | $\dfrac{66-40}{6-3}=8.66$ | $\dfrac{8.66-10.33}{6-1}=-0.334$ | |
| 6 | 190 | $\dfrac{190-58}{6-4}=66$ | $\dfrac{182.5-66}{10-4}=19.42$ | $\dfrac{19.42-8.66}{10-3}=1.5371$ | $\dfrac{1.5271+0.334}{10-1}=0.2079$ |
| 10 | 920 | $\dfrac{920-190}{10-6}=182.5$ | | | |

**Example:** Find the third divided difference with arguments 2, 4, 9, 10 of the function $f(x) = x^3 - 2x$.

**Solution:**

| $x$ | $y$ | first | second | third |
|---|---|---|---|---|
| 2 | 4 | | | |
| | | $\dfrac{56-4}{4-2} = 26$ | | |
| 4 | 56 | | $\dfrac{131-26}{9-2} = 15$ | |
| | | $\dfrac{711-56}{9-4} = 131$ | | $\dfrac{23-15}{10-2} = 1$ |
| 9 | 711 | | $\dfrac{269-131}{10-4} = 23$ | |
| | | $\dfrac{980-711}{10-9} = 269$ | | |
| 10 | 980 | | | |

Hence, the third divided difference is 1.

### 4.8.4  Newton's Divided Difference Formula

**Theorem:** Let $f(x)$ be a function taking values $y_0, y_1, \cdots, y_n$ corresponding to the values $x_0, x_1, \cdots, x_n$. Then the Newton's divided difference formula is given by

$$f(x) = f(x_0) + (x - x_0)f(x_0, x_1) + (x - x_0)(x - x_1)f(x_0, x_1, x_2)$$
$$+ \cdots + (x - x_0)(x - x_1) \cdots (x - x_{n-1}) f(x_0, x_1, x_2, \cdots, x_n)$$

**Proof:** Consider

$$f(x, x_0) = \frac{f(x) - f(x_0)}{x - x_0}$$
$$\Rightarrow f(x) = f(x_0) + (x - x_0)f(x, x_0) \quad \cdots \quad (1)$$

Also

$$f(x, x_0, x_1) = \frac{f(x, x_0) - f(x_0, x_1)}{x - x_1}$$
$$\Rightarrow f(x, x_0) = f(x_0, x_1) + (x - x_1)f(x, x_0, x_1) \quad \cdots \quad (2)$$

Substituting (2) in (1), we have

$$f(x) = f(x_0) + (x - x_0)\big[f(x_0, x_1) + (x - x_1)f(x, x_0, x_1)\big]$$
$$= f(x_0) + (x - x_0)f(x_0, x_1) + (x - x_0)(x - x_1)f(x, x_0, x_1) \quad \cdots \text{ (3)}$$

Also
$$f(x, x_0, x_1, x_2) = \frac{f(x, x_0, x_1) - f(x_0, x_1, x_2)}{x - x_2}$$
$$\Rightarrow f(x, x_0, x_1) = f(x_0, x_1, x_2) + (x - x_2)f(x, x_0, x_1, x_2) \quad \cdots \text{ (4)}$$

Substituting (4) in (3), we have

$$f(x) = f(x_0) + (x - x_0)f(x_0, x_1) + (x - x_0)(x - x_1)\big[f(x_0, x_1, x_2) + (x - x_2)f(x, x_0, x_1, x_2)\big]$$
$$= f(x_0) + (x - x_0)f(x_0, x_1) + (x - x_0)(x - x_1)f(x_0, x_1, x_2)$$
$$+ (x - x_0)(x - x_1)(x - x_2)f(x, x_0, x_1, x_2)$$

Continuing this way,

$$f(x) = f(x_0) + (x - x_0)f(x_0, x_1) + (x - x_0)(x - x_1)f(x_0, x_1, x_2)$$
$$+ \cdots + (x - x_0)(x - x_1)\cdots(x - x_{n-1})f(x_0, x_1, x_2, \cdots, x_n)$$

This is the **Newton's divided difference formula.**

**Note:** If $f(x)$ is a polynomial of degree $n$ or $n$th order divided difference is a constant then
$$f(x, x_0, x_1, \cdots, x_n) = 0.$$

**Example:** Obtain the Newton's divided difference interpolation polynomial and hence find $f(6)$: (Dec-2020)

| $x$:    | 3   | 7   | 9  | 10 |
|---------|-----|-----|----|----|
| $f(x)$: | 168 | 120 | 72 | 63 |

**Solution:** Construct the following divided difference table:

| $x$ | $y$ | first div. diff | second div. diff | third div. diff |
|---|---|---|---|---|
| 3 | 168 | | | |
| | | $\dfrac{120-168}{7-3} = \boxed{-12}$ | | |
| 7 | 120 | | $\dfrac{-24+12}{9-3} = \boxed{-2}$ | |
| | | $\dfrac{71-120}{9-7} = -24$ | | $\dfrac{5+2}{10-3} = \boxed{1}$ |
| 9 | 72 | | $\dfrac{-9+24}{10-7} = 5$ | |
| | | $\dfrac{63-72}{10-9} = -9$ | | |
| 10 | 63 | | | |

Apply Newton's divided difference formula, we get

$$f(x) = f(x_0) + (x-x_0)f(x_0,x_1) + (x-x_0)(x-x_1)f(x_0,x_1,x_2)$$
$$+ (x-x_0)(x-x_1)(x-x_2)f(x_0,x_1,x_2,x_3)$$
$$= 168 + (x-3)(-12) + (x-3)(x-7)(-2) + (x-3)(x-7)(x-9)(1)$$
$$= x^3 - 21x^2 + 119x - 27$$

At $x = 6$, $f(6) = 147$.

**Example:** Find the polynomial the lowest possible degree which assumes the values $3, 12, 15, -21$ when $x$ takes the values $3, 2, 1, -1$ respectively.

**Solution:** Construct the following divided difference table:

| $x$ | $y$ | first div. diff | second div. diff | third div. diff |
|---|---|---|---|---|
| -1 | -21 | | | |
| | | $\dfrac{15+21}{1+1} = \boxed{18}$ | | |
| 1 | 15 | | $\dfrac{-3-18}{2+1} = \boxed{-7}$ | |
| | | $\dfrac{12-15}{2-1} = -3$ | | $\dfrac{-3+7}{3+1} = \boxed{1}$ |
| 2 | 12 | | $\dfrac{-9+3}{3-1} = -3$ | |
| | | $\dfrac{3+2}{3-2} = -9$ | | |
| 3 | 3 | | | |

Apply Newton's divided difference formula, we get

$$f(x) = f(x_0) + (x - x_0)f(x_0, x_1) + (x - x_0)(x - x_1)f(x_0, x_1, x_2)$$
$$+ (x - x_0)(x - x_1)(x - x_2)f(x_0, x_1, x_2, x_3)$$
$$= -21 + (x + 1)(18) + (x + 1)(x - 1)(-7) + (x + 1)(x - 1)(x - 2)(1)$$
$$= x^3 - 9x^2 + 17x + 6.$$

**Example:** Find the equation of the cubic curve which passes through the points $(4, -43)$, $(7, 83)$ and $(12, 1053)$. Hence find $f(10)$.

**Solution:** We have to find the equation of the cubic curve from the following data

| $x$: | 4 | 7 | 9 | 12 |
|---|---|---|---|---|
| $f(x)$: | -43 | 83 | 327 | 1053 |

The divided difference table is

| $x$ | $y$ | first div. diff | second div. diff | third div. diff |
|---|---|---|---|---|
| 4 | -43 | | | |
| | | 42 | | |
| 7 | 83 | | 16 | |
| | | 122 | | 1 |
| 9 | 327 | | 24 | |
| | | 242 | | |
| 12 | 1053 | | | |

We apply Newton's divided difference formula

$$f(x) = f(x_0) + (x - x_0)f(x_0, x_1) + (x - x_0)(x - x_1)f(x_0, x_1, x_2)$$
$$+ (x - x_0)(x - x_1)(x - x_2)f(x_0, x_1, x_2, x_3)$$
$$= -43 + (x - 4)(42) + (x - 4)(x - 7)(16) + (x - 4)(x - 7)(x - 9)(1)$$
$$= x^3 - 4x^2 - 7x - 15.$$

$$\therefore f(10) = 1000 - 400 - 70 - 15 = 515$$

**Example:** Find Newton's divided difference polynomial for the data in the table below. Also find $f(2.5)$.

| $x$:    | $-3$ | $-1$ | $0$  | $3$ | $5$  |
|---------|------|------|------|-----|------|
| $f(x)$: | $-30$ | $-22$ | $-12$ | $330$ | $3458$ |

**Solution:** The divided difference table is:

| $x$ | $y$ | first div. diff | second div. diff | third div. diff | fourth div. diff |
|---|---|---|---|---|---|
| $-3$ | $-30$ | | | | |
| | | $\dfrac{-22+30}{-1+3} = \boxed{4}$ | | | |
| $-1$ | $-22$ | | $\dfrac{10-4}{0+3} = \boxed{2}$ | | |
| | | $\dfrac{-12+22}{0+1} = 10$ | | $\dfrac{26-2}{3+3} = \boxed{4}$ | |
| $0$ | $-12$ | | $\dfrac{114-10}{3+1} = 26$ | | $\dfrac{44-4}{5+3} = \boxed{5}$ |
| | | $\dfrac{330+12}{3-0} = 114$ | | $\dfrac{290-26}{3+1} = 44$ | |
| $3$ | $330$ | | $\dfrac{1564-114}{5-0} = 290$ | | |
| | | $\dfrac{3458-330}{3-5} = 1564$ | | | |
| $5$ | $3458$ | | | | |

Here $x_0 = -3, x_1 = -1, x_2 = 0, x_3 = 3, x_4 = 5$ and $y_0 = -30, y_1 = -22, y_2 = -12, y_3 = 330, y_4 = 3458$. The divided differences are

$f(x_0, x_1) = 4$, $f(x_1, x_2) = 10$, $f(x_2, x_3) = 114$, $f(x_3, x_4) = 1564$

$f(x_0, x_1, x_2) = \dfrac{f(x_1, x_2) - f(x_0, x_1)}{x_2 - x_0} = \dfrac{10 - 4}{0 + 3} = 2,$

$f(x_1, x_2, x_3) = 26$, $f(x_2, x_3, x_4) = 290$,

$f(x_0, x_1, x_2, x_3) = \dfrac{f(x_1, x_2, x_3) - f(x_0, x_1, x_2)}{x_3 - x_0} = \dfrac{26 - 2}{3 + 3} = 4$, $f(x_1, x_2, x_3, x_4) = 44$

and finally

$f(x_0, x_1, x_2, x_3, x_4) = \dfrac{f(x_1, x_2, x_3, x_4) - f(x_0, x_1, x_2, x_3)}{x_4 - x_0} = \dfrac{44 - 4}{5 + 3} = 5.$

The Newton's divided difference formula, we get

$$f(x) = f(x_0) + (x - x_0)f(x_0, x_1) + (x - x_0)(x - x_1)f(x_0, x_1, x_2)$$
$$+ (x - x_0)(x - x_1)(x - x_2)f(x_0, x_1, x_2, x_3)$$
$$+ (x - x_0)(x - x_1)(x - x_2)(x - x_3)f(x_0, x_1, x_2, x_3, x_4)$$
$$= -230 + (x + 3)(4) + (x + 3)(x + 1)(2) + (x + 3)(x + 1)(x - 0)(4)$$
$$+ (x + 3)(x + 1)(x - 0)(x - 3)(5)$$
$$= 5x^4 + 9x^3 - 27x^2 - 21x - 12.$$

It is the required 4th degree polynomial. Now when $x = 2.5$, $y(2.5) = 102.6785$.

**Example:** Determine $f(x)$ as a polynomial in $x$ or the the following data using Newton's divided formula and hence find $f(-3)$ and $f(9)$ **(Dec-2020)**

| $x$: | $-4$ | $-1$ | 0 | 2 | 5 |
|---|---|---|---|---|---|
| $f(x)$: | 1245 | 33 | 5 | 9 | 1335 |

**Solution:** The divided difference table is:

| $x$ | $y$ | first div. diff | second div. diff | third div. diff | fourth div. diff |
|---|---|---|---|---|---|
| $-4$ | 1245 | | | | |
| | | $\frac{33 - 1245}{-1 - (-4)} = \boxed{-404}$ | | | |
| $-1$ | 33 | | $\frac{-28 + 404}{0 + 4} = \boxed{94}$ | | |
| | | $\frac{5 - 33}{0 - (-1)} = -28$ | | $\frac{10 - 94}{2 + 4} = \boxed{-14}$ | |
| 0 | 5 | | $\frac{2 + 28}{2 + 1} = 10$ | | $\frac{13 + 14}{5 + 4} = \boxed{3}$ |
| | | $\frac{9 - 5}{0 - 2} = 2$ | | $\frac{88 - 10}{5 + 1} = 13$ | |
| 2 | 9 | | $\frac{442 - 2}{5 - 0} = 88$ | | |
| | | $\frac{1335 - 9}{5 - 2} = 442$ | | | |
| 5 | 1335 | | | | |

The Newton's divided difference formula, we get

$$f(x) = f(x_0) + (x-x_0)f(x_0,x_1) + (x-x_0)(x-x_1)f(x_0,x_1,x_2)$$
$$+ (x-x_0)(x-x_1)(x-x_2)f(x_0,x_1,x_2,x_3)$$
$$+ (x-x_0)(x-x_1)(x-x_2)(x-x_3)f(x_0,x_1,x_2,x_3,x_4)$$
$$= 1245 + (x+4)(-404) + (x+3)(x+1)(94) - (x+4)(x+1)x(14)$$
$$+ (x+4)(x+1)x(x-2)(3)$$
$$= 3x^4 - 5x^3 + 6x^2 - 14x + 5.$$

Now when $x = -3$, $f(-3) = 479$
and when $x = 9$, $f(9) = 16403$.

**Example:** By means of Newton's divided difference formula find the value of $f(8)$ and $f(15)$ from the following table: **(April-2022)**

| $x$:    | 4  | 5   | 7   | 10  | 11   | 13   |
|---------|----|-----|-----|-----|------|------|
| $f(x)$: | 48 | 100 | 294 | 900 | 1210 | 2028 |

**Solution:** Here $x_0 = 4, x_1 = 5, x_2 = 7, x_3 = 10, x_4 = 11, x_5 = 13$.
The divided difference table is given below:

| $x$ | $y$ | first div. diff | second div. diff | third div. diff | fourth div. diff |
|---|---|---|---|---|---|
| 4 | 48 | | | | |
| | | $\dfrac{100-48}{5-4} = \boxed{52}$ | | | |
| 5 | 100 | | $\dfrac{97-52}{7-4} = \boxed{15}$ | | |
| | | $\dfrac{294-100}{7-5} = 97$ | | $\dfrac{21-15}{10-4} = \boxed{1}$ | |
| 7 | 294 | | $\dfrac{202-97}{10-5} = 21$ | | $\boxed{0}$ |
| | | $\dfrac{900-294}{10-7} = 202$ | | $\dfrac{27-21}{11-5} = 1$ | |
| 10 | 900 | | $\dfrac{310-202}{11-7} = 27$ | | 0 |
| | | $\dfrac{1210-900}{11-10} = 310$ | | $\dfrac{33-27}{13-7} = 1$ | |
| 11 | 1210 | | $\dfrac{409-310}{13-10} = 33$ | | |
| | | $\dfrac{2028-1210}{13-11} = 409$ | | | |
| 13 | 2028 | | | | |

Now, using Newton's divided difference formula, we get

$$f(x) = f(x_0) + (x-x_0)f(x_0, x_1) + (x-x_0)(x-x_1)f(x_0, x_1, x_2)$$
$$+ (x-x_0)(x-x_1)(x-x_2)f(x_0, x_1, x_2, x_3)$$
$$+ (x-x_0)(x-x_1)(x-x_2)(x-x_3)f(x_0, x_1, x_2, x_3, x_4)$$
$$= 48 + 52(x-4) + 15(x-4)(x-5) + (1)(x+4)(x-5)(x-7)$$
$$= x^2(x-1).$$

Now when $x = 8$, $f(8) = 448$ and when $x = 15$, $f(15) = 3150$.

**Example:** Given that $f(0) = 8, f(1) = 68$ and $f(5) = 123$, construct a divided difference table. Using the table, determine the value of $f(2)$.

**Solution:** The divided difference table is:

| $x$ | $y$ | first div. diff | second div. diff |
|---|---|---|---|
| 0 | 8 | | |
| | | $\dfrac{68-8}{1-0} = 60$ | |
| 1 | 68 | | $\dfrac{13.75-60}{5-0} = -9.25$ |
| | | $\dfrac{123-68}{5-1} = 13.75$ | |
| 5 | 120 | | |

Now making use of Newton's divided difference for unequal intervals, we get

$$f(x) = f(x_0) + (x-x_0)f(x_0,x_1) + (x-x_0)(x-x_1)f(x_0,x_1,x_2)$$
$$\Rightarrow f(2) = 8 + (2-0)(60) + (2-0)(2-1)(-9.25)$$
$$= 109.50.$$

**Example:** Given $\log_{10} 654 = 2.8156$, $\log_{10} 658 = 2.8182$, $\log_{10} 659 = 2.8189$, $\log_{10} 661 = 2.8202$, find the value of $\log_{10} 656$ by the divided difference formula.

**Solution:** Shift the origin to 654, then we can consider the data given for $x = 0, 4, 5$ and 7 and are required to find $f(x)$ at $x = 2$.

The divided difference table is as under:

| $x$ | $y$ | first div. diff | second div. diff | third div. diff |
|---|---|---|---|---|
| 0 | 2.8156 | | | |
| | | $\dfrac{2.8182 - 2.856}{4 - 0} = \boxed{0.00065}$ | | |
| 4 | 2.8182 | | $\dfrac{0.00070 - 0.00065}{5 - 0} = \boxed{0.00001}$ | |
| | | $\dfrac{2.8189 - 2.8182}{5 - 4} = 0.00070$ | | $\dfrac{-0.00007 - 0.00001}{7 - 0} = \boxed{-0.000004}$ |
| 5 | 2.8189 | | $\dfrac{0.00065 - 0.00070}{7 - 4} = -0.000017$ | |
| | | $\dfrac{2.8202 - 2.8189}{7 - 5} = 0.00065$ | | |
| 7 | 2.8202 | | | |

Applying Newton's divided difference formula, we get

$$f(x) = f(x_0) + (x - x_0)f(x_0, x_1) + (x - x_0)(x - x_1)f(x_0, x_1, x_2)$$
$$+ (x - x_0)(x - x_1)(x - x_2)f(x_0, x_1, x_2, x_3)$$

$$\Rightarrow f(2) = 2.8156 + (x - 0)(0.00065) + (x - 0)(x - 4)(0.00001)$$
$$+ (x - 0)(x - 4)(x - 5)(-0.000004)$$

$$\Rightarrow \log_{10} 656 = 2.8156 + 0.0013 - 0.00004 - 0.0000048$$
$$= 2.8168 \text{ approximately.}$$

**Example:** Given the values, find $f(9)$ using divided difference formula

| $x$:    | 5   | 7   | 11   | 13   | 17   |
|---------|-----|-----|------|------|------|
| $f(x)$: | 150 | 392 | 1452 | 2366 | 5202 |

using Newton's divided difference formula. **(April-2022)**

**Solution:** Construct the following divided difference table:

| $x$ | $y$ | first div. diff | second div. diff | third div. diff |
|-----|-----|-----------------|------------------|-----------------|
| 5   | 150 |                 |                  |                 |
|     |     | $\dfrac{392 - 150}{7 - 5} = \boxed{121}$ |                  |                 |
| 7   | 392 |                 | $\dfrac{265 - 121}{11 - 5} = \boxed{24}$ |                 |
|     |     | $\dfrac{1452 - 392}{11 - 7} = 265$ |                  | $\dfrac{32 - 24}{13 - 5} = \boxed{1}$ |
| 11  | 1452 |                | $\dfrac{457 - 265}{13 - 7} = 32$ |                 |
|     |     | $\dfrac{2366 - 1452}{13 - 11} = 457$ |                  | $\dfrac{42 - 32}{17 - 7} = 1$ |
| 13  | 2366 |                | $\dfrac{709 - 457}{17 - 11} = 42$ |                 |
|     |     | $\dfrac{5202 - 2366}{17 - 13} = 709$ |                  |                 |
| 17  | 5202 |                 |                  |                 |

Apply Newton's divided difference formula, we get

$$f(x) = f(x_0) + (x - x_0)f(x_0, x_1) + (x - x_0)(x - x_1)f(x_0, x_1, x_2)$$
$$+ (x - x_0)(x - x_1)(x - x_2)f(x_0, x_1, x_2, x_3)$$
$$= 150 + (x - 5)(121) + (x - 5)(x - 7)(24) + (x - 5)(x - 7)(x - 11)(1)$$
$$\Rightarrow f(9) = 150 + 484 + 192 - 16$$
$$= 810.$$

**Example:** Using Newton's divided difference formula, find the missing value from the table:

| $x$:    | 1  | 2  | 4 | 5   | 6 |
|---------|----|----|---|-----|---|
| $f(x)$: | 14 | 15 | 5 | ... | 9 |

**Solution:** The divided difference table is

| $x$ | $y$ | first div. diff | second div. diff | third div. diff |
|---|---|---|---|---|
| 1 | 14 | | | |
| | | $\dfrac{15-14}{2-1} = \boxed{1}$ | | |
| 2 | 15 | | $\dfrac{-5-1}{4-1} = \boxed{-2}$ | |
| | | $\dfrac{5-15}{4-2} = -5$ | | $\dfrac{7/4 + 2}{6-1} = \boxed{\dfrac{3}{4}}$ |
| 4 | 5 | | $\dfrac{2+5}{6-2} = \dfrac{7}{4}$ | |
| | | $\dfrac{9-5}{6-4} = 2$ | | |
| 6 | 9 | | | |

Apply Newton's divided difference formula, we get

$$f(x) = f(x_0) + (x - x_0)f(x_0, x_1) + (x - x_0)(x - x_1)f(x_0, x_1, x_2)$$
$$+ (x - x_0)(x - x_1)(x - x_2)f(x_0, x_1, x_2, x_3)$$
$$= 14 + (x-1)(1) + (x-1)(x-2)(-2) + (x-1)(x-2)(x-4)\left(\dfrac{3}{4}\right)$$

Putting $x = 5$, we get

$$y(5) = 14 + 4 + 4(3)(-2) + (4)(3)(1)\left(\frac{3}{4}\right) = 3.$$

Hence missing value is 3.

**EXERCISE**

1. Find the missing term in the following using Newton's divided difference formula

| $x$: | 0 | 1 | 2 | 3 | 4 |
|---|---|---|---|---|---|
| $f(x)$: | 1 | 3 | 9 | $\cdots$ | 81 |

**Ans:** 31

2. Fit a polynomial and find $f(1)$ and $f(8)$

| $x$: | $-1$ | 0 | 3 | 6 | 7 |
|---|---|---|---|---|---|
| $f(x)$: | 3 | $-6$ | 39 | 822 | 1611 |

**Ans:** Divided differences are $-9, 6, 5, 1$
$f(x) = x^4 - 3x^3 + 5x^2 - 6$, $f(1) = -3, f(8) = 2874$.

3. Find a cubic polynomial from the following table using Newton's divided difference formula

| $x$: | 0 | 1 | 2 | 5 |
|---|---|---|---|---|
| $f(x)$: | 2 | 3 | 12 | 147 |

**Ans:** Divided differences are $1, 4, 7$, $f(x) = x^3 + x^2 - x + 2$.

4. Find the function $f(x)$ from the following table

| $x$: | 0 | 1 | 4 | 5 |
|---|---|---|---|---|
| $f(x)$: | 8 | 11 | 78 | 123 |

**Ans:** $x^3 - x^2 + 3x + 8$.

5. Find the polynomial satisfied by $(0, 8)$, $(1, 11)$, $(4, 68)$ and $(5, 123)$ using Newton's divided difference formula.

**Ans:** $x^3 - x^2 - 3x + 8$.

6. Determine $f(x)$ as a polynomial in $x$ for the following data using Newton's divided difference formula **(Dec-2020)**

| $x$:    | 0 | 2 | 3 | 5 | 8  |
|---------|---|---|---|---|----|
| $f(x)$: | 1 | 3 | 5 | 9 | 11 |

**Ans:** $\dfrac{x^4}{360} - \dfrac{17x^3}{180} + \dfrac{271x^2}{360} - \dfrac{3x}{20} + 1$

7. Determine $f(x)$ as a polynomial in $x$ for the following data using Newton's divided difference formula **(Dec-2020)**

| $x$:    | 1  | 3  | 6  | 9  | 11  |
|---------|----|----|----|----|-----|
| $f(x)$: | 12 | 33 | 41 | 55 | 133 |

**Ans:** $\dfrac{139}{2400}x^4 - \dfrac{2071}{2400}x^3 + \dfrac{6803}{2400}x^2 + \dfrac{19351}{2400}x + \dfrac{763}{400}$

8. Find a polynomial by using Newtons divided difference formula for the data. **(Aug-2021)**

| $x$:    | $-2$ | 0   | 1   | 3   | 6   |
|---------|------|-----|-----|-----|-----|
| $f(x)$: | 121  | 135 | 189 | 200 | 225 |

**Ans:** $\dfrac{103}{90}x^4 - \dfrac{779}{90}x^3 + \dfrac{161}{45}x^2 + \dfrac{869}{15}x + 135$

9. Find $y(5)$ from the table using Newtons divided differences **(Aug-2021)**

| $x$:    | 0 | 1 | 3  | 8   |
|---------|---|---|----|-----|
| $f(x)$: | 1 | 3 | 13 | 128 |

**Ans:** 38.8571

10. Using Newton's divided difference formula find $\log_{10} 310$ from the data given below:

| $x$:          | 300    | 304    | 305    | 307    |
|---------------|--------|--------|--------|--------|
| $\log_{10} x$: | 2.4771 | 2.4829 | 2.4843 | 2.4871 |

**Ans:** 2.4786

11. Find the unique polynomial $P(x)$ of degree 2 or less such that $P(1) = 1, P(3) = 27, P(4) = 64$ using Newton's divided difference formula.

**Ans:** $8x^2 - 19x + 12$

12. Find $y(10)$ from the table using Newton's divided differences  (**Mar-2022**)

| $x:$ | 5  | 6  | 9  | 11 |
|------|----|----|----|----|
| $y:$ | 12 | 13 | 14 | 16 |

**Ans:** 14.6666

13. Using Newton's divided difference formula, find $y(8)$ from the following table

(**April-2022**)

| $x:$ | 5  | 6  | 9  | 10 |
|------|----|----|----|----|
| $y:$ | 12 | 13 | 14 | 16 |

**Ans:** 13.3

14. Find the polynomial for the following data using Newton's divided difference formula  (**Mar-2022**)

| $x:$ | 0   | 2   | 3   | 6   |
|------|-----|-----|-----|-----|
| $y:$ | 648 | 704 | 729 | 792 |

**Ans:** $-x^2 + 30x + 648$

15. Using Newton's divided difference formula, find $y(8)$ from the following table

(**April-2022**)

| $x:$ | 3  | 5  | 9  | 11 |
|------|----|----|----|----|
| $y:$ | 10 | 13 | 12 | 18 |

**Ans:** 11.5625

16. Find $y(4)$ using Newton's divided difference formula  (**Jan-2020**)

| $x:$ | 3 | 6  | 8  | 9  |
|------|---|----|----|----|
| $y:$ | 2 | 13 | 18 | 23 |

**Ans:** 7.55

17. Find $f(2.1)$ defined by the set of values $(2,2), (6,3), (9,4), (10,6)$ using Newton's divided difference formula. **(Jan-2020)**

**Ans:** 2.15651

18. Fit the polynomial defined by the set of values $(5,12), (6,13), (9,14), (11,16)$ using Newton's divided difference formula. **(Jan-2020)**

**Ans:** $\dfrac{x^3}{20} - \dfrac{7}{6}x^2 + \dfrac{557}{60}x - \dfrac{23}{2}$

19. Find the interpolating polynomial $f(x)$ from the table using Newton's divided differences **(Aug-2022)**

| x | 3 | 5 | 7 | 9 | 11 |
|---|---|---|---|---|---|
| y | 6 | 24 | 58 | 108 | 74 |

20. Using Newton's divided difference formula evaluate $f(140)$ from the following data **(Aug-2022)**

| x | 110 | 130 | 160 | 190 |
|---|---|---|---|---|
| y | 10.8 | 8.1 | 5.5 | 4.8 |

21. Using Newton's divided difference formula evaluate $f(15)$ from the following data **(Aug-2022)**

| x: | 5 | 6 | 9 | 11 |
|---|---|---|---|---|
| y: | 12 | 13 | 14 | 16 |

22. Using Newton's divided difference formula find $f(x)$ from the following data **(Aug-2022)**

| x | −1 | 0 | 2 | 3 |
|---|---|---|---|---|
| y | −8 | 3 | 1 | 2 |

23. Using Newton's divided difference formula evaluate $f(x)$ from the following data **(Aug-2022)**

| x | 0 | 1 | 3 | 4 |
|---|---|---|---|---|
| y | −12 | 0 | 6 | 12 |

# Chapter 5

# Numerical Integration and Solution of Ordinary Differential Equations

## 5.1 Numerical Integration

Numerical integration is used to obtain approximate answers for definite integrals that cannot be solved analytically.

The process of evaluating a definite integral from a set of tabulated values of the integrand $f(x)$ which is not known explicitly is called **numerical integration**. In this section, we derive several formulae for numerical integration.

**Newton-Cote's quadrature formula:**

Let $I = \int\limits_{a}^{b} f(x)dx$, where $f(x)$ takes the values $y_0, y_1, \cdots, y_n$ for $x = x_0, x_1, \cdots, x_n$.

Let us divide the interval $[a, b]$ into $n$ equal subintervals of width $h$ so that $x_0 = a, x_1 = x_0 + h, x_2 = x_0 + 2h, \cdots, x_n = x_0 + nh = b$.

Newton's forward difference formula is

$$y(x) = y(x_0 + ph) = y_0 + p\Delta y_0 + \frac{p(p-1)}{2!}\Delta^2 y_0 + \frac{p(p-1)(p-2)}{3!}\Delta^3 y_0 + \cdots$$

where $p = \dfrac{x - x_0}{h}$.

$$\therefore I = \int_a^b f(x)dx$$

$$= \int_{x_0}^{x_n} f(x)dx$$

$$= \int_{x_0}^{x_0+nh} f(x)dx$$

$$= \int_{x_0}^{x_0+nh} \left[ y_0 + p\Delta y_0 + \frac{p(p-1)}{2!}\Delta^2 y_0 + \frac{p(p-1)(p-2)}{3!}\Delta^3 y_0 + \cdots \right] dx$$

Since $x = x_0 + ph$, $dx = hdp$ and $p$ varies from 0 to $n$ hence above integral becomes

$$\therefore I = h \int_0^n \left[ y_0 + p\Delta y_0 + \frac{p(p-1)}{2!}\Delta^2 y_0 + \frac{p(p-1)(p-2)}{3!}\Delta^3 y_0 + \cdots \right] dp$$

$$= h \left[ py_0 + \frac{p^2}{2}\Delta y_0 + \frac{1}{2}\left(\frac{p^3}{3} - \frac{p^2}{2}\right)\Delta^2 y_0 + \frac{1}{6}\left(\frac{p^4}{4} - p^3 + p^2\right)\Delta^3 y_0 + \cdots \right]_0^n$$

$$= h \left[ ny_0 + \frac{n^2}{2}\Delta y_0 + \frac{1}{2}\left(\frac{n^3}{3} - \frac{n^2}{2}\right)\Delta^2 y_0 + \frac{1}{6}\left(\frac{n^4}{4} - n^3 + n^2\right)\Delta^3 y_0 + \cdots \right]$$

$$\boxed{\therefore \int_{x_0}^{x_0+nh} f(x)dx = h\left[ny_0 + \frac{n^2}{2}\Delta y_0 + \frac{1}{2}\left(\frac{n^3}{3} - \frac{n^2}{2}\right)\Delta^2 y_0 + \cdots\right]}$$

This is called **Newton-Cote's quadrature formula**. It is a general quadrature formula from which we can derive various special formulae giving different values for $n$.

**Trapezoidal Rule:**

Put $n = 1$ in the quadrature formula. Then all differences higher than the first will become zero.

$$\therefore \int_{x_0}^{x_1} f(x)dx = \int_{x_0}^{x_0+h} f(x)dx = h\left[y_0 + \frac{1}{2}\Delta y_0\right] = h\left[y_0 + \frac{1}{2}(y_1 - y_0)\right] = \frac{h}{2}(y_0 + y_1).$$

Similarly

$$\int_{x_1}^{x_2} f(x)dx = \int_{x_0+h}^{x_0+2h} f(x)dx = h\left[y_1 + \frac{1}{2}\Delta y_1\right] = h\left[y_1 + \frac{1}{2}(y_2 - y_1)\right] = \frac{h}{2}(y_1 + y_2)$$

$$\int_{x_2}^{x_3} f(x)dx = \int_{x_0+2h}^{x_0+3h} f(x)dx = \frac{h}{2}(y_2 + y_3)$$

$$\cdots \qquad \cdots \qquad \cdots$$

$$\int_{x_{n-1}}^{x_n} f(x)dx = \int_{x_0+(n-1)h}^{x_0+nh} f(x)dx = \frac{h}{2}(y_{n-1} + y_n).$$

Adding these $n$ integrals, we obtain

$$\int_{x_0}^{x_n} f(x)dx = \int_{x_0}^{x_0+h} f(x)dx + \int_{x_0+h}^{x_0+2h} f(x)dx + \cdots + \int_{x_{n-1}}^{x_n} f(x)dx$$

$$= \frac{h}{2}(y_0 + y_1) + \frac{h}{2}(y_1 + y_2) + \cdots + \frac{h}{2}(y_{n-1} + y_n)$$

$$\boxed{\therefore \int_{x_0}^{x_n} f(x)dx = \frac{h}{2}\left[(y_0 + y_n) + 2(y_1 + y_2 + \cdots + y_{n-1})\right]}$$

$$\therefore \int_{x_0}^{x_n} f(x)dx = \frac{h}{2}\left[\text{sum of the first and last ordinates} + 2(\text{sum of the remaining ordinates})\right]$$

This is known as the **Trapezoidal rule**.

**Note:** Trapezoidal rule can be applied to any number of subintervals, odd or even.

**Simpson's $\frac{1}{3}^{rd}$ Rule:**

Put $n = 2$ in Newton-Cotes's quadrature formula. Since $x$ takes one of the three values $x_0$, $x_1$ or $x_2$, all the differences of third and higher order become zero.

$$\therefore \int_{x_0}^{x_2} f(x)dx = h\left[2y_0 + \frac{4}{2}\Delta y_0 + \frac{1}{2}\left(\frac{8}{3} - \frac{4}{2}\right)\Delta^2 y_0\right]$$

$$= h\left[2y_0 + 2(y_1 - y_0) + \frac{1}{3}\Delta^2 y_0\right]$$

$$= h\left[2y_0 + 2(y_1 - y_0) + \frac{1}{3}(E - 1)^2 y_0\right]$$

$$= h\left[2y_0 + 2(y_1 - y_0) + \frac{1}{3}(E^2 - 2E + 1)y_0\right]$$

$$= h\left[2y_1 + \frac{1}{3}(y_2 - 2y_1 + y_0)\right]$$

$$= h\left[\frac{1}{3}y_2 + \frac{4}{3}y_1 + \frac{1}{3}y_0\right]$$

Thus $\int_{x_0}^{x_2} f(x)dx = \frac{h}{3}[y_0 + 4y_1 + y_2]$.

Similarly, $\int_{x_2}^{x_4} f(x)dx = \frac{h}{3}[y_2 + 4y_3 + y_4]$.

Finally, $\int_{x_{n-2}}^{x_n} f(x)dx = \frac{h}{3}[y_{n-2} + 4y_{n-1} + y_n]$, $n$ being even.

Adding all these integrals, we have when $n$ is even

$$\int_{x_0}^{x_n} f(x)dx = \int_{x_0}^{x_2} f(x)dx + \int_{x_2}^{x_4} f(x)dx + \cdots + \int_{x_{n-2}}^{x_n} f(x)dx$$

$$= \frac{h}{3}\left[(y_0 + 4y_1 + y_2) + (y_2 + 4y_3 + y_4) + \cdots + (y_{n-2} + 4y_{n-1} + y_n)\right]$$

$$\therefore \int_{x_0}^{x_n} f(x)dx = \frac{h}{3}\left[(y_0 + y_n) + 4(y_1 + y_3 + y_5 + \cdots) + 2(y_2 + y_4 + y_6 + \cdots)\right]$$

$$\therefore \int_{x_0}^{x_n} f(x)dx = \frac{h}{3}\left[\text{sum of the first and last ordinates}\right.$$

$$+ 4(\text{sum of the odd ordinates})$$

$$\left. + 2(\text{sum of the remaining even ordinates})\right]$$

This is known as the Simpson's $\frac{1}{3}^{rd}$ rule or simply Simpson's rule and is most commonly used.. This rule requires the given interval must be divided into an even number of equal subintervals of width $h$.

**Simpson's $\frac{3}{8}^{th}$ Rule:**

$$\int_{x_0}^{x_n} f(x)dx = \frac{3h}{8}\Big[(y_0 + y_n) + 3(y_1 + y_2 + y_4 + \cdots) + 2(y_3 + y_6 + y_9 + \cdots)\Big]$$

This formula is known as the Simpson's $\frac{3}{8}^{th}$. This rule is employed by dividing the interval into $n$ equal sub-intervals when $n$ is a multiple of 3.

**Example:** The table below shows the temperature $f(t)$ as a function of time:

| $t$ | 1 | 2 | 3 | 4 | 5 | 6 | 7 |
|---|---|---|---|---|---|---|---|
| $f(t)$ | 81 | 75 | 80 | 83 | 78 | 70 | 60 |

Use Simpson's $\frac{1}{3}$ method to estimate $\int_{1}^{7} f(t)dt$.

**Solution:** From the table, $y_0 = 81, y_1 = 75, y_2 = 80, y_3 = 83, y_4 = 78, y_5 = 70, y_6 = 60$ and $h = 1$.

Using Simpson's rule, we have

$$\int_{x_0}^{x_n} f(x)dx = \frac{h}{3}\Big[(y_0 + y_n) + 4(y_1 + y_3 + y_5 + \cdots) + 2(y_2 + y_4 + y_6 + \cdots)\Big]$$

$$\Rightarrow \int_{1}^{7} f(t)dt = \frac{h}{3}\Big[(y_0 + y_6) + 4(y_1 + y_3 + y_5) + 2(y_2 + y_4)\Big]$$

$$= \frac{1}{3}\Big[(81 + 60) + 4(75 + 83 + 70) + 2(80 + 78)\Big]$$

$$= \frac{1}{3}[141 + 912 + 316]$$

$$= \frac{1}{3}[1369]$$

$$= 456.333$$

**Example:** Evaluate $\int_{0.6}^{2} y\,dx$, using Trapezoidal rule:

| $x$ | 0.6  | 0.8  | 1.0  | 1.2  | 1.4  | 1.6  | 1.8   | 2.0   |
|-----|------|------|------|------|------|------|-------|-------|
| $y$ | 1.23 | 1.58 | 2.03 | 4.32 | 6.25 | 8.38 | 10.23 | 12.45 |

**Solution:** From the table,

$y_0 = 1.23, y_1 = 1.58, y_2 = 2.03, y_3 = 4.32, y_4 = 6.25, y_5 = 8.38, y_6 = 10.23$ and $y_7 = 12.45$. Given $h = 0.2$.

By Trapezoidal rule,

$$\int_{x_0}^{x_n} f(x)\,dx = \frac{h}{2}\Big[(y_0 + y_n) + 2(y_1 + y_2 + y_3 + \cdots)\Big]$$

$$\Rightarrow \int_{0.6}^{2} y\,dx = \frac{h}{2}\Big[(y_0 + y_7) + 2(y_1 + y_2 + y_3 + y_4 + y_5 + y_6)\Big]$$

$$= \frac{0.2}{2}\Big[(1.23 + 12.45) + 2(1.58 + 2.03 + 4.32 + 6.25 + 8.38 + 10.23)\Big]$$

$$= 0.1\big[13.68 + 65.58\big]$$

$$= 0.1\big[79.26\big]$$

$$= 7.926$$

**Example:** Given that  (Dec-2021)

| $x$ | 4.0 | 4.2 | 4.4 | 4.6 | 4.8 | 5.0 | 5.2 |
|-----|-----|-----|-----|-----|-----|-----|-----|
| $y = \log x$ | 1.3863 | 1.4351 | 1.4816 | 1.5261 | 1.5686 | 1.6094 | 1.6487 |

Evaluate $\int_{4}^{5.2} \log x\,dx$ by (a) Trapezoidal rule (b) Simpson's 1/3rd rule (c) Simpson's 3/8th rule.

**Solution:** Given $h = 0.2, y_0 = 1.3863, y_1 = 1.4351, y_2 = 1.4816, y_3 = 1.5261, y_4 = 1.5686, y_5 = 1.6094, y_6 = 1.6487$.

(a) By the Trapezoidal rule, we have

$$\int_{4}^{5.2} \log x \, dx = \frac{h}{2}\Big[(y_0 + y_6) + 2(y_1 + y_2 + y_3 + y_4 + y_5)\Big]$$

$$= \frac{0.2}{2}\Big[3.035 + 2(7.6208)\Big]$$

$$= 0.1(18.2766)$$

$$= 1.82766$$

(b) By the Simpson's 1/3rd rule, we have

$$\int_{4}^{5.2} \log x \, dx = \frac{h}{3}\Big[(y_0 + y_6) + 4(y_1 + y_3 + y_5) + 2(y_2 + y_4)\Big]$$

$$= \frac{0.2}{3}\Big[3.035 + 4(4.5706) + 2(3.0502)\Big]$$

$$= 1.82785$$

(c) By the Simpson's 3/8th rule, we have

$$\int_{4}^{5.2} \log x \, dx = \frac{3h}{8}\Big[(y_0 + y_6) + 3(y_1 + y_2 + y_4 + y_5) + 2y_3\Big]$$

$$= \frac{0.6}{8}\Big[3.035 + 3(6.0947) + 2(1.5261)\Big]$$

$$= 1.82785$$

**Example:** The following Table gives 7 values of an independent variable $x$ and the corresponding values of $y = f(x)$.

| $x$ | 0 | 1 | 2 | 3 | 4 | 5 | 6 |
|---|---|---|---|---|---|---|---|
| $y = f(x)$ | 0.146 | 0.161 | 0.176 | 0.190 | 0.204 | 0.217 | 0.230 |

Evaluate $\int_{0}^{6} f(x) \, dx$, by using Simpson's $(3/8)^{th}$ rule.

**Solution:** Here the given 7 values of $x$ are $x_0 = 0, x_1 = 1, x_2, \cdots, x_6 = 6$ and the corresponding values of $y$ are $y_0 = 0.146, y_1 = 0.161, y_2 = 0.176, \cdots, y_6 = 0.230$. Also

$h = 1$ and $n = 6$ which is a multiple of 3.

Using Simpson's $(3/8)^{th}$ rule, we have

$$\int_0^6 f(x)dx = \frac{3h}{8}\Big[(y_0 + y_6) + 3(y_1 + y_2 + y_4 + y_5) + 2y_3\Big]$$

$$= \frac{3}{8}\Big[(0.146 + 0.230) + 3(0.161 + 0.176 + 0.204 + 0.217) + 2(0.190)\Big]$$

$$= \frac{3}{8}\Big[0.376 + 2.274 + 0.38\Big]$$

$$= 1.3625.$$

**Example:** Evaluate $\int_0^1 \sqrt{1+x^3}dx$, taking $h = 0.1$ using

(i) Trapezoidal rule  (ii) Simpson's $\frac{1}{3}^{rd}$ rule

Solution: Let $f(x) = (1+x^3)^{1/2}$.

Given $x_0 = 0$, $x_n = 1$ and $h = 0.1$

| $x$ | 0 | 0.1 | 0.2 | 0.3 | 0.4 | 0.5 | 0.6 |
|---|---|---|---|---|---|---|---|
| $y$ | $1(y_0)$ | $1.0005(y_1)$ | $1.004(y_2)$ | $1.0134(y_3)$ | $1.0315(y_4)$ | $1.0607(y_5)$ | $1.1027(y_6)$ |

| 0.7 | 0.8 | 0.9 | 1.0 |
|---|---|---|---|
| $1.1588(y_7)$ | $1.2296(y_8)$ | $1.3149(y_9)$ | $1.4142(y_{10})$ |

Trapezoidal rule is given by

$$\int_{x_0}^{x_n} f(x)dx = \frac{h}{2}\Big[(y_0 + y_n) + 2(y_1 + y_2 + y_3 + \cdots)\Big]$$

$$\Rightarrow \int_0^1 (1+x^3)^{1/2}dx = \frac{h}{2}\Big[(y_0 + y_{10}) + 2(y_1 + y_2 + y_3 + y_4 + y_5 + y_6 + y_7 + y_8 + y_9)\Big]$$

$$= \frac{0.1}{2}\Big[(1 + 1.4142) + 2(1.0005 + 1.004 + 1.0134 + 1.0315 + 1.0607$$
$$+ 1.1027 + 1.1588 + 1.2296 + 1.3149)\Big]$$

$$= (0.05)\big[2.4142 + 2(9.9161)\big]$$

$$= (0.05)\big[22.2466\big]$$

$$= 1.11232$$

Simpson's $\frac{1}{3}^{rd}$ rule is given by

$$\int_{x_0}^{x_n} f(x)dx = \frac{h}{3}\Big[(y_0 + y_n) + 4(y_1 + y_3 + y_5 + \cdots) + 2(y_2 + y_4 + y_6 + \cdots)\Big]$$

$$\Rightarrow \int_0^1 (1+x^3)^{1/2}dx = \frac{h}{3}\Big[(y_0 + y_{10}) + 4(y_1 + y_3 + y_5 + y_7 + y_9) + 2(y_2 + y_4 + y_6 + y_8)\Big]$$

$$= \frac{0.1}{3}\Big[(1 + 1.4142) + 4(1.0005 + 1.0134 + 1.0607 + 1.1588 + 1.3149$$
$$+ 2(1.004 + 1.0315 + 1.1027 + 1.2296)\Big]$$

$$= \frac{0.1}{3}\big[2.4142 + 22.1932 + 8.7356\big]$$

$$= \frac{0.1}{3}\big[33.343\big]$$

$$= 1.1114.$$

**Example:** By using Simpson's $\frac{3}{8}^{th}$ rule, evaluate $\int_0^1 \sqrt{1+x^4}dx.$

**Solution:** Let $f(x) = \sqrt{1+x^4}.$ (Dec-2020), (July-2021), (Dec-2022)

Given $x_0 = 0$, $x_n = 1$ and and take $n = 6$.
$$\therefore h = \frac{x_n - x_0}{n} = \frac{1-0}{6} = \frac{1}{6}.$$

| $x$ | 0 | $\frac{1}{6}$ | $\frac{2}{6}$ | $\frac{3}{6}$ | $\frac{4}{6}$ | $\frac{5}{6}$ | $\frac{6}{6} = 1$ |
|---|---|---|---|---|---|---|---|
| $y$ | $1(y_0)$ | $1.0003(y_1)$ | $1.0062(y_2)$ | $1.0307(y_3)$ | $1.0943(y_4)$ | $1.2175(y_5)$ | $1.4142(y_6)$ |

Simpson's $\frac{3}{8}^{th}$ rule is given by

$$\int_{x_0}^{x_n} f(x)dx = \frac{3h}{8}\left[(y_0 + y_n) + 3(y_1 + y_2 + y_4 + y_5 + \cdots) + 2(y_3 + y_6 + y_9 + \cdots)\right]$$

$$\Rightarrow \int_0^1 \sqrt{1+x^4}\,dx = \frac{3h}{8}\left[(y_0 + y_6) + 3(y_1 + y_2 + y_4 + y_5) + 2y_3\right]$$

$$= \frac{3}{48}\left[(1 + 1.4142) + 3(1.0003 + 1.0062 + 1.0943 + 1.2175) + 2(1.0307)\right]$$

$$= \frac{1}{16}[2.4142 + 3(4.3183) + 2(1.0307)]$$

$$= \frac{1}{16}[17.4305]$$

$$= 1.0894.$$

**Example:** Evaluate $\int_0^6 \frac{dx}{1+x^2}$ by using (i) Trapezoidal rule, (ii) Simpson's $\frac{1}{3}^{rd}$ rule (iii) Simpson's $\frac{3}{8}^{th}$ rule.

**Solution:** We divide the interval $(0,6)$ into six equal parts. Here $f(x) = \frac{dx}{1+x^2}$, $x_0 = 0, x_n = 6, n = 6$.
$$\therefore h = \frac{x_n - x_0}{n} = \frac{6-0}{6} = 1.$$
The values of $f(x)$ are given below:

| $x$ | 0 | 1 | 2 | 3 | 4 | 5 | 6 |
|---|---|---|---|---|---|---|---|
| $y$ | 1 | 0.5 | 0.2 | 0.1 | 0.058824 | 0.03846 | 0.027027 |
|   | $y_0$ | $y_1$ | $y_2$ | $y_3$ | $y_4$ | $y_5$ | $y_6$ |

**Trapezoidal rule:**

$$\int_{x_0}^{x_n} f(x)dx = \frac{h}{2}\Big[(y_0 + y_n) + 2(y_1 + y_2 + y_3 + \cdots)\Big]$$

$$\Rightarrow \int_0^6 \frac{dx}{1+x^2} = \frac{h}{2}\Big[(y_0 + y_6) + 2(y_1 + y_2 + y_3 + y_4 + y_5)\Big]$$

$$= \frac{1}{2}\Big[(1 + 0.027027) + 2(0.5 + 0.2 + 0.1 + 0.058824 + 0.03846)\Big]$$

$$= (0.5)\big[1.027027 + 2(0.897284)\big]$$

$$= (0.5)\big[2.821595\big]$$

$$= 1.4108$$

**Simpson's $\frac{1}{3}^{rd}$ rule:**

$$\int_{x_0}^{x_n} f(x)dx = \frac{h}{3}\Big[(y_0 + y_n) + 4(y_1 + y_3 + y_5 + \cdots) + 2(y_2 + y_4 + y_6 + \cdots)\Big]$$

$$\Rightarrow \int_0^6 \frac{dx}{1+x^2} = \frac{h}{3}\Big[(y_0 + y_6) + 4(y_1 + y_3 + y_5) + 2(y_2 + y_4)\Big]$$

$$= \frac{1}{3}\Big[(1 + 0.027027) + 4(0.5 + 0.1 + 0.03846) + 2(0.2 + 0.058824)\Big]$$

$$= \frac{1}{3}\Big[1.027027 + 4(0.63846) + 2(0.258824)\Big]$$

$$= 1.3662.$$

**Simpson's $\frac{3}{8}^{\text{th}}$ rule:**

$$\int_{x_0}^{x_n} f(x)dx = \frac{3h}{8}\Big[(y_0 + y_n) + 3(y_1 + y_2 + y_4 + \cdots) + 2(y_3 + y_6 + y_9 + \cdots)\Big]$$

$$\Rightarrow \int_0^6 \frac{dx}{1+x^2} = \frac{3h}{8}\Big[(y_0 + y_6) + 3(y_1 + y_2 + y_4 + y_5) + 2y_3\Big]$$

$$= \frac{3}{8}\Big[(1 + 0.027027) + 3(0.5 + 0.2 + 0.058824 + 0.03846) + 0.2\Big]$$

$$= 0.375\Big[1.027027 + 2.3918 + 0.2\Big]$$

$$= 1.3571.$$

**Example:** Evaluate $\int_0^1 \frac{dx}{1+x^2}$ by using Simpson's $\frac{1}{3}^{\text{rd}}$ rule and $\frac{3}{8}^{\text{th}}$ rule. Hence obtain the approximate value of $\pi$ in each case.

**Solution:** Divide the range into six equal parts each of with $h = \frac{1}{6}$ and compute the value of $f(x) = \frac{dx}{1+x^2}$ at each point of subdivision.
These values are given below:

| $x$: | 0 | $\frac{1}{6}$ | $\frac{2}{6}$ | $\frac{3}{6}$ | $\frac{4}{6}$ | $\frac{5}{6}$ | $\frac{6}{6} = 1$ |
|---|---|---|---|---|---|---|---|
| $y$: | $1(y_0)$ | $0.97297(y_1)$ | $0.9(y_2)$ | $0.8(y_3)$ | $0.69231(y_4)$ | $0.59016(y_5)$ | $0.5(y_6)$ |

By Simpson's $\frac{1}{3}^{\text{rd}}$ rule, we have

$$\int_{x_0}^{x_n} f(x)dx = \frac{h}{3}\Big[(y_0 + y_n) + 4(y_1 + y_3 + y_5 + \cdots) + 2(y_2 + y_4 + y_6 + \cdots)\Big]$$

$$\Rightarrow \int_0^1 \frac{dx}{1+x^2} = \frac{h}{3}\Big[(y_0 + y_6) + 4(y_1 + y_3 + y_5) + 2(y_2 + y_4)\Big]$$

$$= \frac{1}{18}\Big[(1 + 0.5) + 4(0.97297 + 0.8 + 0.59016) + 2(0.9 + 0.06923)\Big]$$

$$= 0.785395 \quad \cdots\cdots \quad (1)$$

By Simpson's $\frac{3}{8}^{th}$ rule, we have

$$\int_{x_0}^{x_n} f(x)dx = \frac{3h}{8}\Big[(y_0 + y_n) + 3(y_1 + y_2 + y_4 + \cdots) + 2(y_3 + y_6 + y_9 + \cdots)\Big]$$

$$\Rightarrow \int_0^1 \frac{dx}{1+x^2} = \frac{3h}{8}\Big[(y_0 + y_6) + 3(y_1 + y_2 + y_4 + y_5) + 2y_3\Big]$$

$$= \frac{1}{16}\Big[(1+0.5) + 3(0.07297 + 0.9 + 0.069231 + 0.59016) + 2(0.8)\Big]$$

$$= 0.785395 \quad \cdots \cdots \quad (2)$$

By actual integration, we have

$$\int_0^\pi \frac{dx}{1+x^2} = \Big[\tan^{-1} x\Big]_0^1 = \pi/4 \quad \cdots \cdots \quad (3)$$

From (1) and (2), we get $\pi = 3.14188$ and from (2) and (3), we get $\pi = 3.141580$.

**Example:** Calculate an approximate value of $\int_0^{\pi/2} \sin x \, dx$ by (a) Trapezoidal rule (b) Simpson's rule using 11 ordinates.

**Solution:** Divide the range of integration into ten equal parts.

$$\therefore h = \frac{x_n - x_0}{n} = \frac{\pi/2 - 0}{10} = \pi/20.$$

| $x$ | 0 | $\frac{\pi}{20}$ | $\frac{2\pi}{20}$ | $\frac{3\pi}{20}$ | $\frac{4\pi}{20}$ | $\frac{5\pi}{20}$ |
|---|---|---|---|---|---|---|
| $y = \sin x$ | $0(y_0)$ | $0.15643\,(y_1)$ | $0.30902(y_2)$ | $0.45399(y_3)$ | $0.58778\,(y_4)$ | $0.70711(y_5)$ |

| $\frac{6\pi}{20}$ | $\frac{7\pi}{20}$ | $\frac{8\pi}{20}$ | $\frac{9\pi}{20}$ | $\frac{10\pi}{20} = \frac{\pi}{2}$ |
|---|---|---|---|---|
| $0.80901(y_6)$ | $0.89101\,(y_7)$ | $0.95106(y_8)$ | $0.98769(y_9)$ | $1\,(y_{10})$ |

(a) By the Trapezoidal rule, we have

$$\int_0^{\pi/2} \sin x\, dx = \frac{h}{2}\Big[(y_0 + y_{10}) + 2(y_1 + y_2 + \cdots + y_9)\Big]$$

$$= \frac{\pi/20}{2}\Big[1 + 2(5.8531)\Big]$$

$$= \frac{\pi}{40}[12.7062]$$

$$= 0.9979$$

(b) By the Simpson's 1/3rd rule, we have

$$\int_0^{\pi/2} \sin x\, dx = \frac{h}{3}\Big[(y_0 + y_{10}) + 4(y_1 + y_3 + y_5 + y_7 + y_9) + 2(y_2 + y_4 + y_6 + y_8)\Big]$$

$$= \frac{\pi/20}{3}\Big[1 + 4(3.19623) + 2(2.65687)\Big]$$

$$= \frac{\pi}{40}[19.09866]$$

$$= 1.$$

**Example:** Evaluate $\int_0^{\pi/2} e^{\sin x}\, dx$ correct to four decimal places by Simpson's $\frac{3}{8}$ rule.

**Solution:** Let $f(x) = e^{\sin x}$ and $n = 6$. Then $h = \dfrac{\pi/2 - 0}{6} = \dfrac{\pi}{12}$.

| $x$ | 0 | $\frac{\pi}{12}$ | $\frac{2\pi}{12}$ | $\frac{3\pi}{12}$ | $\frac{4\pi}{12}$ | $\frac{5\pi}{12}$ | $\frac{6\pi}{12} = \frac{\pi}{2}$ |
|---|---|---|---|---|---|---|---|
| $y$ | $1(y_0)$ | $1.2954\ (y_1)$ | $1.6487(y_2)$ | $2.0281(y_3)$ | $2.3774\ (y_4)$ | $2.6272(y_5)$ | $2.7183(y_6)$ |

Simpson's $\frac{3}{8}^{th}$ rule is given by

$$\int_{x_0}^{x_n} f(x)dx = \frac{3h}{8}\Big[(y_0 + y_n) + 3(y_1 + y_2 + y_4 + \cdots) + 2(y_3 + y_6 + y_9 + \cdots)\Big]$$

$$\Rightarrow \int_{0}^{\pi/2} e^{\sin x} dx = \frac{3h}{8}\Big[(y_0 + y_6) + 3(y_1 + y_2 + y_4 + y_5) + 2y_3\Big]$$

$$= \frac{3(\pi/12)}{8}\Big[(1 + 2.7183) + 3(1.2954 + 1.6487 + 2.3774 + 2.6272) + 2(2.0281)\Big]$$

$$= \frac{\pi}{32}[3.7183 + 23.8461 + 4.0562]$$

$$= \frac{\pi}{32}[31.6205]$$

$$= 3.1043$$

**Example:** Evaluate $\int_{0}^{\pi} \sin x \, dx$ taking $n = 10$ by (i) trapezoidal rule (ii) Simpson's $\frac{1}{3}$ rule.

**Solution:** Let $f(x) = \sin x$. Here, $n = 10$ and $h = \frac{\pi - 0}{10} = \frac{\pi}{10}$.

| $x$ | 0 | $\frac{\pi}{10}$ | $\frac{2\pi}{10}$ | $\frac{3\pi}{10}$ | $\frac{4\pi}{10}$ | $\frac{5\pi}{10}$ |
|---|---|---|---|---|---|---|
| $y = \sin x$ | $0(y_0)$ | $0.3090(y_1)$ | $0.5878(y_2)$ | $0.8090(y_3)$ | $0.9511(y_4)$ | $1(y_5)$ |

| $\frac{6\pi}{10}$ | $\frac{7\pi}{10}$ | $\frac{8\pi}{10}$ | $\frac{9\pi}{10}$ | $\frac{10\pi}{10} = \pi$ |
|---|---|---|---|---|
| $0.9511(y_6)$ | $0.8090(y_7)$ | $0.5878(y_8)$ | $0.3090(y_9)$ | $0(y_{10})$ |

(i) By Trapezoidal rule,

$$\int_{x_0}^{x_n} f(x)dx = \frac{h}{2}\Big[(y_0 + y_n) + 2(y_1 + y_2 + y_3 + \cdots)\Big]$$

$$\Rightarrow \int_0^{\pi} \sin x\, dx = \frac{h}{2}\Big[(y_0 + y_{10}) + 2(y_1 + y_2 + y_3 + y_4 + y_5 + y_6 + y_7 + y_8 + y_9)\Big]$$

$$= \frac{\pi/10}{2}\Big[(0+0) + 2(0.3090 + 0.5878 + 0.8090 + 0.9511 + 1 + 0.9511$$
$$+ 0.8090 + 0.5878 + 0.3090)\Big]$$

$$= \frac{\pi}{20}[12.6276]$$

$$= 1.9835$$

(ii) By Simpson's $\frac{1}{3}^{\text{rd}}$ rule,

$$\int_{x_0}^{x_n} f(x)dx = \frac{h}{3}\Big[(y_0 + y_n) + 4(y_1 + y_3 + y_5 + \cdots) + 2(y_2 + y_4 + y_6 + \cdots)\Big]$$

$$\Rightarrow \int_0^{\pi} \sin x\, dx = \frac{h}{3}\Big[(y_0 + y_6) + 4(y_1 + y_3 + y_5 + y_7 + y_9) + 2(y_2 + y_4 + y_6 + y_8)\Big]$$

$$= \frac{\pi/10}{3}\Big[(0+0) + 4(0.3090 + 0.8090 + 1.0 + 0.8090 + 0.3090)$$
$$+ 2(0.5878 + 0.9511 + 0.9511 + 0.5878)\Big]$$

$$= \frac{\pi}{30}[12.944 + 6.1556]$$

$$= \frac{\pi}{30}[19.0996]$$

$$= 2.0001$$

**Example:** Evaluate $\int_0^{\pi} \frac{\sin x}{x}dx$ using (i) Trapezoidal rule (ii) Simpson's $\frac{1}{3}^{\text{rd}}$ rule (iii) Simpson's $\frac{3}{8}^{\text{th}}$ rule.

**Solution:** All the formulae are applicable if the number of subintervals is a multiple of 6. So, we divide the $(0, \pi)$ into 6 equal parts.

$$\therefore h = \frac{\pi - 0}{6} = \frac{\pi}{6}.$$

| $x$ | 0 | $\frac{\pi}{6}$ | $\frac{2\pi}{6}$ | $\frac{3\pi}{6}$ | $\frac{4\pi}{6}$ | $\frac{5\pi}{6}$ | $\frac{6\pi}{6}=\pi$ |
|---|---|---|---|---|---|---|---|
| $y$ | $1(y_0)$ | $0.9549(y_1)$ | $0.8269(y_2)$ | $0.6366(y_3)$ | $0.4134(y_4)$ | $0.19090(y_5)$ | $0(y_6)$ |

Trapezoidal rule is given by

$$\int_{x_0}^{x_n} f(x)dx = \frac{h}{2}\Big[(y_0+y_n)+2(y_1+y_2+y_3+\cdots)\Big]$$

$$\Rightarrow \int_0^{\pi} \frac{\sin x}{x}dx = \frac{h}{2}\Big[(y_0+y_6)+2(y_1+y_2+y_3+y_4+y_5)\Big]$$

$$= \frac{\pi}{12}\Big[(1+0)+2(0.9549+0.8269+0.6366+0.4134+0.19090)\Big]$$

$$= \frac{\pi}{12}[1+6.0454]$$

$$= (0.2618)[7.0454]$$

$$= 1.84448$$

Simpson's $\frac{1}{3}^{rd}$ rule is given by

$$\int_{x_0}^{x_n} f(x)dx = \frac{h}{3}\Big[(y_0+y_n)+4(y_1+y_3+y_5+\cdots)+2(y_2+y_4+y_6+\cdots)\Big]$$

$$\Rightarrow \int_0^{\pi} \frac{\sin x}{x}dx = \frac{h}{3}\Big[(y_0+y_6)+4(y_1+y_3+y_5)+2(y_2+y_4)\Big]$$

$$= \frac{\pi}{18}\Big[(1+0)+4(0.9549+0.6366+0.19090)+2(0.8269+0.4134)\Big]$$

$$= \frac{\pi}{18}[1+7.1296+2.4806]$$

$$= \frac{\pi}{18}[10.6102]$$

$$= 1.8518$$

Simpson's $\frac{3}{8}^{th}$ rule is given by

$$\int_{x_0}^{x_n} f(x)dx = \frac{3h}{8}\left[(y_0 + y_n) + 3(y_1 + y_2 + y_4 + \cdots) + 2(y_3 + y_6 + y_9 + \cdots)\right]$$

$$\Rightarrow \int_0^\pi \frac{\sin x}{x}dx = \frac{3h}{8}\left[(y_0 + y_6) + 3(y_1 + y_2 + y_4) + 2y_3\right]$$

$$= \frac{3(\pi/6)}{8}\left[(1 + 0) + 3(0.9549 + 0.8269 + 0.4134 + 0.19090) + 2(0.6366)\right]$$

$$= \frac{\pi}{16}[1 + 7.1583 + 1.2732]$$

$$= \frac{\pi}{16}[9.4315]$$

$$= 1.8518.$$

**Example:** Evaluate $\int_0^2 e^{-x^2} dx$, using Simpson's $\frac{1}{3}$ rule by taking $h = 0.25$.

**Solution:** Let $f(x) = e^{-x^2}$. Here $x_0 = 0$, $x_n = 2$ and $h = 0.25$.

(Dec-2020), (Apr-2022)

| $x$ | 0 | 0.25 | 0.5 | 0.75 | 1.0 |
|---|---|---|---|---|---|
| $y = e^{-x^2}$ | $1(y_0)$ | $0.9394(y_1)$ | $0.7788(y_2)$ | $0.5698(y_3)$ | $0.3678\ (y_4)$ |

| 1.25 | 1.50 | 1.75 | 2.0 |
|---|---|---|---|
| $0.2096(y_5)$ | $0.1054\ (y_6)$ | $0.04677(y_7)$ | $0.0183(y_8)$ |

By Simpson's $\frac{1}{3}^{rd}$ rule,

$$\int_{x_0}^{x_n} f(x)dx = \frac{h}{3}\Big[(y_0 + y_n) + 4(y_1 + y_3 + y_5 + \cdots) + 2(y_2 + y_4 + y_6 + \cdots)\Big]$$

$$\Rightarrow \int_0^2 e^{-x^2} dx = \frac{h}{3}\Big[(y_0 + y_8) + 4(y_1 + y_3 + y_5 + y_7) + 2(y_2 + y_4 + y_6)\Big]$$

$$= \frac{0.25}{3}\Big[(1 + 0.0183) + 4(0.9394 + 0.5698 + 0.2096 + 0.04677)$$

$$+ 2(0.7788 + 0.3678 + 0.1054)\Big]$$

$$= 0.0833\big[1.0183 + 7.06228 + 2.504\big]$$

$$= 0.0833\big[10.58458\big]$$

$$= 0.881696$$

**Example:** Use Simpson's $(1/3)^{rd}$ rule to find $\int_0^{0.6} e^{-x^2} dx$ by taking seven ordinates.

**Solution:** Let $f(x) = e^{-x^2}$. Here $x_0 = 0, x_n = 0.6$. Divide the interval $(0, 0.6)$ into six equal parts.

$$\therefore h = \frac{x_n - x_0}{n} = \frac{0.6 - 0}{6} = 0.1.$$

The values of $f(x)$ are given below:

| $x$ | 0 | 0.1 | 0.2 | 0.3 | 0.4 | 0.5 | 0.6 |
|---|---|---|---|---|---|---|---|
| $y = f(x)$ | 1 | 0.9900 | 0.9608 | 0.9139 | 0.8521 | 0.7788 | 0.6977 |
|  | $y_0$ | $y_1$ | $y_2$ | $y_3$ | $y_4$ | $y_5$ | $y_6$ |

By Simpson's $\frac{1}{3}^{rd}$ rule, we have

$$\int_{x_0}^{x_n} f(x)dx = \frac{h}{3}\Big[(y_0+y_n) + 4(y_1+y_3+y_5+\cdots) + 2(y_2+y_4+y_6+\cdots)\Big]$$

$$\Rightarrow \int_0^{0.6} e^{-x^2}dx = \frac{h}{3}\Big[(y_0+y_6) + 4(y_1+y_3+y_5) + 2(y_2+y_4)\Big]$$

$$= \frac{0.1}{3}\Big[(1+0.6977) + 4(0.99+0.9139+0.7788) + 2(0.9608+0.8521)\Big]$$

$$= \frac{0.1}{3}\big[1.6977 + 10.7308 + 3.6258\big]$$

$$= \frac{0.1}{3}\big[16.0543\big]$$

$$= 0.5351.$$

**Example:** Evaluate $\int_0^{\pi/2} \sqrt{\sin\theta}\,d\theta$ using (i) Simpson's $\frac{1}{3}$ rule (ii) Simpson's $\frac{3}{8}$ rule.

**Solution:** Let $f(\theta) = \sqrt{\sin\theta}$. Here $a=0, b=\frac{\pi}{2}$ and $h = \frac{\pi/2-0}{6} = \frac{\pi}{12}$.

| $\theta$ | 0 | $\frac{\pi}{12}$ | $\frac{2\pi}{12}$ | $\frac{3\pi}{12}$ | $\frac{4\pi}{12}$ |
|---|---|---|---|---|---|
| $y = \sqrt{\sin\theta}$ | $0(y_0)$ | $0.5087(y_1)$ | $0.7071(y_2)$ | $0.8409(y_3)$ | $0.9306\ (y_4)$ |

| $\frac{5\pi}{12}$ | $\frac{6\pi}{12}=\frac{\pi}{2}$ |
|---|---|
| $0.9828(y_5)$ | $1(y_6)$ |

By Simpson's $\frac{1}{3}^{rd}$ rule,

$$\int_{x_0}^{x_n} f(x)dx = \frac{h}{3}\Big[(y_0 + y_n) + 4(y_1 + y_3 + y_5 + \cdots) + 2(y_2 + y_4 + y_6 + \cdots)\Big]$$

$$\Rightarrow \int_0^{\pi/2} \sqrt{\sin\theta}\,d\theta = \frac{h}{3}\Big[(y_0 + y_6) + 4(y_1 + y_3 + y_5) + 2(y_2 + y_4)\Big]$$

$$= \frac{\pi/12}{3}\Big[(0+1) + 4(0.5087 + 0.8409 + 0.9828)$$
$$+ 2(0.7071 + 0.9306)\Big]$$

$$= \frac{\pi}{36}[1 + 3.2754 + 9.3296]$$

$$= \frac{\pi}{36}[13.605]$$

$$= 1.1872$$

By Simpson's $\frac{3}{8}^{th}$ rule,

$$\int_{x_0}^{x_n} f(x)dx = \frac{3h}{8}\Big[(y_0 + y_n) + 3(y_1 + y_2 + y_4 + \cdots) + 2(y_3 + y_6 + \cdots)\Big]$$

$$\Rightarrow \int_0^{\pi/2} \sqrt{\sin\theta}\,d\theta = \frac{3}{8}\left(\frac{\pi}{12}\right)\Big[(y_0 + y_6) + 3(y_1 + y_2 + y_4 + y_5) + 2(y_3)\Big]$$

$$= \frac{\pi}{32}\Big[(0+1) + 3(0.5087 + 0.7071 + 0.9306 + 0.9826) + 2(0.8409)\Big]$$

$$= \frac{\pi}{32}[1 + 9.3876 + 1.6818]$$

$$= \frac{\pi}{36}[12.0694]$$

$$= 1.18491$$

**Example:** Compute the value of $\int_{0.2}^{1.4} (\sin x - \log x + e^x)dx$ using Simpson's $\frac{3}{8}^{th}$ rule.

**Solution:** Let $f(x) = \sin x - \log x + e^x$.  (May-2017), (Apr-2022)

Here $x_0 = 0.2, x_n = 1.4$ and $n = 6$.

$$\therefore h = \frac{x_n - x_0}{n} = \frac{1.4 - 0.2}{6} = 0.2$$

The values of $y$ are as given below:

| $x$ | 0.2 | 0.4 | 0.6 | 0.8 | 1.0 | 1.2 | 1.4 |
|---|---|---|---|---|---|---|---|
| $y$ | 3.0295 | 2.7975 | 2.8976 | 3.1660 | 3.5598 | 4.0698 | 4.7042 |
|  | $y_0$ | $y_1$ | $y_2$ | $y_3$ | $y_4$ | $y_5$ | $y_6$ |

By Simpson's $\frac{3}{8}^{th}$ rule, we have

$$\int_{x_0}^{x_n} f(x)dx = \frac{3h}{8}\left[(y_0 + y_n) + 3(y_1 + y_2 + y_4 + \cdots) + 2(y_3 + y_6 + \cdots)\right]$$

$$\Rightarrow \int_{0.2}^{1.4} (\sin x - \log x + e^x)dx = \frac{0.6}{8}\left[(y_0 + y_6) + 3(y_1 + y_2 + y_4 + y_5) + 2(y_3)\right]$$

$$= \frac{0.6}{8}\left[(3.0295 + 4.7042) + 3(2.7975 + 2.8976 + 3.5598 + 4.0698) + 2(3.1660)\right]$$

$$= \frac{0.6}{8}\left[7.7336 + 3(13.3247) + 2(3.1660)\right]$$

$$= \frac{0.6}{8}[54.0398]$$

$$= 4.05298$$

**Example:** Evaluate $\int_0^1 \frac{dx}{1+x^2}$ using Trapezoidal rule with $h = 0.2$. Hence determine the value of $\pi$.

**Solution:** Let $f(x) = \frac{1}{1+x^2}$. Here $h = 0.2$.

The values of $x$ and $f(x)$ are tabulated below:

| $x$ | 0 | 0.2 | 0.4 | 0.6 | 0.8 | 1.0 |
|---|---|---|---|---|---|---|
| $y = f(x)$ | 1($y_0$) | 0.9615($y_1$) | 0.8621($y_2$) | 0.7353($y_3$) | 0.6098 ($y_4$) | 0.5 ($y_5$) |

Using Trapezoidal rule,

$$\int_{x_0}^{x_n} f(x)dx = \frac{h}{2}\Big[(y_0 + y_n) + 2(y_1 + y_2 + y_3 + \cdots)\Big]$$

$$\Rightarrow \int_0^1 \frac{dx}{1+x^2} = \frac{h}{2}\Big[(y_0 + y_5) + 2(y_1 + y_2 + y_3 + y_4)\Big]$$

$$= \frac{0.2}{2}\Big[(1+0.5) + 2(0.9615 + 0.8621 + 0.7353 + 0.6098)\Big]$$

$$= 0.1[1.5 + 6.3374]$$

$$= 0.1[7.8374]$$

$$= 0.78374 \quad \cdots\cdots \quad (1)$$

**To find the value of $\pi$:**

By actual integration

$$\int_0^1 \frac{dx}{1+x^2} = \tan^{-1} x \Big|_0^1 = \tan^{-1} 1 - \tan^{-1} 0 = \frac{\pi}{4} - 0 = \frac{\pi}{4} \quad \cdots\cdots \quad (2)$$

From (1) and (2), we have

$$\frac{\pi}{4} = 0.78374$$

$$\Rightarrow \pi = 3.13496.$$

**Example:** Evaluate $\int_0^1 \frac{dx}{1+x}$ using the Simpson's $(3/8)^{th}$ rule by dividing the interval into six equal parts. Hence deduce an approximate value of $\log 2$.

**Solution:** Let $f(x) = \frac{1}{1+x}$. When we divide the given interval $[0,1]$ into six equal parts we have $n = 6$.

$$\therefore h = \frac{1-0}{6} = \frac{1}{6}.$$

The values of $x$ and $f(x)$ are shown in the following Table:

| $x$ | 0 | $\frac{1}{6}$ | $\frac{2}{6}$ | $\frac{3}{6}$ | $\frac{4}{6}$ | $\frac{5}{6}$ | $\frac{6}{6} = 1$ |
|---|---|---|---|---|---|---|---|
| $y = f(x)$ | 1 | 0.8571 | 0.75 | 0.6667 | 0.6 | 0.5455 | 0.5 |
| | $y_0$ | $y_1$ | $y_2$ | $y_3$ | $y_4$ | $y_5$ | $y_6$ |

The Simpson's $(3/8)^{th}$ rule, we have

$$\int_{x_0}^{x_n} f(x)dx = \frac{3h}{8}\Big[(y_0 + y_n) + 3(y_1 + y_2 + +y_4 + \cdots) + 2(y_3 + y_6 + \cdots)\Big]$$

$$\Rightarrow \int_0^1 \frac{dx}{1+x} = \frac{3h}{8}\Big[(y_0 + y_6) + 3(y_1 + y_2 + y_4 + y_5) + 2y_3\Big]$$

$$= \frac{3}{8} \times \frac{1}{6}\Big[(1 + 0.5) + 3(0.8571 + 0.75 + 0.6 + 0.5455) + 2(0.6667)\Big]$$

$$= \frac{1}{16}[1.5 + 8.2578 + 1.3334]$$

$$= 0.6932 \quad \cdots\cdots \quad (1)$$

**To find the value of $\log 2$:**

By actual integration, we find

$$\int_0^1 \frac{dx}{1+x} = \Big[\log(1+x)\Big]_0^1 = \log 2 \quad \cdots\cdots \quad (2)$$

From (1) and (2), we find that $\log 2 = 0.6932$.

**Example:** Find the value of $\log 2^{1/3}$ from $\int_0^1 \frac{x^2}{1+x^3} dx$ using Simpson's 1/3 rd rule with $h = 0.25$.

**Solution:** Let $f(x) = \frac{x^2}{1+x^3}$. Here $h = 0.25$.
The values of $x$ and $f(x)$ are tabulated below:

| $x$ | 0 | 0.25 | 0.5 | 0.75 | 1 |
|---|---|---|---|---|---|
| $y = f(x)$ | $0(y_0)$ | $0.0615(y_1)$ | $0.2222(y_2)$ | $0.3956(y_3)$ | $0.5(y_4)$ |

Using Simpson's 1/3 rd rule, we have

$$\int_{x_0}^{x_n} f(x)dx = \frac{h}{3}\Big[(y_0 + y_n) + 2(y_2 + y_4 + \cdots) + 4(y_1 + y_3 + \cdots)\Big]$$

$$\Rightarrow \int_0^1 \frac{x^2}{1+x^3}dx = \frac{h}{3}\Big[(y_0 + y_4) + 2y_2 + 4(y_1 + y_3)\Big]$$

$$= \frac{0.25}{3}\Big[(0 + 0.5) + 2(0.2222) + 4(0.0615 + 0.3956)\Big]$$

$$= 0.2311 \quad \cdots\cdots \quad (1)$$

**To find the value of** $\log 2^{1/3}$:

By actual integration

$$\int_0^1 \frac{x^2}{1+x^3}dx = \frac{1}{3}\int_0^1 \frac{3x^2}{1+x^3}dx$$

$$= \frac{1}{3}\Big[\log(1+x^3)\Big]_0^1$$

$$= \frac{1}{3}\log 2$$

$$= \log 2^{1/3} \quad \cdots\cdots \quad (2)$$

From (1) and (2), we have

$$\log 2^{1/3} = 0.2311.$$

**Example:** Evaluate $\int_0^1 \frac{x}{1+x^2}dx$ by using the Simpson's $(3/8)^{th}$, dividing the interval into 3 equal parts. Hence find an approximate value of $\log \sqrt{2}$.

**Solution:** Here, the interval is $[0,1]$ and it has to be divided into 3 equal parts. Hence we take $h = 1/3$ so that $x_0 = 0, x_1 = 1/3, x_2 = 2/3$ and $x_3 = 1$. We compute the corresponding values of $y = f(x) = \frac{x}{1+x^2}$, and the values are tabulated below:

| $x$: | $x_0 = 0$ | $x_1 = 1/3$ | $x_2 = 2/3$ | $x_3 = 1$ |
|---|---|---|---|---|
| $y = f(x)$: | $y_0 = 0$ | $y_1 = 3/10$ | $y_2 = 6/13$ | $y_3 = 1/2$ |

The Simpson's $(3/8)^{th}$ rule, we have

$$\int_{x_0}^{x_n} f(x)dx = \frac{3h}{8}\Big[(y_0 + y_n) + 3(y_1 + y_2 + + y_4 + \cdots) + 2(y_3 + y_6 + \cdots)\Big]$$

$$\Rightarrow \int_0^1 \frac{x}{1+x^2}dx = \frac{3h}{8}\Big[(y_0 + y_3) + 3(y_1 + y_2)\Big]$$

$$= \frac{1}{8}\Big[0 + \frac{1}{2} + 3\Big(\frac{3}{10} + \frac{6}{13}\Big)\Big]$$

$$= 0.348077 \quad \cdots\cdots \quad (1)$$

**To find the value of** $\log\sqrt{2}$:

By actual integration, we find

$$\int_0^1 \frac{x}{1+x^2}dx = \frac{1}{2}\int_0^1 \frac{2x}{1+x^2}dx$$

$$= \frac{1}{2}\Big[\log(1+x^2)\Big]_0^1$$

$$= \frac{1}{2}\log 2$$

$$= \log\sqrt{2} \quad \cdots\cdots \quad (2)$$

From (1) and (2), we find that $\log\sqrt{2} = 0.348077$.

**Example:** The velocity $v$ of a particle at distance $s$ from a point on its linear path is given by the following table:

| s in meters | 0 | 10 | 20 | 30 | 40 | 50 | 60 |
|---|---|---|---|---|---|---|---|
| v in m/sec | 47 | 58 | 64 | 65 | 61 | 52 | 38 |

Estimate the time taken to travel 60 meters by using Simpson's $\frac{1}{3}$rd rule.

**Solution:** Here $h = 10$.

If $t$ sec be the time taken to travel the distance $s(m)$ then $v = \frac{ds}{dt}$. Hence $dt = \frac{ds}{v}$.

To find the time taken to travel 60 meters, we have to evaluate $\int_0^{60} dt = \int_0^{60} \frac{ds}{v}$.

Let $y = \frac{1}{v}$. The table values for $y$ for different values of $s$ are given below:

| s | 0 | 10 | 20 | 30 | 40 | 50 | 60 |
|---|---|---|---|---|---|---|---|
| $y = \dfrac{1}{v}$ | 0.0213 | 0.0172 | 0.0156 | 0.0154 | 0.0164 | 0.0192 | 0.0263 |

By Simpson's $\frac{1}{3}$rd rule,

$$\int_0^{60} y\,ds = \frac{h}{3}\Big[(y_0 + y_6) + 2(y_2 + y_4) + 4(y_1 + y_3 + y_5)\Big]$$

$$\Rightarrow \int_0^{60} dt = \frac{10}{3}\Big[(0.0213 + 0.0263) + 2(0.0156 + 0.0164) + 4(0.0172 + 0.0154 + 0.0192)\Big]$$

$$= 1.0627$$

∴ Time taken to travel 60 meters = 1.0627 sec.

**Example:** The velocity $v(km/min)$ of a train which starts from rest is given at fixed intervals of time $t(min)$ as follows:

| t | 2 | 4 | 6 | 8 | 10 | 12 | 14 | 16 | 18 | 20 |
|---|---|---|---|---|---|---|---|---|---|---|
| v | 10 | 18 | 25 | 29 | 32 | 20 | 11 | 5 | 2 | 0 |

Estimate approximately the distance covered in 20 minutes using Simpson's $\frac{1}{3}$rd rule.

**Solution:** Here $h = 2$.

If $s(km)$ be the distance covered in $t(min)$, then $v = \dfrac{ds}{dt} \Rightarrow ds = v\,dt$.

To find the distance covered in 20 minutes, we have to evaluate $\int_0^{20} ds = \int_0^{20} v\,dt$.

The train starts from rest, ∴ the velocity $v = 0$ when $t = 0$

The given table of velocities can be written as

| t | 0 | 2 | 4 | 6 | 8 | 10 | 12 | 14 | 16 | 18 | 20 |
|---|---|---|---|---|---|---|---|---|---|---|---|
| v | 0 | 10 | 18 | 25 | 29 | 32 | 20 | 11 | 5 | 2 | 0 |
|   | $v_0$ | $v_1$ | $v_2$ | $v_3$ | $v_4$ | $v_5$ | $v_6$ | $v_7$ | $v_8$ | $v_9$ | $v_{10}$ |

By Simpson's $\frac{1}{3}$rd rule,

$$\int_0^{20} v\,dt = \frac{h}{3}\Big[(v_0 + v_{10}) + 2(v_2 + v_4 + v_6 + v_8) + 4(v_1 + v_3 + v_5 + v_7 + v_9)\Big]$$

$$\Rightarrow \int_0^{20} ds = \frac{2}{3}\Big[(0+0) + 2(18 + 29 + 20 + 5) + 4(10 + 25 + 32 + 11 + 2)\Big]$$

$$= 309.3$$

Hence, the distance covered by the train in 20 minutes = 309.3 $km$.

**Example:** A solid of revolution is formed by rotating about the $x$-axis, the area between the $x$-axis, the lines $x = 0$, $x = 1$ and a curve through the points with the following co-ordinates:

| $x$ | 0.00 | 0.25 | 0.50 | 0.75 | 1.00 |
|---|---|---|---|---|---|
| $y$ | 1.0000 | 0.9896 | 0.9589 | 0.9089 | 0.8415 |

Estimate the volume of the solid formed using Simpson's rule.

**Solution:** Here $h = 0.25$, $y_0 = 1$, $y_1 = 0.9896$, $y_2 = 0.9589$, $y_3 = 0.9089$, $y_4 = 0.8415$.
Then, the required volume of the solid generated

$$= \int_0^1 \pi y^2\,dx = \pi \cdot \frac{h}{3}\Big[(y_0^2 + y_4^2) + 4(y_1^2 + y_3^2) + 2y_2^2\Big]$$

$$= \frac{0.25\pi}{3}\Big[\{1 + (0.8415)^2\} + 4\{(0.9896)^2 + (0.9089)^2\} + 2(0.9589)^2\Big]$$

$$= \frac{0.25\pi}{3}\Big[1.7081 + 7.2216 + 1.8389\Big]$$

$$= 0.2618(10.7686)$$

$$= 2.81922.$$

**Example:** A rocket is launched from the ground. Its acceleration measured every 5 seconds is tabulated below. Find the velocity and the position of the rocket at $t = 40$ seconds. Use trapezoidal rule as well as Simpson's rule

| t: | 0 | 5 | 10 | 15 | 20 | 25 | 30 | 35 | 40 |
|---|---|---|---|---|---|---|---|---|---|
| $a(t)$ : | 40.0 | 45.25 | 48.50 | 51.25 | 54.35 | 59.48 | 61.5 | 64.3 | 68.7 |

**Solution:** Let $s$ be distance traveled by the rocket with velocity $v(t)$ and acceleration $a(t)$ at time $t$.

Then $a = \dfrac{dv}{dt}$.

Integrating from $t = 0$ to $t = 40$ seconds, we get

$$v\Big|_{t=0}^{40} = \int_0^{40} \frac{dv}{dt} dt = \int_0^{40} a(t)\,dt.$$

Here $h = 5$, $a_0 = 40.0$, $a_1 = 45.25$, $a_2 = 48.50$, $a_3 = 51.25$, $a_4 = 54.35$, $a_5 = 59.48$, $a_6 = 61.5$, $a_7 = 64.3$ and $a_8 = 68.7$.

By Trapezoidal rule, we have

The required velocity $= \int_0^{40} a(t)\,dt$

$= \dfrac{h}{2}\Big[(a_0 + a_8) + 2(a_1 + a_2 + a_3 + a_4 + a_5 + a_6 + a_7)\Big]$

$= \dfrac{5}{2}\Big[(400 + 68.7) + 2(45.25 + 48.50 + 51.25 + 54.35 + 59.48 + 61.5 + 64.3)\Big]$

$= \dfrac{5}{2}\Big[108.7 + 2(384.63)\Big]$

$= \dfrac{5}{2}[877.96]$

$= 2194.9$

Position of the rocket at $t = 40$ seconds$=(2194.9)(40) = 87796$.

By Simpson's rule, we have

The required velocity $= \int_0^{40} a(t)dt$

$= \dfrac{h}{3}\Big[(a_0 + a_8) + 2(a_2 + a_4 + a_6) + 4(a_1 + a_3 + a_5 + a_7)\Big]$

$= \dfrac{5}{2}\Big[(400 + 68.7) + 2(45.25 + 48.50 + 51.25 + 54.35 + 59.48$

$\qquad + 61.5 + 64.3) + 4(45.25 + 51.25 + 59.48 + 64.3)\Big]$

$= \dfrac{5}{3}\Big[108.7 + 328.7 + 881.123\Big]$

$= 2197.5$

Position of the rocket at $t = 40$ seconds $= (2197.5)(40) = 87900$.

## EXERCISE

1. Evaluate $\int_0^1 \dfrac{x}{x^3+5}dx$ using the Simpson's $(1/3)^{rd}$ rule. **(Dec-2020)**
Ans: 0.10385

2. Evaluate $\int_0^1 \dfrac{x}{x^3+5}dx$ using Trapezoidal rule. **(Dec-2020)**
Ans: 0.1124

3. Using Simpson's $(3/8)^{th}$ rule evaluate $\int_0^2 e^{x^2} dx$. **(July-2021)**

4. Using Trapezoidal rule evaluate $\int_0^2 \dfrac{dx}{1+x^2}$. **(July-2021)**

5. Evaluate $\int_0^\pi \sin x\, dx$ using Trapezoidal rule, Simpson's 1/3rd and 3/8th rule.
**(Jan-2020), (Mar-2022)**

6. Evaluate $\int_0^1 x^3 dx$ using Simpson's 1/3rd and Simpson's 3/8th rules. **(July-2021)**

7. Using Trapezoidal rule evaluate $\int_0^1 (1+2x+e^x)dx$. **(Nov-2021)**

8. Using Simpson's $(1/3)^{rd}$ rule evaluate $\int_0^\pi \dfrac{1}{1+2x}dx$. **(Nov-2021)**

9. Using Simpson's $(1/3)^{rd}$ rule evaluate $\int_{1}^{2} \dfrac{dx}{x}$.  (Mar-2022)

Ans: 0.69325

10. Using Simpson's $(3/8)^{th}$ rule evaluate $\int_{0}^{\pi} e^{2x} dx$.  (Mar-2022)

11. Evaluate $\int_{0}^{2} e^{-x^2} dx$, using Trapezoidal rule by taking $h = 0.25$.  (Apr-2022)

## 5.2 Numerical Solution of Ordinary Differential Equations

In this section we develop numerical methods for finding a solution of ordinary differential equation of first order and first degree which is of the form

$$\frac{dy}{dx} = f(x, y) \quad \cdots \cdots \quad (1)$$

with the initial condition $y(x_0) = y_0$, which is called the initial value problem.

The general solution of equation (1) will be obtained in any one of the two forms given below:

(1). A series for $y$ in terms of powers of $x$, from which the value of $y$ can be obtained by direct substitution. The methods *Taylor and Picard* belong to this type.

(2). Given a set of tabulated values of $x$ and $y$, we obtain $y$ by iterative process. The methods of Euler and Runge-Kutta belong to this type.

### 5.2.1 Taylor's Series Method

Consider the first order differential equation

$$\frac{dy}{dx} = f(x, y), \quad y(x_0) = y_0 \quad \cdots \cdots \quad (1)$$

Let $y = y(x)$ be the exact solution of this problem. Expanding $y(x)$ in Taylor's series abut the point $x_0$, we get

$$y(x) = y(x_0) + (x - x_0)y'(x_0) + \frac{(x - x_0)^2}{2!}y''(x_0) + \frac{(x - x_0)^3}{3!}y'''(x_0) + \cdots \cdots \quad (2)$$

The expression (2) is the *Taylor Series Solution* of the problem (1) at a point $x$ in a neighborhood of the point $x_0$. For a specified value of $x$ in a neighborhood of $x_0$, this expression yields the corresponding value of $y$.

Thus, for a specified value $x_1$ of $x$ in a neighborhood of $x_0$, the solution at $x_1$ is given by

$$y(x_1) = y(x_0) + (x_1 - x_0)y'(x_0) + \frac{(x_1 - x_0)^2}{2!} y''(x_0) + \frac{(x_1 - x_0)^3}{3!} y'''(x_0) + \cdots \cdots \quad (3)$$

If we need to find the solution at a point $x$ in a neighborhood of $x_1$, we replace $x_0$ by $x_1$ and $y_0$ by $y_1 = y(x_1)$ in the solution (2). Thus, in a neighborhood of $x_1$, the solution is given by

$$y(x) = y(x_1) + (x - x_1)y'(x_1) + \frac{(x - x_1)^2}{2!} y''(x_1) + \frac{(x - x_1)^3}{3!} y'''(x_1) + \cdots \cdots \quad (4)$$

For a specified value $x_2$ of $x$ in a neighborhood of $x_1$, the solution at $x_2$ is got by changing $x$ to $x_2$ in the solution (4).

Similar procedure is adopted for obtaining the solution in a neighborhood of $x_2$.

The method of obtaining an approximate solution of the problem (1) on the basis of the Taylor's series as described above is known as the **Taylor's Series method**.

**Example:** Using Taylor's series method to solve $\dfrac{dy}{dx} = 1 + xy$ with $y(0) = 1$.

Hence find $y(0.1)$. (Dec-2020)

**Solution:** The given differential equation is $y' = 1 + xy$ and $y(0) = 1$.

Here $x_0 = 0$. Therefore, the Taylor's series solution for $y = f(x)$ in a neighborhood of $x_0 = 0$ is

$$y(x) = y(x_0) + (x - x_0)y'(x_0) + \frac{(x - x_0)^2}{2!} y''(x_0) + \frac{(x - x_0)^3}{3!} y'''(x_0) + \frac{(x - x_0)^4}{4!} y''''(x_0) + \cdots$$

$$= y(0) + xy'(0) + \frac{x^2}{2!} y''(0) + \frac{x^3}{3!} y'''(0) + \frac{x^4}{4!} y''''(0) + \cdots \cdots \quad (1)$$

with $y(0) = 1$.

From the given differential equation, we get

$$y' = 1 + xy$$
$$y'' = xy' + y$$
$$y''' = xy'' + y' + y' = xy'' + 2y'$$
$$y'''' = xy''' + y'' + 2y'' = xy''' + 3y''$$
and so on.

Using the given initial condition that $y(0) = 1$ in these, we obtain

$$y'(0) = 1 + 0y(0) = 1 + 0 = 1$$
$$y''(0) = 0y'(0) + y(0) = 0 + 1 = 1$$
$$y'''(0) = 0y''(0) + 2y'(0) = 0(1) + 2(1) = 2$$
$$y''''(0) = xy'''(0) + 3y''(0) = 0(2) + 3(1) = 3$$

Putting these and the condition $y(0) = 1$ in expression (1), we get

$$y(x) = y(0) + xy'(0) + \frac{x^2}{2}y''(0) + \frac{x^3}{6}y'''(0) + \frac{x^4}{24}y''''(0) + \cdots$$
$$= 1 + x + \frac{x^2}{2} + \frac{x^3}{3} + \frac{x^4}{8} + \cdots$$

Taking $x = x_1 = 0.1$ in this expression, we get

$$y(0.1) = 1 + (0.1) + \frac{(0.1)^2}{2} + \frac{(0.1)^3}{3} + \frac{(0.1)^4}{8} + \cdots$$
$$\approx 1.1053 \text{ correct to four decimal places}$$

This is the required solution.

**Example:** Solve $y' = x - y^2, y(0) = 1$ by Taylor's series method and find $y(0.1)$ correct to four decimal places. **(Dec-2020), (July-2021)**

**Solution:** The given differential equation is $y' = x - y^2$ and $y(0) = 1$.
Here $x_0 = 0$. Hence, the Taylor's series solution for the given problem in a neighborhood of $x_0 = 0$ is

$$y(x) = y(0) + xy'(0) + \frac{x^2}{2!}y''(0) + \frac{x^3}{3!}y'''(0) + \frac{x^4}{4!}y''''(0) + \cdots \cdots \quad (1)$$

with $y(0) = 1$.

From the given differential equation, we get

$$y' = x - y^2$$
$$y'' = 1 - 2yy'$$
$$y''' = -2[yy'' + (y')^2]$$
$$y'''' = -2[yy''' + y'y'' + 2y'y''] = -2[yy''' + 3y'y'']$$

and so on. Using the given initial condition that $y(0) = 1$ in these, we obtain

$y'(0) = 0 - [y(0)]^2 = 0 - 1^2 = -1$

$y''(0) = 1 - 2y(0)y'(0) = 1 - 2(1)(-1) = 1 + 2 = 3$

$y'''(0) = -2[y(0)y''(0) + (y'(0))^2] = -2[(1)(3) + (-1)^2] = -2(4) = -8$

$y''''(0) = -2[y(0)y'''(0) + 3y'(0)y''(0)] = -2[(1)(-8) + 3(-1)(3)] = -2[-8 - 9] = 34.$

Putting these and the condition $y(0) = 1$ in (1), we get

$$y(x) = y(0) + xy'(0) + \frac{x^2}{2}y''(0) + \frac{x^3}{6}y'''(0) + \frac{x^4}{24}y''''(0) + \cdots$$

$$= 1 + x(-1) + \frac{x^2}{2}(3) + \frac{x^3}{6}(-8) + \frac{x^4}{24}(34) + \cdots$$

$$= 1 - x + \frac{3}{2}x^2 - \frac{4}{3}x^3 + \frac{17}{12}x^4 + \cdots$$

Taking $x = x_1 = 0.1$ in this expression, we get

$$y(x_1) = y(0.1) = 1 - (0.1) + \frac{3}{2}(0.1)^2 - \frac{4}{3}(0.1)^3 + \frac{17}{12}(0.1)^4 + \cdots$$

$$= 1 - 0.1 + 0.015 - 0.001333 + 0.000146$$

$$\approx 0.9138.$$

This is the required solution.

**Example:** Solve $y' = x + y$ given $y(1) = 0$. Hence find $y(1.1)$ and $y(1.2)$ by Taylor's series method. Compare with its exact value. (Apr-2022)

**Solution:** The given differential equation is $y' = x + y$ and $y(1) = 0$.

Here $x_0 = 1$. Hence, the Taylor's series solution for the given problem in a neighborhood of $x_0 = 1$ is

$$y(x) = y(1) + (x-1)y'(1) + \frac{(x-1)^2}{2!}y''(1) + \frac{(x-1)^3}{3!}y'''(1) + \frac{(x-1)^4}{4!}y''''(1) + \cdots \cdots \quad (1)$$

From the given differential equation, we find

$$\left.\begin{array}{c} y' = x + y \\ y'' = 1 + y' \\ y''' = y'' \\ y'''' = y''' \end{array}\right\} \quad \cdots \cdots \quad (2)$$

and so on.

Using the given initial condition that $y(1) = 0$ in these, we obtain

$$y'(1) = 1 + y(1) = 1 + 0 = 1$$
$$y''(1) = 1 + y'(1) = 1 + 1 = 2$$
$$y'''(1) = y''(1) = 2$$
$$y''''(1) = y'''(1) = 2 \text{ etc.}$$

Using these and the condition $y(1) = 0$ in expression (1), we get

$$y(x) = 0 + (x-1)(1) + \frac{(x-1)^2}{2!}(2) + \frac{(x-1)^3}{3!}(2) + \frac{(x-1)^4}{4!}y'''(2) + \cdots \cdots$$
$$= (x-1) + (x-1)^2 + \frac{1}{3}(x-1)^3 + \frac{1}{12}(x-1)^4 + \cdots \cdots$$

**Step 1:** To find $y(1.1)$

For $x = x_1 = 1.1$ this becomes

$$y(1.1) = 0.1 + (0.1)^2 + \frac{1}{3}(0.1)^3 + \frac{1}{12}(0.1)^4 + \cdots$$
$$\approx 0.11033.$$

This is an approximate solution of the given problem at the point $x_1 = 1.1$.

**Step 2:** Now we will find $y(1.2)$.

The Taylor's series solution at a point $x$ in a neighborhood of $x_1 = 1.1$ is given by

$$y(x) = y(1.1) + (x - 1.1)y'(1.1) + \frac{(x-1.1)^2}{2!}y''(1.1) + \frac{(x-1.1)^3}{3!}y'''(1.1)$$
$$+ \frac{(x-1.1)^4}{4!}y''''(1.1) + \cdots \cdots \quad (3)$$

Now, from expressions (2), we find

$$y'(1.1) = 1.1 + y(1.1) = 1.1 + 0.1103 = 1.2103$$
$$y''(1.1) = 1 + y'(1.1) = 1 + 1.2.2103 = 2.2103$$
$$y'''(1.1) = y''(1.1) = 2.2103$$
$$y''''(1.1) = y'''(1.1) = 2.2103 \text{ etc.}$$

Using these in expression (3), we get

$$y(x) = y(1.1) + (x-1.1)(1.2103) + \frac{(x-1.1)^2}{2!}(2.2103) +$$
$$\frac{(x-1.1)^3}{3!}(2.2103) + \frac{(x-1.1)^4}{4!}(2.2103) + \cdots \cdots$$

For $x = 1.2$, this becomes

$$y(1.2) = 0.11033 + (0.1)(1.2103) + \frac{1}{2}(0.01)(2.2103)$$
$$+ \frac{1}{6}(0.001)(2.2103) + \frac{1}{24}(0.0001)(2.2103)$$
$$\approx 0.2427$$

This is an approximate solution of the given problem at the point $x_2 = 1.2$.

### Analytical Solution:

The given equation is $\dfrac{dy}{dx} - y = x$.

This is a linear differential equation of the form $\dfrac{dy}{dx} + P(x)y = Q(x)$

Here $P(x) = -1$ and $Q(x) = x$.

$$\therefore \text{I.F} = e^{\int P dx} = e^{\int -dx} = e^{-x}.$$

∴ The general solution is

$$y(I.F) = \int Q(I.F)dx + x$$

$$\Rightarrow ye^{-x} = \int xe^{-x}dx + c$$

$$= x\left(\frac{e^{-x}}{-1}\right) - (1)\left(\frac{e^{-x}}{(-1)^2}\right) + c$$

$$= -(x+1)e^{-x} + c$$

$$\Rightarrow y = -(x+1) + ce^x.$$

We have $y(1) = 0 \Rightarrow 0 = -2 + ce \Rightarrow c = 2e^{-1}$

Hence the solution is $y = -(x+1) + 2e^{-1}e^x = -x - 1 + 2e^{x-1}$.

Thus $y(1.1) = -1.1 - 1 + 2e^{0.1} = 0.11034$

$y(1.2) = -1.2 - 1 + 2e^{0.2} = 0.2428$.

We can tabulate the values as follows:

| $x$ | Taylor's series method($y$) | Exact solution($y$) |
|---|---|---|
| 1.1 | 0.1103 | 0.11034 |
| 1.2 | 0.2427 | 0.2428 |

**Example:** Using Taylor's series method find the values of $y$ at $x = 0.1$ and $x = 0.2$ correct to five decimals places from $\dfrac{dy}{dx} = x^2y - 1$, $y(0) = 1$.

**Solution:** The given differential equation is $y' = x^2y - 1$ and $y(0) = 1$.

Here $x_0 = 0$. Hence, the Taylor's series solution for the given problem in a neighborhood of $x_0 = 0$ is

$$y(x) = y(0) + xy'(0) + \frac{x^2}{2!}y''(0) + \frac{x^3}{3!}y'''(0) + \frac{x^4}{4!}y''''(0) + \cdots \cdots \quad (1)$$

with $y(0) = 1$.

From the given differential equation, we get

$$\left.\begin{array}{l} y' = x^2 y - 1 \\ y'' = 2xy + x^2 y' \\ y''' = 2xy' + 2y + 2xy' + x^2 y'' = 2y + 4xy' + x^2 y'' \\ y'''' = 2y' + 4xy'' + 4y' + x^2 y''' + 2xy'' = 6y' + 6xy'' + x^2 y''' \end{array}\right\} \quad \ldots \ldots \quad (2)$$

and so on.

Using the given initial condition that $y(0) = 1$ in these, we obtain

$$y'(0) = 0^2 y(0) - 1 = (0)^2(1) - 1 = -1$$

$$y''(0) = 2(0)y(0) + (0)^2 y'(0) = 2(0)(1) + (0)^2(-1) = 0$$

$$y'''(0) = 2y(0) + 4(0)y'(0) + (0)^2 y''(0) = 2(1) + 4(0)(-1) + (0)^2(0) = 2$$

$$y''''(0) = 6y'(0) + 6(0)y''(0) + (0)^2 y'''(0) = 6(-1) + 6(0)(0) + (0)^2(2) = -6$$

The Taylor's series solution for the given problem in a neighborhood of $x_0 = 0$ is

$$y(x) = y(0) + xy'(0) + \frac{x^2}{2}y''(0) + \frac{x^3}{6}y'''(0) + \frac{x^4}{24}y''''(0) + \cdots$$

$$= 1 + x(-1) + \frac{x^2}{2}(0) + \frac{x^3}{6}(2) + \frac{x^4}{24}(-6) + \cdots$$

$$= 1 - x + \frac{x^3}{3} - \frac{x^4}{4} + \cdots$$

Setting $x = x_1 = 0.1$ in this expression, we get the solution at $x_1 = 0.1$ as

$$y(0.1) = 1 - (0.1) + \frac{(0.1)^3}{3} - \frac{(0.1)^4}{4} + \cdots = 0.9003 \quad \ldots \ldots \quad (3)$$

Next, the solution in a neighborhood of $x_1 = 0.1$ is

$$y(x) = y(0.1) + (x - 0.1)y'(0.1) + \frac{(x - 0.1)^2}{2!}y''(0.1)$$

$$+ \frac{(x - 0.1)^3}{3!}y'''(0.1) + + \frac{(x - 0.1)^4}{4!}y''''(0.1) + \cdots \quad \ldots \ldots \quad (4)$$

From expression (2) and (3), we find that

$$y'(0.1) = (0.1)^2 y(0.1) - 1 = (0.01)(0.9003) - 1 = -0.991$$

$$y''(0.1) = 2(0.1)y(0.1) + (0.1)^2 y'(0.1) = (0.2)(0.9003) + (0.01)(-0.991) = 0.1702$$

$$y'''(0.1) = 2y(0.1) + 4(0.1)y'(0.1) + (0.1)^2 y''(0.1)$$
$$= 2(0.9003) + (0.4)(-0.991) + (0.01)(0.1702) = 1.4059$$

$$y''''(0.1) = 6y'(0.1) + 6(0.1)y''(0.1) + (0.1)^2 y'''(0.1)$$
$$= 6(-0.991) + (0.6)(0.1702) + (0.01)(1.4059) = -5.8298$$

Using these and (3) in (4) and setting $x = 0.2$, we get the solution at $x_2 = 0.2$ as

$$y(0.2) = 0.9903 + (0.1)(-0.991) + \frac{1}{2}(0.01)(0.1702)$$
$$+ \frac{1}{6}(0.001)(1.059) + \frac{1}{24}(0.0001)(-5.8298)$$
$$\approx 0.8023$$

**Example:** Using Taylor's series method, find $y(0.1)$ correct to 3 decimals places given that $\dfrac{dy}{dx} = e^x - y^2$, $y(0) = 1$.

**Solution:** The given differential equation is $y' = e^x - y^2$ and $y(0) = 1$. Here $x_0 = 0$. Hence Taylor's series solution of the problem in a neighborhood of $x_0 = 0$ is

$$y(x) = y(0) + xy'(0) + \frac{x^2}{2} y''(0) + \frac{x^3}{6} y'''(0) + \frac{x^4}{24} y''''(0) + \cdots \quad (1)$$

with $y(0) = 1$.

From the given differential equation, we get

$$y' = e^x - y^2$$
$$y'' = e^x - 2yy'$$
$$y''' = e^x - 2[yy'' + (y')^2]$$
$$y'''' = e^x - 2[yy''' + y'y'' + 2y'y'']$$
$$= e^x - 2[yy''' + 3y'y'']$$

and so on. Using the initial condition that $y(0) = 1$ in these, we obtain

$$y'(0) = e^0 - (y(0))^2 = 1 - 1 = 0$$
$$y''(0) = e^0 - 2y(0)y'(0) = 1 - 0 = 1$$
$$y'''(0) = e^0 - 2[y(0)y''(0) + (y')^2(0)] = 1 - 2 = -1$$
$$y''''(0) = e^0 - 2[y(0)y'''(0) + 3y'(0)y''(0)] = 1 - 2[-1+0] = 1+2 = 3.$$

Putting these and the condition $y(0) = 1$ in (1), we get

$$y(x) = y(0) + xy'(0) + \frac{x^2}{2}y''(0) + \frac{x^3}{6}y'''(0) + \frac{x^4}{24}y''''(0) + \cdots$$
$$= 1 + x(0) + \frac{x^2}{2}(1) + \frac{x^3}{6}(-1) + \frac{x^4}{24}(3) + \cdots$$
$$= 1 + \frac{x^2}{2} - \frac{x^3}{6} + \frac{x^4}{8} + \cdots$$

Taking $x = x_1 = 0.1$ in this expression, we get

$$y(x_1) = y(0.1) = 1 + \frac{(0.1)^2}{2} - \frac{(0.1)^3}{3} + \frac{(0.1)^4}{8} + \cdots$$
$$= 1 + 0.005 - 0.00016 + 0.0000125$$
$$\approx 1.0048525$$

**Example:** Solve $y' = 3x + y^2$, $y(0) = 1$, using Taylor's series method and compute $y(0.1)$.

**Solution:** The given differential equation is $y' = 3x + y^2$ and $y(0) = 1$.
Here $x_0 = 0$. Hence Taylor's series solution of the problem in a neighborhood of $x_0 = 0$ is

$$y(x) = y(0) + xy'(0) + \frac{x^2}{2}y''(0) + \frac{x^3}{6}y'''(0) + \frac{x^4}{24}y''''(0) + \cdots \qquad (1)$$

with $y(0) = 1$.

From the given differential equation, we get

$$y' = 3x + y^2$$
$$y'' = 3 + 2yy'$$
$$y''' = 2[yy'' + (y')^2]$$
$$y'''' = 2[yy''' + y'y'' + 2y'y''] = 2[yy''' + 3y'y'']$$

and so on. Using the initial condition that $y(0) = 1$ in these, we obtain

$$y'(0) = 3(0) + [y(0)]^2 = 1$$
$$y''(0) = 3 + 2y(0)y'(0) = 3 + 2(1)(1) = 3 + 2 = 5$$
$$y'''(0) = 2[y(0)y''(0) + (y'(0))^2] = 2(5 + 1) = 12$$
$$y''''(0) = 2[y(0)y'''(0) + 3y'(0)y''(0)] = 2(12 + 15) = 54.$$

Putting these and the condition $y(0) = 1$ in (1), we get

$$y(x) = y(0) + xy'(0) + \frac{x^2}{2}y''(0) + \frac{x^3}{6}y'''(0) + \frac{x^4}{24}y''''(0) + \cdots$$
$$= 1 + x(1) + \frac{x^2}{2}(5) + \frac{x^3}{6}(12) + \frac{x^4}{24}(54) + \cdots$$
$$= 1 + x + \frac{5}{2}x^2 + 2x^3 + \frac{27}{12}x^4 + \cdots$$

Taking $x = x_1 = 0.1$ in this expression, we get

$$y(x_1) = y(0.1) = 1 + (0.1) + \frac{5}{2}(0.1)^2 + 2(0.1)^3 + \frac{27}{12}(0.1)^4 + \cdots$$
$$= 1 + 0.1 + 0.025 + 0.002 + 0.000225$$
$$\approx 1.127225.$$

**Example:** Find the solutions of the initial value problem

$$y' = x + y^2, \quad y(0) = 1$$

at the points $x = 0.1, x_2 = 0.2$ and $x_3 = 0.3$, using Taylor's series method.

**Solution:** The given differential equation is $y' = y^2 + x$ and $y(0) = 1$.

From this, we find

$$\left.\begin{array}{l} y' = x + y^2 \\ y'' = 2yy' + 1 \\ y''' = 2yy'' + 2(y')^2 \\ y'''' = 2yy''' + 2y'y'' + 4y'y'' = 2[yy''' + 3y'y''] \end{array}\right\} \quad \ldots\ldots \quad (1)$$

and so on.

The Taylor's series solution of the problem in a neighborhood of the point $x_0 = 0$ is

$$y(x) = y(x_0) + xy'(x_0) + \frac{x^2}{2!}y''(x_0) + \frac{x^3}{3!}y'''(x_0) + +\frac{x^4}{4!}y''''(x_0)\cdots$$

$$= y(0) + xy'(0) + \frac{x^2}{2}y''(0) + \frac{x^3}{6}y'''(0) + \frac{x^4}{24}y''''(0) + \cdots$$

with $y(0) = 1$. From (1), we find that

$$y'(0) = [y(0)]^2 + 0 = 1^2 + 0 = 1$$

$$y''(0) = 2y(0)y'(0) + 1 = 2(1)(1) + 1 = 3$$

$$y'''(0) = 2y(0)y''(0) + 2(y'(0))^2 = 2(1)(3) + 2(1)^2 = 8$$

$$y''''(0) = 2[y(0)y'''(0) + 3y'(0)y''(0) = 2(1)(8) + 6(1)(3) = 16 + 18 = 34$$

Using these and $y(0) = 1$ and setting $x = x_1 = 0.1$, we get

$$y(0.1) = 1 + (0.1) \times 1 + \frac{1}{2}(0.1)^2(3) + \frac{1}{6}(0.1)^3(8) + \frac{1}{24}(0.1)^4(34)$$

$$= 1.1164 \quad \cdots \quad (2)$$

This is required solution at the point $x_1 = 0.1$.

Next, the solution in a neighborhood of the point $x_1 = 0.1$ is

$$y(x) = y(0.1) + (x - 0.1)y'(0.1) + \frac{(x - 0.1)^2}{2!}y''(0.1)$$

$$+ \frac{(x - 0.1)^3}{3!}y'''(0.1) + +\frac{(x - 0.1)^4}{4!}y''''(0.1) + \cdots \quad (3)$$

From (1) and (2), we find that

$y'(0.1) = [y(0.1)]^2 + 0.1 = (1.1164)^2 + 0.1 = 1.3463$

$y''(0.1) = 2y(0.1)y'(0.1) + 1 = 2(1.1164)(1.3468) + 1 = 4.0071$

$y'''(0.1) = 2y(0.1)y''(0.1) + 2(y'(0.1))^2 = 2(1.1164)(4.0071) + 2(1.3463)^2 = 12.5721$

$y''''(0.1) = 2[y(0.1)y'''(0.1) + 3y'(0.1)y''(0.1)]$

$\quad = 2(1.1164)(12.5721) + 6(1.3463)(4.0071) = 60.4395$

Using these in (3) and setting $x = x_2 = 0.2$, we get

$y(0.2) = 1.1164 + (0.1)(1.3463) + \dfrac{1}{2}(0.1)^2(4.0071) + \dfrac{1}{6}(0.1)^3(12.5721) + \dfrac{1}{24}(0.1)^4(60.4395)$

$\quad = 1.2734 \qquad \cdots \quad (4)$

Lastly, the solution in a neighborhood of the point $x_2 = 0.2$ is

$$y(x) = y(0.2) + (x - 0.2)y'(0.2) + \dfrac{(x-0.2)^2}{2!}y''(0.2)$$
$$+ \dfrac{(x-0.2)^3}{3!}y'''(0.2) + +\dfrac{(x-0.2)^4}{4!}y''''(0.2) + \cdots \qquad (5)$$

From (1) and (4), we find that

$y'(0.2) = [y(0.2)]^2 + 0.2 = (1.2734)^2 + 0.2 = 1.8215$

$y''(0.2) = 2y(0.2)y'(0.2) + 1 = 2(1.2734)(1.8215) + 1 = 5.639$

$y'''(0.2) = 2y(0.2)y''(0.2) + 2(y'(0.2))^2 = 2(1.2734)(5.639) + 2(1.8215)^2 = 20.9971$

$y''''(0.2) = 2[y(0.2)y'''(0.2) + 3y'(0.2)y''(0.2) = 2(1.2734)(20.9971) + 6(1.8215)(5.639) = 115.104$

Using these and (4) in (5) and setting $x = x_3 = 0.3$, we get

$y(0.3) = 1.2734 + (0.1)(1.8215) + \dfrac{1}{2}(0.1)^2(5.639) + \dfrac{1}{6}(0.1)^3(20.9971) + \dfrac{1}{24}(0.1)^4(115.104)$

$\quad = 1.4877$

This is the required solution at the point $x_3 = 0.3$.

**Example:** Solve $y' = x - y^2$, $y(0) = 1$ and $y'(0) = 1$ by Taylor's series method.

Hence find the values of $y$ at $x = 0.2$ and $x = 0.4$.

**Solution:** The given differential equation is $y' = x - y^2$, $y(0) = 1$ and $y'(0) = 1$.
The Taylor's series solution for the given problem in a neighborhood of $x_0 = 0$ is

$$y(x) = y(x_0) + xy'(x_0) + \frac{x^2}{2!}y''(x_0) + \frac{x^3}{3!}y'''(x_0) + +\frac{x^4}{4!}y''''(x_0) + \cdots$$

$$= y(0) + xy'(0) + \frac{x^2}{2}y''(0) + \frac{x^3}{6}y'''(0) + \frac{x^4}{24}y''''(0) + \cdots \qquad (1)$$

From the given differential equation, we find

$$\left.\begin{array}{l} y'' = 1 - 2yy' \\ y''' = -2[yy'' + (y')^2] \\ y'''' = -2[yy''' + y'y'' + 2y'y''] = -2[yy''' + 3y'y''] \end{array}\right\} \quad \cdots \cdots \quad (2)$$

and so on. From these and the given initial conditions $y(0) = 1$ and $y'(0) = 1$, we find that

$$y''(0) = 1 - 2y(0)y'(0) = 1 - 2(1)(1) = -1$$

$$y'''(0) = -2[y(0)y''(0) + (y'(0))^2] = -2[1(-1) + 1^2] = 0$$

$$y''''(0) = -2[y(0)y'''(0) + 3y'(0)y''(0)] = -2[1(0) + 3(1)(-1)] = 6$$

Using these and $y(0) = 1$ and $y'(0) = 1$ in expression (1), we get

$$y(x) = 1 + x(1) + \frac{x^2}{2}(-1) + \frac{x^3}{6}(0) + \frac{x^4}{24}(6) + \cdots$$

$$= 1 + x - \frac{1}{2}x^2 + 0 + \frac{1}{4}x^4 + \cdots$$

$$= 1 + x - \frac{x^2}{2} + \frac{x^4}{4} + \cdots$$

For $x = x_1 = 0.2$, this becomes

$$y(0.2) = 1 + (0.2) - \frac{(0.2)^2}{2} + \frac{(0.2)^4}{4} + \cdots$$

$$= 1 + 0.2 - 0.02 + 0.0004 \approx 1.1804 \qquad \cdots \quad (3)$$

This is an approximate solution of the given problem at the point $x_1 = 0.2$.
Next, the Taylor's series solution at a point $x$ in a neighborhood of $x_1 = 0.2$ is given

by

$$y(x) = y(x_1) + (x - x_1)y'(x_1) + \frac{(x-x_1)^2}{2!}y''(x_1) + \frac{(x-x_1)^3}{3!}y'''(x_1)$$
$$+ \frac{(x-x_1)^4}{4!}y''''(x_1) + \cdots$$
$$= y(0.2) + (x - 0.2)y'(0.2) + \frac{(x-0.2)^2}{2!}y''(0.2) + \frac{(x-0.2)^3}{3!}y'''(0.2)$$
$$+ \frac{(x-0.2)^4}{4!}y''''(0.2) + \cdots \tag{4}$$

Now from expression (2), we find

$$y'(0.2) = 0.2 - [y(0.2)]^2 = 0.2 - (1.1804)^2 = -1.1933$$

$$y''(0.2) = 1 - 2y(0.2)y'(0.2) = 1 - 2(1.1804)(-1.1933) = 3.8171$$

$$y'''(0.2) = -2[y(0.2)y''(0.2) + (y'(0.2))^2]$$
$$= -2[(1.1804)(3.8171) + (-1.1933)^2] = -11.8593$$

$$y''''(0.2) = -2[y(0.2)y'''(0.2) + 3y'(0.2)y''(0.2)]$$
$$= -2[(1.1804)(-11.8593) + 3(-1.1933)(3.8171)] = 55.3271$$

Using these in expression (4), we get

$$y(x) = 1.1804 + (x - 0.2)(-1.1933) + \frac{(x-0.2)^2}{2}(3.8171) + \frac{(x-0.2)^3}{6}(-11.8593)$$
$$+ \frac{(x-0.2)^4}{24}(55.3271) + \cdots$$

For $x = 0.4$, this becomes

$$y(0.4) = 1.1804 + (0.2)(-1.1933) + (0.02)(3.8171) + (0.001333)(-11.8593)$$
$$+ (0.000066)(55.3271) + \cdots$$
$$= 1.005925$$

**Example:** Using Taylor's series method, find the approximate solutions for the following initial-value problem at the points $x_1 = 0.1$ and for $x_2 = 0.2$:

$$\frac{dy}{dx} = 2y + 3e^x, \quad y(0) = 0.$$

Compare the numerical solution obtained with exact solution.
**Solution:** The given differential equation is

$$y' = 2y + 3e^x. \qquad \cdots \quad (1)$$

Here $x_0 = 0$. Hence, the Taylor's series solution for the given problem in a neighborhood of $x_0 = 0$ is

$$y(x) = y(0) + xy'(0) + \frac{x^2}{2!}y''(0) + \frac{x^3}{3!}y'''(0) + + \frac{x^4}{4!}y''''(0) + \cdots \qquad (2)$$

with $y(0) = 0$.
From the given differential equation, we get

$$\left.\begin{array}{l} y' = 2y + 3e^x \\ y'' = 2y' + 3e^x \\ y''' = 2y'' + 3e^x \\ y'''' = 2y''' + 3e^x \\ y''''' = 2y'''' + 3e^x \end{array}\right\} \qquad \cdots \cdots \quad (3)$$

and so on. Using the given initial condition that $y(0) = 0$ in these, we obtain

$$y'(0) = 2y(0) + 3e^0 = 2(0) + 3 = 3$$
$$y''(0) = 2y'(0) + 3e^0 = 2(3) + 3 = 9$$
$$y'''(0) = 2y''(0) + 3e^0 = 2(9) + 3 = 21$$
$$y''''(0) = 2y'''(0) + 3e^0 = 2(21) + 3 = 45$$

Putting these and the condition $y(0) = 0$ in (2), we get

$$y(x) = y(0) + xy'(0) + \frac{x^2}{2}y''(0) + \frac{x^3}{6}y'''(0) + \frac{x^4}{24}y''''(0) + \cdots$$
$$= 0 + x(3) + \frac{x^2}{2}(9) + \frac{x^3}{6}(21) + \frac{x^4}{24}(45) + \frac{x^5}{24}(93) + \cdots$$
$$= 3x + \frac{9}{2}x^2 + \frac{7}{2}x^3 + \frac{15}{8}x^4 + \cdots$$

Taking $x = x_1 = 0.1$ in this expression, we get the solution at $x_1 = 0.1$ as

$$y(0.1) = 3(0.1) + \frac{9}{2}(0.1)^2 + \frac{7}{2}(0.1)^3 + \frac{15}{8}(0.1)^4 + \cdots = 0.3487. \quad \cdots \quad (4)$$

Next, the solution at a point in a neighborhood of $x_1 = 0.1$ is given by

$$y(x) = y(0.1) + (x - 0.1)y'(0.1) + \frac{(x - 0.1)^2}{2!}y''(0.1) + \frac{(x - 0.1)^3}{3!}y'''(0.1)$$
$$+ \frac{(x - 0.1)^4}{4!}y''''(0.1) + \cdots \quad (5)$$

Using $y(0.1) \approx 0.3487$, we find, from expressions (3),

$y'(0.1) = 2y(0.1) + 3e^{0.1} = 2(0.3487) + 3(1.1052) = 4.013,$

$y''(0.1) = 2y'(0.1) + 3e^{0.1} = 2(4.013) + 3(1.1052) = 11.3416,$

$y'''(0.1) = 2y''(0.1) + 3e^{0.1} = 2(11.3416) + 3(1.1052) = 25.9988,$

$y''''(0.1) = 2y'''(0.1) + 3e^{0.1} = 2(25.9988) + 3(1.1052) = 55.3132.$

Using these in (5) and setting $x = x_2 = 0.2$, we get the solution at $x = 0.2$ as

$$y(0.2) \approx 0.3487 + (0.1)(4.013) + \frac{1}{2}(0.01)(11.3416) + \frac{1}{6}(0.001)(25.9988)$$
$$+ \frac{1}{24}(0.0001)(55.3132) \approx 0.81127.$$

**Analytical Solution:**

The given equation is $\frac{dy}{dx} - 2y = 3e^x$.

This is a linear differential equation of the form $\frac{dy}{dx} + P(x)y = Q(x)$

Here $P(x) = -2$ and $Q(x) = 3e^x$.

$$\therefore \text{I.F} = e^{\int P dx} = e^{\int -2 dx} = e^{-2x}.$$

∴ The general solution is

$$y(I.F) = \int Q(I.F)dx + c$$
$$\Rightarrow ye^{-2x} = \int 3e^x e^{-2x} dx + c$$
$$= \int 3e^{-x} dx + c$$
$$= -3e^{-x} + c$$
$$\Rightarrow y = -3e^x + ce^{2x}.$$

We have $y(0) = 0 \Rightarrow 0 = -3 + c \Rightarrow c = 3$
Hence the solution is $y = -3e^x + 3e^{2x} = 3(e^{2x} - e^x)$.
Thus $y(0.1) = 3(e^{0.2} - e^{0.1}) = 0.3486$
$$y(0.2) = 3(e^{0.4} - e^{0.2}) = 0.81126.$$
We can tabulate the values as follows:

| $x$ | Taylor's series method($y$) | Exact solution($y$) |
|---|---|---|
| 0.1 | 0.3487 | 0.3486 |
| 0.2 | 0.81127 | 0.81126 |

**Example:** Use Taylor's series method to find $y$ at the points $x_1 = 0.1$ and $x_2 = 0.2$, given

$$\frac{dy}{dx} = x^2 + y^2, \quad y(0) = 1.$$

**Solution:** From the given differential equation, we find

$$\left. \begin{array}{l} y' = x^2 + y^2 \\ y'' = 2x + 2yy' \\ y''' = 2 + 2[yy'' + (y')^2] \\ y'''' = 2[yy''' + y'y'' + 2y'y''] = 2[yy''' + 3y'y''] \end{array} \right\} \quad \cdots\cdots \quad (1)$$

and so on. Using the given condition $y(0) = 1$ in these we obtain

$$\left.\begin{array}{l} y'(0) = 0^2 + [y(0)]^2 = 0 + 1^2 = 1 \\ y''(0) = 2(0) + 2y(0)y'(0) = 2 \\ y'''(0) = 2 + 2[y(0)y''(0) + (y'(0))^2] = 8 \\ y''''(0) = 2[y(0)y'''(0) + 3y'(0)y''(0)] = 28 \end{array}\right\} \quad \ldots\ldots \quad (2)$$

The Taylor's series solution of the given problem in a neighborhood of $x_0 = 0$ is

$$y(x) = y(0) + xy'(0) + \frac{x^2}{2!}y''(0) + \frac{x^3}{3!}y'''(0) + \frac{x^4}{4!}y''''(0) + \cdots \quad (3)$$

with $y(0) = 1$.

Using (2) and the condition $y(0) = 1$ in (3), we get

$$y(x) = 1 + x + \frac{x^2}{2}(2) + \frac{x^3}{6}(8) + \frac{x^4}{24}(28) + \cdots$$

$$= 1 + x + x^2 + \frac{4}{3}x^3 + \frac{7}{6}x^4 + \cdots \quad (4)$$

For $x = x_1 = 0.1$, expression (4) yields the solution at $x_1 = 0.1$ as

$$y(0.1) = 1 + (0.1) + (0.1)^2 + \frac{4}{3}(0.1)^3 + \frac{7}{6}(0.1)^4 + \cdots \approx 0.1114. \quad \cdots \quad (5)$$

Next, the solution at a point $x$ in a neighborhood of $x_1 = 0.1$ is given by

$$y(x) = y(0.1) + (x - 0.1)y'(0.1) + \frac{(x-0.1)^2}{2!}y''(0.1) + \frac{(x-0.1)^3}{3!}y'''(0.1)$$

$$+ \frac{(x-0.1)^4}{4!}y''''(0.1) + \cdots \quad (6)$$

Using the expressions (1) and (5), we find that

$$y'(0.1) = (0.1)^2 + [y(0.1)]^2 = (0.1)^2 + (1.1114)^2 = 1.2452,$$

$$y''(0.1) = 2(0.1) + 2y(0.1)y'(0.1) = 0.2 + 2(1.1114)(1.2452) \approx 2.9678$$

$$y'''(0.1) = 2 + 2[y(0.1)y''(0.1) + (y'(0.1))^2]$$

$$= 2 + 2[(1.1114)(2.9678) + (1.2452)^2] = 11.6979$$

$$y''''(0.1) = 2[y(0.1)y'''(0.1) + 3y'(0.1)y''(0.1)]$$

$$= 2[(1.1114)(11.6979) + 3(1.2452)(2.9678)] = 48.1751$$

Using these in (6), we get

$$y(x) = 1.1114 + (x - 0.1)(1.2452) + \frac{1}{2}(2.9678)(x - 0.1)^2 + \frac{1}{6}(11.6979)(x - 0.1)^3$$
$$+ \frac{1}{24}(48.1751)(x - 0.1)^4 + \cdots$$

For $x = 0.2$, this yields the solution at $x_2 = 0.2$ as

$$y(0.2) = 1.1114 + (1.2452)(0.1) + \frac{1}{2}(1.4839)(0.1)^2 + (1.94965)(0.1)^3$$
$$+ (2.0073)(0.1)^4 + \cdots$$
$$\approx 1.2529.$$

**Example:** Using Taylor's series method, solve the problem $\dfrac{dy}{dx} = xy + y^2$, $y(0) = 1$
at the points $x_1 = 0.1, x_2 = 0.2$ and $x_3 = 0.3$.

**Solution:** The given differential equation is $y' = xy + y^2$.
From this, we find (by successive differentiation)

$$\left.\begin{aligned}
y' &= xy + y^2 \\
y'' &= y + xy' + 2yy' = y + (x + 2y)y' \\
y''' &= y' + (x + 2y)y'' + (1 + 2y')y' \\
&= (x + 2y)y'' + 2(1 + y')y' \\
y'''' &= (x + 2y)y''' + (1 + 2y')y'' + 2\big[(1 + y')y'' + y''y'\big] \\
&= (x + 2y)y''' + 3y'' + 2(y + 2)y'y''
\end{aligned}\right\} \quad \cdots\cdots \quad (1)$$

and so on.

The Taylor's series solution of the problem in a neighborhood of the point $x_0 = 0$ is

$$y(x) = y(x_0) + xy'(x_0) + \frac{x^2}{2!}y''(x_0) + \frac{x^3}{3!}y'''(x_0) + +\frac{x^4}{4!}y''''(x_0) \cdots$$
$$= y(0) + xy'(0) + \frac{x^2}{2}y''(0) + \frac{x^3}{6}y'''(0) + \frac{x^4}{24}y''''(0) + \cdots \qquad (2)$$

with $y(0) = 1$. From (1), we find that

$$y'(0) = (0)y(0) + [y(0)]^2 = 0 + 1^2 = 1,$$
$$y''(0) = y(0) + (0 + 2y(0))y'(0) = 1 + (2)(1) = 3,$$
$$y'''(0) = (0 + 2y(0))y''(0) + 2(1 + y'(0))y'(0) = (2)(3) + 2(1+1)1 = 10,$$
$$y''''(0) = (0 + 2y(0))y'''(0) + 3y''(0) + 2(y(0) + 2)y'(0)y''(0)$$
$$= 2(10) + 3(3) + (2)(3)(1)(3) = 47.$$

Using these and $y(0) = 1$ and setting $x = x_1 = 0.1$ in (2), we get

$$y(0.1) = 1 + (0.1) \times 1 + \frac{1}{2}(0.1)^2(3) + \frac{1}{6}(0.1)^3(10) + \frac{1}{24}(0.1)^4(47)$$
$$\approx 1.1169 \qquad \cdots \quad (3)$$

This is required solution at the point $x_1 = 0.1$.

Next, the solution in a neighborhood of the point $x_1 = 0.1$ is

$$y(x) = y(0.1) + (x - 0.1)y'(0.1) + \frac{(x-0.1)^2}{2!}y''(0.1)$$
$$+ \frac{(x-0.1)^3}{3!}y'''(0.1) + + \frac{(x-0.1)^4}{4!}y''''(0.1) + \cdots \qquad (4)$$

From (1) and (3), we find that

$$y'(0.1) = (0.1)y(0.1) + [y(0.1)]^2 = (0.1)(1.1169) + (1.1169)^2 = 1.3592,$$
$$y''(0.1) = y(0.1) + [0.1 + 2y(0.1)]y'(0.1) = 1.1169 + [0.1 + 2(1.1169)](1.3592) = 4.289,$$
$$y'''(0.1) = [0.1 + 2y(0.1)]y''(0.1) + 2[1 + y'(0.1)]y'(0.1)$$
$$= [0.1 + 2(1.1169)](4.289) + 2[1 + (1.3592)](1.3592) = 16.4229,$$
$$y''''(0.1) = [0.1 + 2y(0.1)]y'''(0.1) + 3y''(0.1) + 2[y(0.1) + 2]y'(0.1)y''(0.1)$$
$$= [0.1 + 2(1.1169)](16.4229) + 3(4.289) + 2[1.1169 + 2](1.3592)(4.289) = 87.5353.$$

Using these in (4) and setting $x = x_2 = 0.2$, we get

$$y(0.2) = 1.1169 + (0.1)(1.3592) + \frac{1}{2}(0.1)^2(4.289) + \frac{1}{6}(0.1)^3(16.4229) + \frac{1}{24}(0.1)^4(87.5353)$$
$$\approx 1.2774 \qquad \cdots \quad (5)$$

This is required solution at the point $x_2 = 0.2$.

Lastly, the solution in a neighborhood of the point $x_2 = 0.2$ is

$$y(x) = y(0.2) + (x - 0.2)y'(0.2) + \frac{(x - 0.2)^2}{2!}y''(0.2)$$
$$+ \frac{(x - 0.2)^3}{3!}y'''(0.2) + \frac{(x - 0.2)^4}{4!}y''''(0.2) + \cdots \qquad (6)$$

From (1) and (5), we find that

$y'(0.2) = (0.2)y(0.2) + [y(0.2)]^2 = (0.2)(1.2774) + (1.2774)^2 = 1.8872,$

$y''(0.2) = y(0.2) + [0.2 + 2y(0.2)]y'(0.2) = 1.2774 + [0.2 + 2(1.2774)](1.8872) = 6.4763,$

$y'''(0.2) = [0.2 + 2y(0.2)]y''(0.2) + 2[1 + y'(0.2)]y'(0.2)$
$= [0.2 + 2(1.2774)](6.4763) + 2[1 + (1.8872)](1.8872) = 28.7383,$

$y''''(0.2) = [0.2 + 2y(0.2)]y'''(0.2) + 3y''(0.2) + 2[y(0.2) + 2]y'(0.2)y''(0.2)$
$= [0.2 + 2(1.2774)](28.7383) + 3(6.4763) + 2[1.2774 + 2](1.8872)(6.4763) = 178.7104.$

Using these and (5) in (6) and setting $x = x_3 = 0.3$, we get

$$y(0.3) = 1.2774 + (0.1)(1.8872) + \frac{1}{2}(0.1)^2(6.4763) + \frac{1}{6}(0.1)^3(28.7383) + \frac{1}{24}(0.1)^4(178.7104)$$

$$\approx 1.504$$

This is the required solution at the point $x_3 = 0.3$.

**Example:** Using Taylor's series method to find $y(4.1)$ and $y(4.2)$, given that

$$\frac{dy}{dx} = \frac{1}{x^2 + y}, \quad y(4) = 4.$$

Consider terms upto third degree.

**Solution:** The given differential equation is $y' = \frac{1}{x^2 + y}$.

From this, we find (by successive differentiation)

$$\left.\begin{array}{l}y' = \dfrac{1}{x^2+y}\\[4pt]y'' = \dfrac{-1}{(x^2+y)^2}(2x+y') = -\dfrac{2x+y'}{(x^2+y)^2}\\[4pt]y''' = -\dfrac{(x^2+y)^2(2+y'') - (2x+y')2(x^2+y)(2x+y')}{(x^2+y)^4}\\[4pt]\phantom{y'''} = \dfrac{2(2x+y')^2 - (x^2+y)(2+y'')}{(x^2+y)^3}\end{array}\right\} \quad \cdots\cdots \quad (1)$$

From these and the given initial condition $y(4) = 4$, we find that

$$y'(4) = \dfrac{1}{4^2 + y(4)} = \dfrac{1}{16+4} = \dfrac{1}{20} = 0.05,$$

$$y''(4) = -\dfrac{2(4) + y'(4)}{(4^2 + y(4))^2} = -\dfrac{8 + 0.05}{(16+4)^2} = -0.020125$$

$$y'''(4) = \dfrac{2[2(4) + y'(4)]^2 - \left(4^2 + y(4)\right)\left(2 + y''(4)\right)}{[4^2 + y(4)]^3} = 0.0112509$$

The Taylor's series solution for $y$ in a neighborhood of the point $x_0 = 4$ is

$$y(x) = y(4) + (x-4)y'(4) + \dfrac{(x-4)^2}{2}y''(4) + \dfrac{(x-4)^3}{6}y'''(4) + \cdots \quad (2)$$

Substituting the values of $y(4), y'(4), y''(4)$ and $y'''(4)$ in this expression, we get

$$y(x) \approx 4 + (x-4)(0.05) - \dfrac{(x-4)^2}{2}(0.020125) + \dfrac{(x-4)^3}{6}(0.0112509)$$

Setting $x = x_1 = 4.1$ in this expression, we get

$$y(x) \approx 4 + (0.1)(0.05) - \dfrac{(0.1)^2}{2}(0.020125) + \dfrac{(0.1)^3}{6}(0.0112509) \approx 4.0049 \quad \cdots \quad (3)$$

Next, the solution in a neighborhood of the point $x_1 = 4.1$ is given by

$$y(x) = y(4.1) + (x-4.1)y'(4.1) + \dfrac{(x-4.1)^2}{2!}y''(4.1) + \dfrac{(x-4.1)^3}{3!}y'''(4.1) + \cdots \quad (4)$$

From (1) and (3), we find that

$$y'(4.1) = \frac{1}{4.1^2 + y(4.1)} = \frac{1}{16.81 + 4.0049} = 0.04804,$$

$$y''(4.1) = -\frac{2(4.1) + y'(4.1)}{(4.1^2 + y(4.1))^2} = -\frac{8.2 + 0.04804}{(16.81 + 4.0049)^2} = -0.01904$$

$$y'''(4.1) = \frac{2[2(4.1) + y'(4.1)]^2 - (4.1^2 + y(4.1))(2 + y''(4.1))}{[4.1^2 + y(4.1)]^3} = 0.01052$$

Using these and (3) in (4) and setting $x = x_2 = 4.2$ in (4), we get

$$y(4.2) = 4.0049 + (0.1)(0.04804) + \frac{1}{2}(0.1)^2(0.01052) + \frac{1}{6}(0.1)^3(0.01052)$$

$$\approx 4.0096.$$

This is required solution at the point $x_2 = 4.2$.

## EXERCISE

1. Solve $y' = x + y, y(0) = 1$ by Taylor's series method. Hence find the values of $y$ at $x = 0.1$ and $x = 0.2$.
   **Ans:** $y(x) = 1 + x + x^2 + \frac{x^3}{3} + \frac{x^4}{6} + \cdots$, $y(0.1) = 1.1103, y(0.2) = 1.2427$.

2. Tabulate the values of $y(0.1), y(0.2), y(0.3)$ and $y(0.4)$ using Taylor series given that $\frac{dy}{dx} = x^2 - y, y(0) = 1$.
   **Ans:** $y(0.1) = 0.905167, y(0.2) = 0.821227, y(0.3) = 0.749147, y(0.4) = 0.689652$.

3. Find $y(0.1), y(0.2)$ using Taylor's series method if $\frac{dy}{dx} = e^x - 2y$, $y(0) = 1$.
   (Sep-2021)

4. Solve $y' = y - x^2$, $y(0) = 1$, by Taylor's series method up to fourth approximation. Hence, find the value of $y(0.1)$. (Apr-2022)

### 5.2.2 Picard's Method of Successive Approximation

Consider the first order differential equation

$$\frac{dy}{dx} = f(x, y), \qquad \cdots \cdots \quad (1)$$

with initial condition $y(x_0) = y_0$.

Let us rewrite the differential equation of the problem in the separable form $dy = f(x, y)dx$ and integrate both sides (w.r.t $x$) from $x_0$ to $x$ to get

$$\int_{x_0}^{x} dy = \int_{x_0}^{x} f(x, y)dx$$

$$\Rightarrow y(x) - y(x_0) = \int_{x_0}^{x} f(x, y)dx$$

Since $y(x_0) = y_0$ by the initial condition of the problem, we obtain

$$y(x) = y_0 + \int_{x_0}^{x} f(x, y)dx \qquad \cdots\cdots \quad (2)$$

In equation (2), the unknown function $y$ appears under the integral sign, such type of equation is called integral equation.

The first approximation $y_1$ is obtained by replacing $y$ by $y_0$ in $f(x, y)$ in equation (2), we have

$$y_1 = y_0 + \int_{x_0}^{x} f(x, y_0)dx \qquad \cdots\cdots \quad (3)$$

The second approximation $y_2$ is obtained by replacing $y$ by $y_1$ in $f(x, y)$ in equation (2), we have

$$y_2 = y_0 + \int_{x_0}^{x} f(x, y_1)dx \qquad \cdots\cdots \quad (4)$$

Continuing this process the $n$th approximation is given by

$$\boxed{y_n = y_0 + \int_{x_0}^{x} f(x, y_{n-1})dx, \quad n = 1, 2, \cdots, n}$$

This is known as **Picard's iteration formula**. The process of iteration is stopped when any two values of iteration are approximately the same.

**Note:** Picard's method gives a sequence of approximations $y_1, y_2, \cdots$ each giving a better result than the preceding one. But this can be applied only to equations in which the successive integration can be obtained easily.

**Example:** Use Picard's method, find the value of $y$ for $x = 0.4$ given that

$$y' = x^2 + y^2, \quad y(0) = 0.$$

**Solution:** The Picard's iteration formula for the differential equation $\dfrac{dy}{dx} = f(x, y)$ is

$$y_n = y_0 + \int_{x_0}^{x} f(x, y_{n-1}) dx, \quad n = 1, 2, \cdots, n.$$

Here $f(x, y) = x^2 + y^2$, $x_0 = 0$ and $y_0 = 0$.
The first approximation i.e., for $n = 1$ is

$$y_1 = y_0 + \int_{x_0}^{x} f(x, y_0) dx$$

$$= 0 + \int_{0}^{x} (x^2 + y_0^2) dx = \int_{0}^{x} (x^2 + 0) dx = \frac{x^3}{3}$$

The second approximation i.e., for $n = 2$ is

$$y_2 = y_0 + \int_{x_0}^{x} f(x, y_1) dx$$

$$= 0 + \int_{0}^{x} (x^2 + y_1^2) dx = \int_{0}^{x} \left( x^2 + \left(\frac{x^3}{3}\right)^2 \right) dx = \int_{0}^{x} \left( x^2 + \frac{x^6}{9} \right) = \frac{x^3}{3} + \frac{x^7}{63}$$

Calculating further would be tedious.

∴ We limit to second approximation.
Assuming a value of $x = 0.4$, we have

$$y(0.4) = \frac{(0.4)^3}{3} + \frac{(0.4)^7}{63} = 0.0123 + 2.6 \times 10^{-5} = 0.021326.$$

**Example:** Obtain $y(0.1)$, given $y' = \dfrac{y - x}{y + x}$, $y(0) = 1$ by Picard's method.

or

Find the value of $y$ for $x = 0.1$ by Picard's method, given that **(Apr-2022)**

$$\frac{dy}{dx} = \frac{y-x}{y+x}, \quad y(0) = 1.$$

**Solution:** The Picard's iteration formula for the differential equation $\frac{dy}{dx} = f(x,y)$ is

$$y_n = y_0 + \int_{x_0}^{x} f(x, y_{n-1})dx, \quad n = 1, 2, \cdots, n.$$

Here $f(x,y) = \frac{y-x}{y+x}$, $x_0 = 0$ and $y_0 = 1$.

∴ The first approximation i.e., for $n = 1$ is

$$y_1 = y_0 + \int_{x_0}^{x} f(x, y_0)dx$$

$$= 1 + \int_{0}^{x} \left(\frac{y_0 - x}{y_0 + x}\right)dx$$

$$= 1 + \int_{0}^{x} \frac{1-x}{1+x}dx$$

$$= 1 + \int_{0}^{x} \left(-1 + \frac{2}{1+x}\right)dx \quad \text{(By partial fractions)}$$

$$= 1 + \Big[-x + 2\log(1+x)\Big]_{0}^{x}$$

$$= 1 + \Big[-x + 2\log(1+x)\Big] - \Big[0 + 2\log(1+0)\Big]$$

$$= 1 - x + 2\log(1+x)$$

The second approximation i.e., for $n = 2$ is

$$y_2 = y_0 + \int_{x_0}^{x} f(x, y_1) dx$$

$$= 1 + \int_{0}^{x} \left(\frac{y_1 - x}{y_1 + x}\right) dx$$

$$= 1 + \int_{0}^{x} \frac{1 - x + 2\log(1 + x) - x}{1 - x + 2\log(1 + x) + x} dx$$

$$= 1 + \int_{0}^{x} \frac{1 - 2x + 2\log(1 + x)}{1 + 2\log(1 + x)} dx$$

$$= 1 + \int_{0}^{x} \left[1 - \frac{2x}{1 + 2\log(1 + x)}\right] dx$$

$$= 1 + x - 2 \int_{0}^{x} \frac{x}{1 + 2\log(1 + x)} dx$$

which is very difficult to integrate.

Hence we use the first approximation.

$\therefore \quad y(x) = y_1 = 1 - x + 2\log(1 + x)$.

Putting $x = 0.1$, we obtain

$y(0.1) = 1 - 0.1 + 2\log(1.1) = 1.09062$ (correct to 4 decimal places)

**Example:** Solve $y' = y - x^2$, $y(0) = 1$ by Picard's method upto the fourth approximation. Hence, find the values of $y(0.1)$ and $y(0.2)$. **(Apr-2022)**

**Solution:** Here $y' = f(x, y) = y - x^2$, $y(0) = 1$, i.e., $x_0 = 0$, $y_0 = 1$.

The Picard's iteration formula for the differential equation $\frac{dy}{dx} = f(x, y)$ is

$$y_n = y_0 + \int_{x_0}^{x} f(x, y_{n-1}) dx, \quad n = 1, 2, \cdots, n.$$

**First approximation:**

$$y_1 = y_0 + \int_0^x f(x, y_0)\,dx$$

$$= 1 + \int_0^x (y_0 - x^2)\,dx$$

$$= 1 + \int_0^x (1 - x^2)\,dx$$

$$= 1 + \left(x - \frac{x^3}{3}\right)_0^x = 1 + x - \frac{x^3}{3}$$

**Second approximation:**

$$y_2 = y_0 + \int_0^x f(x, y_1)\,dx$$

$$= 1 + \int_0^x (y_1 - x^2)\,dx$$

$$= 1 + \int_0^x \left(1 + x - \frac{x^3}{3} - x^2\right)dx$$

$$= 1 + x + \frac{x^2}{2} - \frac{x^3}{3} - \frac{x^4}{12}$$

### Third approximation:

$$y_3 = y_0 + \int_0^x f(x, y_2)dx$$

$$= 1 + \int_0^x (y_2 - x^2)dx$$

$$= 1 + \int_0^x \left(1 + x + \frac{x^2}{2} - \frac{x^3}{3} - \frac{x^4}{12} - x^2\right)dx$$

$$= 1 + \int_0^x \left(1 + x - \frac{x^2}{2} - \frac{x^3}{3} - \frac{x^4}{12}\right)dx$$

$$= 1 + x + \frac{x^2}{2} - \frac{x^3}{6} - \frac{x^4}{12} - \frac{x^5}{60}$$

### Fourth approximation:

$$y_4 = y_0 + \int_0^x f(x, y_3)dx$$

$$= 1 + \int_0^x (y_3 - x^2)dx$$

$$= 1 + \int_0^x \left(1 + x + \frac{x^2}{2} - \frac{x^3}{6} - \frac{x^4}{12} - \frac{x^5}{60} - x^2\right)dx$$

$$= 1 + \int_0^x \left(1 + x - \frac{x^2}{2} - \frac{x^3}{6} - \frac{x^4}{12} - \frac{x^5}{60}\right)dx$$

$$= 1 + x + \frac{x^2}{2} - \frac{x^3}{6} - \frac{x^4}{24} - \frac{x^5}{60} - \frac{x^6}{360}$$

Putting $x = 0.1$

$$y(0.1) = 1 + 0.1 + \frac{(0.1)^2}{2} - \frac{(0.1)^3}{6} - \frac{(0.1)^4}{24} - \frac{(0.1)^5}{60} - \frac{(0.1)^6}{360}$$

$$= 1 + 0.1 + 0.005 - 0.0001666 - 0.00000416 - 0.000000166 - 0.0000000271$$

$$= 1.104829$$

Putting $x = 0.2$

$$y(0.2) = 1 + 0.2 + \frac{(0.2)^2}{2} - \frac{(0.2)^3}{6} - \frac{(0.2)^4}{24} - \frac{(0.2)^5}{60} - \frac{(0.2)^6}{360}$$

$$= 1 + 0.2 + 0.02 - 0.0013333 - 0.00006666 - 0.000005333 - 0.0000001777$$

$$= 1.21859$$

**Example:** Using Picard's process of successive approximations, obtain a solution upto the fifth approximation of the equation $\frac{dy}{dx} = x + y$ such that $y = 1$ when $x = 0$. Check your answer by finding the exact particular solution.

**Solution:** Here $\frac{dy}{dx} = y' = f(x, y) = x + y$, $x_0 = 0$, $y_0 = 1$.

The Picard's iteration formula for the differential equation $\frac{dy}{dx} = f(x, y)$ is

$$y_n = y_0 + \int_{x_0}^{x} f(x, y_{n-1}) dx, \quad n = 1, 2, \cdots, n.$$

**First approximation:**

$$y_1 = y_0 + \int_{x_0}^{x} f(x, y_0) dx = y_0 + \int_{x_0}^{x} (x + y_0) dx = 1 + \int_{0}^{x} (x + 1) dx = 1 + x + \frac{x^2}{2}.$$

**Second approximation:**

$$y_2 = y_0 + \int_{x_0}^{x} f(x, y_1) dx = y_0 + \int_{x_0}^{x} (x + y_1) dx$$

$$= 1 + \int_{0}^{x} \left\{ x + \left(1 + x + \frac{x^2}{2}\right) \right\} dx$$

$$= 1 + \int_{0}^{x} \left(1 + 2x + \frac{x^2}{2}\right) dx = 1 + x + x^2 + \frac{x^3}{6}.$$

**Third approximation:**

$$y_3 = y_0 + \int_{x_0}^{x} f(x, y_2)dx = y_0 + \int_{x_0}^{x} (x + y_2)dx$$

$$= 1 + \int_{0}^{x} \left\{ x + \left(1 + x + x^2 + \frac{x^3}{6}\right) \right\} dx$$

$$= 1 + \int_{0}^{x} \left(1 + 2x + x^2 + \frac{1}{6}x^3\right) dx$$

$$= 1 + x + x^2 + \frac{x^3}{3} + \frac{x^4}{24}.$$

**Fourth approximation:**

$$y_4 = y_0 + \int_{x_0}^{x} f(x, y_3)dx = y_0 + \int_{x_0}^{x} (x + y_3)dx$$

$$= 1 + \int_{0}^{x} \left\{ x + \left(1 + x + x^2 + \frac{x^3}{3} + \frac{x^4}{24}\right) \right\} dx$$

$$= 1 + \int_{0}^{x} \left(1 + 2x + x^2 + \frac{x^3}{3} + \frac{x^4}{24}\right) dx$$

$$= 1 + x + x^2 + \frac{x^3}{3} + \frac{x^4}{12} + \frac{x^5}{120}.$$

**Fourth approximation:**

$$y_5 = y_0 + \int_{x_0}^{x} f(x, y_4)dx = y_0 + \int_{x_0}^{x} (x + y_4)dx$$

$$= 1 + \int_{0}^{x} \left\{ x + \left(1 + x + x^2 + \frac{x^3}{3} + \frac{x^4}{12} + \frac{x^5}{120}\right) \right\} dx$$

$$= 1 + \int_{0}^{x} \left(1 + 2x + x^2 + \frac{x^3}{3} + \frac{x^4}{12} + \frac{x^5}{120}\right) dx$$

$$= 1 + x + x^2 + \frac{x^3}{3} + \frac{x^4}{12} + \frac{x^5}{60} + \frac{x^6}{720}. \qquad \cdots\cdots \quad (1)$$

The exact solution of $\dfrac{dy}{dx} = x + y$, $y(0) = 1$ can be found as follows:

The equation can be written as $\dfrac{dy}{dx} - y = x$. This is a linear in $x$.

Here $P(x) = -1, Q(x) = x$. $\therefore$ I.F $= e^{\int P dx} = e^{\int (-1) dx} = e^{-x}$.

The general solution is

$$ye^{-x} = \int xe^{-x} dx + c$$

$$= -(1+x)e^{-x} + c$$

$$\Rightarrow y = -x - 1 + ce^{x}.$$

Since $y = 1$, when $x = 0$, we obtain

$$1 = 0 - 1 + c \Rightarrow c = 2.$$

Thus the required particular solution is

$$y = -x - 1 + 2e^{x}. \qquad \cdots\cdots \quad (2)$$

Using the series

$$e^{x} = 1 + x + \dfrac{x^{2}}{2} + \dfrac{x^{3}}{3!} + \dfrac{x^{4}}{4!} + \cdots$$

we get

$$y = 1 + x + x^{2} + \dfrac{x^{3}}{3} + \dfrac{x^{4}}{12} + \dfrac{x^{5}}{60} + \dfrac{x^{6}}{360} \cdots \qquad (3)$$

Comparing (1) and (3), it is clear that (1), approximates to the exact particular solution (3) upto the term in $x^{5}$.

**Example:** Solve $y' = \dfrac{x^{2}}{y^{2}+1}$, $y(0) = 0$ Picard's method of successive approximation. Find $y(1)$.

**Solution:** Here $f(x,y) = \dfrac{x^{2}}{y^{2}+1}, x_{0} = 0, y_{0} = 0$.

By Picard's method, a sequence of successive approximations to $y$ are given by

$$y_{n} = y_{0} + \int_{x_{0}}^{x} f(x, y_{n-1}) dx \Rightarrow y_{n} = \int_{0}^{x} \dfrac{x^{2}}{y_{n-1}^{2}+1} dx \quad \cdots\cdots \quad (1)$$

## First approximation

For $n = 1$, equation (1) becomes

$$y_1 = \int_0^x \frac{x^2}{y_0^2 + 1} dx = \int_0^x \frac{x^2}{0+1} dx = \int_0^x x^2 dx = \frac{x^3}{3}.$$

## Second approximation

For $n = 2$, equation (1) becomes

$$y_2 = \int_0^x \frac{x^2}{y_1^2 + 1} dx = \int_0^x \frac{x^2}{\frac{x^6}{9} + 1} dx = 9\int_0^x \frac{x^2}{x^6 + 9} dx = 3\int_0^x \frac{3x^2}{(x^3)^2 + 3^2} dx = \tan^{-1}\left(\frac{x^3}{3}\right).$$

## Third approximation

For $n = 3$, equation (1) becomes

$$y_3 = \int_0^x \frac{x^2}{y_2^2 + 1} dx = \int_0^x \frac{x^2}{\left(\tan^{-1}\left(\frac{x^3}{3}\right)\right)^2 + 1} dx.$$

The integration is difficult.

Expanding $\tan^{-1}$, we get from second approximation

$$y = \tan^{-1}\left(\frac{x^3}{3}\right) = \frac{x^3}{3} - \frac{1}{3}\left(\frac{x^3}{3}\right)^3 + \frac{1}{5}\left(\frac{x^3}{3}\right)^5 + \cdots = \frac{x^3}{3} - \frac{x^9}{81} + \frac{x^{15}}{1215} + \cdots$$

At $x = 1$, $y(1) = \frac{1}{3} - \frac{1}{81} + \frac{1}{1215} = 0.321810699$.

**Example:** Employ Picard's method to obtain, correct to four decimal places, solution of the differential equation $dy/dx = x^2 + y^2$ for $x = 0.4$, given that $y = 0$ when $x = 0$.

**Solution:** Here $f(x, y) = x^2 + y^2$, $x_0 = 0$, $y_0 = 0$.

By Picard's method, a sequence of successive approximations to $y$ are given by

$$y_n = y_0 + \int_{x_0}^x f(x, y_{n-1}) dx \Rightarrow y_n = \int_0^x (x^2 + y_{n-1}^2) dx, \quad n = 1, 2, 3 \cdots \quad \cdots \quad (1)$$

## First approximation

For $n = 1$, equation (1) becomes

$$y_1 = \int_0^x (x^2 + y_0^2)\,dx = \int_0^x x^2\,dx = \frac{x^3}{3}.$$

## Second approximation

For $n = 2$, equation (1) becomes

$$y_2 = \int_0^x (x^2 + y_1^2)\,dx = \int_0^x \left\{x^2 + \left(\frac{x^3}{3}\right)^2\right\}dx = \frac{x^3}{3} + \frac{x^7}{63}.$$

## Third approximation

For $n = 3$, equation (1) becomes

$$y_3 = \int_0^x (x^2 + y_2^2)\,dx = \int_0^x \left\{x^2 + \left(\frac{x^3}{3} + \frac{x^7}{63}\right)^2\right\}dx$$

$$= \int_0^x \left\{x^2 + \frac{x^6}{9} + \frac{x^{14}}{63^2} + 2\cdot\frac{x^{10}}{189}\right\}dx$$

$$= \frac{x^3}{3} + \frac{x^7}{63} + \frac{x^{15}}{15\times 63^2} + 2\frac{x^{11}}{11\times 189}.$$

From this, we find that

$$y(0.4) = \frac{(0.4)^3}{3} + \frac{(0.4)^7}{63} + \frac{(0.4)^{15}}{15\times 63^2} + 2\frac{(0.4)^{11}}{11\times 189} = 0.0213593.$$

**Example:** Using Picard's solve $dy/dx = 1 + xy$ with $y(0) = 2$. Find $y(0.1)$, $y(0.2)$ and $y(0.3)$.

**Solution:** Here $f(x, y) = 1 + xy$, $x_0 = 0$, $y_0 = 2$.

By Picard's method, a sequence of successive approximations to $y$ are given by

$$y_n = y_0 + \int_{x_0}^x f(x, y_{n-1})\,dx, \quad n = 1, 2, 3\cdots \qquad (1)$$

## First approximation

For $n = 1$, equation (1) becomes

$$y_1 = y_0 + \int_{x_0}^{x} f(x, y_0)dx = 2 + \int_{0}^{x} (1 + xy_0)dx$$

$$= 2 + \int_{0}^{x} (1 + 2x)dx$$

$$= 2 + x + x^2.$$

## Second approximation

For $n = 2$, equation (1) becomes

$$y_2 = y_0 + \int_{x_0}^{x} f(x, y_1)dx = 2 + \int_{0}^{x} (1 + xy_1)dx$$

$$= 2 + \int_{0}^{x} \left[1 + x(2 + x + x^2)\right]dx$$

$$= 2 + x + x^2 + \frac{x^3}{3} + \frac{x^4}{4}.$$

## Third approximation

For $n = 3$, equation (1) becomes

$$y_3 = y_0 + \int_{x_0}^{x} f(x, y_2)dx = 2 + \int_{0}^{x} (1 + xy_2)dx$$

$$= 2 + \int_{0}^{x} \left[1 + x\left(2 + x + x^2 + \frac{x^3}{3} + \frac{x^4}{4}\right)\right]dx$$

$$= 2 + x + x^2 + \frac{x^3}{3} + \frac{x^4}{4} + \frac{x^5}{15} + \frac{x^6}{24}. \quad \cdots \quad (1)$$

Putting $x = 0.1, 0.2$ and $0.3$ in (1), we get

$$y(0.1) = 2 + 0.1 + (0.1)^2 + \frac{(0.1)^3}{3} + \frac{(0.1)^4}{4} + \frac{(0.1)^5}{15} + \frac{(0.1)^6}{24} = 2.11035$$

$$y(0.2) = 2 + 0.2 + (0.2)^2 + \frac{(0.2)^3}{3} + \frac{(0.2)^4}{4} + \frac{(0.2)^5}{15} + \frac{(0.2)^6}{24} = 2.24309$$

$$y(0.3) = 2 + 0.3 + (0.3)^2 + \frac{(0.3)^3}{3} + \frac{(0.3)^4}{4} + \frac{(0.3)^5}{15} + \frac{(0.3)^6}{24} = 2.40122.$$

### EXERCISE

1. Use Picard's method to obtain $y$ for $x = 0.2$ given $\frac{dy}{dx} = x - y$, $y(0) = 1$.

   **Ans:** $y(x) = 1 - x + x^2 - \frac{x^3}{3} + \frac{x^4}{12} - \frac{x^5}{60} + \frac{x^6}{720}$, $y(0.2) = 0.8374$.

2. Find $y(0.2)$ using Picard's method given that $\frac{dy}{dx} = xy$, $y(0) = 1$.

   **Ans:** $y(0.2) = 1.0202$.

3. Given that $\frac{dy}{dx} = 1 + xy$ and $y(0) = 1$. Compute $y(0.1)$ and $y(0.2)$ using Picard's method.

   **Ans:** $y(0.1) = 1.1053, y(0.2) = 1.22282$.

4. Evaluate $y(0.1)$ using Picard's method for $\frac{dy}{dx} = y + x^2$, $y(0) = 1$.   (Nov-2017)

5. Solve $y' + y = e^x$ and $y(0) = 0$ using Picard's method compute $y(0.1)$. (Dec-2020)

6. Solve $y' = 5x - y$ and $y(1) = 1$ by Picard's method compute $y(1.1)$.   (Dec-2020)

## 5.2.3 Euler's Method

Taylor's series method and Picard's method that we have discussed in the previous two sections yield the solution of a differential equation in the form of a power series. We now proceed to describe methods which give the solution in the form of table values at equally spaced points.

Consider the differential equation,

$$\frac{dy}{dx} = f(x, y), \text{ with } y(x_0) = y_0. \quad \cdots \cdots \quad (1)$$

Suppose we want to find the approximate value of $y$ say $y_n$ when $x = x_n$. We divide the interval $[x_0, x_n]$ into $n$-subintervals of equal length say $h$, with the division point $x_0, x_1, \cdots, x_n$, where $x_n = x_0 + nh$, $n = 1, 2, 3, \cdots, n$.

Let us assume that
$$f(x, y) \approx f(x_{n-1}, y_{n-1})$$
in $[x_{n-1}, x_n]$. Integrating (1) in $[x_{n-1}, x_n]$, we get

$$\int_{x_{n-1}}^{x_n} dy = \int_{x_{n-1}}^{x_n} f(x, y) dx$$

$$\Rightarrow [y_n - y_{n-1}] = \int_{x_{n-1}}^{x_n} f(x, y) dx$$

$$\Rightarrow y_n \approx y_{n-1} + f(x_{n-1}, y_{n-1}) \int_{x_{n-1}}^{x_n} dx$$

$$\Rightarrow y_n \approx y_{n-1} + f(x_{n-1}, y_{n-1})(x_n - x_{n-1})$$

$$\Rightarrow y_n \approx y_{n-1} + h f(x_{n-1}, y_{n-1}) \quad \cdots \cdots \quad (2)$$

Equation (2) is called Euler's iteration formula.

Hence
$$\boxed{y_{n+1} = y_n + h f(x_n, y_n), \quad n = 0, 1, 2, 3, \cdots.}$$

### 5.2.4 Examples

**Example:** Using the Euler's method, find $y(0.3)$ given $y' = x + y$, $y(0) = 1$. Take step length $h = 0.1$.

**Solution:** The Euler's formulae for numerical solution of the differential equation $\dfrac{dy}{dx} = f(x, y)$ is

$$y_{n+1} = y_n + h f(x_n, y_n), \quad n = 0, 1, 2, 3, \cdots \quad \cdots \quad (1)$$

Here $f(x, y) = x + y$, $x_0 = 0$, $y_0 = 1$ and $h = 0.1$. Then $x_1 = x_0 + h = 0.1$, $x_2 = x_1 + h = 0.2$ and $x_3 = x_2 + h = 0.3$.

We have to find $y_3 = y(x_3) = y(0.3)$.

Putting $n = 0$ in (1), we get

$$y_1 = y_0 + hf(x_0, y_0) = 1 + (0.1)f(0, 1) = 1 + (0.1)(0 + 1) = 1 + 0.1 = 1.1$$

i.e., $y(0.1) = 1.1$.

Putting $n = 1$ in (1), we get

$$y_2 = y_1 + hf(x_1, y_1) = 1.1 + (0.1)f(0.1, 1.1)$$
$$= 1 + (0.1)(0.1 + 1.1) = 1.1 + 0.12 = 1.22$$

i.e., $y(0.2) = 1.22$.

Putting $n = 2$ in (1), we get

$$y_3 = y_2 + hf(x_2, y_2) = 1.22 + (0.1)f(0.2, 1.22)$$
$$= 1.22 + (0.1)(0.2 + 1.22) = 1.22 + 0.142 = 1.362$$

Thus, $y_3 = y(0.3) = 1.362$ is the required solution.

**Example:** Using the Euler's method, solve for $y$ at $x = 0.1$ from $\dfrac{dy}{dx} = x + y + xy$, $y(0) = 1$, taking step size $h = 0.025$.

**Solution:** The Euler's formulae for numerical solution of the differential equation $\dfrac{dy}{dx} = f(x, y)$ is

$$y_{n+1} = y_n + hf(x_n, y_n), \quad n = 0, 1, 2, 3, \cdots \quad (1)$$

Here $f(x, y) = x + y + xy$, $x_0 = 0$, $y_0 = 1$ and $h = 0.025$. Then $x_1 = x_0 + h = 0.025$, $x_2 = x_1 + h = 0.05$, $x_3 = x_2 + h = 0.075$, $x_4 = x_3 + h = 0.1$.

We have to find $y_4 = y(x_4) = y(0.1)$.

Putting $n = 0$ in (1), we get

$$y_1 = y_0 + hf(x_0, y_0) = y_0 + h(x_0 + y_0 + x_0 y_0) = 1 + (0.025)[0 + 1 + (0)(1)] = 1.025$$

i.e., $y(0.025) = 1.025$.

Putting $n = 1$ in (1), we get

$$y_2 = y_1 + hf(x_1, y_1) = y_1 + h(x_1 + y_1 + x_1y_1)$$
$$= 1.025 + (0.025)\big[0.025 + 1.025 + (0.025)(1.025)\big] = 1.05189$$

i.e., $y(0.2) = 1.05189$.

Putting $n = 2$ in (1), we get

$$y_3 = y_2 + hf(x_2, y_2) = y_2 + h(x_2 + y_2 + x_2y_2)$$
$$= 1.05189 + (0.025)\big[0.05 + 1.05189 + (0.05)(1.05189)\big] = 1.08075$$

i.e., $y(0.3) = 1.08075$.

Putting $n = 3$ in (1), we get

$$y_4 = y_3 + hf(x_3, y_3) = y_3 + h(x_3 + y_3 + x_3y_3)$$
$$= 1.08075 + (0.025)\big[0.075 + 1.08075 + (0.075)(1.08075)\big] = 1.11167$$

Thus $y_4 = y(0.1) = 1.11167$, is the required solution.

**Example:** Given $y' = x^2 - y$, $y(0) = 1$, find $y(0.1)$ and $y(0.2)$ using Euler's method.

**Solution:** The Euler's formulae for numerical solution of the differential equation $\dfrac{dy}{dx} = f(x,y)$ is

$$y_{n+1} = y_n + hf(x_n, y_n), \quad n = 0, 1, 2, 3, \cdots \quad (1)$$

Here $f(x,y) = x^2 - y$, $x_0 = 0$, $y_0 = 1$. Let us take step length $h = 0.1$ and set $x_1 = x_0 + h = 0.1$ and $x_2 = x_1 + h = 0.2$.

We have to find $y_1 = y(x_1) = y(0.1)$ and $y_2 = y(x_2) = y(0.2)$.

Putting $n = 0$ in (1), we get

$$y_1 = y_0 + hf(x_0, y_0) = 1 + (0.1)f(0, 1) = 1 + (0.1)(0^2 - 1) = 1 - 0.1 = 0.9$$

Putting $n = 1$ in (1), we get

$$y_2 = y_1 + hf(x_1, y_1) = 0.9 + (0.1)f(0.1, 0.9)$$
$$= 0.9 + (0.1)[(0.1)^2 - 0.9]$$
$$= 0.9 + (0.1)(0.89)$$
$$= 0.811$$

Thus, the required values are $y(0.1) = 0.9$ and $y(0.2) = 0.811$.

**Example:** Using Euler's method, find $y(0.1)$ and $y(0.2)$ given $y' = x^2 + y^2$, $y(0) = 1$.

**Solution:** The Euler's formulae for numerical solution of the differential equation $\dfrac{dy}{dx} = f(x, y)$ is

$$y_{n+1} = y_n + hf(x_n, y_n), \quad n = 0, 1, 2, 3, \cdots \qquad (1)$$

Here $f(x, y) = x^2 + y^2$, $x_0 = 0$, $y_0 = 1$. Let us take step length $h = 0.1$ and set $x_1 = x_0 + h = 0.1$ and $x_2 = x_1 + h = 0.2$.

We have to find $y_1 = y(x_1) = y(0.1)$ and $y_2 = y(x_2) = y(0.2)$.

Putting $n = 0$ in (1), we get

$$y_1 = y_0 + hf(x_0, y_0) = 1 + (0.1)f(0, 1) = 1 + (0.1)(0^2 + 1^2) = 1 + 0.1 = 1.1$$

Putting $n = 1$ in (1), we get

$$y_2 = y_1 + hf(x_1, y_1) = 1.1 + (0.1)f(0.1, 1.1)$$
$$= 1.1 + (0.1)[(0.1)^2 + (1.1)^2]$$
$$= 1.1 + (0.1)(1.22)$$
$$= 1.222$$

Thus, the required values are $y(0.1) = 1.1$ and $y(0.2) = 1.222$.

**Example:** Using the Euler's method, find an approximate solutions of the problem

$$\dfrac{dy}{dx} = x - y, \quad y(0) = 1$$

at the points 0.2 and 0.4 taking the step length $h = 0.2$.

**Solution:** The Euler's formulae for numerical solution of the differential equation $\dfrac{dy}{dx} = f(x,y)$ is

$$y_{n+1} = y_n + hf(x_n, y_n), \quad n = 0, 1, 2, 3, \cdots \quad \cdots \quad (1)$$

Here $f(x,y) = x - y$, $x_0 = 0$, $y_0 = 1$. Let $x_1 = x_0 + h = 0.2$ and $x_2 = x_1 + h = 0.4$.
We have to find $y_1 = y(x_1) = y(0.2)$ and $y_2 = y(x_2) = y(0.4)$.
Putting $n = 0$ in (1), we get

$$y_1 = y_0 + hf(x_0, y_0) = 1 + (0.2)f(0,1) = 1 + (0.2)(0-1) = 1 - 0.2 = 0.8$$

This is an approximate solution for the given problem at the point $x_1 = 0.2$.
Putting $n = 1$ in (1), we get

$$y_2 = y_1 + hf(x_1, y_1) = 0.8 + (0.2)f(0.2, 0.8)$$
$$= 0.8 + (0.2)\big[0.2 - 0.8\big]$$
$$= 0.8 + (0.2)(-0.6)$$
$$= 0.68$$

This is an approximate solution for the given problem at the point $x_2 = 0.4$.

**Example:** Compute $y$ at $x = 0.25$ by Euler's method in 5 steps given $y' = 2xy$, $y(0) = 1$.

**Solution:** The Euler's formulae for numerical solution of the differential equation $\dfrac{dy}{dx} = f(x,y)$ is

$$y_{n+1} = y_n + hf(x_n, y_n), \quad n = 0, 1, 2, 3, \cdots \quad \cdots \quad (1)$$

Here $f(x,y) = 2xy$, $x_0 = 0$, $y_0 = 1$ and $h = 0.05$.
Then $x_1 = 0.05$, $x_2 = 0.10$, $x_3 = 0.15$, $x_4 = 0.20$, $x_5 = 0.25$.

Putting $n = 0, 1, 2, 3, 4$, we get

$$y_1 = y_0 + hf(x_0, y_0) = 1 + (0.05)f(0, 1) = 1 + (0.05)(2)(0)(1) = 1$$
$$y_2 = y_1 + hf(x_1, y_1) = 1 + (0.05)f(0.05, 1) = 1.005$$
$$y_3 = y_2 + hf(x_2, y_2) = 1.005 + (0.005)f(0.10, 1.005) = 1.01505$$
$$y_4 = y_3 + hf(x_3, y_3) = 1.01505 + (0.05)f(0.15, 1.01505) = 1.03027$$
$$y_5 = y_4 + hf(x_4, y_4) = 1.03027 + (0.05)f(0.20, 1.03027) = 1.05088$$

Therefore, $y(0.25) = 1.05088$.

**Example:** Use Euler's method to find $y(0.1)$, $y(0.2)$ and $y(0.3)$, given

$$y' = (x^3 + xy^2)e^{-x}, \quad y(0) = 1.$$

**Solution:** The Euler's formulae for numerical solution of the differential equation $\dfrac{dy}{dx} = f(x, y)$ is

$$y_{n+1} = y_n + hf(x_n, y_n), \quad n = 0, 1, 2, 3, \cdots \qquad \cdots \quad (1)$$

Here $f(x, y) = (x^3 + xy^2)e^{-x}$, $x_0 = 0$, $y_0 = 1$. Let us take $h = 0.1$ and set $x_1 = x_0 + h = 0.1$, $x_2 = x_1 + h = 0.2$ and $x_2 = x_2 + h = 0.3$. We have to find $y_1 = y(x_1) = y(0.1)$, $y_2 = y(x_2) = y(0.2)$ and $y_3 = y(x_3) = y(0.3)$.
Now $x_1 = x_0 + h = 0 + 0.1 = 0.1$.
Putting $n = 0$ in (1), we get

$$y_1 = y_0 + hf(x_0, y_0) = 1 + (0.1)f(0, 1) = 1 + (0.1)(0 + 0)e^{-0} = 1 + 0 = 1$$

Putting $n = 1$ in (1), we get

$$y_2 = y_1 + hf(x_1, y_1)$$
$$= 1 + (0.1)f(0.1, 1)$$
$$= 1 + (0.1)\left[(0.1)^3 + (0.1)(1)^2\right]e^{-0.1}$$
$$= 1 + (0.1)\left[0.001 + 0.1\right]e^{-0.1}$$
$$= 1 + (0.1)(0.101)e^{-0.1}$$
$$= 1.009138.$$

Putting $n = 2$ in (1), we get

$$y_3 = y_2 + hf(x_2, y_2)$$
$$= 1.009138 + (0.1)f(0.2, 1.009138)$$
$$= 1.009138 + (0.1)\left[(0.2)^3 + (0.2)(1.009138)^2\right]e^{-0.2}$$
$$= 1.009138 + (0.1)\left[0.008 + 0.20336719\right]e^{-0.2}$$
$$= 1.009138 + (0.1)(0.2116719)e^{-0.2}$$
$$= 1.0264682.$$

Thus, the required values are

$$y(0.1) = y_1 = 1, \quad y(0.2) = y_2 = 1.009138, \quad y(0.3) = 1.0264682.$$

**Example:** By the Euler's method, find an approximate solution at $x = 0.2$ of the initial-vale problem $y' = 1 - y$, $y(0) = 0$, by taking the step length $h = 0.1$.

**Solution:** The Euler's formulae for numerical solution of the differential equation $\dfrac{dy}{dx} = f(x, y)$ is

$$y_{n+1} = y_n + hf(x_n, y_n), \quad n = 0, 1, 2, 3, \cdots \quad \cdots \quad (1)$$

Here $f(x, y) = 1 - y$, $x_0 = 0$, $y_0 = 0$. Also $x_1 = x_0 + h = 0.1$ and $x_2 = x_1 + h = 0.2$. We have to find $y_2 = y(x_2) = y(0.2)$.

Putting $n = 0$ in (1), we get

$$y_1 = y_0 + hf(x_0, y_0) = 0 + (0.1)f(0, 1) = 0 + (0.1)(1 - 0) = 0.1$$

Putting $n = 1$ in (1), we get

$$y_2 = y_1 + hf(x_1, y_1)$$
$$= 0.1 + (0.1)f(0.1, 0.1)$$
$$= 0.1 + (0.1)(1 - 0.1)$$
$$= 0.1 + (0.1)(0.9)$$
$$= 0.19.$$

Thus, $y_2 = y(0.2) = 0.19$ is the required solution.

**Example:** Using Euler's method, solve for $y$ at $x = 2$ from $\dfrac{dy}{dx} = 3x^2 + 1$, $y(1) = 2$ taking step size (i) $h = 0.5$ (ii) $h = 0.25$. **(Dec-2020)**

**Solution:** The Euler's formulae for numerical solution of the differential equation $\dfrac{dy}{dx} = f(x, y)$ is

$$y_{n+1} = y_n + hf(x_n, y_n), \quad n = 0, 1, 2, 3, \cdots \qquad \cdots \quad (1)$$

Here $f(x, y) = 3x^2 + 1$, $x_0 = 1$, $y_0 = 2$.

**To find $y(2)$ with $h = 0.5$:**

We have to find $y_2 = y(x_2) = y(2)$.

Now $x_1 = x_0 + h = 1 + 0.5 = 1.5$.

Putting $n = 0$ in (1), we get

$$y(1.5) = y_1 = y_0 + hf(x_0, y_0) = 2 + (0.5)f(1, 2) = 2 + (0.5)(3 + 1) = 2 + 2 = 4$$

Now $x_2 = x_1 + h = 1.5 + 0.5 = 2$.

Putting $n = 1$ in (1), we get

$$y(2) = y_2 = y_1 + hf(x_1, y_1) = 4 + (0.5)f(1.5, 4) = 4 + (0.5)[3(1.5)^2 + 1] = 7.875.$$

Thus, $y_2 = y(2) = 7.875$ is the required solution.

**To find $y(2)$ with $h = 0.25$:**

We have to find $y_4 = y(x_4) = y(2)$.

Now $x_1 = 1.25, x_2 = 1.5, x_3 = 1.75, x_4 = 2$.

Putting $n = 0, 1, 2, 3$ in (1), we get

$$y_1 = y(1.25) = y_0 + hf(x_0, y_0) = 2 + (0.25)f(1, 2) = 3$$

$$y_2 = y(1.5) = y_1 + hf(x_1, y_1) = 3 + (0.25)f(1.25, 3) = 4.42188$$

$$y_3 = y(1.75) = y_2 + hf(x_2, y_2) = 4.4218 + (0.25)f(1.5, 4.4218) = 6.35938$$

$$y_4 = y(2) = y_3 + hf(x_3, y_3) = 6.3593 + (0.25)f(2, 6.3593) = 8.90626$$

Thus, $y_2 = y(2) = 8.90626$ is the required solution.

**Example:** Using Euler's method find the approximate solution of $y$ at $x = 0.1$ given $y' = \dfrac{y - x}{y + x}$, $y(0) = 1$ by taking $h = 0.02$.

**Solution:** The Euler's formulae for numerical solution of the differential equation $\dfrac{dy}{dx} = f(x, y)$ is

$$y_{n+1} = y_n + hf(x_n, y_n), \quad n = 0, 1, 2, 3, \cdots \quad \cdots \quad (1)$$

Here $f(x, y) = \dfrac{y - x}{y + x}$, $x_0 = 0, y_0 = 1$ and $h = 0.02$. Also,

$x_1 = x_0 + h = 0.02$, $x_2 = x_1 + h = 0.04$, $x_3 = x_2 + h = 0.06$, $x_4 = x_3 + h = 0.08$,

$x_5 = x_4 + h = 0.1$.

Putting $n = 0$ in (1), we get

$$y(0.02) = y_1 = y_0 + hf(x_0, y_0) = 1 + (0.02)f(0, 1) = 1 + (0.02)\left(\frac{1 - 0}{1 + 0}\right) = 1.0200$$

Putting $n = 1$ in (1), we get

$$y(0.04) = y_2 = y_1 + hf(x_1, y_1) = 1.0200 + (0.02)f(0.02, 1.0200)$$

$$= 1.0200 + (0.02)\left(\frac{1.0200 - 0.02}{1.0200 + 0.02}\right) = 1.0392$$

Putting $n = 2$ in (1), we get

$$y(0.06) = y_3 = y_2 + hf(x_2, y_2) = 1.0392 + (0.02)f(0.04, 1.0392)$$
$$= 1.0392 + (0.02)\left(\frac{1.0392 - 0.04}{1.0392 + 0.04}\right) = 1.0577$$

Putting $n = 3$ in (1), we get

$$y(0.08) = y_4 = y_3 + hf(x_3, y_3) = 1.0577 + (0.02)f(0.06, 1.0577)$$
$$= 1.0577 + (0.02)\left(\frac{1.0577 - 0.06}{1.0577 + 0.06}\right) = 1.0756$$

Putting $n = 4$ in (1), we get

$$y(0.1) = y_5 = y_4 + hf(x_4, y_4) = 1.0756 + (0.02)f(0.08, 1.0756)$$
$$= 1.0756 + (0.02)\left(\frac{1.0756 - 0.08}{1.0756 + 0.08}\right) = 1.0928$$

Thus, $y_5 = y(0.1) = 1.0928$ is the required solution.

**Example:** Find $y(1.01), y(1.02), y(1.03)$ and $y(1.04)$ by solving the problem $y' = xy^{1/3}$, $y(1) = 1$.

**Solution:** The Euler's formulae for numerical solution of the differential equation $\frac{dy}{dx} = f(x, y)$ is

$$y_{n+1} = y_n + hf(x_n, y_n), \quad n = 0, 1, 2, 3, \cdots \quad \cdots \quad (1)$$

Here $f(x, y) = xy^{1/3}$, $x_0 = 1$, $y_0 = 1$ and $h = 0.01$. Also,
$x_1 = x_0 + h = 1.01$, $x_2 = x_1 + h = 1.02$, $x_3 = x_2 + h = 1.03$, $x_4 = x_3 + h = 1.04$.
We have to find $y_1, y_2, y_3$ and $y_4$.

Putting $n = 0$ in (1), we get

$$y(1.01) = y_1 = y_0 + hf(x_0, y_0)$$
$$= 1 + (0.01)f(1, 1)$$
$$= 1 + (0.01)1(1^{1/3})$$
$$= 1.01$$

Putting $n = 1$ in (1), we get

$$y(1.02) = y_2 = y_1 + hf(x_1, y_1)$$
$$= 1.01 + (0.01)f(1.01, 1.01)$$
$$= 1.01 + (0.01)(1.01)(1.01)^{1/3}$$
$$= 1.0201$$

Putting $n = 2$ in (1), we get

$$y(1.03) = y_3 = y_2 + hf(x_2, y_2)$$
$$= 1.0201 + (0.01)f(1.03, 1.0201)$$
$$= 1.0201 + (0.01)(1.03)(1.0201)^{1/3}$$
$$= 1.0304$$

Putting $n = 3$ in (1), we get

$$y(1.04) = y_4 = y_3 + hf(x_3, y_3)$$
$$= 1.0304 + (0.01)f(1.04, 1.0304)$$
$$= 1.0304 + (0.01)(1.04)(1.0304)^{1/3}$$
$$= 1.0409$$

Thus, the required values are $y(1.01) = 1.01, y(1.02) = 1.0201, y(1.03) = 1.0304$ and $y(1.04) = 1.0409$.

## EXERCISE

1. If $y' = x - y^2, y(0) = 1$, find $y(0.2)$ by using Euler's method when $h = 0.1$.
   **Ans:** $y(0.1) = 0.9, y(0.2) = 0.829$.
2. Using Euler's method, find $y(0.1)$ and $y(0.2)$ given $y' = 1 + xy, y(0) = 1$.
   **Ans:** $y(0.1) = 1.1, y(0.2) = 1.211$.
3. Evaluate $y(0.1)$ using Euler's method for $\dfrac{dy}{dx} = x - ye^x, y(0) = 1$.

4. Using Euler's method, solve $y' = y^2 + x$, $y(0) = 1$, compute $y(0.1), y(0.2)$.
   
   **(Dec-2020), (July-2021)**

5. Using Euler's method, find $y(0.2)$ and $y(0.4)$ given $y' = x + y, y(0) = 1$.
   
   **(Dec-2020)**

6. Evaluate $y(1)$ by Euler's method for $\dfrac{dy}{dx} = \dfrac{x+y}{y-x}, y(0) = 1$ by taking $h = 0.2$.
   
   **(Dec-2021), (Mar-2022)**

7. Evaluate $y(0.1)$ by Euler's method for $\dfrac{dy}{dx} = \dfrac{x+y}{y-x}, y(0) = 1$.
   
   **(Jan-2020)**

8. Find the solution of $\dfrac{dy}{dx} = x - y, y(0) = 1$ at $x = 0.1, 0.2, 0.3, 0.4 \& 0.5$ using Euler's method.
   
   **(Sep-2021)**

9. Using Euler's method find $y(0.2), y(0.4)$ and $y(0.6)$ given $y' = y + e^x, y(0) = 0$.
   
   **Ans:** $y(0.2) = 0.24678, y(0.4) = 0.60302$ **(Apr-2022)**

10. Given $y' = x + \sin y, y(0) = 1$, compute $y(0.2)$ using Euler's method taking $h = 0.05$.
    
    **(Apr-2022)**
    
    **Ans:** 1.19725

## 5.2.5 Modified Euler's Method

Consider the differential equation

$$\frac{dy}{dx} = f(x,y) \quad \text{with initial condition} \quad y(x_0) = y_0$$

$$\Rightarrow dy = f(x,y)dx.$$

Integrating the above equation in $\left[x_{r-1}, x_r\right]$

$$\int_{x_{r-1}}^{x_r} dy = \int_{x_{r-1}}^{x_r} f(x,y)dx.$$

Using Trapezoidal rule, we get

$$\left(y\right)_{x_{r-1}}^{x_r} = \frac{h}{2}\left[f\left(x_{r-1}, y_{r-1}\right) + f\left(x_r, y_r\right)\right]$$

$$\Rightarrow y(x_r) - y(x_{r-1}) = \frac{h}{2}\left[f\left(x_{r-1}, y_{r-1}\right) + f\left(x_r, y_r\right)\right]$$

$$\Rightarrow y_r = y_{r-1} + \frac{h}{2}\left[f\left(x_{r-1}, y_{r-1}\right) + f\left(x_r, y_r\right)\right]$$

Replacing $f(x_r, y_r)$ by its approximate value $f(x_r, y_r^{(0)})$, we get

$$y_r^{(1)} = y_{r-1} + \frac{h}{2}\left[f\left(x_{r-1}, y_{r-1}\right) + f\left(x_r, y_r^{(0)}\right)\right]$$

Here $y_r^{(1)}$ is the first approximation to $y_r = y(x_r)$.
Proceeding as above, we get the iteration formula

$$y_r^{(n)} = y_{r-1} + \frac{h}{2}\left[f\left(x_{r-1}, y_{r-1}\right) + f\left(x_r, y_r^{(n-1)}\right)\right] \quad \cdots\cdots \quad (1)$$

where $y_r^{(n)}$ denote the $n$th approximation to $y_r$.
At the point $x_1 = x_0 + h$, the solution got by the use of the Euler's formula is given by

$$y_1^{(0)} = y_0 + hf(x_0, y_0).$$

The modified values of $y_1$ are

$$y_1^{(1)} = y_0 + \frac{h}{2}\left[f(x_0, y_0) + f(x_1, y_1^{(0)})\right]$$

$$y_1^{(2)} = y_0 + \frac{h}{2}\left[f(x_0, y_0) + f(x_1, y_1^{(1)})\right]$$

$$\vdots$$

$$\vdots$$

We proceed in this way until two successive approximations are approximately equal.

Next, if we wish to find the solution at the point $x_2 = x_1 + h = x_0 + 2h$. At the point $x_2$, the solution got by the use of the Euler's formula is given by

$$y_2^{(0)} = y_1 + hf(x_1, y_1).$$

The modified values of $y_2$ are

$$y_2^{(1)} = y_1 + \frac{h}{2}\left[f(x_1, y_1) + f(x_2, y_2^{(0)})\right]$$
$$y_2^{(2)} = y_1 + \frac{h}{2}\left[f(x_1, y_1) + f(x_2, y_2^{(1)})\right]$$
$$\vdots$$
$$\vdots$$

We proceed in this way until two successive approximations are approximately equal.

### 5.2.6 Examples

**Example:** Find $y(0.2)$ and $y(0.4)$ given that $\frac{dy}{dx} = x - y^2$, $y(0) = 1$ using Euler's modified formula.

**Solution:** Here $f(x,y) = x - y^2$, $x_0 = 0$, $y_0 = 1$. Take $h = 0.2$. Then $x_1 = x_0 + h = 0.2$ and $x_2 = x_1 + h = 0.4$.

Using Euler's formula,

$$\begin{aligned}y_1^{(0)} &= y_0 + hf(x_0, y_0)\\ &= 1 + (0.2)f(0, 1)\\ &= 1 + (0.2)(0 - 1^2)\\ &= 1 - 0.2\\ &= 0.8\end{aligned}$$

**To find $y_1$ i. e., $y(0.2)$**

$$y_1^{(1)} = y_0 + \frac{h}{2}\left[f(x_0, y_0) + f(x_1, y_1^{(0)})\right]$$
$$= 1 + \frac{0.2}{2}\left[f(0, 1) + f(0.2, 0.8)\right]$$
$$= 1 + (0.1)\left[(0 - 1^2) + (0.2 - 0.8^2)\right]$$
$$= 1 + (0.1)\left[-1 - 0.44\right]$$
$$= 1 - 0.144$$
$$= 0.856$$

$$y_1^{(2)} = y_0 + \frac{h}{2}\left[f(x_0, y_0) + f(x_1, y_1^{(1)})\right]$$
$$= 1 + \frac{0.2}{2}\left[f(0, 1) + f(0.2, 0.856)\right]$$
$$= 1 + (0.1)\left[(0 - 1^2) + (0.2 - 0.856^2)\right]$$
$$= 1 + (0.1)\left[-1 - 0.5327\right]$$
$$= 1 - 0.15327$$
$$= 0.8467$$

$$y_1^{(3)} = y_0 + \frac{h}{2}\left[f(x_0, y_0) + f(x_1, y_1^{(2)})\right]$$
$$= 1 + \frac{0.2}{2}\left[f(0, 1) + f(0.2, 0.8467)\right]$$
$$= 1 + (0.1)\left[(0 - 1^2) + (0.2 - 0.8467^2)\right]$$
$$= 1 + (0.1)\left[-1 - 0.5169\right]$$
$$= 1 - 0.15169$$
$$= 0.8483$$

$$y_1^{(4)} = y_0 + \frac{h}{2}\left[f(x_0, y_0) + f(x_1, y_1^{(3)})\right]$$
$$= 1 + \frac{0.2}{2}\left[f(0,1) + f(0.2, 0.8483)\right]$$
$$= 1 + (0.1)\left[(0 - 1^2) + (0.2 - 0.8483^2)\right]$$
$$= 1 + (0.1)\left[-1 - 0.5196\right]$$
$$= 1 - 0.15196$$
$$= 0.84804 \text{ correct to 3 decimal places}$$

Since the values of $y_1^{(3)}$ and $y_1^{(4)}$ are equal, we take
$$y_1 = y(0.2) = 0.84804.$$

**To find** $y_2$ i. e., $y(0.4)$

We take $x_1 = 0.2$ and $y_1 = 0.84804$ and $x_2 = 0.4, h = 0.2$

$$f(x_1, y_1) = f(0.2, 0.84804) = 0.2 - 0.84804^2 = -0.5192.$$

Euler's formula gives

$$y_2^{(0)} = y_1 + hf(x_1, y_1)$$
$$= 0.84804 + (0.2)f(0.2, 0.84804)$$
$$= 0.84804 - 0.10384$$
$$= 0.7442.$$

$$y_2^{(1)} = y_1 + \frac{h}{2}\left[f(x_1, y_1) + f(x_2, y_2^{(0)})\right]$$
$$= 0.84804 + \frac{0.2}{2}\left[f(0.2, 0.84804) + f(0.4, 0.7442)\right]$$
$$= 0.84804 + (0.1)\left[-0.5192 + (0.4 - 0.7442^2)\right]$$
$$= 0.84804 + (0.1)\left[-0.5192 - 0.1538\right]$$
$$= 0.78074$$

$$y_2^{(2)} = y_1 + \frac{h}{2}\left[f(x_1, y_1) + f(x_2, y_2^{(1)})\right]$$

$$= 0.84804 + \frac{0.2}{2}\left[f(0.2, 0.84804) + f(0.4, 0.78074)\right]$$

$$= 0.84804 + (0.1)\left[-0.5192 + (0.4 - 0.78074^2)\right]$$

$$= 0.84804 + (0.1)\left[-0.5192 - 0.2096\right]$$

$$= 0.77516$$

$$y_2^{(3)} = y_1 + \frac{h}{2}\left[f(x_1, y_1) + f(x_2, y_2^{(2)})\right]$$

$$= 0.84804 + \frac{0.2}{2}\left[f(0.2, 0.84804) + f(0.4, 0.77516)\right]$$

$$= 0.84804 + (0.1)\left[-0.5192 + (0.4 - 0.77516^2)\right]$$

$$= 0.84804 + (0.1)\left[-0.5192 - 0.2009\right]$$

$$= 0.77603$$

$$y_2^{(4)} = y_1 + \frac{h}{2}\left[f(x_1, y_1) + f(x_2, y_2^{(3)})\right]$$

$$= 0.84804 + \frac{0.2}{2}\left[f(0.2, 0.84804) + f(0.4, 0.77603)\right]$$

$$= 0.84804 + (0.1)\left[-0.5192 + (0.4 - 0.77603^2)\right]$$

$$= 0.84804 + (0.1)\left[-0.5192 - 0.2022\right]$$

$$= 0.7759$$

$$y_2^{(5)} = y_1 + \frac{h}{2}\left[f(x_1, y_1) + f(x_2, y_2^{(4)})\right]$$

$$= 0.2468 + \frac{0.2}{2}\left[f(0,0) + f(0.4, 0.7759)\right]$$

$$= 0.84804 + (0.1)\left[-0.5192 + (0.4 - 0.7759^2)\right]$$

$$= 0.84804 + (0.1)\left[-0.5192 - 0.2020\right]$$

$$= 0.7759$$

Since the values of $y_2^{(4)}$ and $y_2^{(5)}$ are equal, we take

$$y_2 = y(0.4) = 0.7759.$$

**Example:** Using modified Euler's method, find $y(0.2)$, $y(0.4)$ and $y(0.6)$ given that $y' = y + e^x$, $y(0) = 0$.

**Solution:** Here $f(x,y) = y + e^x$, $x_0 = 0, y_0 = 0$. Also, $h = 0.2$. We take $x_1 = x_0 + h = 0.2$, $x_2 = x_1 + h = 0.4$, $x_3 = x_2 + h = 0.6$.

We have to find $y_1 = y(x_1) = y(0.2)$, $y_2 = y(x_2) = y(0.4)$ and $y_3 = y(x_3) = y(0.6)$.

To find $y_1$ i. e., $y(0.2)$

Using Euler's formula

$$y_1^{(0)} = y_0 + hf(x_0, y_0)$$
$$= 0 + (0.2)f(0,0)$$
$$= (0.2)(0 + e^0)$$
$$= 0.2$$

$$y_1^{(1)} = y_0 + \frac{h}{2}\left[f(x_0, y_0) + f(x_1, y_1^{(0)})\right]$$
$$= 0 + \frac{0.2}{2}\left[f(0,0) + f(0.2, 0.2)\right]$$
$$= (0.1)\left[(0 + e^0) + (0.2 + e^{0.2})\right]$$
$$= (0.1)\left[1 + 1.42140\right]$$
$$= (0.1)(2.42140)$$
$$= 0.24214$$

$$y_1^{(2)} = y_0 + \frac{h}{2}\left[f(x_0, y_0) + f(x_1, y_1^{(1)})\right]$$
$$= 0 + \frac{0.2}{2}\left[f(0,0) + f(0.2, 0.24214)\right]$$
$$= (0.1)\left[1 + (0.24214 + e^{0.2})\right]$$
$$= 0.2463$$

$$y_1^{(3)} = y_0 + \frac{h}{2}\left[f(x_0, y_0) + f(x_1, y_1^{(2)})\right]$$
$$= 0 + \frac{0.2}{2}\left[f(0,0) + f(0.2, 0.2463)\right]$$
$$= (0.1)\left[1 + (0.2463 + e^{0.2})\right]$$
$$= 0.2468$$

$$y_1^{(4)} = y_0 + \frac{h}{2}\left[f(x_0, y_0) + f(x_1, y_1^{(3)})\right]$$
$$= 0 + \frac{0.2}{2}\left[f(0,0) + f(0.2, 0.2468)\right]$$
$$= (0.1)\left[1 + (0.2468 + e^{0.2})\right]$$
$$= 0.2468 \text{ correct to 4 decimal places}$$

Since the values of $y_1^{(3)}$ and $y_1^{(4)}$ are equal, we take
$$y_1 = y(0.2) = 0.2468.$$

**To find $y_2$ i. e., $y(0.4)$**

We take $x_1 = 0.2$ and $y_1 = 0.2468$ and $x_2 = 0.4, h = 0.2$

$$f(x_1, y_1) = f(0.2, 0.2468) = 0.2468 + e^{0.2} = 0.2468 + 1.2214 = 1.4682.$$

Euler's formula gives

$$y_2^{(0)} = y_1 + hf(x_1, y_1)$$
$$= 0.2468 + (0.2)f(0.2, 1.4682)$$
$$= 0.5404$$

$$y_2^{(1)} = y_1 + \frac{h}{2}\left[f(x_1, y_1) + f(x_2, y_2^{(0)})\right]$$
$$= 0.2468 + \frac{0.2}{2}\left[f(0.2, 0.2468) + f(0.4, 0.5404)\right]$$
$$= 0.2468 + (0.1)\left[1.4682 + (0.5404 + e^{0.4})\right]$$
$$= 0.2468 + (0.1)\left[1.4682 + (0.5404 + 1.4918)\right]$$
$$= 0.5968$$

$$y_2^{(2)} = y_1 + \frac{h}{2}\left[f(x_1, y_1) + f(x_2, y_2^{(1)})\right]$$

$$= 0.2468 + \frac{0.2}{2}\left[f(0.2, 0.2468) + f(0.4, 0.5968)\right]$$

$$= 0.2468 + (0.1)\left[1.4682 + (0.5968 + e^{0.4})\right]$$

$$= 0.2468 + (0.1)\left[1.4682 + (0.5968 + 1.4918)\right]$$

$$= 0.6025$$

$$y_2^{(3)} = y_1 + \frac{h}{2}\left[f(x_1, y_1) + f(x_2, y_2^{(2)})\right]$$

$$= 0.2468 + \frac{0.2}{2}\left[f(0.2, 0.2468) + f(0.4, 0.6025)\right]$$

$$= 0.2468 + (0.1)\left[1.4682 + (0.6025 + e^{0.4})\right]$$

$$= 0.2468 + (0.1)\left[1.4682 + (0.6025 + 1.4918)\right]$$

$$= 0.603$$

$$y_2^{(4)} = y_1 + \frac{h}{2}\left[f(x_1, y_1) + f(x_2, y_2^{(3)})\right]$$

$$= 0.2468 + \frac{0.2}{2}\left[f(0.2, 0.2468) + f(0.4, 0.603)\right]$$

$$= 0.2468 + (0.1)\left[1.4682 + (0.603 + e^{0.4})\right]$$

$$= 0.2468 + (0.1)\left[1.4682 + (0.603 + 1.4918)\right]$$

$$= 0.6031$$

$$y_2^{(5)} = y_1 + \frac{h}{2}\left[f(x_1, y_1) + f(x_2, y_2^{(4)})\right]$$

$$= 0.2468 + \frac{0.2}{2}\left[f(0.2, 0.2468) + f(0.4, 0.6031)\right]$$

$$= 0.2468 + (0.1)\left[1.4682 + (0.6031 + e^{0.4})\right]$$

$$= 0.2468 + (0.1)\left[1.4682 + (0.6031 + 1.4918)\right]$$

$$= 0.6031$$

Since the values of $y_2^{(4)}$ and $y_2^{(5)}$ are equal, we take

$$y_2 = y(0.4) = 0.6031.$$

**To find** $y_3$ i. e., $y(0.6)$

We take $x_2 = 0.4$ and $y_2 = 0.6031$ and $x_2 = 0.6, h = 0.2$

$$f(x_2, y_2) = f(0.4, 0.6031) = 0.6031 + e^{0.4} = 0.6031 + 1.4918 = 2.0949.$$

Euler's formula gives

$$y_3^{(0)} = y_2 + hf(x_2, y_2)$$
$$= 0.6031 + (0.2)(2.0949)$$
$$= 1.02208$$

$$y_3^{(1)} = y_2 + \frac{h}{2}\left[f(x_2, y_2) + f(x_3, y_3^{(0)})\right]$$
$$= 0.6031 + \frac{0.2}{2}\left[f(0.4, 0.6031) + f(0.6, 1.02208)\right]$$
$$= 0.6031 + (0.1)\left[2.0949 + (1.02208 + e^{0.6})\right]$$
$$= 0.6031 + (0.1)\left[2.0949 + (1.02208 + 1.8221)\right]$$
$$= 1.097$$

$$y_3^{(2)} = y_2 + \frac{h}{2}\left[f(x_2, y_2) + f(x_3, y_3^{(1)})\right]$$
$$= 0.6031 + \frac{0.2}{2}\left[f(0.4, 0.6031) + f(0.6, 1.097)\right]$$
$$= 0.6031 + (0.1)\left[2.0949 + (0.5968 + e^{0.6})\right]$$
$$= 0.6031 + (0.1)\left[2.0949 + (1.097 + 1.8221)\right]$$
$$= 1.1045$$

$$y_3^{(3)} = y_2 + \frac{h}{2}\left[f(x_2, y_2) + f(x_3, y_3^{(2)})\right]$$
$$= 0.6031 + \frac{0.2}{2}\left[f(0.4, 0.6031) + f(0.6, 1.1045)\right]$$
$$= 0.6031 + (0.1)\left[2.0949 + (1.1045 + e^{0.6})\right]$$
$$= 0.6031 + (0.1)\left[2.0949 + (1.1045 + 1.8221)\right]$$
$$= 1.1052$$

$$y_3^{(4)} = y_2 + \frac{h}{2}\left[f(x_2, y_2) + f(x_3, y_3^{(3)})\right]$$

$$= 0.6031 + \frac{0.2}{2}\left[f(0.4, 0.6031) + f(0.4, 0.603)\right]$$

$$= 0.6031 + (0.1)\left[2.0949 + (1.1052 + e^{0.6}]\right.$$

$$= 0.6031 + (0.1)\left[2.0949 + (1.1052 + 1.8221)\right]$$

$$= 1.1053$$

$$y_3^{(5)} = y_2 + \frac{h}{2}\left[f(x_1, y_1) + f(x_2, y_2^{(4)})\right]$$

$$= 0.6031 + \frac{0.2}{2}\left[f(0.4, 0.6031) + f(0.4, 1.1053)\right]$$

$$= 0.6031 + (0.1)\left[2.0949 + (1.1053 + e^{0.6}]\right.$$

$$= 0.6031 + (0.1)\left[2.0949 + (1.1053 + 1.8221)\right]$$

$$= 1.1053$$

Since the values of $y_3^{(4)}$ and $y_3^{(5)}$ are equal, we take

$$y_3 = y(0.6) = 1.1053.$$

**Example:** Given $y' = x + \sin y$, $y(0) = 1$. Compute $y(0.2)$ and $y(0.4)$ with $h = 0.2$ using Euler's modified method.

**Solution:** Here $f(x, y) = x + \sin y$, $x_0 = 0, y_0 = 1, h = 0.2$. We set $x_1 = x_0 + h = 0.2$ and $x_2 = x_1 + h0.4$.

**To find $y_1$ i. e., $y(0.2)$** Using Euler's formula,

$$y_1^{(0)} = y_0 + hf(x_0, y_0)$$

$$= 1 + (0.2)f(0, 1)$$

$$= 1 + (0.2)(0 + \sin 1)$$

$$= 1 + 0.1683$$

$$= 1.1683$$

$$y_1^{(1)} = y_0 + \frac{h}{2}\left[f(x_0, y_0) + f(x_1, y_1^{(0)})\right]$$

$$= 1 + \frac{0.2}{2}\left[f(0, 1) + f(0.2, 1.1683)\right]$$

$$= 1 + (0.1)\left[(0 + \sin 1) + (0.2 + \sin 1.1683)\right]$$

$$= 1 + (0.1)\left[0.8414 + 1.1201\right]$$

$$= 1 + (0.1)(1.9615)$$

$$= 1.19615$$

$$y_1^{(2)} = y_0 + \frac{h}{2}\left[f(x_0, y_0) + f(x_1, y_1^{(1)})\right]$$

$$= 1 + \frac{0.2}{2}\left[f(0, 1) + f(0.2, 1.19615)\right]$$

$$= 1 + (0.1)\left[(0 + \sin 1) + (0.2 + \sin 1.19615)\right]$$

$$= 1 + (0.1)\left[0.8414 + 1.1306\right]$$

$$= 1 + (0.1)(1.972)$$

$$= 1.1972$$

$$y_1^{(3)} = y_0 + \frac{h}{2}\left[f(x_0, y_0) + f(x_1, y_1^{(2)})\right]$$

$$= 1 + \frac{0.2}{2}\left[f(0, 1) + f(0.2, 1.1972)\right]$$

$$= 1 + (0.1)\left[(0 + \sin 1) + (0.2 + \sin 1.1972)\right]$$

$$= 1 + (0.1)\left[0.8414 + 1.1310\right]$$

$$= 1 + (0.1)(1.9724)$$

$$= 1.19724$$

$$y_1^{(4)} = y_0 + \frac{h}{2}\left[f(x_0, y_0) + f(x_1, y_1^{(3)})\right]$$
$$= 1 + \frac{0.2}{2}\left[f(0, 1) + f(0.2, 1.19724)\right]$$
$$= 1 + (0.1)\left[(0 + \sin 1) + (0.2 + \sin 1.19724)\right]$$
$$= 1 + (0.1)\left[0.8414 + 1.1310\right]$$
$$= 1 + (0.1)(1.9724)$$
$$= 1.19724$$

Since the values of $y_1^{(3)}$ and $y_1^{(4)}$ are equal, we take

$$y_1 = y(0.2) = 1.19724.$$

**To find** $y_2$ **i. e.,** $y(0.4)$

We take $x_1 = 0.2$ and $y_1 = 1.19724$ and $x_2 = 0.4, h = 0.2$

$$f(x_1, y_1) = f(0.2, 1.19724) = 0.2 + \sin 1.19724 = 0.2 + 0.9310 = 1.131.$$

Euler's formula gives

$$y_2^{(0)} = y_1 + hf(x_1, y_1)$$
$$= 1.19724 + (0.2)f(0.2, 1.19724)$$
$$= 1.42344$$

$$y_2^{(1)} = y_1 + \frac{h}{2}\left[f(x_1, y_1) + f(x_2, y_2^{(0)})\right]$$
$$= 1.19724 + \frac{0.2}{2}\left[1.131 + f(0.4, 1.42344)\right]$$
$$= 1.19724 + (0.1)\left[1.131 + (0.4 + \sin 1.42344)\right]$$
$$= 1.19724 + (0.1)\left[1.131 + 1.3891\right]$$
$$= 1.44925$$

$$y_2^{(2)} = y_1 + \frac{h}{2}\left[f(x_1, y_1) + f(x_2, y_2^{(1)})\right]$$

$$= 1.19724 + \frac{0.2}{2}\left[f(0.2, 1.19724) + f(0.4, 1.44925)\right]$$

$$= 1.19724 + (0.1)\left[1.131 + (0.4 + \sin 1.44925)\right]$$

$$= 1.19724 + (0.1)\left[1.131 + 1.3926\right]$$

$$= 1.4496$$

$$y_2^{(3)} = y_1 + \frac{h}{2}\left[f(x_1, y_1) + f(x_2, y_2^{(2)})\right]$$

$$= 1.19724 + \frac{0.2}{2}\left[f(0.2, 1.19724) + f(0.4, 1.4496)\right]$$

$$= 1.19724 + (0.1)\left[1.131 + (0.4 + \sin 1.4496)\right]$$

$$= 1.19724 + (0.1)\left[1.131 + 1.3927\right]$$

$$= 1.4496$$

Since the values of $y_2^{(2)}$ and $y_2^{(3)}$ are equal, we take

$$y_2 = y(0.4) = 1.4496.$$

**EXERCISE**

1. Solve $y' = 2x - y$ and $y(1) = 3$ by modified Euler's method and compute $y(1.1)$.
**Ans:** $y(1.1) = 2.91428$

2. Using modified Euler's method evaluate $y(0.1), y(0.2)$ and $y(0.3)$ given $y' = x + y, y(0) = 1$.
**Ans:** $y(0.1) = 1.1105, y(0.2) = 1.2432, y(0.3) = 1.3951$.

3. Find $y(0.1)$ and $y(0.2)$ given that $\frac{dy}{dx} = x^2 - y$, $y(0) = 1$ using Euler's modified formula.
**Ans:** $y(0.1) = 0.9052, y(0.2) = 0.82137$.

4. By modified Euler's formula find $y(0.3)$ given that $\frac{dy}{dx} = x^2 - y^2$, $y(0) = 1$.
(July-2021)

5. By modified Euler's formula find $y(0.2)$ given that $\frac{dy}{dx} = 2x + y^2$, $y(0) = 1$.

6. Find the solution of $\dfrac{dy}{dx} = x - y$, $y(0) = 1$ at $x = 0.1, 0.2$ using modified Euler's method.

(Mar-2022)

(Mar-2022)

### 5.2.7 Runge-Kutta Methods

The Taylor's series method of solving differential equations numerically is handicapped by the problem of finding the higher order derivatives. Euler's method is less efficient in practical problems since it requires $h$ to be small for obtaining reasonable accuracy.

The Runge-Kutta methods do not require the calculations of higher order derivatives and they are designed to give greater accuracy with the advantage of requiring only the function values at some selected points on the sub-interval. These methods agree with Taylor's series solution upto the terms of $h^r$ where $r$ is the order of the Runge-Kutta (R. K. method).

### 5.2.8 R-K Second Order Method

Consider $\dfrac{dy}{dx} = f(x, y)$, $y(x_0) = y_0$. To find $y_1 = y(x_0 + h)$. The second order Runge-Kutta formula is

$$y_1 = y_0 + \dfrac{1}{2}[k_1 + k_2]$$

where $k_1 = hf(x_0, y_0)$ and $k_2 = hf(x_0 + h, y_0 + k_1)$.

**Example:** Apply second order R-K method to find $y(0.1)$ and $y(0.2)$, where $\dfrac{dy}{dx} = y - x$ and $y(0) = 2$.

**Solution:** Second-order R-K formula is

$$y_1 = y_0 + \dfrac{1}{2}(k_1 + k_2)$$

where $k_1 = hf(x_0, y_0)$, $k_2 = hf(x_0 + h, y_0 + k_1)$.

Here $f(x, y) = y - x$, $x_0 = 0$ and $y_0 = 2$ and we take $h = 0.1$.

$$\therefore k_1 = hf(x_0, y_0)$$
$$= hf(0, 2)$$
$$= (0.1)(2 - 0)$$
$$= 0.2$$

$$k_2 = hf(x_0 + h, y_0 + k_1)$$
$$= (0.1)f(0.1, 2 + 0.2)$$
$$= (0.1)f(0.1, 2.2)$$
$$= (0.1)[2.2 - 0.1]$$
$$= 0.21$$

Hence $y_1 = y_0 + \dfrac{1}{2}(k_1 + k_2)$

$\Rightarrow y_1 = y(0.1) = 2 + \dfrac{1}{2}(0.20 + 0.21) = 2 + 0.2050 = 2.2050$.

To determine $y_2 = y(0.2)$, we note that $x_1 = 0.1$, $y_1 = 2.2050$.

$$k_1 = hf(x_1, y_1)$$
$$= (0.1)f(0.1, 2.205)$$
$$= (0.1)(2.205 - 0.1)$$
$$= 0.2105$$

$$k_2 = hf(x_1 + h, y_1 + k_1)$$
$$= (0.1)f(0.2, 2.205 + 0.2105)$$
$$= (0.1)f(0.2, 2.4155)$$
$$= (0.1)(2.4155 - 0.2)$$
$$= (0.1)(2.2155)$$
$$= 0.22155.$$

$$\therefore y_2 = y(0.2) = y_1 + \frac{1}{2}(k_1 + k_2)$$
$$= 2.2050 + \frac{1}{2}(0.2105 + 0.22155)$$
$$= 2.4210.$$

**Example:** Using R-K method of second order, compute $y(2.5)$ from $\frac{dy}{dx} = \frac{x+y}{x}$, $y(2) = 2$, taking $h = 0.25$.

**Solution:** Here $f(x,y) = \frac{x+y}{x}$.

**Step 1:** $x_0 = 2$, $y_0 = 2$ and $h = 0.25$.

$$k_1 = hf(x_0, y_0) = (0.25)f(2,2) = (0.25)\left(\frac{2+2}{2}\right) = 0.5$$

$$k_2 = hf(x_0 + h, y_0 + k_1) = (0.25)f(2.25, 2.5) = (0.25)\left(\frac{2.25 + 2.5}{2.25}\right) = 0.5278$$

Hence,
$$y_1 = y(2.25) = y_0 + \frac{1}{2}(k_1 + k_2)$$
$$= 2 + \frac{1}{2}(0.5 + 0.5278)$$
$$= 2.5139.$$

**Step 2:** Now starting from $(x_1, y_1)$, we get $(x_2, y_2)$,

Here $x_1 = x_0 + h = 2 + 0.25 = 2.25$, $y_1 = 2.514$ and $h = 0.25$.

$$k_1 = hf(x_1, y_1) = (0.25)f(2.25, 2.5139) = (0.25)\left(\frac{2.25 + 2.5139}{2.25}\right) = 0.5293$$

$$k_2 = hf(x_1 + h, y_1 + k_1) = (0.25)f(2.25 + 0.25, 2.5139 + 0.5293)$$
$$= (0.25)f(2.5, 3.0432) = (0.25)\left(\frac{2.5 + 3.0432}{2.5}\right) = 0.55432$$

Hence,
$$y_2 = y(2.5) = y_1 + \frac{1}{2}(k_1 + k_2)$$
$$= 2.5139 + \frac{1}{2}(0.5293 + 0.55432)$$
$$= 3.05301.$$

**Example:** Given $\dfrac{dy}{dx} = -y$, $y(0) = 1$, using R-K method of second order find the values of $y$ at $x = 0.1$ and $x = 0.2$.

**Solution:** Second-order R-K formula is

$$y_1 = y_0 + \frac{1}{2}(k_1 + k_2)$$

where $k_1 = hf(x_0, y_0)$, $k_2 = hf(x_0 + h, y_0 + k_1)$.

Here $f(x, y) = -y$, $x_0 = 0$ and $y_0 = 1$ and we take $h = 0.1$.

To find $y_1$ i.e., $y(0.1)$

$$k_1 = hf(x_0, y_0)$$
$$= hf(0, 1)$$
$$= (0.1)(-1)$$
$$= -0.1$$

$$k_2 = hf(x_0 + h, y_0 + k_1)$$
$$= (0.1)f(0.1, 1 - 0.1)$$
$$= (0.1)f(0.1, 0.9)$$
$$= (0.1)\left[-0.9\right]$$
$$= -0.09$$

Hence $y_1 = y_0 + \dfrac{1}{2}(k_1 + k_2)$

$\Rightarrow y_1 = y(0.1) = 1 + \dfrac{1}{2}(-0.1 - 0.09) = 1 - 0.09 = 0.905$.

To find $y_2 = y(0.2)$

$x_1 = x_0 + h = 0.1$, $y_1 = 0.905$, $h = 0.1$.

$$k_1 = hf(x_1, y_1)$$
$$= (0.1)f(0.1, 0.905)$$
$$= (0.1)(-0.905)$$
$$= -0.0905$$

$$k_2 = hf(x_1 + h, y_1 + k_1)$$
$$= (0.1)f(0.2, 0.905 - 0.0905)$$
$$= (0.1)f(0.2, 0.8145)$$
$$= (0.1)(-0.8145)$$
$$= -0.08145.$$
$$\therefore y_2 = y(0.2) = y_1 + \frac{1}{2}(k_1 + k_2)$$
$$= 0.905 + \frac{1}{2}(-0.0905 - 0.08145)$$
$$= 0.819025.$$

**EXERCISE**

1. Given $y' = \frac{1}{2}(1+x)y^2$ and $y(0) = 1$, using second order R-K method find $y(0.3)$ by taking $h = 0.1$.
**Ans:** $y(0.3) = 1.2073$

2. Apply second order R-K method to find $y(0.2)$, where $\frac{dy}{dx} = x + \sqrt{y}$ and $y(0) = 1$.

3. Using Runge-Kutta second order formula solve the equation $\frac{dy}{dx} = 2 + \sqrt{xy}$ with $y(1) = 1$ for $x = 1(0.2)1.6$.

4. Using Runge-Kutta second order formula solve the equation $\frac{dy}{dx} = (1+x^2)y$ with $y(0) = 1$ for $x = 0(0.2)0.6$.
**Ans:** $y(0.2) = 1.2248$, $y(0.4) = 1.5238$, $(0.6) = 1.95575$.

5. Find $y(0.1), y(0.2)$ given that $\frac{dy}{dx} = 1 - 2xy^2$, $y(0) = 1$ by R-K method of second order if $h = 0.1$. **(July-2021)**

## 5.2.9 Runge-Kutta Fourth-Order Method

This method is most commonly used and is refereed as the Runge-Kutta method. The working rule for solving the initial value problem

$$\frac{dy}{dx} = f(x, y), \quad y(x_0) = y_0$$

by fourth order Runge-Kutta method as follows:

Calculate successively:

$$k_1 = hf(x_0, y_0)$$
$$k_2 = hf\left(x_0 + \frac{h}{2}, y_0 + \frac{k_1}{2}\right)$$
$$k_3 = hf\left(x_0 + \frac{h}{2}, y_0 + \frac{k_2}{2}\right)$$
$$k_4 = hf(x_0 + h, y_0 + k_3)$$

Then the required approximate values is given by

$$\boxed{y_1 = y_0 + \frac{1}{6}(k_1 + 2k_2 + 2k_3 + k_4).}$$

Similarly, the value of $y$ in the second interval is obtained by replacing $x_0$ by $x_1$ and $y_0$ by $y_1$ in the above set of formulae and we obtain $y_2$.

In general to find $y_n$ substitute $x_{n-1}, y_{n-1}$ in the expression for $k_1, k_2$ etc.

**Note 1.** The operation is identical whether the differential equation is linear or nonlinear.

**Note 2.** To evaluate $y_{n+1}$ we need information only at the point $y_n$. Information at the points $y_{n-1}, y_{n-2}$ etc. are not directly required. Hence R. K methods are steps methods.

**Example:** Solve the initial value problem

$$\frac{dy}{dx} = y - x, \quad y(0) = 2$$

at the point $x = 0.2$, by using the fourth order Runge-Kutta method. Take step length $h = 0.2$.

**Solution:** Here $f(x, y) = y - x$, $x_0 = 0, y_0 = 2$ and $h = 0.2$. Then $x_1 = x_0 + h = 0.2$. We have to find $y_1 = y(x_1) = y(0.2)$.

**To find $y(0.2)$ i.e., $y_1$**

Using the definitions of $k_1, k_2, k_3$ and $k_4$, we find

$$k_1 = hf(x_0, y_0)$$
$$= (0.2)f(0, 2)$$
$$= (0.2)(2 - 0)$$
$$= 0.4$$

$$k_2 = hf\left(x_0 + \frac{h}{2}, y_0 + \frac{k_1}{2}\right)$$
$$= (0.2)f\left(0 + \frac{0.2}{2}, 2 + \frac{0.4}{2}\right)$$
$$= (0.2)f(0.1, 2.2)$$
$$= (0.2)[2.2 - 0.1]$$
$$= 0.42$$

$$k_3 = hf\left(x_0 + \frac{h}{2}, y_0 + \frac{k_2}{2}\right)$$
$$= (0.2)f\left(0 + \frac{0.2}{2}, 2 + \frac{0.42}{2}\right)$$
$$= (0.2)f(0.1, 2.21)$$
$$= (0.2)[2.21 - 0.1]$$
$$= 0.422$$

$$k_4 = hf(x_0 + h, y_0 + k_3)$$
$$= (0.2)f(0 + 0.2, 2 + 0.422)$$
$$= (0.2)f(0.2, 2.422)$$
$$= (0.2)[2.422 - 0.2]$$
$$= 0.4444$$

Hence,

$$y_1 = y(0.2) = y_0 + \frac{1}{6}(k_1 + 2k_2 + 2k_3 + k_4)$$

$$= 2 + \frac{1}{6}(0.4 + 2(0.42) + 2(0.422) + 0.4444)$$

$$= 2 + 0.4214$$

$$= 2.4214.$$

**Example:** Find $y(0.1)$ and $y(0.2)$, using Runge-Kutta fourth order formula, given that $y' = x^2 - y$ and $y(0) = 1$. **(Jan-2020)**

**Solution:** Here $f(x, y) = x^2 - y$, $x_0 = 0$, $y_0 = 1$ and $h = 0.1$, so that $x_1 = 0.1$ and $x_2 = 0.2$. We have to find $y_1 = y(x_1) = y(0.1)$ and $y_2 = y(x_2) = y(0.2)$.

**Step 1:** To find $y(0.1)$ i.e., $y_1$.

$$k_1 = hf(x_0, y_0)$$
$$= (0.1)f(0, 1)$$
$$= (0.1)(0 - 1)$$
$$= -0.1$$

$$k_2 = hf\left(x_0 + \frac{h}{2}, y_0 + \frac{k_1}{2}\right)$$
$$= (0.1)f\left(0 + \frac{0.1}{2}, 1 + \frac{-0.1}{2}\right)$$
$$= (0.1)f(0.05, 0.95)$$
$$= (0.1)\left[(0.05)^2 - 0.95\right]$$
$$= -0.09475$$

$$k_3 = hf\left(x_0 + \frac{h}{2}, y_0 + \frac{k_2}{2}\right)$$
$$= (0.1)f\left(0 + \frac{0.1}{2}, 1 - \frac{0.09475}{2}\right)$$
$$= (0.1)f(0.05, 0.952625)$$
$$= (0.1)\left[(0.05)^2 - 0.952625\right]$$
$$= -0.0950125$$

$$k_4 = hf(x_0 + h, y_0 + k_3)$$
$$= (0.1)f(0 + 0.1, 1 - 0.0950125)$$
$$= (0.1)f(0.1, 0.9049875)$$
$$= (0.1)\left[(0.1)^2 - 0.905\right]$$
$$= -0.0895$$

Hence,

$$y_1 = y(0.1) = y_0 + \frac{1}{6}(k_1 + 2k_2 + 2k_3 + k_4)$$
$$= 1 + \frac{1}{6}(-0.1 - 0.1895 - 0.190025 - 0.0895)$$
$$= 0.90516$$

**Step 2:** To find $y(0.2)$ i.e., $y_2$.

Now, we have $x_1 = x_0 + h = 0.1$, $y_1 = 0.90516$ and $h = 0.1$.

$$k_1 = hf(x_1, y_1)$$
$$= (0.1)f(0.1, 0.90516)$$
$$= (0.1)\left[(0.1)^2 - 0.90516\right]$$
$$= -0.089516$$

$$k_2 = hf\left(x_1 + \frac{h}{2}, y_1 + \frac{k_1}{2}\right)$$
$$= (0.1)f\left(0.1 + \frac{0.1}{2}, 0.90516 - \frac{0.08952}{2}\right)$$
$$= (0.1)f(0.15, 0.8604)$$
$$= (0.1)\left[(0.15)^2 - 0.8604\right]$$
$$= -0.08379$$

$$k_3 = hf\left(x_1 + \frac{h}{2}, y_1 + \frac{k_2}{2}\right)$$
$$= (0.1)f\left(0.1 + \frac{0.1}{2}, 0.90516 - \frac{0.08379}{2}\right)$$
$$= (0.1)f(0.15, 0.863265)$$
$$= (0.1)\left[(0.15)^2 - 0.863265\right]$$
$$= -0.08408$$

$$k_4 = hf(x_1 + h, y_1 + k_3)$$
$$= (0.1)f(0.1 + 0.1, 0.90516 - 0.08408)$$
$$= (0.1)f(0.2, 0.82108)$$
$$= (0.1)\left[(0.2)^2 - 0.82108\right]$$
$$= -0.078108$$

Hence,

$$y_2 = y(0.2) = y_1 + \frac{1}{6}(k_1 + 2k_2 + 2k_3 + k_4)$$
$$= 0.90516 + \frac{1}{6}(-0.089516 - 0.16758 - 0.16816 - 0.078108)$$
$$= 0.82127$$

**Example:** Use Runge-Kutta method to find $y$ when $x = 1.2$ in steps of $0.1$, given that $dy/dx = x^2 + y^2$ and $y(1) = 1.5$. **(Apr-2022)**

**Solution:** Here $f(x,y) = x^2 + y^2$, $x_0 = 1$, $y_0 = 1.5$ and $h = 0.1$.

**Step 1:** To find $y(1.1)$, ie., $y_1$.

$$k_1 = hf(x_0, y_0)$$
$$= (0.1)f(1, 1.5)$$
$$= (0.1)\left[1^2 + (1.5)^2\right]$$
$$= 0.325$$

$$k_2 = hf\left(x_0 + \frac{h}{2}, y_0 + \frac{k_1}{2}\right)$$
$$= (0.1)f\left(1 + \frac{0.1}{2}, 1.5 + \frac{0.325}{2}\right)$$
$$= (0.1)f(1.05, 1.6625)$$
$$= (0.1)\left[(1.05)^2 + (1.6625)^2\right]$$
$$= 0.3866$$

$$k_3 = hf\left(x_0 + \frac{h}{2}, y_0 + \frac{k_2}{2}\right)$$
$$= (0.1)f\left(1 + \frac{0.1}{2}, 1.5 + \frac{0.3866}{2}\right)$$
$$= (0.1)f(1.05, 1.6933)$$
$$= (0.1)\left[(1.05)^2 + (1.6933)^2\right]$$
$$= 0.39698$$

$$k_4 = hf(x_0 + h, y_0 + k_3)$$
$$= (0.1)f(1 + 0.1, 1.5 + 0.39698)$$
$$= (0.1)f(1.1, 1.89698)$$
$$= (0.1)\left[(1.1)^2 + (1.89698)^2\right]$$
$$= 0.4808$$

By R-K fourth order formula

$$y_1 = y(1.1) = y_0 + \frac{1}{6}(k_1 + 2k_2 + 2k_3 + k_4)$$
$$= 1.5 + \frac{1}{6}(0.325 + 0.7732 + 0.79396 + 0.4808)$$
$$= 1.5 + 0.39548$$
$$= 1.89548 \simeq 1.8955$$

**Step 2:** To find $y(1.2)$, i.e., $y_2$.

Now, we have $x_1 = x_0 + h = 1.1$, $y_1 = 1.8955$ and $h = 0.1$.

$$k_1 = hf(x_1, y_1)$$
$$= (0.1)f(1.1, 1.8955)$$
$$= (0.1)[(1.1)^2 + (1.8955)^2]$$
$$= 0.4803$$

$$k_2 = hf\left(x_1 + \frac{h}{2}, y_1 + \frac{k_1}{2}\right)$$
$$= (0.1)f\left(1.1 + \frac{0.1}{2}, 1.8955 + \frac{0.4803}{2}\right)$$
$$= (0.1)f(1.15, 2.13565)$$
$$= (0.1)[(1.15)^2 + (2.13565)^2]$$
$$= 0.58835$$

$$k_3 = hf\left(x_1 + \frac{h}{2}, y_1 + \frac{k_2}{2}\right)$$
$$= (0.1)f\left(1.1 + \frac{0.1}{2}, 1.8955 + \frac{58835}{2}\right)$$
$$= (0.1)f(1.15, 2.189675)$$
$$= (0.1)\left[(0.15)^2 + (2.189675)^2\right]$$
$$= 0.6117$$

$$k_4 = hf(x_1 + h, y_1 + k_3)$$
$$= (0.1)f(1.1 + 0.1, 1.8955 + 0.6117)$$
$$= (0.1)f(1.2, 2.5072)$$
$$= (0.1)\left[(1.2)^2 + (2.5072)^2\right]$$
$$= 0.7726$$

Hence,

$$y_2 = y(1.2) = y_1 + \frac{1}{6}(k_1 + 2k_2 + 2k_3 + k_4)$$
$$= 1.8955 + \frac{1}{6}(0.4803 + 1.1767 + 1.2234 + 0.7726)$$
$$= 1.8955 + \frac{1}{6}(3.653)$$
$$= 11.8955 + 0.6088$$
$$= 2.5043$$

**Example:** Find $y(0.1)$ and $y(0.2)$ by Runge-Kutta fourth order method given that $y' = xy + y^2$, $y(0) = 1$. **(Dec-2016)**

**Solution:** Here $f(x, y) = xy + y^2$, $x_0 = 0, y_0 = 1$ and $h = 0.1$.

**Step 1:** To find $y(0.1)$ i.e., $y_1$

Now
$$k_1 = hf(x_0, y_0)$$
$$= (0.1)f(0, 1)$$
$$= (0.1)[(0)(1) + (1)^2]$$
$$= 0.1$$

$$k_2 = hf\left(x_0 + \frac{h}{2}, y_0 + \frac{k_1}{2}\right)$$
$$= (0.1)f\left(0 + \frac{0.1}{2}, 1 + \frac{0.1}{2}\right)$$
$$= (0.1)f(0.05, 1.05)$$
$$= (0.1)[(0.05)(1.05) + (1.05)^2]$$
$$= 0.1155$$

$$k_3 = hf\left(x_0 + \frac{h}{2}, y_0 + \frac{k_2}{2}\right)$$
$$= (0.1)f\left(0 + \frac{0.1}{2}, 1 + \frac{0.1155}{2}\right)$$
$$= (0.1)f(0.05, 1.05775)$$
$$= (0.1)[(0.05)(1.05775) + (1.05775)^2)]$$
$$= 0.1172$$

$$k_4 = hf(x_0 + h, y_0 + k_3)$$
$$= (0.1)f(0 + 0.1, 1 + 0.1172)$$
$$= (0.1)f(0.1, 1.1172)$$
$$= (0.1)[(0.1)(1.1172) + (1.1172)^2]$$
$$= 0.1360$$

By R-K fourth order formula
$$y_1 = y(0.1) = y_0 + \frac{1}{6}(k_1 + 2k_2 + 2k_3 + k_4)$$
$$= 1 + \frac{1}{6}(0.1 + 0.231 + 0.2344 + 0.1360)$$
$$= 1.1169$$

**Step 2:** To find $y(0.2)$, i.e., $y_2$.

Now, we have $x_1 = x_0 + h = 0.1, y_1 = 1.1169$ and $h = 0.1$.

$$k_1 = hf(x_1, y_1)$$
$$= (0.1)f(0.1, 1.1169)$$
$$= (0.1)\left[(0.1)(1.1169) + (1.11692)^2\right]$$
$$= 0.1359$$

$$k_2 = hf\left(x_1 + \frac{h}{2}, y_1 + \frac{k_1}{2}\right)$$
$$= (0.1)f\left(0.1 + \frac{0.1}{2}, 1.1169 + \frac{0.1359}{2}\right)$$
$$= (0.1)f(0.15, 1.18495)$$
$$= (0.1)\left[(0.15)(1.188495) + (1.18495)^2\right]$$
$$= 0.1582$$

$$k_3 = hf\left(x_1 + \frac{h}{2}, y_1 + \frac{k_2}{2}\right)$$
$$= (0.1)f\left(0.1 + \frac{0.1}{2}, 1.1169 + \frac{0.1582}{2}\right)$$
$$= (0.1)f(0.15, 1.916)$$
$$= (0.1)\left[(0.15)(1.196) + (1.196)^2\right]$$
$$= 0.1609$$

$$k_4 = hf(x_1 + h, y_1 + k_3)$$
$$= (0.1)f(0.1 + 0.1, 1.1169 + 0.1609)$$
$$= (0.1)f(0.2, 1.2778)$$
$$= (0.1)\left[(0.2)(1.2778) + (1.2778)^2\right]$$
$$= 0.18883$$

Hence,

$$y_2 = y(0.2) = y_1 + \frac{1}{6}(k_1 + 2k_2 + 2k_3 + k_4)$$
$$= 1.1169 + \frac{1}{6}(0.1359 + 2(0.1582) + 2(0.1609) + 0.18883)$$
$$= 1.27738$$

**Example:** Using Runge-Kutta fourth order formula, find $y(0.1)$ and $y(0.2)$ for the equation $y' = \dfrac{y-x}{y+x}, y(0) = 1$. **(Dec-2016), (Dec-2020), (July-2021)**

**Solution:** Here $f(x,y) = \dfrac{y-x}{y+x}$, $x_0 = 0, y_0 = 1$ and $h = 0.1$ so that $x_1 = 0.1$ and $x_2 = 0.2$. We have to find $y_1 = y(x_1) = 0.1$ and $y_2 = y(x_2) = y(0.2)$.

**Step 1:** To find $y(0.1)$ i.e., $y_1$

Now
$$k_1 = hf(x_0, y_0)$$
$$= (0.1)f(0,1)$$
$$= (0.1)\left[\frac{1-0}{1+0}\right]$$
$$= 0.1$$

$$k_2 = hf\left(x_0 + \frac{h}{2}, y_0 + \frac{k_1}{2}\right)$$
$$= (0.1)f\left(0 + \frac{0.1}{2}, 1 + 0.05\right)$$
$$= (0.1)f(0.05, 1.05)$$
$$= (0.1)\left[\frac{1.05 - 0.05}{1.05 + 0.05}\right]$$
$$= 0.0909$$

$$k_3 = hf\left(x_0 + \frac{h}{2}, y_0 + \frac{k_2}{2}\right)$$
$$= (0.1)f\left(0 + \frac{0.1}{2}, 1 + 0.0455\right)$$
$$= (0.1)f(0.05, 1.04545)$$
$$= (0.1)\left[\frac{1.04545 - 0.05}{1.04545 + 0.05}\right]$$
$$= 0.09087$$

$$k_4 = hf(x_0 + h, y_0 + k_3)$$
$$= (0.1)f(0 + 0.1, 1 + 0.09087)$$
$$= (0.1)f(0.1, 1.09087)$$
$$= (0.1)\left[\frac{1.09087 - 0.1}{1.09087 + 0.1}\right]$$
$$= 0.08321$$

By R-K fourth order formula

$$y_1 = y(0.1) = y_0 + \frac{1}{6}(k_1 + 2k_2 + 2k_3 + k_4)$$
$$= 1 + \frac{1}{6}[0.1 + 2(0.0900) + 2(0.09087) + 0.08321]$$
$$= 1.091125$$

**Step 2:** To find $y(0.2)$, i.e., $y_2$.

Here $x_1 = x_0 + h = 0.1, y_1 = 1.091125$ and $h = 0.1$.

$$k_1 = hf(x_1, y_1) = (0.1)f(0.1, 1.091125)$$
$$= (0.1)\left[\frac{1.091125 - 0.1}{1.091125 + 0.1}\right]$$
$$= 0.083209$$

$$k_2 = hf\left(x_1 + \frac{h}{2}, y_1 + \frac{k_1}{2}\right)$$
$$= (0.1)f(0.15, 1.13273)$$
$$= (0.1)\left[\frac{1.13273 - 0.15}{1.13273 + 0.15}\right]$$
$$= 0.07661$$

$$k_3 = hf\left(x_1 + \frac{h}{2}, y_1 + \frac{k_2}{2}\right)$$
$$= (0.1)f(0.15, 1.12943)$$
$$= (0.1)\left[\frac{1.12943 - 0.15}{1.12943 + 0.15}\right]$$
$$= 0.07655$$

$$k_4 = hf(x_1 + h, y_1 + k_3)$$
$$= (0.1)f(0.2, 1.167675)$$
$$= (0.1)\left[\frac{1.167675 - 0.2}{1.167675 + 0.2}\right]$$
$$= 0.070753$$

Hence,

$$y_2 = y(0.2) = y_1 + \frac{1}{6}(k_1 + 2k_2 + 2k_3 + k_4)$$
$$= 1.091125 + \frac{1}{6}[0.083209 + 2(0.07661) + 2(0.07655) + 0.070753]$$
$$= 1.167835.$$

**Example:** Using Runge-Kutta fourth order formula, solve $\dfrac{dy}{dx} = \dfrac{y^2 - x^2}{y^2 + x^2}$ with $y(0) = 1$ at $x = 0.2, 0.4$.

**Solution:** Here,
$$f(x, y) = \frac{y^2 - x^2}{y^2 + x^2}, \quad x_0 = 0, y_0 = 1, h = 0.2,$$

so that $x_1 = 0.2$ and $x_2 = 0.4$. We have to find $y_1 = y(x_1) = y(0.2)$ and $y_2 = y(x_2) = y(0.4)$.

**To find $y_1$**

$$k_1 = hf(x_0, y_0)$$
$$= (0.2)f(0, 1)$$
$$= (0.2)\left[\frac{1^2 - 0^2}{1^2 + 0^2}\right]$$
$$= 0.2$$

$$k_2 = hf\left(x_0 + \frac{h}{2}, y_0 + \frac{k_1}{2}\right)$$
$$= (0.2)f(0.1, 1.1)$$
$$= (0.2)\left[\frac{(1.1)^2 - (0.1)^2}{(1.1)^2 + (0.1)^2}\right]$$
$$= 0.19672$$

$$k_3 = hf\left(x_0 + \frac{h}{2}, y_0 + \frac{k_2}{2}\right)$$
$$= (0.2)f(0.1, 1.09836)$$
$$= (0.2)\left[\frac{(1.09836)^2 - (0.1)^2}{(1.09836)^2 + (0.1)^2}\right]$$
$$= 0.19671$$

$$k_4 = hf(x_0 + h, y_0 + k_3)$$
$$= (0.2)f(0.2, 1.19671)$$
$$= (0.2)\left[\frac{(1.19671)^2 - (0.1)^2}{(1.19671)^2 + (0.1)^2}\right]$$
$$= 0.18913$$

Hence, by R-K fourth order formula

$$y_1 = y(0.1) = y_0 + \frac{1}{6}(k_1 + 2k_2 + 2k_3 + k_4)$$
$$= 1 + \frac{1}{6}(0.2 + 0.39344 + 0.39342 + 0.18913)$$
$$= 1.196$$

**To find $y(0.4)$, i.e., $y_2$.**

Here $x_1 = x_0 + h = 0.2, y_1 = 1.196$ and $h = 0.2$.

$$k_1 = hf(x_1, y_1)$$
$$= (0.2)f(0.2, 1.196)$$
$$= (0.2)\left[\frac{(1.196)^2 - (0.2)^2}{(1.196)^2 + (0.2)^2}\right]$$
$$= 0.18912$$

$$k_2 = hf\left(x_1 + \frac{h}{2}, y_1 + \frac{k_1}{2}\right)$$
$$= (0.2)f(0.3, 1.29056)$$
$$= (0.2)\left[\frac{(1.29056)^2 - (0.2)^2}{(1.29056)^2 + (0.3)^2}\right]$$
$$= 0.17949$$

$$k_3 = hf\left(x_1 + \frac{h}{2}, y_1 + \frac{k_2}{2}\right)$$
$$= (0.2)f(0.3, 1.28575)$$
$$= (0.2)\left[\frac{(1.28575)^2 - (0.3)^2}{(1.28575)^2 + (0.3)^2}\right]$$
$$= 0.17935$$

$$k_4 = hf(x_1 + h, y_1 + k_3)$$
$$= (0.2)f(0.4, 1.37535)$$
$$= (0.2)\left[\frac{(1.37535)^2 - (0.4)^2}{(1.37535)^2 + (0.4)^2}\right]$$
$$= 0.1688$$

Hence,

$$y_2 = y(0.4) = y_1 + \frac{1}{6}(k_1 + 2k_2 + 2k_3 + k_4)$$
$$= 1.196 + \frac{1}{6}(0.18912 + 0.35898 + 0.3587 + 0.1688)$$
$$= 1.3753.$$

**Example:** Using Runge-Kutta method of fourth order, solve for $y$ at $x = 1.2$ and $x = 1.4$ from $\dfrac{dy}{dx} = \dfrac{2xy + e^x}{x^2 + xe^x}$, given $x_0 = 1, y_0 = 0$.

**Solution:** We have $f(x,y) = \dfrac{2xy + e^x}{x^2 + xe^x}$, $x_0 = 1, y_0 = 0$ and $h = 0.2$ so that $x_1 = 1.2$ and $x_2 = 1.4$. We have to find $y_1 = y(x_1) = 1.2$ and $y_2 = y(x_2) = y(1.4)$.

**Step 1:** To find $y(1.2)$ i.e., $y_1$

Now

$$k_1 = hf(x_0, y_0)$$
$$= (0.2)f(1, 0)$$
$$= (0.2)\left[\dfrac{2(1)(0) + e^1}{1^2 + (1)e^1}\right]$$
$$= (0.2)\left[\dfrac{e}{1+e}\right]$$
$$= 0.1462$$

$$k_2 = hf\left(x_0 + \dfrac{h}{2}, y_0 + \dfrac{k_1}{2}\right)$$
$$= (0.2)f\left(1 + \dfrac{0.2}{2}, 0 + \dfrac{0.1462}{2}\right)$$
$$= (0.2)f(1.1, 0.0731)$$
$$= (0.2)\left[\dfrac{2(1.1)(0.0731) + e^{1.1}}{(1.1)^2 + (1.1)e^{1.1}}\right]$$
$$= (0.2)\left[\dfrac{3.1650}{4.5146}\right]$$
$$= 0.1402$$

$$k_3 = hf\left(x_0 + \frac{h}{2}, y_0 + \frac{k_2}{2}\right)$$
$$= (0.2)f\left(1 + \frac{0.2}{2}, 0 + \frac{0.1402}{2}\right)$$
$$= (0.2)f(1.1, 0.0701)$$
$$= (0.2)\left[\frac{2(1.1)(0.0701) + e^{1.1}}{(1.1)^2 + (1.1)e^{1.1}}\right]$$
$$= (0.2)\left[\frac{3.1584}{4.5146}\right]$$
$$= 0.1399$$

$$k_4 = hf(x_0 + h, y_0 + k_3)$$
$$= (0.2)f(1 + 0.2, 0 + 0.1399)$$
$$= (0.2)f(1.2, 0.1399)$$
$$= (0.2)\left[\frac{2(1.2)(0.1399) + e^{1.2}}{(1.2)^2 + (1.2)e^{1.2}}\right]$$
$$= (0.2)\left[\frac{3.6559}{5.4241}\right]$$
$$= 0.1348$$

By R-K fourth order formula
$$y_1 = y(1.2) = y_0 + \frac{1}{6}(k_1 + 2k_2 + 2k_3 + k_4)$$
$$= 0 + \frac{1}{6}[0.1462 + 0.2804 + 0.2795 + 0.1348]$$
$$= 0.1402$$

**Step 2:** To find $y(1.4)$ i.e., $y_2$

Now

$$k_1 = hf(x_1, y_1)$$
$$= (0.2)f(1.2, 0.1402)$$
$$= (0.2)\left[\frac{2(1.2)(0.1402) + e^{1.2}}{(1.2)^2 + (1.2)e^{1.2}}\right]$$
$$= (0.2)\left[\frac{3.6566}{5.4241}\right]$$
$$= 0.1348$$

$$k_2 = hf\left(x_1 + \frac{h}{2}, y_1 + \frac{k_1}{2}\right)$$
$$= (0.2)f\left(1.2 + \frac{0.2}{2}, 0.1402 + \frac{0.1348}{2}\right)$$
$$= (0.2)f(1.3, 0.2076)$$
$$= (0.2)\left[\frac{2(1.3)(0.2076) + e^{1.3}}{(1.3)^2 + (1.3)e^{1.3}}\right]$$
$$= (0.2)\left[\frac{4.2091}{6.4601}\right]$$
$$= 0.1303$$

$$k_3 = hf\left(x_1 + \frac{h}{2}, y_1 + \frac{k_2}{2}\right)$$
$$= (0.2)f\left(1.2 + \frac{0.2}{2}, 0.1402 + \frac{0.1303}{2}\right)$$
$$= (0.2)f(1.30.2054)$$
$$= (0.2)\left[\frac{2(1.3)(0.2054) + e^{1.3}}{(1.3)^2 + (1.3)e^{1.3}}\right]$$
$$= (0.2)\left[\frac{4.2033}{6.4601}\right]$$
$$= 0.1301$$

$$k_4 = hf(x_1 + h, y_1 + k_3)$$
$$= (0.2)f(1.2 + 0.2, 0.1402 + 0.1301)$$
$$= (0.2)f(1.4, 0.2703)$$
$$= (0.2)\left[\frac{2(1.4)(0.2703) + e^{1.4}}{(1.4)^2 + (1.4)e^{1.4}}\right]$$
$$= (0.2)\left[\frac{4.8120}{7.6373}\right]$$
$$= 0.1260$$

By R-K fourth order formula
$$y_2 = y(1.4) = y_1 + \frac{1}{6}(k_1 + 2k_2 + 2k_3 + k_4)$$
$$= 0.1402 + \frac{1}{6}[0.1348 + 0.2606 + 0.2602 + 0.1260]$$
$$= 0.1402 + 0.1303$$
$$= 0.2705$$

**Example:** Evaluate $y(0.8)$ using R-K method given $y' = (x+y)^{1/2}, y(0.4) = 0.41$.

**Solution:** Here $f(x,y) = (x+y)^{1/2}$.

**To find $y(0.8)$ i.e., $y_1$**

Here $x_0 = 0.4$, $y_0 = 0.41$, $h = 0.4$ and $x_1 = x_0 + h = 0.8$. Now

$$k_1 = hf(x_0, y_0) = (0.4)f(0.4, 0.41) = 0.36$$
$$k_2 = hf\left(x_0 + \frac{h}{2}, y_0 + \frac{k_1}{2}\right) = (0.4)f(0.6, 0.59) = 0.4363$$
$$k_3 = hf\left(x_0 + \frac{h}{2}, y_0 + \frac{k_2}{2}\right) = (0.4)f(0.6, 0.6282) = 0.4433$$
$$k_4 = hf(x_0 + h, y_0 + k_3) = (0.4)f(0.8, 0.8533) = 0.5143$$

Hence, by R-K fourth order formula
$$y_1 = y(0.8) = y_0 + \frac{1}{6}(k_1 + 2k_2 + 2k_3 + k_4)$$
$$= 0.41 + \frac{1}{6}(0.36 + 0.8726 + 0.8866 + 0.5143)$$
$$= 0.8489.$$

**Example:** Find $y(0.1)$ and $y(0.2)$ using Runge-Kutta fourth order formula given that $y' = x + y$, $y(0) = 1$.

**Solution:** Here $f(x, y) = x + y$, $x_0 = 0$, $y_0 = 1$ and $h = 0.1$. Then $x_1 = x_0 + h = 0.1$ and $x_2 = x_1 + h = 0.2$. We have to find $y_1 = y(x_1) = y(0.1)$ and $y_2 = y(x_2) = y(0.2)$.

**To find $y(0.)$ i.e., $y_1$**

Now $x_0 = 0, y_0 = 1$ and $h = 0.1$.

$k_1 = hf(x_0, y_0) = (0.1)f(0, 1) = (0.1)(0 + 1) = 0.1$

$k_2 = hf\left(x_0 + \dfrac{h}{2}, y_0 + \dfrac{k_1}{2}\right) = (0.1))f(0.05, 1.05) = (0.1)(0.05 + 1.05) = 0.11$

$k_3 = hf\left(x_0 + \dfrac{h}{2}, y_0 + \dfrac{k_2}{2}\right) = (0.1)f(0.05, 1.055) = (0.1)(0.05 + 1.055) = 0.1105$

$k_4 = hf(x_0 + h, y_0 + k_3) = (0.1)f(0.1, 1.1105) = (0.1)(0.1 + 1.1105) = 0.12105$

Hence, by R-K fourth order formula

$$y_1 = y(0.1) = y_0 + \dfrac{1}{6}(k_1 + 2k_2 + 2k_3 + k_4)$$

$$= 1 + \dfrac{1}{6}(0.1 + 0.22 + 0.2210 + 0.12105)$$

$$= 1.11034$$

**To find $y(0.2)$, i.e., $y_2$.**

Here $x_1 = x_0 + h = 0 + 0.1 = 0.1$, $y_1 = 1.11034$ and $h = 0.1$.

$k_1 = hf(x_1, y_1) = (0.1)f(0.1, 1.11034) = (0.1)(0.1 + 1.11034) = 0.121034$

$k_2 = hf\left(x_1 + \dfrac{h}{2}, y_1 + \dfrac{k_1}{2}\right) = (0.1)f(0.15, 1.170857) = (0.1)(0.15 + 1.170857) = 0.1320857$

$k_3 = hf\left(x_1 + \dfrac{h}{2}, y_1 + \dfrac{k_2}{2}\right) = (0.1)f(0.15, 1.1763829) = (0.1)(0.15 + 1.1763829) = 0.1326382$

$k_4 = hf(x_1 + h, y_1 + k_3) = (0.1)f(0.2, 1.2429783) = (0.1)(0.2 + 1.242783) = 0.1442978$

Hence,

$$y_2 = y(0.2) = y_1 + \dfrac{1}{6}(k_1 + 2k_2 + 2k_3 + k_4)$$

$$= 1.11034 + \dfrac{1}{6}(0.121034 + 0.2641714 + 0.2652764 + 0.1442978)$$

$$= 1.242803.$$

**Example:** Apply Runge-Kutta method to find an approximate value of $y$ for $x = 0.2$ in step of 0.1, if $dy/dx = x + y^2$ given that $y = 1$, where $x = 0$.

**Solution:** Here $f(x, y) = x + y^2$, $x_0 = 0, y_0 = 1$.

Let us choose $h = 0.1$ and calculate various values as follows:

**Step 1:** Here $x_0 = 0, y_0 = 1$ and $h = 0.1$.

$$k_1 = hf(x_0, y_0) = (0.1)f(0, 1) = 0.1000$$

$$k_2 = hf\left(x_0 + \frac{h}{2}, y_0 + \frac{k_1}{2}\right) = (0.1))f(0.05, 1.1) = 0.1152$$

$$k_3 = hf\left(x_0 + \frac{h}{2}, y_0 + \frac{k_2}{2}\right) = (0.1)f(0.05, 1.1152) = 0.1168$$

$$k_4 = hf(x_0 + h, y_0 + k_3) = (0.1)f(0.1, 1.1168) = 0.1347$$

Hence, by R-K fourth order formula

$$y_1 = y(0.1) = y_0 + \frac{1}{6}(k_1 + 2k_2 + 2k_3 + k_4)$$

$$= 1 + \frac{1}{6}(0.1000 + 0.2304 + 0.2336 + 0.1347)$$

$$= 1.1165$$

**Step 2:** Here $x_1 = x_0 + h = 0 + 0.1 = 0.1, y_1 = 1.1165$ and $h = 0.1$.

$$k_1 = hf(x_1, y_1) = (0.1)f(0.1, 1.1165) = 0.1347$$

$$k_2 = hf\left(x_1 + \frac{h}{2}, y_1 + \frac{k_1}{2}\right) = (0.1)f(0.15, 1.1838) = 0.1551$$

$$k_3 = hf\left(x_1 + \frac{h}{2}, y_1 + \frac{k_2}{2}\right) = (0.1)f(0.15, 1.194) = 0.1576$$

$$k_4 = hf(x_1 + h, y_1 + k_3) = (0.1)f(0.2, 1.1576) = 0.1823$$

Hence,

$$y_2 = y(0.2) = y_1 + \frac{1}{6}(k_1 + 2k_2 + 2k_3 + k_4)$$

$$= 1.1165 + 0.1571$$

$$= 1.2736.$$

**Example:** Use Runge-Kutta method of order 4, compute $y(0.2)$ and $y(0.4)$ from $10\dfrac{dy}{dx} = x^2 + y^2$, $y(0) = 1$, taking $h = 0.2$.

**Solution:** Here $f(x, y) = \dfrac{x^2 + y^2}{10}$, $x_0 = 0, y_0 = 1, h = 0.2$.

**Step 1:** Here $x_0 = 0, y_0 = 1$ and $h = 0.2$.

$$k_1 = hf(x_0, y_0) = \dfrac{h}{10}(x_0^2 + y_0^2) = \dfrac{0.2}{10}(0 + 1) = 0.02$$

$$k_2 = hf\left(x_0 + \dfrac{h}{2}, y_0 + \dfrac{k_1}{2}\right)$$

$$= \dfrac{h}{10}\left[\left(x_0 + \dfrac{h}{2}\right)^2 + \left(y_0 + \dfrac{k_1}{2}\right)^2\right]$$

$$= \dfrac{0.2}{10}\left[\left(0 + \dfrac{0.2}{2}\right)^2 + \left(1 + \dfrac{0.02}{2}\right)^2\right]$$

$$= 0.0206$$

$$k_3 = hf\left(x_0 + \dfrac{h}{2}, y_0 + \dfrac{k_2}{2}\right)$$

$$= \dfrac{h}{10}\left[\left(x_0 + \dfrac{h}{2}\right)^2 + \left(y_0 + \dfrac{k_2}{2}\right)^2\right]$$

$$= \dfrac{0.2}{10}\left[\left(0 + \dfrac{0.2}{2}\right)^2 + \left(1 + \dfrac{0.0206}{2}\right)^2\right]$$

$$= 0.0206$$

$$k_4 = hf(x_0 + h, y_0 + k_3)$$

$$= \dfrac{h}{10}\left[\left(x_0 + h\right)^2 + \left(y_0 + k_3\right)^2\right]$$

$$= \dfrac{0.2}{10}\left[(0 + 0.2)^2 + (1 + 0.0206)^2\right]$$

$$= 0.0216$$

Hence, by R-K fourth order formula

$$y_1 = y(0.1) = y_0 + \dfrac{1}{6}(k_1 + 2k_2 + 2k_3 + k_4)$$

$$= 1 + \dfrac{1}{6}\Big(0.02 + 2(0.0206) + 2(0.0206) + 0.0216\Big)$$

$$= 1.1 + 0.0206$$

$$= 1.0206$$

**Step 2:** Here $x_1 = x_0 + h = 0.2 + 0.2 = 0.4$, $y_0 = 1.0206$ and $h = 0.2$.

$$k_1 = hf(x_1, y_1) = \frac{h}{10}(x_1^2 + y_1^2) = \frac{0.2}{10}(0.2^2 + 1.0206^2) = 0.0216$$

$$k_2 = hf\left(x_0 + \frac{h}{2}, y_0 + \frac{k_1}{2}\right)$$
$$= \frac{h}{10}\left[\left(x_1 + \frac{h}{2}\right)^2 + \left(y_1 + \frac{k_1}{2}\right)^2\right]$$
$$= \frac{0.2}{10}\left[\left(0.2 + \frac{0.2}{2}\right)^2 + \left(1.0206 + \frac{0.0216}{2}\right)^2\right]$$
$$= 0.0230$$

$$k_3 = hf\left(x_1 + \frac{h}{2}, y_1 + \frac{k_2}{2}\right)$$
$$= \frac{h}{10}\left[\left(x_1 + \frac{h}{2}\right)^2 + \left(y_1 + \frac{k_2}{2}\right)^2\right]$$
$$= \frac{0.2}{10}\left[\left(0.2 + \frac{0.2}{2}\right)^2 + \left(1.0206 + \frac{0.023}{2}\right)^2\right]$$
$$= 0.0231$$

$$k_4 = hf(x_0 + h, y_0 + k_3)$$
$$= \frac{h}{10}\left[\left(x_1 + h\right)^2 + \left(y_1 + k_3\right)^2\right]$$
$$= \frac{0.2}{10}\left[(0.2 + 0.2)^2 + (1.0206 + 0.0201)^2\right]$$
$$= 0.0249$$

Hence, by R-K fourth order formula

$$y_2 = y(0.2) = y_0 + \frac{1}{6}(k_1 + 2k_2 + 2k_3 + k_4)$$
$$= 1.0206 + \frac{1}{6}\left(0.0216 + 2(0.023) + 2(0.0231) + 0.0249\right)$$
$$= 1.0206 + 0.0231$$
$$= 1.0437$$

**Example:** By using the Runge-Kutta method of order 4, solve the equation $\frac{dy}{dx} = 3x + \frac{y}{2}$ with $y(0) = 1$ at the points $x = 0.1$ and $x = 0.2$, taking step length $h = 0.1$.

**Solution:** Here $f(x, y) = 3x + \frac{y}{2}$, $x_0 = 0$, $y_0 = 1$ and $h = 0.1$, so that $x_1 =$

$x_0 + h = 0.1$, $x_2 = x_1 + h = 0.2$. We are required to find $y_1 = y(x_1) = y(0.1)$ and $y_2 = y(x_2) = y(0.2)$.

**Step 1:** Here $x_0 = 0, y_0 = 1$ and $h = 0.1$.

$$k_1 = hf(x_0, y_0) = (0.1)f(0,1) = (0.1)\left[3(0) + \frac{1}{2}\right] = 0.05$$

$$k_2 = hf\left(x_0 + \frac{h}{2}, y_0 + \frac{k_1}{2}\right) = (0.1))f(0.05, 1.025) = (0.1)\left\{3(0.05) + \frac{1.025}{2}\right\} = 0.06625$$

$$k_3 = hf\left(x_0 + \frac{h}{2}, y_0 + \frac{k_2}{2}\right) = (0.1)f(0.05, 1.033125) = (0.1)\left\{3(0.05) + \frac{1.033125}{2}\right\} = 0.0666$$

$$k_4 = hf(x_0 + h, y_0 + k_3) = (0.1)f(0.1, 1.0666) = (0.1)\left\{3(0.1) + \frac{1.06666}{2}\right\} = 0.08333$$

Hence, by R-K fourth order formula

$$y_1 = y(0.1) = y_0 + \frac{1}{6}(k_1 + 2k_2 + 2k_3 + k_4)$$

$$= 1 + \frac{1}{6}(0.05 + 0.1325 + 0.13332 + 0.08333)$$

$$= 1.06653$$

**Step 2:** Here $x_1 = 0.1, y_1 = 1.06653$ and $h = 0.1$.

$$k_1 = hf(x_1, y_1) = (0.1)f(0.1, 1.06653) = (0.1)\left[3(0.1) + \frac{1.06653}{2}\right] = 0.08331$$

$$k_2 = hf\left(x_0 + \frac{h}{2}, y_1 + \frac{k_1}{2}\right) = (0.1))f(0.15, 1.1082) = (0.1)\left\{3(0.15) + \frac{1.1082}{2}\right\} = 0.10041$$

$$k_3 = hf\left(x_1 + \frac{h}{2}, y_1 + \frac{k_2}{2}\right) = (0.1)f(0.15, 1.11674) = (0.1)\left\{3(0.15) + \frac{1.11674}{2}\right\} = 0.10084$$

$$k_4 = hf(x_1 + h, y_1 + k_3) = (0.1)f(0.2, 1.16737) = (0.1)\left\{3(0.2) + \frac{1.6737}{2}\right\} = 0.11837$$

Hence, by R-K fourth order formula

$$y_2 = y(0.2) = y_0 + \frac{1}{6}(k_1 + 2k_2 + 2k_3 + k_4)$$

$$= 1.06653 + \frac{1}{6}(0.08333 + 0.20082 + 0.20168 + 0.11837)$$

$$= 1.06533 + 0.1007$$

$$= 1.16723.$$

**Example:** Use Runge-Kutta method of the fourth order to find $y(0.1)$, given that
$y' = \dfrac{1}{x+y}$, $y(0) = 1$.

**Solution:** Here $f(x,y) = \dfrac{1}{x+y}$, $x_0 = 0, y_0 = 1$ and $h = 0.1$. Then $x_1 = x_0 + h = 0.1$.
We are required to find $y_1 = y(x_1) = y(0.1)$.
Here $x_0 = 0, y_0 = 1$ and $h = 0.1$.

$$k_1 = hf(x_0, y_0)$$
$$= (0.1)f(0,1)$$
$$= (0.1)\left[\dfrac{1}{1+0}\right]$$
$$= 0.1$$

$$k_2 = hf\left(x_0 + \dfrac{h}{2}, y_0 + \dfrac{k_1}{2}\right)$$
$$= (0.1))f(0.1, 1.05)$$
$$= (0.1)\left[\dfrac{1}{0.5 + 1.05}\right]$$
$$= (0.1)\left[\dfrac{1}{1.1}\right]$$
$$= 0.0909$$

$$k_3 = hf\left(x_0 + \dfrac{h}{2}, y_0 + \dfrac{k_2}{2}\right)$$
$$= (0.1)f(0.05, 1.045)$$
$$= (0.1)\left[\dfrac{1}{0.05 + 1.045}\right]$$
$$= 0.0913$$

$$k_4 = hf(x_0 + h, y_0 + k_3)$$
$$= (0.1)f(0.1, 1.0913)$$
$$= (0.1)\left[\dfrac{1}{0.1 + 1.0913}\right]$$
$$= 0.0839$$

Hence, by R-K fourth order formula

$$y_1 = y(0.1) = y_0 + \frac{1}{6}(k_1 + 2k_2 + 2k_3 + k_4)$$
$$= 1 + \frac{1}{6}[0.1 + 2(0.0909) + 2(0.0913) + 0.0839]$$
$$= 1 + 0.0914$$
$$= 1.0914$$

**Example:** Evaluate $y(1.1)$ by using Runge-Kutta method, given

$$\frac{dy}{dx} + \frac{y}{x} = \frac{1}{x^2}; \quad y(1) = 1.$$

**Solution:** Here $f(x,y) = \frac{1}{x^2} - \frac{y}{x}$, $x_0 = 1$, $y_0 = 1$ and $h = 0.1$. Then $x_1 = x_0 + h = 1.1$. We are required to find $y_1 = y(x_1) = y(1.1)$.

Here $x_0 = 1, y_0 = 1$ and $h = 0.1$.

$$k_1 = hf(x_0, y_0)$$
$$= (0.1)f(1,1)$$
$$= (0.1)\left[\frac{1}{1^2} - \frac{1}{1}\right]$$
$$= 0$$

$$k_2 = hf\left(x_0 + \frac{h}{2}, y_0 + \frac{k_1}{2}\right)$$
$$= (0.1)f(1.05, 1)$$
$$= (0.1)\left[\frac{1}{(1.05)^2} - \frac{1}{1.05}\right]$$
$$= (0.1)[0.90703 - 0.95238]$$
$$= -0.004535$$

$$k_3 = hf\left(x_0 + \frac{h}{2}, y_0 + \frac{k_2}{2}\right)$$
$$= (0.1)f(1.05, 0.99773)$$
$$= (0.1)\left[\frac{1}{(1.05)^2} - \frac{0.99773}{1.05}\right]$$
$$= (0.1)\left[0.90703 - 0.95022\right]$$
$$= -0.004319$$

$$k_4 = hf(x_0 + h, y_0 + k_3)$$
$$= (0.1)f(1.1, 0.99568)$$
$$= (0.1)\left[\frac{1}{(1.1)^2} - \frac{0.99568}{1.1}\right]$$
$$= (0.1)\left[0.82645 - 0.90516\right]$$
$$= -0.007871$$

Hence, by R-K fourth order formula

$$y_1 = y(0.1) = y_0 + \frac{1}{6}(k_1 + 2k_2 + 2k_3 + k_4)$$
$$= 1 + \frac{1}{6}[0 + 2(-0.004535) + 2(-0.004319) - 0.007871]$$
$$= 1 - 0.0042632$$
$$= 0.9958$$

**Example:** Find $y(0.1)$ and $y(0.2)$ using Runge-Kutta fourth order formula given that $y' = x + x^2 y$, $y(0) = 1$. **(Dec-2020), (Apr-2022)**

**Solution:** Here $f(x, y) = x + x^2 y$, $x_0 = 0, y_0 = 1$ and $h = 0.1$, so that $x_1 = x_0 + h = 0.1$, $x_2 = x_1 + h = 0.2$. We are required to find $y_1 = y(x_1) = y(0.1)$ and $y_2 = y(x_2) = y(0.2)$.

**Step 1:** Here $x_0 = 0, y_0 = 1$ and $h = 0.1$.

$$k_1 = hf(x_0, y_0) = 0$$
$$k_2 = hf\left(x_0 + \frac{h}{2}, y_0 + \frac{k_1}{2}\right) = 0.0525$$
$$k_3 = hf\left(x_0 + \frac{h}{2}, y_0 + \frac{k_2}{2}\right) = 0.0053$$
$$k_4 = hf(x_0 + h, y_0 + k_3) = 0.011$$

Hence, by R-K fourth order formula

$$y_1 = y(0.1) = y_0 + \frac{1}{6}(k_1 + 2k_2 + 2k_3 + k_4)$$
$$= 1 + \frac{1}{6}(0.5 + 0.0105 + 0.0106 + 0.011)$$
$$= 1.0054$$

**Step 2:** Here $x_1 = 0.1, y_1 = 1.0054$ and $h = 0.1$.

$$k_1 = hf(x_1, y_1) = 0.011$$
$$k_2 = hf\left(x_0 + \frac{h}{2}, y_1 + \frac{k_1}{2}\right) = 0.0173$$
$$k_3 = hf\left(x_1 + \frac{h}{2}, y_1 + \frac{k_2}{2}\right) = 0.0173$$
$$k_4 = hf(x_1 + h, y_1 + k_3) = 0.0241$$

Hence, by R-K fourth order formula

$$y_2 = y(0.2) = y_0 + \frac{1}{6}(k_1 + 2k_2 + 2k_3 + k_4) = 1.0228.$$

### EXERCISE

1. Find by Runge-Kutta method, the approximate value of $y$ at $x = 0.6$ and $0.8$, given that $y = 0.41$ when $x = 0.4$ and $dy/dx = \sqrt{x+y}$. Take $h = 0.2$.
   **Ans:** 0.61035, 0.84899

2. Given that $dy/dx = (y^2 - 2x)/(y^2 + x)$ and $y = 1$ at $x = 0$, find $y$ for $x = 0.1, 0.2$ and $0.3$ by using the Runge-Kutta method.

**Ans:** 1.09111, 1.1677, 1.2352.

3. 4. Find $y(0.1)$ and $y(0.2)$ using R-K fourth order formula given that $y' = x + x^2$, $y(0) = 1$. **(Dec-2020)**

4. Find $y(0.1)$ using R-K method of fourth order if $\dfrac{dy}{dx} = 2e^x + y$, $y(0) = 1$. **(Mar-2022)**

5. Using Runge-Kutta method of fourth order, find $y(0.3)$ given that $\dfrac{dy}{dx} = \dfrac{1}{2}(1 + x)y^2$, $y(0) = 1$. **(Apr-2022)**

www.ingramcontent.com/pod-product-compliance
Lightning Source LLC
Chambersburg PA
CBHW070925220526
45470CB00014B/1404